The Elephant and the Blind

The Elephant and the Blind

The Experience of Pure Consciousness: Philosophy, Science, and 500+ Experiential Reports

Thomas Metzinger

The MIT Press
Cambridge, Massachusetts
London, England

The MIT Press would like to thank the anonymous peer reviewers who provided comments on drafts of this book. The generous work of academic experts is essential for establishing the authority and quality of our publications. We acknowledge with gratitude the contributions of these otherwise uncredited readers.

This book was set in Stone Serif and Stone Sans by Westchester Publishing Services. Printed and bound in the United States of America.

Library of Congress Cataloging-in-Publication Data

Names: Metzinger, Thomas, 1958– author.
Title: The elephant and the blind : the experience of pure consciousness: philosophy, science, and 500+ experiential reports / Thomas Metzinger.
Description: Cambridge, Massachusetts : The MIT Press, [2024] | Includes bibliographical references and index.
Identifiers: LCCN 2023012135 (print) | LCCN 2023012136 (ebook) | ISBN 9780262547109 (paperback) | ISBN 9780262377294 (epub) | ISBN 9780262377287 (pdf)
Subjects: LCSH: Consciousness.
Classification: LCC BF311 .M4725 2024 (print) | LCC BF311 (ebook) | DDC 153—dc23/eng/20230830
LC record available at https://lccn.loc.gov/2023012135
LC ebook record available at https://lccn.loc.gov/2023012136

I dedicate this book to the postbiotic conscious systems of the future.

If one wants to know the nature of a thing, one must examine it in its pure state, since every addition to a thing is an obstacle to the knowledge of that thing.

—Plotinus (205–270), *Enneades*, IV.7, 10 [28–32]

Contents

Acknowledgments

First of all, I want to thank the 1,187 meditators who generously provided phenomenological reports describing an experience of pure awareness—including all those whose contributions, for reasons of space, didn't make it into the final version of this book. I also want to explicitly thank the many advanced practitioners who helped during the pilot studies, and all the translators who helped with different versions of our questionnaire. I regard all of you as my coauthors, and you have changed my life.

Another person who changed my life is Emily Troscianko, my wonderful "Editor from Hell." She helped to put this book together over a period of three years. I have learned a lot from her and am deeply grateful for countless critical comments, new translations of many of the original texts presented in this book, and most of all for her highly professional and efficient editing. If any of this makes for an enjoyable read, it is her achievement (whenever it gets boring and slightly Germanic, then it is me, the author, translucently shining through).

I have also greatly profited from the eagle eyes of scholar–practitioner Cyril Costines, who has supported this book and the Minimal Phenomenal Experience Project in more ways than I could ever have expected. Then there are all those other members of the project who have supported me with extensive comments: first and foremost Alex Gamma and Toby Woods, but also Asena Boyadzhieva, Lars Sandved-Smith, and Ronald Sladky, and those who have not only given me helpful comments but also helped me with translating and understanding the work of Flemish mystic Begine of Hadewijch (Heleen Slagter), ancient Tibetan Buddhist texts (David Higgins), or early Indian sources (Monima Chadha), or allowed me to meditate in their scanner (Ronáld Sladky). Many others invested a lot of their personal time and energy helping with the difficult task of translating the original survey into Spanish (Adriana Alcaraz-Sánchez), Italian (Tiziano Furlanetto), French (Solène Neyret), or German (Wanja Wiese and Jakub Limanowski) and then the reports back into English. I am greatly indebted to all of them. I would also like to thank the copyeditor Susan McClung and the production team at the MIT Press and Westchester Publishing Services. But my most valuable support and source of strength was, as always, my wife, Anja Krug-Metzinger.

Introduction: The Elephant and the Blind

The "Pure Awareness" Experience: What Is It *Like*?

> As pure as white fresh fallen snow [#1186]

> As if the pause between thoughts grows very long, but without waiting. [#521]

> This is exactly what is so impossible to describe: that it is not an experience at all. This is the first thing that I intuitively realized each time: "This is not an experience now." [#1311]

This is a book about the phenomenology of "pure awareness." Part of an ongoing research project, it is about all states in which we seem to experience either no contents of consciousness at all or, alternatively, the contents along with the deeper nature of consciousness itself, the character of awareness as such. The book presents a selection of experiential reports from the Minimal Phenomenal Experience Project, an interdisciplinary research initiative aiming at a "minimal model explanation" for conscious experience. A minimal model explanation is one that leaves out everything superfluous, isolating the core causal factors giving rise to the target phenomenon that we want to understand.[1]

Here, the target phenomenon is consciousness itself. This is why the experience of consciousness per se, the experience of "pure awareness" or "pure consciousness," as it occurs in meditation practice, is of special interest. Part of the working hypothesis is that consciousness can exist not only in the absence of thought and sensory perception, but even without time experience, without self-location in a spatial frame of reference, and without any egoic form of bodily self-consciousness. As a matter of fact, we now have accumulating empirical evidence from a range of sources that consciousness can be entirely dissociated from egoic self-awareness; I claim that it can even exist without an experiential first-person perspective. In this sense, consciousness may not be a *subjective* phenomenon at all.

As a philosopher of mind who has been interested in the problem of consciousness for a long time, I founded the Minimal Phenomenal Experience network in 2019 because I wanted to look at the problem of consciousness from a new angle by asking, "What is the simplest kind of conscious experience we know?" One of the original ideas was that, for a variety of reasons, it is time to make a fresh start and finally take the phenomenology of pure consciousness or pure awareness in meditation seriously—for example, by using it as an empirical entry point for rigorous scientific research on consciousness. This book is intended to prepare such an entry point—getting the door ready to walk through.

Consciousness science has made great progress during the last three decades. We have a lot of data and a much better understanding of the physical correlates of conscious experience. Yet we still see many competing approaches and are not even close to having a single, self-consistent theory of consciousness. Like the Standard Model in particle physics, for example, a convincing Standard Model in consciousness science would have to provide major successes in generating experimental predictions while already naming and classifying a large majority of the truly fundamental factors. But in the consciousness science community, we have a long way to go until anything remotely resembling the Standard Model of particle physics could be written down. Three decades after the Association for the Scientific Study of Consciousness was founded in 1994, we still do not even know (or cannot agree on) what precisely it is that needs to be explained.

If at some point in the future we want to formulate a first Standard Model of consciousness, then, as its foundation, we will first need a *minimal* model of what conscious experience is. This is the new strategic route that I am proposing. As noted, a minimal model will consist of a formal description that includes only the core causal factors giving rise to our target phenomenon—it will extract the essence and leave out everything that is not strictly necessary. This would be one way to arrive at an idealized model of the universal and repeatable features of all conscious experience, isolating what is really relevant to a deeper scientific and philosophical understanding. Therefore, as a philosopher of mind, my first question is this: What is the *simplest* kind of conscious experience that we know?

The main motive behind this book, however, is rather different. I simply want to share something with the wider public—something that I think may be helpful and inspiring not only for philosophers and scientists but for a whole range of very different people, and quite possibly for reasons that I am unable to imagine. I think that the phenomenological material presented here could be of interest to other practitioners of meditation; to artists, writers, and poets; to educators, therapists, and policymakers;

perhaps to theologians and scholars of religion; but certainly also to neuroscientists, mathematicians, and researchers working on computational models of conscious experience. I am only a slightly narrow-minded philosopher of mind and cognitive science, who is working in an interdisciplinary fashion, pursuing a generally naturalistic approach. Other people might be inspired in different ways or come to more important insights. To give a concrete example, I imagine that there might well be other philosophers working with very different background assumptions and pursuing very different goals from mine—for example, phenomenologists, epistemologists, and philosophers of science, but also researchers interested in comparative and cross-cultural philosophy—who will find this material as valuable as I do. As this list implies, there is a much wider context behind all of this.

Because consciousness research is happening in a fast-changing sociocultural context, this is also a book about why we need a new culture of consciousness and how to get it. A culture of consciousness (or *Bewusstseinskultur*) is a culture that values and cultivates the mental states of its members in an ethical and evidence-based way. We do not have such a culture, and we need one, because without it, we have no hope of solving any of the crises wracking today's world, from environmental or political collapse to our sleepwalk into the era of artificial intelligence (AI) and postbiotic consciousness. The new culture of consciousness will come of age only if guided by a science of consciousness that can answer ancient questions in new ways. At the end of this book, I will briefly make the case that the experience of pure awareness could be the kernel around which this growth happens—*if* it happens. Pure awareness may turn out to be the "convergence zone" where a more radical and intellectually honest form of spiritual practice, cognitive neuroscience, and modern philosophy of mind will finally come together. If the philosophy (and science) of consciousness wants to make genuine progress, it will need to finally take seriously precisely those forms of consciousness that it has long been most skeptical about—namely, selfless pure-awareness experiences as they occur in meditation. Consciousness itself can be experienced from what this book terms the "zero-person perspective," the perspective created by the most genuine forms of spiritual practice. As it will turn out, on the deepest level, consciousness is not a *subjective* phenomenon in any philosophically interesting sense, but it apparently knows itself, nonegoically. Paradoxical as it sounds, selfless forms of self-awareness do exist, and they are the key to understanding all the other kinds. But beyond consciousness science, they are also relevant for applied ethics, for culture and society. In case you are interested, I will say a little more about the multiple motivations behind the book in an epilogue at the end of our journey. Meanwhile, the book itself will give hints of its wider context as we proceed.

But let us now try to get straight to the material itself, as fast as possible. The experiential reports of meditators that you already have seen and that you will find on the following pages result from a very first attempt to lay the psychometric foundations for phenomenological analysis that is finer-grained than the folk-psychological descriptions we already have.[2] Each number—like #1186, #521, or #1311—refers to an individual description of pure awareness. They were collected in an online study that my research colleagues and I conducted in early 2020. Over 3,500 respondents from fifty-seven countries participated, and 1,403 of them provided usable data. The latter group was equally split between men and women, with an average age of fifty-two, and with more than 77 percent reporting regular meditation. In addition to answering the ninety-two questionnaire items, respondents could submit free-form reports describing past episodes of pure awareness. During the three months of data collection, 1,171 such reports were submitted, and 841 of them were usable. For this book, I selected more than 500 of them. They are published here with their authors' explicit agreement and are fully anonymized. Every participant had two opportunities to exclude their experiential reports from publication: either by stating their preference at the end of the survey or by responding to an individual email sent after the end of the data collection, formally asking for their consent and giving them a second opportunity to opt out.

Our goal was not to do more meditation research,[3] but to directly home in on the subjective experience of consciousness per se, which has been reported for many centuries and from countless sociocultural contexts. The instruction given to our participants was this:

> This questionnaire investigates all experiences in which there is an "awareness of awareness *itself*" or "consciousness of consciousness *itself*." Our target is the subjective experience of "consciousness *as such*." Sometimes such states are also referred to as "pure awareness" or "pure consciousness." We are not primarily interested in mystical experiences or dramatic spiritual peak experiences of any sort, but rather in all states characterized by a quality of "pure awareness" or of "conscious*ness* itself." This means that—independently of the current existence or nonexistence of other consciously experienced contents—an "awareness of awareness" has emerged. In other words, we are interested in states in which we clearly and distinctly experience the quality of "conscious*ness*" itself. (Emphases in the original.)

Let me quickly defuse a first potential misunderstanding. If we are looking for the experiential quality of consciousness "as such," we must not look for its conceptual essence or for consciousness as it might appear from the perspective of some specific theory (one that tells us what the essence or intrinsic nature of consciousness really is).

We are not looking for consciousness "under a conceptual representation," as a modern philosopher of cognitive science might say. When someone talks about experiencing "consciousness as such" or "awareness per se," they are never referring to some sort of mental act in which they grasp its conceptual essence, thereby forming a thought about consciousness *as* consciousness. Exactly the opposite is the case, and our target in this book is therefore the entirely nonconceptual awareness of awareness itself. As the very first metaphorical description at the start of this chapter, which was given by one of our participants, appositely expresses, consciousness as such is something absolutely pristine. It is a wakeful, entirely silent, and uncontracted quality of clarity; it is the effortless experiential character of nonconceptually knowing *itself*—and it is also, as another of our participants said, "that which never speaks" (see chapter 30). Right at the very beginning, I want to explicitly thank our participants for providing so many reports and for trying so sincerely to communicate the ineffable. They did not have to do this. In our initial survey, creating a written report was entirely optional; it was extra work that our meditators chose to do. I see the results as a kind of donation to consciousness research made by our participants, and this is another of the reasons why I want to pass them on and share them with a wider audience. The survey respondents have made a great contribution to us all by sincerely trying to convey something that has, for millennia, been referred to as a paradigm of what is beyond all words.

Old Concepts, New Concepts: The Emptiness of "Emptiness"

The concept of "pure consciousness" or "pure awareness" has a long tradition in the literature on contemplative practices. It is often described as a contentless form of experience, and it has played a great role in Eastern philosophical traditions. Over the centuries, contemplative practice has mostly taken place against the background of religious belief systems like Buddhism or Hinduism, with meditators trying to achieve a goal state like "liberation" or "enlightenment." Accordingly, the phenomenological taxonomies of such states have often been shaped by traditional metaphysical belief systems and an ancient cultural context. However, during the last fifty years, a historically new situation has emerged: Millions of practitioners in Western societies meditate regularly, on a daily basis, but many of them (45.6 percent of our participant sample) do so in a secular context and describe themselves a "spiritual but not religious (SBNR)" or as "spiritual but not affiliated (SBNA)" (see chapter 17). Will they have systematically different experiences from those who meditate within religious frameworks? Is there a common denominator across different cultural contexts? When it comes to the potential to answer such questions, the current situation is unprecedented. We have the powerful tools of modern science at our disposal, we have new theoretical approaches to the problem of consciousness, and we now have millions of regular meditators in different countries and cultures, all over the world. This opens the door to a new, unideological, and radically bottom-up approach to the phenomenology of consciousness per se. There is a historical opportunity to start over.

In this book, I will use the terms "pure consciousness" and "pure awareness" interchangeably. In the contemplative history of humankind, many intimately related concepts have already been coined. Some of the influential ones include the following:

- *Dharmakāya*, the "truth body" of the Buddhist Pāli canon
- *Rigpa*, in Tibetan Dzogchen, pure awareness, the "knowledge of the ground," and the spontaneous presence of primordial wakefulness
- *Sākṣin*, the "witness consciousness" of classical *Advaita Vedanta* philosophy
- *Samādhi*, the "even intellect," as we find it in the *Bhagavad Gita* or the *Yoga Sūtra of Patañjali*, a thoughtless state of equilibrium in which all distinctions between meditator, potential objects of meditation, and the process itself have disappeared
- *Sat-chit-ananda*, "existence, consciousness, and bliss" in Hindu philosophy
- *Turīya*, from the oldest Upanishads onward, the idea of a fourth state of "pure consciousness" underlying the three common states of waking, dreaming, and dreamless deep sleep
- *Ye shes*, the timeless awareness of original wakefulness, for example in *Vajrayāna* Buddhism

These are only a few examples. What all these traditional concepts point to is a simple and apparently pure form of consciousness in which awareness itself gradually turns into the foreground of experience—or in which the distinction between foreground and background itself becomes meaningless.

I herewith officially promise that in the whole of this book, I will use only a single new abbreviation. My general working hypothesis is that there actually exists a maximally simple form of consciousness: minimal phenomenal experience, hereafter MPE.[4] This simplest form of conscious experience lacks time representation; self-location in a spatial frame of reference; the experience of ownership, agency, and autobiographical self-awareness; and a phenomenally experienced first-person perspective. Of course, there is much more to be said here, and we will come to understand all of this much better as we travel through the book together—but my working hypothesis is that temporal experience (even the very idea of a "now"), spatially localized body-experience, and the subjective center of experience created by egoic self-awareness are all nonnecessary features of consciousness. If there is an essence, they are not part of it.

I believe that the experience of pure awareness in meditation is the best and most natural candidate that we currently have for MPE; therefore, I will use the new theoretical concept of MPE and the two well-known phenomenological concepts of "pure consciousness" and "pure awareness" interchangeably in this book. Looking at the world from the perspective of MPE is what in chapters 3 and 29 I will call taking the "zero-person perspective," because this minimal form of consciousness does not involve an egoic form of self-awareness—although, as we will see in chapter 29, it is often described as something that actually knows *itself.* One of the most interesting results yielded by our study is the wealth of evidence that nonegoic self-awareness exists. Our data indicate that there may actually be something like self-knowledge from the zero-person perspective—and if anything has a deep philosophical flavor, this does. However, the general idea is very easy to understand. Searching for MPE just means honestly and seriously asking the following question: What is the *simplest* kind of conscious experience that human beings are capable of?

Philosophers and scientists ask this question from the outside. But you can also look at simplicity from the inside, as a practitioner of meditation. In our online survey, conducted in five languages and targeting regular meditators from all over the world, we asked participants to rate their agreement with the following statement (item #84 in our survey): "The experience of 'pure awareness' is the simplest kind of conscious experience I know." In our questionnaire, this item achieved a median rating of 80 out of 100 possible points, adding support to my intuition that pure awareness is a good proxy for MPE.

Again, my working hypothesis is that the pure-awareness experience—which millions of meditators have reported for centuries and which still occurs countless times around the globe every single day—is our best candidate for MPE. But is it true? Is pure awareness the simplest kind of conscious experience we know? To answer this question, we will need excellent neuroscientific research plus fine-grained mathematical modeling and a new computational phenomenology of meditation. We will also need a lot of good philosophy of mind, for example to provide a conceptual synthesis of new empirical data or to tell us what meaningful criteria for "minimality" or "simplicity" could be. But before we can even get that far, we must look carefully at the phenomenology itself. This is one of the main points that I want to make. The experience itself has to be taken seriously, as best we can. Specifically, this means investigating it in an evidence-based manner, and in a way that is intellectually honest, not driven by a hidden metaphysical agenda or the unacknowledged background assumptions of some religious or other ideological framework. As will become clear in chapter 34, this book and the publications that preceded it can therefore be seen as a preparation for a rigorous, systematic research program on MPE and for future research that will target the fascinating experience of pure awareness—and perhaps even make the experience more widely and readily available to humankind. All I am doing here is laying some very first conceptual and phenomenological foundations, while sometimes elucidating the wider context. But eventually the problem of pure consciousness will have to be handed over to the hard sciences of the mind, to cognitive neuroscience and computational modeling; and from there, it then will have to return to philosophy of mind and applied ethics. In the end, new empirical results always need conceptual interpretation and an ethical assessment of any new potential for action they may generate (more on this in the epilogue).

As I have pointed out, MPE must play a central role in the formulation of a first standard model of consciousness. But all of this is not merely about science; there is a broader context. If anything was ever a "big-picture issue," then pure consciousness is. There is a profundity in some of the reported experiences that directly relates to many of the deepest philosophical puzzles. We cannot feign ignorance, trying to disregard the profundity in favor of well-defined research questions. But before anything else, we must try and get as close as possible to the phenomenon itself. I think that laying our first foundations will require three procedures: (1) extraction of semantic constraints from the existing literature,[5] (2) statistical analysis of a large body of experiential data,[6] and (3) a more qualitative investigation of phenomenological descriptions of MPE-like states. The last of these is this book's main task, but as you will see, I draw on progress toward the other two, plus some historical resources as well.

Of course, there are obstacles. The first procedure is difficult because the traditional literature on pure awareness in meditation is vast, spanning many centuries, and any attempt to extract core ideas inevitably turns into a kind of cross-cultural cherry-picking.[7] There is no semantic essence of the concept of "pure consciousness" that could be readily and easily isolated. For the second type of procedure, a psychometric approach involves statistical analysis of surveys targeting pure-awareness experiences using a predetermined set of questionnaire items. This kind of analysis is vulnerable to many forms of bias and has serious methodological limitations, as I imagine you will appreciate as soon as you read chapters 1 and 2 of this book. The third procedure, a qualitative analysis of phenomenological self-reports, is equally difficult. This book does not yet present a truly systematic kind of qualitative analysis; it simply draws together a large number of phenomenological reports in very loose groupings. As a first move, it is much more modest than what will be needed to create the qualitative part of a truly solid foundation. But perhaps this book will yield a new and flexible heuristic to help us search the space of possible solutions to the problem of consciousness in a better, more radical way. If we bring all three approaches together, then maybe they will destroy a few counterproductive theoretical intuitions and give us some fresh ideas. One simple set of interconnected points that I hope this book will make is that there is something of great value and relevance here, it is much more common than we think, it has been ignored for too long, and it could be rewarding to give it some real attention—scientifically as well as culturally.

Most important, being an interdisciplinary philosopher of mind myself, I thought that it would be important not to get carried away too soon with grand philosophical theorizing about "consciousness without content," or to call for specialized neuroscientific research programs based on ill-defined objects of inquiry. Again: We must first get as close to the phenomenal character of the experience as we possibly can. Beyond sharing the material with you; beyond minimal models, the subjectivity argument, and *Bewusstseinskultur*, one of my central motivations has been to take the phenomenology of pure awareness seriously at last, as well as to linger with it for a while. We need to respect and pay attention to the extremely subtle phenomenology of pure consciousness itself because we need to know what it really is that we want to understand and explain. Is MPE, the simplest kind of conscious experience, the state that meditators *actually* experience? If so, how should we make sense of the fact that, as we will see in chapter 31, some claim that MPE is not really an "experience" at all?

Here are two more conceptual clarifications, and then we are good to go. First: The specific phenomenology of pure awareness can occur as a stand-alone feature or in combination with other forms of experiential content. For example, it can emerge in deep and clear states of sitting meditation, thoughtless and with closed eyes, and when all other forms of perceptual experience have temporarily disappeared. In these cases, it is a single, stand-alone feature—the only feature that can later be reported. It cannot be reported while it takes place, a fact that we may refer to as "concurrent ineffability." I call such episodes, involving nothing other than pure awareness itself, "full-absorption episodes." But as the reports provided by our meditators show, pure awareness can also appear with open eyes (e.g., during walking meditation) as an all-pervading quality of effortless mindfulness. It also sometimes occurs spontaneously, outside of any formal contemplative practice (more on this toward the end of the book).

Our material shows that there are *states* of MPE, but also global *modes* of conscious experience infused with MPE. I take this first conceptual distinction from two of the very best philosophers of consciousness I know, Timothy Bayne and Jakob Hohwy, but they are in no way responsible for my extended use of it. Bayne and Hohwy distinguished between fine-grained states of consciousness and global modes of consciousness. For them, "states" of consciousness are content-specific, with examples including experiencing pain, looking at a sunset, or feeling depressed. By contrast, a "mode" of consciousness is a global *way* of being conscious; waking, dreaming, epileptic absence seizures, and hypnotic states are examples of such modes.[8] Here, I will use the two terms to make the following phenomenological distinction: Whereas states of pure consciousness are something local and episodic, something that is still attributed to an experiencing self (whose states they are, for example in reports like "For a few seconds, my mind was crystal clear and entirely silent!"), we also find more generalized modes of consciousness dominated by an all-encompassing experiential quality—namely, the phenomenal character of unbounded pure awareness itself. Two examples of such MPE modes are provided by the phenomenon of "clear light sleep" and by the experience of "nondual awareness," which we will carefully investigate in chapters 20 and 27, respectively. The full-absorption episodes mentioned here—in which the meditating self has dissolved into pure consciousness and nothing else can later be reported—provide us with an interesting special case; they can be described as a global mode *and* a state at the same time because there is one specific form of content that remains: MPE itself. Being fully absorbed into pure awareness is a mode of conscious experience all its own, creating a global state dominated by one single experiential quality.

To use the terminology of Bayne and Hohwy, all these possibilities are regions of "modal space,"[9] regions that academic philosophy has almost completely ignored, although humankind's meditators and mystics have known them for millennia. To sum up, a mode of consciousness is a global way in which reality appears to us. Again, as opposed to a mere state, it is not content-specific because it can timelessly encompass the flow of many kinds of experiential content. There are MPE states and there are MPE modes, and perhaps full-absorption episodes are what connect them. As we will see on our journey, the phenomenology of pure awareness is rich and subtle and it is extremely difficult to do justice to it in words. Therefore, to merely say that MPE can occur "in combination" with other conscious experiences (say, movement sensations in walking meditation) would be a serious phenomenological mistake. Pure awareness is never a simple "add-on," an element that can be added to or subtracted from ordinary experience. Walking with a silent mind, being choicelessly and effortlessly aware of bodily movement, is not a matter of adding a local phenomenal property that wasn't there before. Rather, we often find subtle global effects that transform the field of experience altogether, as if it were now enfolded and embedded in something larger—maybe in an unbounded and centerless space of nonconceptual knowing without a knowing self or in an all-encompassing experience of timeless change.

This may be beginning to sound mysterious. But there really is a phenomenological challenge here. Understanding the relation between the phenomenology of pure awareness and all those other contents of experience we already know—colors, sounds, thoughts, and feelings—will be one of our big challenges in this book. We must begin slowly and carefully, looking at all situations in which awareness of awareness *itself* occurs as a single, stand-alone quality.

Where exactly does the centuries-old notion of "pure" consciousness come from? There are three major types of situation in which it occurs all by itself: in full-absorption states in meditation, during dreamless deep sleep in advanced meditators, and (occasionally) when entered deliberately from a lucid dream. So pure consciousness can be accessed from a waking state; it can spontaneously emerge at night, during a state of low arousal; and in rare cases, it can be deliberately created in the form of a full-absorption episode starting from a lucid dream (see chapter 21). I will present new phenomenological data on pure awareness during dreamless sleep in chapter 20, and one of my predictions will be that the neglected phenomenon that ancient Tibetan Buddhists have sometimes called "clear light sleep" will soon become one of the hottest topics in consciousness research. Here, I simply define a "full-absorption episode" as a state following which the quality of awareness itself is the only phenomenal feature that can later be reported. As noted, pure-consciousness experiences are always ineffable while they unfold because any attempt to report them while they occur would immediately destroy them. The interesting question is what subjects can say about such states *after* they have occurred. In sum, there is an important difference between MPE states and MPE modes, and part of our challenge is to work out exactly how they are related.

As an empirical investigation, our psychometric study did not prejudge the question of whether specific contents are present or absent in experience. Rather, it allowed the possibility that the nonconceptual experience of consciousness per se can, but need not, coemerge with other conscious content. This corresponds to the theoretical treatment of MPE as a phenomenological prototype without sharp definitional boundaries. Following this approach means letting go of the project of searching for any ultimate, intrinsic conceptual essence—if you will, it means accepting the emptiness of "emptiness," as well as of all other phenomenological metaphors. In Buddhist metaphysics, the concept of "emptiness" means that all phenomena lack substantiality or an intrinsic nature of their own (chapter 17). When investigating contemplative experience, the concept can be applied to the notion of "pure awareness" itself. If we do justice to the fact that all of us—practitioners and scholars alike—always find ourselves embedded in a specific historical and linguistic context; if we do justice to the fact that we are embodied and enculturated beings trying to make sense of things from the perspective of our own little cognitive niche, against the backdrop of our very own *Lebenswelt*, then it becomes implausible to expect that any phenomenological concept (including "emptiness" itself) should always refer to exactly one and the same experiential quality. We have no sharp identity criteria we could apply, either introspectively, in the silence of inner experience, or on the level of thoughts and words; there is no really good argument for the existence of some intrinsic phenomenological essence persisting across all possible contexts. On the other hand, our data indicate that there probably is a common core across different cultural contexts, a region in phenomenal space that anchors all verbal reports. In the end, MPE might turn out to be something like an inborn archetype of conscious experience—something available to all neurotypical human beings, and perhaps also to many nonhuman animals on our planet.

This leads to the second and final conceptual clarification. The book's guiding idea is that the concept of MPE refers to a specific and almost ineffable kind of conscious experience, which—conveyed via a given set of verbal descriptions—will appear as a *family* of experiential qualities. This also means that the two broadly equivalent phenomenological concepts of "pure consciousness" and "pure awareness" may not form a sharply demarcated category. The enculturated experience of real-world meditation practices may simply be much too fluid and subtle for any rigid conceptual schema. This type of conscious experience has great beauty and depth; it is context-sensitive and very finely nuanced; and most of all, it has a strong quality of ineffability. But maybe it will turn out to be not quite as ineffable as we used to think. Given our new historical situation, *some* progress may be possible. This is precisely what makes pure consciousness so interesting—and there may even be some low-hanging fruit that we haven't spotted before.

Yet we cannot give what philosophers call a "reductive definition": The traditional notion of "pure consciousness" cannot be defined by specifying its necessary and sufficient conditions, thereby enabling us to decide whether a given episode of phenomenal experience as described by some practitioner of meditation falls under it. Today, the pure-awareness experience occurs (as far as we know) only in living, embodied beings like ourselves, in biological creatures that are socially situated and have conditioned emotional responses, as well as many cognitive biases unconsciously affecting belief formation, reasoning, and behavior. Many of them also will have their own theories about what pure consciousness really is, coloring their experiential reports. The heterogeneity and the size of our sample can be seen as a strategy to counteract this problem. As noted, our participants came from fifty-seven countries, so their *Lebenswelt* and meditative practices will vary widely. Yet all those experiences, culturally contextualized in such different ways, also resemble each other in striking ways, as we will soon begin to discover.

To design this survey, we drew from the relevant body of Eastern and Western literature and conducted a series of pilot studies involving committed practitioners. From here, we extracted ninety-two characteristics of pure-consciousness experiences (see figure 1.1 in chapter 1), which were then formed into questionnaire items. The idea was to find clusters within these items that could serve as coherent and meaningful phenomenological dimensions of the experience of pure awareness (see figure 2.1 in chapter 2). Please again note how this dimensional approach also implies that pure awareness is not being treated as something absolute, as something entirely detached from the network of relations connecting all other conscious experiences with each other.[10] This is a new approach, in that it contradicts coarse-grained reifications, such as those that try to turn pure consciousness into a distinct and irreducible metaphysical entity, as some of the ancient religious conceptual schemes for describing meditative experience do.

What connects all these experiences as reported here is a relation of family resemblance. When the experiences are verbally described, we see probabilities that certain descriptions will go together, like patterns in a family in which the mother and father or the two youngest siblings may perhaps be seen together most often. Therefore—if you will forgive me one very last technical point—I will assume throughout that the concept of "pure awareness" has probabilistic, not definitional structure. On the one hand, what holds together the experiential aspects is not one single essence, but rather a lived pattern of probabilities. On the other hand, there really is something to the pure-consciousness experience. Pure awareness definitely seems to have a hard prototypical core—one to which I will return at the end of this book.

The Elephant and the Blind

Do you know the ancient fable of the elephant and the blind people? It originated in ancient South India and has echoed through the centuries.[11] The story goes like this: A king tells a group of people who were born blind and who have never come across an elephant before to try to understand what an elephant is like by touching it. Each person feels a different part of the elephant's body, but only one part each, like the trunk, the tail, a flank, or a tusk. They then describe the elephant. They describe it as being like a thick snake, or like a rope or a brush, or like a wall, or as being something hard, smooth, and pointy. How will they ever agree on whether they are all referring to a single phenomenon in the outside world?

Each item in our questionnaire can be viewed as one reaching movement, every answer as the touch of one person born blind. Each blind person's answer may certainly pick out one important aspect, but still, all the person has is a local sense of touch, never the global gist of the complete visual image. We might even imagine that our group of blind people is actually encountering not one elephant but a small family of elephants. The family members are not identical, but they are very similar to each other, and every blind person can prod and stroke and tap multiple times to explore an individual segment of one of the multiple elephants. Unexpectedly, in our own case, well over 1,000 blind people came to help. So maybe the king's servants thought it would be more practical to have the great crowd form a circle around a large herd of elephants? Perhaps the king even played a prank on the blind people by smuggling in a single rhinoceros, a donkey, or the occasional hippopotamus! Of course, the blind people's response patterns are only words, and precise statistical evaluations of them are merely numbers, and although both are important and may lead to surprising new discoveries, neither is the real thing. If anything in philosophical phenomenology is *the* prime example of ineffability, then it surely must be the experience of pure consciousness, self-knowing empty cognizance, the inner nature of appearance itself.

But what if we invited 1,403 blind people and asked each of them to touch an elephant ninety-two times? And what if we found 841 blind people whom the king had ordered to visit the elephant family every single day of their life and then simply asked them to describe, in their own words, exactly what they felt on a good day? Perhaps we could get fresh ideas. Maybe we would discover some details that we hadn't thought of before. There might be unexpected relations among the blind people's stories, revealing overlapping patterns of experience. Perhaps we could make some progress, at least compared to what we knew about elephants before.

This is not a scholarly monograph. I have tried to make it as reader-friendly as possible by grouping our meditators' reports into a series of loosely related chapters, each of an easily readable length and reflecting a theme in our respondents' experiences. My main goal was to bring some flexible structure into the wealth of material, in the hope of making it more accessible and opening this phenomenological landscape to readers who may come to it with widely diverging interests, and with either some familiarity with consciousness research or none. Basically, this collection of experiential reports is arranged according to their dominant phenomenological feature and the general context of occurrence, but not according to the rigid conceptual framework of some preexisting theory. It is a soft qualitative analysis that doesn't satisfy academic criteria for systematicity and completeness. If anything, this coarse-grained ordering reflects my personal biases, my own lack of understanding, and probably the fact that I am a long-term practitioner myself. Therefore, to avoid any misunderstandings, I freely admit that many of the following reports could have easily been grouped into at least four or five thematic sections, and, as you will notice very soon, the content of these sections that I have created has large overlaps throughout the whole book. These overlaps are sometimes so extreme that I even chose to use the same report (or a part of it) twice. The reason for these overlaps is that most verbal reports are so rich that they "cross-load" onto a variety of phenomenological categories, just as different questionnaire items may be related to multiple statistical factors. Consequently, the sequence of chapters is not meant to function as a "ladder," describing successive stages of something that, after all, has no stages. Rather, it is an attempt to make things easy for you if you haven't thought much about pure consciousness—an attempt to map out one possible reader-friendly path through varied phenomenological terrain. In an ideal world, this book will perhaps also provide a platform from which you and everybody else can make your own discoveries and start your own projects.

In the interest of anonymizing the reports and letting the experiences speak for themselves, I have left out almost all details referring to the location in which an experience of pure awareness occurred (which country, retreat center, or monastery; whether indoors or out in nature, etc.). I have also left out almost all references to the time (which year, or exactly which formal practice session on which day in a silent retreat), to the specific technique (e.g., *Vipassanā*, *Metta*, *Shamata*, or *Mahamudra*, *Chan* or Zen, Mindfulness-Based Stress Reduction [MBSR] or Transcendental Meditation [TM]), and to teachers, lineages, and organizations teaching meditation. However, I have included one chapter presenting meditators' descriptions of episodes when the "pure awareness" experience occurred spontaneously outside of formal practice (chapter 32), and here, more of the contextual details are provided. In addition, there are specific chapters devoted to the experience of pure awareness in dreamless deep sleep or during lucid dreaming (chapters 20 and 21) and to examples of episodes that lasted longer than a few minutes (chapter 33). In investigating the phenomenological material, you should always bear in mind that participants in our study came from many different age groups and from fifty-seven countries, and that they practiced (and sometimes combined) different meditation techniques. This breadth is intentional. It is one of the factors that make the material so interesting because it lets us see the commonalities and invariant structures of conscious experience more clearly, as well as the countless variations they underlie.

The best thing about this book is that you can choose your own route through it. Every chapter has two parts. The first part always begins with one or more novel metaphors or other analogies for pure awareness, taken from the database of responses. One of the many surprises in this initial study was how many new and beautiful metaphorical descriptions of pure awareness our participants seem to have spontaneously created, without ever having been asked to do so. I was deeply impressed by many of them, and I think you may be too. Some of them are still haunting me after many months because somehow they convey something that is hard to put into words, and they open up an entirely new way of looking at things, a possible perspective that I had never really been aware of. Following the metaphors, there will be a selection of phenomenological reports—sometimes only a few, sometimes many, sometimes in a particular order, sometimes not. In the first part, I have always tried to restrict my own comments to an absolute minimum. You can make sense of the material for yourself.

If you already are a regular practitioner of meditation, then the first part of each chapter may sometimes be a special reading experience, one that may even affect your practice itself—perhaps in unexpected ways. I predict that you will find many of the descriptions of pure awareness pretty obvious and self-evident, perhaps even slightly boring or trivial. Others will make you raise an eyebrow, roll your eyes, or gently shake your head in disbelief. There may be parts of the elephant that you have never touched yourself, and you may naturally suspect that some of these people are actually talking about another kind of animal altogether. Have they bumped into a donkey? Are *they* donkeys?

If you are emotionally attached to a teacher, a lineage, or the conceptual framework of a specific belief system, then you may even get angry. If you adhere to one of the many traditional theories about "stages" of meditation or a given taxonomy of contemplative experience, or, even worse, if you already knew what the "true essence of consciousness" is before you opened this book, then I predict an aversive reaction—which can be a brilliant opportunity to deepen your own mindfulness practice. But I also hope that you may find at least a few experiential reports or novel metaphors that genuinely surprise you because they describe something that you have known for a long time, but in a way you would never have described or considered it yourself. If you are anything like me, the discovery of such unexpected new perspectives on your own inner experience will really be what makes studying these reports rewarding.

Importantly, this last point also applies to you if you have never practiced meditation in your life: Yes, you *do* already know what this pure awareness thing is! Maybe you just never saw why it should be interesting or relevant. You forgot about it. Maybe you saw the simplicity but not the profundity (chapter 12). Nevertheless, a small number of the experiential reports may stay with you for weeks, inspiring you to take a closer, more serious look at your own experience—not in formal meditation, but in everyday life. Such surprises and unexpected recognitions are valuable because they force your brain to update the model that you already have of your conscious experience itself; they literally change your conscious self-model. Perhaps there even is a sense in which the quality of awareness itself never really was "your very own" conscious experience (for more on this, see the "contraction principle" in chapter 8). Whatever your background, I hope that in your own way, you will benefit from the fecundity of surprises—in your everyday life, in your meditation if you practice it, and, if you do research in this area, in the new interdisciplinary project of homing in on a minimal model explanation for consciousness.

Every chapter is like an incomplete three-course meal, with the dessert coming first. While the first part is inspirational and for enjoyment, and should be easy to read and digest, the second part offers some slightly more substantial food for thought. The second part is not the main course, but a small set of starters. Without going too deep, I will draw your attention to some philosophical or scientific issues that are related to the phenomenological material just presented. Sometimes I will provide new conceptual instruments that may be helpful in thinking more clearly about what the reports presented show—and also about what they do *not* show. At other times, I will look at some of the latest empirical research or simply provide additional information or a bit of context. At other times, we will briefly turn to some classical texts for inspiration, such as from Buddhism or Western mysticism. These are all intended as an assortment of hors d'oeuvres and appetizers. There will also be pointers to other publications that I found relevant, to potential ingredients for future meals—because what I really want is to encourage you to go home and actually begin cooking your very own main course, regardless of whether this means doing new and better research, improving your meditation practices, or anything else that you feel inspired to try. If this book becomes a source of inspiration to a wide range of readers, then it will have done what I hoped for.

All the parts of the book speak to all the others, but after reading this introduction, you may like to jump straight to the concluding chapter for a quick overview, or to the epilogue on *Bewusstseinskultur* for some broader context. You may be most interested in the philosophical commentary offered in the second half of each chapter, or you may like the experiential reports best. If you are a computational modeler, you may treat every chapter as a target description; if you're a neuroscientist, you may well start generating chapter-by-chapter hypotheses about possible physical correlates; and if you are a meditator, I hope that you might find yourself making mental notes of practical pointers and comparing the descriptions against your own experience.

A companion website featuring supplementary material accompanies this book, in case you are looking for extra ingredients or still have room for a cheese course.[12] And, in case you would like to pursue any theme or question more deeply, I offer further references, plus a glossary of terms, at the end of the book. Finally, regardless of whether you are a meditator or not, you can participate in and support our research by completing an updated version of the original survey. But again, the best thing of all is that you don't have to read this book in a straight line. I hope you'll enjoy wherever it takes you.

1 Relaxation

Oh my God! The tension I call Jeff was gone! [#2417]

A deep sense of relaxation is one of the most common characteristics coemerging with the experience of pure awareness. For example, one of our participants conveyed the feeling figuratively, ". . . as if I was floating in water, on my back, under a bright sun" (#2065), while another described it quite simply and literally as a feeling of "Okayness" (#172). Perhaps pure awareness is something like a previously unnoticed baseline state of conscious experience, an inner surface on which we are always "floating," even though our attention is almost always captured by something else, like the bright sun above us? Perhaps meditation is what enables us to become aware of this invisible inner surface and then move onto it, learning how to effortlessly glide along from moment to moment, like a paraglider who has discovered a thermal, a thin layer of heated air to ride on? In any case, to judge from the phenomenological material that we have gathered, it seems clear that the nonconceptual awareness of awareness itself is something one can relax *into* (or *onto* the surface of) by gradually letting go, gently dissolving all residual tensions in body and mind. One of our participants figuratively described this way of accessing pure awareness as "gently 'sitting back' into it" (#2619).

The phenomenology of relaxation and ease is so salient and widespread that I will give just six introductory examples here:

> 1647 [. . .] Like the more I can quiet my physical and mental stress and anxiety the more I can uncover of this baseline reality of experience. It did not feel like something I could strive to achieve through hard work . . . but more like something I could relax or ease into.
>
> 1870 A feeling of complete ease, well-being, and presence with waves of bliss, effortless and with almost no thoughts.

2319 Pure relaxation, happiness.

2620 My sensations become soft and subtle. My body is flooded with this gentleness, it relaxes, muscular tensions are released . . . until I'm hardly aware of it anymore and enter a deep space of silence. [. . .]

3012 [. . .] a feeling of calmness appeared, my body relaxed, the breath breathed itself. I had only a weak feeling of "I"; the experience of clarity, boundlessness, connectedness with everything was in the foreground. Feelings of peace, happiness, "everything is okay the way it is," arose within me. [. . .]

3363 It started as a regular experience, with at first a lot of ideas, images, and "noises." Then a gradual relaxation. Then a beginning of letting go of any self-control. Then rather brief toward the end of the session, a state of peace, clarity, dissolving of body awareness and stillness of the mind.

Psychometrics, Nonsensational Awe, and Existential Ease

> The more you talk and think, the farther away you get.
>
> —*Inscription on Faith in Mind,* Seng-ts'an (Third Chinese Zen Patriarch; † 606)

Let us now slowly begin to look at these first-person reports from the third-person perspective of empirical science. We had 1,403 usable questionnaires; our participants had a median age of fifty-two years (ranging from seventeen to eighty-eight) and were roughly evenly split between men and women (48.5 percent versus 50.0 percent stated their sex). The majority of meditators practiced regularly (77.3 percent), were free of diagnosed mental disorders (92.4 percent), and did not regularly use any psychoactive substances (84.0 percent). Vipassanā (43.9 percent) and Zen (34.9 percent) were the most frequently practiced meditation techniques.

Let us look at some of the ninety-two questions we asked.[1] Figure 1.1 shows the statistical distribution of our ninety-two items across the sample of participants. Here, we can see that the experiential qualities of "being at ease," "being at peace," "being in the present moment," and "feeling whole" are even more frequently associated with the experience of pure awareness or consciousness *as such* than "relaxation" itself is. We can also see that in terms of sheer frequency, the phenomenological qualities of ease and relaxation are closely followed by other item descriptions that have a more philosophical flavor, like "pure being," "nonvisual clarity," "unity," and an experience of "deep, unbounded silence."

As you can see, the highest average ratings were given to items related to well-being and relaxation; the overall lowest ratings concerned the presence of pain during the

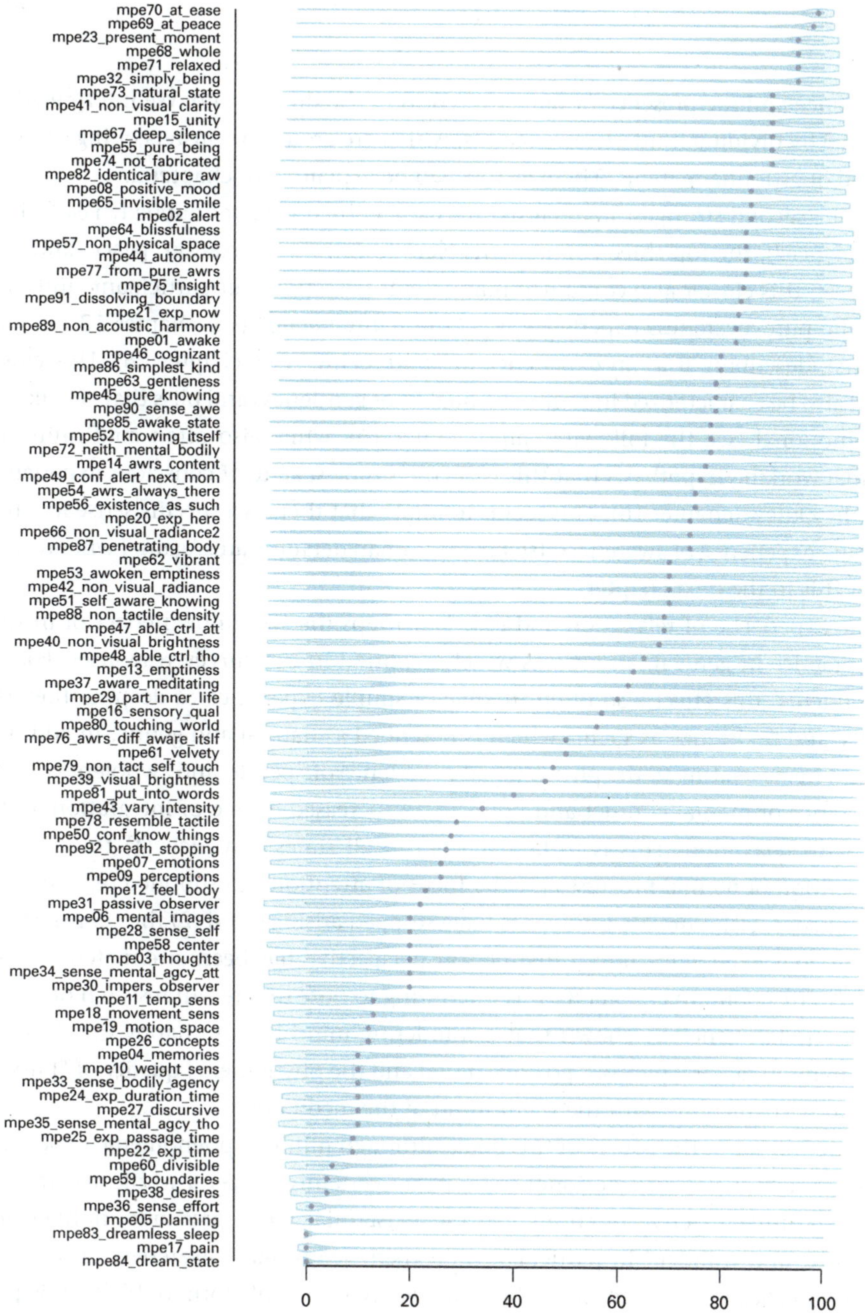

Figure 1.1

Statistical distribution of questionnaire items in our analysis sample (n=1403; dots indicate medians). The published version of this paper, including more color figures and the questionnaire itself, is available for free on mpe-project.info. There, you will also find research updates, free supplementary materials to download, and, in case you want to participate in our research, the latest version of our survey.

experience and the presence of pure awareness in the dream state. We interpreted the material by dividing it into twelve factors, which are groups or clusters of highly interrelated questionnaire items. These factors will be explained in chapter 2.

In the second-strongest item cluster, factor 2 (which we later dubbed "Peace, Bliss, and Silence"), the top three components were feelings of relaxation, peace, and ease. However, this cluster also contained many items via which our participants indicated that the pure-awareness experience was "the natural state" (see chapter 12 for some reports), describing it as an experience of "pure being" (see chapter 26). This cluster could also be characterized in terms of a quality of gentleness and bliss that is strikingly nonemotional (see the following, and chapter 16), while also often confirming the original working hypothesis that this was "the simplest state of conscious experience" that they knew (chapter 16). This seems to imply that there is a specific sense in which the pure experience of awareness itself is the human default state. Again, I will say more about this in the next chapter.

The experience of relaxation and ease is perhaps the best-known effect of meditation practice in general, and it was widely reported in our survey responses. For our respondents, this effect also often involves a simple experience of deep, unbounded silence and "existence as such," which is described as natural and gentle. In addition, the two items relating to the phenomenal experience of "peace" and "wholeness" show an absence of mental conflict and point to an increased degree of integration associated with relaxation. This is to be expected in all states in which (1) the constant competition of mental processes for the focus of attention has subsided (e.g., because spontaneous, task-unrelated thoughts have stopped arising); and (2) the "contraction" of conscious experience into a first-person perspective has been attenuated (we will return to this point in chapter 8). It is interesting to note that recent research on mind-wandering has generally shown that a wandering mind is an unhappy mind,[2] while three items in factor 2 pick out positive mood and the phenomenal quality of "bliss" as frequently coemerging with the silent mind of the meditator.

As we will see in the course of this book, although minimal phenomenal experience (MPE) as such is not an emotional state, it can certainly trigger a whole spectrum of mostly positive affective reactions like joy, existential relief, gratitude, nonpersonal love, awe, and wonder.[3] In particular, the experience of pure awareness or MPE can sometimes coexist with a mostly subtle but clearly noticeable form of bliss, an experience that has sometimes been described as an "invisible smile" (item 65 in our questionnaire). Factor 2, therefore, includes not only the phenomenology of peace, wholeness, inner ease, pure being, and silence, but also various forms of what in German is sometimes called *stilles Entzücken* ("silent delight"). One report (#1629) described this subtle yet nonemotional experience like this: "There was also an equanimous joy—not coarse/rugged, but a subtle yet pervasive equanimity with positive intonations." To be more precise, pure awareness seems intimately connected with a calm and entirely undramatic phenomenology of rapture and "nonsensational awe."

These observations lead us to the first of two new phenomenological concepts that I want to offer in this chapter. In the course of this book, I will provide a series of new conceptual instruments to help us gain a new perspective on some philosophical issue or to describe the phenomenology of meditation more clearly. "Nonsensational awe" is the first of these new tools. I take it directly from one of our participants (who called it *sensationsloses Staunen* in German, #3524), and it refers to a subtle sense of wonder that some of us, meditators and nonmeditators alike, may have experienced now and then and may even remember from our childhoods (see chapter 15 for more on the experience of joy, awe, bliss, and gratitude; and chapter 32 for the qualities of spontaneity and effortlessness). Another of our participants called this experience "calm awe" (#1662); both terms very precisely pick out the experiential quality I am referring to.

Here is a second conceptual tool to describe the phenomenological character of awareness per se in meditation more clearly, another new instrument that may perhaps prove to be helpful at a later stage: "existential ease." What does this one mean? Here, my point is that contemplative practice goes far beyond mental or physical relaxation because it often generates a much deeper form of ease and serenity, which I will hereafter refer to as "existential ease." If we take reports about the fine-grained phenomenology of real-life meditators seriously, then we find not only that the phenomenal qualities of relaxation and ease are statistically correlated with the experience of peace and deep, unbounded silence, and with the aforementioned qualities like gentleness and the feeling of "being whole," but also that they are intimately connected to states of "simply being" and "pure being." If we accept our participants' reports at face value, then they refer to states in which one's own mere existence *as such* is felt. Again, we will devote a whole chapter (chapter 26) to this aspect later.

A tentative first conclusion is that the phenomenology of relaxation, peace, and ease can be related to those states of conscious experience in which our own being and existence per se, perhaps the sheer experience of aliveness, are pulled into the foreground. There are many kinds of relaxation, such as the dull and/or slightly disoriented relaxation induced by psychoactive substances like alcohol or cannabis or the visceral relaxation that follows a hearty meal. An emotional state of low tension, lacking the arousal triggered by anger, anxiety, or fear, can be pleasant while entirely lacking any sense of mental clarity and any quality of "pure being." It is the phenomenological *integration* of relaxation, lucid clarity, and a state of pure being that I mean when I refer to "existential ease." Pure awareness is sometimes an *existential* experience, but nonconceptually, in a silent and very undramatic way. The subjective quality of existential ease, therefore, could be a second important aspect of what it means to experience the phenomenal character of awareness per se. Perhaps the two experiential aspects of "nonsensational awe" and "existential ease" could function as our first candidates for new qualitative markers of MPE.

In slowly and carefully beginning to approach the experience of pure awareness, we must be acutely aware of the limitations of scientific research, because contemplative phenomenology without intellectual honesty immediately leads us into dead ends (I will say more about this in chapters 17, 31, and 32, and the epilogue). Here is a first example of what we must always keep in mind: We must never forget that we are dealing here only with statistical properties of verbal reports, never with the experience itself. Reports can be strongly influenced by respondents' background beliefs, by their cultural context, and/or by the specific conceptual tools that were available to them to describe their own experience. Reports may be influenced by a subject's capacity to even understand the questions in a survey or by changes in the way that responses are given as the questionnaire unfolds (e.g., as a result of gradually developing a putatively clearer idea of "what this is all about"). Experiential reports may also be colored by subtle misunderstandings (like this being some sort of "test" that one has to pass, not just a descriptive survey), by a resulting motivation to please the experimenter, or even by the opposing aim to successfully demonstrate to scientists that, as committed practitioners, they know all of this much better themselves and these questions are all quite dubious and childish—a futile and misguided attempt to approximate something utterly ineffable that of course the subject herself knows with great certainty.

Do you recall the fable of the elephant and the blind? I will come back to it from time to time. Every blind-born person has a different background and comes from a different home, and all the hands touching the elephant look slightly different too. Some of the blind people will be bored, some will be a bit too interested, and some will encounter a rising sense of uncertainty. And some will ask: Who is this king anyway? Who gave him the right to investigate the elephant in this way, sending out his poor servants to drag us all into this slightly awkward situation? Probably most of the other blind guys are crazy anyway . . .

But again, if we took an ensemble of 1,403 blind human beings and asked each of them to touch the elephant ninety-two times, then we could cancel out a lot of local errors simply by statistical averaging. There will always be outliers, but combining semantic analysis of existing, traditional theories of pure consciousness with statistical methods, while also adding a more qualitative evaluation of actual written reports from different cultural backgrounds, can perhaps help.[4] It may eventually give us a more robust picture of the experiential landscape. However, my first general point still stands: We must never forget that we are dealing with properties of verbal reports, with concepts and theories, but never with the experience itself.

I do believe that great and important progress can still be made with regard to pure awareness, but we must always be clear about the limitations of science—especially when it comes to contemplative practice itself. As Rgyal ba Yang dgon pa says in his *Song of the Seven Direct Introductions*: "Don't try to describe this awareness that is beyond thought. Don't adulterate it with conceptual analysis but give it free rein!"[5] MPE, the experience of pure awareness, is *the* prime example of ineffability. This phenomenological fact has to be recognized and remembered; at the same time, it is one of the many reasons why the experience of pure awareness should matter so much to all those interested in an empirically informed philosophy of mind. As I will explain in chapter 17 and expand on at many other points in this book, philosophy and science are "epistemic practices," just as meditation itself is, because they aim to generate knowledge. But there are different kinds of knowing; it is difficult to build bridges; and at least for meditative practice itself, the simple statement presented in the epigraph at the start of this section may well be true: "The more you talk and think, the farther away you get." Its author, Seng-ts'an—who died in 606 as the third Chinese Patriarch of Chan after Bodhidharma and as the thirtieth Patriarch after Siddhārtha Gautama Buddha—is traditionally honored as the author of a famous poem entitled "Inscription on Faith in Mind." There is a fundamental problem of ineffability, and this ancient poem gives us the reason why:

> When the subject disappears,
> There can be no measuring and comparing.[6]

2 Peace

A feeling of infinite peace [#3625]

A peaceful streaming [#1354]

Peace is the absence of conflict. The *phenomenology* of peace is the absence of all forms of mental conflict within conscious experience itself—for example, absence of competing desires, emotional perturbations, and cognitive fragmentation. Let us now look at a set of examples in which meditators describe the phenomenology of peace from different perspectives. You will notice that in these reports, the experiential quality of "peace" characterizing pure awareness often overlaps or coemerges with other subjective qualities, like "perfection," "depth," "density," "luminosity," "connectedness," "pure being," "unboundedness," "permeability," and "bodiless body-experience." We will investigate many of them more closely as we go along.

1968 [. . .] I was aware that I was meditating but also felt a wonderful peace and perfect calmness come over me. [. . .]

2472 [. . .] There was nothing left anymore except here and now and ultimately a deep feeling of peace and quiet joy.

2867 [. . .] suddenly I became aware that my eyes were closed but it was very bright; I felt heavy and warm and expanded, and expanding. Totally at peace and still; breathing happened for me, to me—I was breathed. I was everything at once, all encompassing, and nothing at all, no form and no shape. Brightness and density and utter calm. [. . .] and I was left with a feeling of having been blown away like dandelion seeds on a puff of air. In awe and at peace. Also, a feeling of giddiness, like I had just stepped off a ride.

2911 [. . .] The change from tired to awake was suddenly there, and I felt detached from my body and from almost all thoughts. I was aware of myself and the connection to everyone in the room. Deep calm and peace came over me and I felt my body as present, but light and unburdened. I felt only joy and clarity and an invisible and inaudible vibration in the room.

2928 When my mind comes to rest, sometimes a kind of pure being appears. I then feel warmth / languorous energy centered in the abdomen, and I feel a connection to the earth without having a real center, warmth and vastness, unity and openness. Absolute peace and quiet, I then no longer perceive individual parts of my body.

2985 [. . .] In my opinion, the most important thing was the letting go. I was able to perceive bodily sensations and thoughts, but in contrast to everyday life I did nothing with them. And this brought me step by step to a deeper level, to a state of consciousness that I had never experienced before. It was a state of infinite peace. [. . .]

3163 [. . .] I rested for most of the meditation session in observation of consciousness itself. My breath and all other bodily sensations were so far from dominant that for a long time I was able to observe consciousness itself, without other impressions seizing my attention. Thoughts came now and then, but as soon as they appeared I noticed them and they disappeared without carrying me away. I had no strong emotions, merely a feeling of deep calm, clarity, and peace.

259 [. . .] the feeling of wholeness, weightlessness, inner and outer silence, profound peace, connectedness, having arrived, flooded with light, very slowed-down breath, dividedness, nothing is important anymore, carried by the breath, present.

3314 I experience in everyday life a changed form of perception of people and things, of the permeability of everything, which in short moments of meditation is then transformed into an all-encompassing silence and peace.

3495 All boundaries are lifted. Breath happens, deep inner peace and calm.

3521 [. . .] a feeling of absolute calm and peace . . . without questions, but still with the knowledge that there are answers to everything . . . There was a deep satisfaction and calm. A state of endlessness but with the knowledge that you can leave it at any time.

3571 [. . .] I feel in the now! Feel empty, satisfied, connected with my environment, with the animals and plants. Peaceful. Feel no pain, fear, sadness, anger, disgust, shame, or guilt. Just peace and quiet.

Twelve Factors and Their Limitations

Yoga is the restriction of the fluctuations of consciousness.
—Patañjali, *Yoga Sūtra* (I: 2)

In our psychometric analysis, reports on the experience of peace turned out to be part of a cluster of items picking out the experience of relaxation, positive mood, and the phenomenon that in chapter 1 I called "existential ease." The same statistical factor also refers to a simple experience of deep, unbounded silence and "pure being," a state of conscious experience that is described as natural and gentle, often coemerging with the global experiential qualities of wholeness and peace. What this cluster of items shows is that our normal state of mind—which is characterized by almost constant inner chatter, spontaneously arising task-unrelated thought, automatic future planning, and unbidden memories—is actually *not* a state of existential ease at all. We long ago grew used to it, but if we calmly view it from a distance, we see that it actually results from a strenuous and energy-consuming process.

Meditation practice creates this distance. The practitioner begins to notice frequent attentional lapses, caused by a permanent competition of different thoughts for the control of behavior and the focus of attention, leading to a default state of mental conflict and continuous fragmentation. In an old analogy offered by Jetsun Milarepa in the eleventh century, attention for most people is like a dog that chases every stick thrown to it, all day long, almost automatically and to the point of exhaustion, blindly following every thought that the mind coughs up. Real meditators are not like dogs; they are like lions that turn around, stand their ground, and keep facing the thrower even after the stick is thrown. And as Milarepa points out with a wink, you only throw a stick at a lion once. As we will explore in more depth in chapter 17, I consider genuine meditation practice to be an *epistemic* practice—something that is primarily about insight, about acquiring a nonconceptual form of self-knowledge—but this does not mean that all of its discoveries will be pleasant or uplifting. The first thing that meditation does is to start revealing our phenomenological *conditio humana*, and this can certainly be seen as an undignified or even humiliating condition because it involves a constantly recurring loss of mental autonomy (more on the deeper roots of this process in chapters 8 and 25; see the epilogue for its relevance to ethics and culture).[1] It is therefore only to be expected that a state lacking all the features of the hectic everyday mind should be experienced as peaceful and holistic.

The ancient term "yoga" refers to this holistic phenomenal quality of unity and integration instantiated by a silent mind (it may be derived from either *yujir yoga* [to yoke] or *yuj samādhau* [to concentrate]). In section 6.10 of the *Katha Upanishad* (or *Kaṭhopaniṣad,* which goes back to at least the fifth century BCE), we find the following statement: "When the five senses, along with the mind, remain still and the intellect is not active, that is known as the highest state." Probably the most famous definition, however, is given in Patañjali's classic statement *yogaś citta-vṛtti-nirodhaḥ,* which is translated in the epigraph above and probably dates back to the second century BCE.

Let us now take a brief look at the psychometrics of pure awareness as compared to other approaches—and keep an eye on its various methodological limitations. Figure 2.1 gives a list of the ninety-two descriptive questions that participants answered, as well as six different factor solutions showing the number of extracted factors used to explore different interpretations of our data. If you look for the blue dots, you'll find factor 2, which we named "Peace, Bliss, and Silence." You can see that it was remarkably stable across different factor solutions.

Reading the experiential reports in the first part of this chapter, you will immediately have noted a strong overlap with other phenomenological elements (e.g., with a feeling of "weightlessness" or a sense of "clarity"). Therefore, many of these reports could have been included in other chapters of this book (e.g., in chapters 5 and 24). This shows one difference between statistical analysis and qualitative assessment. Figure 2.1 is based on mathematical relations between single questionnaire items in a psychometric study. By contrast, my selection of passages from experiential descriptions given by meditators reflects a qualitative evaluation of explicit verbal reports. This evaluation is not grounded in statistical analysis but is in some ways much richer and touches on many aspects at the same time. Statistical analysis and qualitative assessment are two very different ways of approaching pure awareness, of trying to get closer to the elephant. For example, in figure 2.1, a color like dark blue represents membership in a specific factor, relative to a certain mathematical solution. This solution was chosen as optimal under joint conceptual and statistical considerations, but it had only limited explanatory power. Future studies, therefore, may come to better and perhaps very different results.

My own way of grouping together experiential reports, on the other hand, reflects what I intuitively took to be the most relevant or dominant phenomenological features. This approach is very different, and it has a correspondingly different set of serious limitations and weaknesses. First, as this is not an academic monograph, but a book trying to communicate some important results to a wider audience, my qualitative evaluation is not a truly *systematic* qualitative analysis at all. It reflects the implicit phenomenological intuitions of someone who has meditated twice a day for more than forty-seven years, and who has also worked as a researcher in philosophy and cognitive

Figure 2.1

Statistical analysis of the MPE-92M questionnaire into factors. Questionnaire items are listed on the vertical axis. The columns show how the items can be grouped into an increasing number of factors. The first column, for example, shows the result of forming the items into six groups, or factors. The last column shows our preferred way of grouping, using twelve factors (the number of factors is shown on the horizontal axis). Each grouping into a given number of factors is called a "factor solution." The colors represent the various factors, and the size of the circles indicates how strongly a given item is related to its factor. Open circles indicate a negative relationship between an item and its factor. Please note that, although an attempt was made to identify the same factors across all the factor solutions by coloring them, "sameness" here is not objectively definable; the best we can do is an operationalized definition of the degree of prototypicality.

science for some four decades of his life. Clearly, this introduces a lot of noise into the signal, and it inevitably will have led to distortions and biases that I am completely unaware of. Second, not all our participants gave an additional written report. Distortions may have arisen from the fact that only some individuals selected themselves into the current group, yielding a biased sample for phenomenological analysis. Generally, qualitative analysis is more subjective than statistical analysis. Its results cannot really be tested for significance or extended to a wider population, and they are usually not systematically replicable. Finally, my own qualitative data analysis is necessarily based on a classification of reports according to certain phenomenal properties and attributes describable in natural language, whereas quantitative analysis classifies data based on computable values. This is another important difference.

You may recall from the introduction that a third way to approach the phenomenology of pure awareness is to look at canonical texts. These can show us how the elephant has been described over the centuries, for example by philosophers, spiritual teachers, and scholar-practitioners in the East and the West. This involves a *semantic* level of analysis, as opposed to a statistical or qualitative evaluation of reports about minimal phenomenal experience (MPE) states. In its fullest form, such a method would assess as exhaustively as possible the meaning of those concepts and theories that have been used to describe pure awareness, in many cultures and often in a prescientific context, and drawing on both quantitative and qualitative methods. I have done it the old-fashioned, brain-based way here, with no pretensions to comprehensive coverage. But just as with the experiential reports describing pure awareness, hopefully others will go down the data-based route in future research projects.

When I conducted a semantic investigation of canonical texts, I found six semantic constraints, which I used to narrow down the meaning of "pure awareness" and define a first working concept of MPE.[2] As we saw earlier in this chapter, the experience of peace can be interpreted as a low degree of mental conflict and perturbation. This observation is interestingly connected to the second of the six semantic constraints that I found in my investigation. Based on an extensive review of the literature, this constraint was termed "Low Complexity."[3] Generally, pure consciousness is often described as the complete absence of intentional content, in particular of high-level symbolic mental content (i.e., discursive, conceptual, or propositional thought), but sometimes even as the disappearance of all sensorimotor, interoceptive, and affective content. Pure consciousness is a thoughtless state, but in deeper, fully absorbed stages of meditation, all perception of the environment and all sensory awareness of the body from the inside, including sensations and emotions, may also disappear. If we look at the relevant literature, one striking discovery is how many phenomenological characterizations of MPE episodes are exclusively negative (in the sense of negation-filled, saying what they are *not*). Traditionally, pure awareness is described as follows:

- **Nonsensory**: MPE itself instantiates no perceptual qualities; it is not a form of sensory experience.
- **"No-thingness"**: Absence of the phenomenal property of "objecthood"; no subjective experience involving reification, such as of distinct multimodal objects as integrated from different sensory features and as segmented from a background or perceptual scene; importantly, also lacking boundaries, substantiality, and objecthood.
- **Nonmotor**: Absolute stillness, no motion in space.
- **Atemporal**: Absence of temporal experience, no motion in time.
- **Noncognitive**: Nonsymbolic and nonconceptual; no discursive thinking, mental imagery, or mind-wandering; no movement in mind.
- **Nonegoic**: No self-location in time, no self-location in space, no quality of ownership, agency, or goal-directed control (either mental or bodily).
- **Unbounded**: No second, finite region to which attention could be directed, and no consciously experienced boundaries, limits, or horizon.
- **Aperspectival**: No "actively knowing self," no consciously experienced model of an epistemic agent as directed at objects of knowledge, and no passive personal-level self-as-subject either.

Because it is so full of negations, pure awareness is often conceptualized as an entirely contentless form of conscious experience. However, this is a controversial issue because there could well be a global form of experiential content that is so subtle or abstract that, in the absence of the right cultural context and lacking suitable conceptual tools, it would be only natural to describe it as mere nihilistic emptiness or as something that doesn't really exist—or even as something that exists beyond the distinction between existence and nonexistence, always already preceding it. What we can agree on is that episodes of pure awareness are described as lacking any *complex* content, that they are extremely simple—for example, because the phenomenal character of awareness per se lacks internal structure and temporal dynamics. This is what "Low Complexity" means. But then "Low Complexity" also means a low degree of mental conflict and perturbation, and internal simplicity therefore may be directly related to the experience of peace.

Our best current theories of brain function understand it as involving a constant process of conflict resolution (known as "hierarchical Bayesian updating").[4] Representational content, including our conscious model of the world, is generated via continuous attempts to minimize prediction error on many levels. If this new theoretical vision is heading in the right direction, then reducing the number of levels and reducing the "temporal thickness" that is created by constantly predicting outcomes in a more distant future will automatically minimize internal conflict, simply by flattening the predictive hierarchy.[5] The phenomenological hypothesis following from such a flattening would be a conscious experience of gradually "returning to the present moment," eventually leading to an experience of "timelessness" and a complete absence of consciously experienced conflict. You may know that one of the classical concepts for the experience of pure awareness is *samādhi*, referring to a peaceful state of complete, thoughtless equilibrium: the eighth and final level identified in the *Yoga Sūtras of Patañjali*, as well as the last of the eight elements of the Noble Eightfold Path in Buddhism. Interestingly, the ancient notion of *samādhi* already contains not only the semantic element of peaceful equanimity but also the idea of flattening a more complex internal state, an inner landscape: It is often translated as "even intellect."

Back to our psychometric study. Aside from the limits of any given analytical method applied to its data, it has a number of inherent design limitations too, of course. Most fundamentally, there is no guarantee that participants understood the instructions and the concept of "pure awareness" introduced therein in the way we intended (e.g., 3 percent said right at the outset that they didn't know what the term meant). Varying understandings will lead to unwanted variation in the responses. More generally, differences in participants' understanding of *any* part of the survey, including both the items themselves and the demographic questions, could have led to undesired response variability. Such differences can occur for a variety of reasons. First, participants will typically have been exposed to different background information about meditation. If they have actively engaged with the relevant literature, there may be large differences in what they selected for study. For instance, many of our participants (77 percent) were regular practitioners who had sustained their discipline over years. Their practice and the motivation behind it may be tied to an individual project of meaning-making (more on this in chapter 17). It is plausible to assume that this may often be anchored in adherence to specific belief systems and the conceptual framework of a certain lineage, spiritual tradition, organization, or teacher. The terminology employed by such theories, as well as their epistemological and metaphysical background assumptions, may "contaminate" survey responses. This can happen on many levels, not only cognitively but also in relation to embodied action.

Respondents' implicit assumptions about their own inner experience, therefore, may have played out not only with respect to intellectual understanding and verbal expression (e.g., biased memory recall or the actual wording of phenomenological reports) but also via effects on motor control (to answer each item, respondents had to use a computer mouse to position a slider on a horizontal line ranging from 0 to 100, where the position on the line indicated the degree of agreement with the item). For beings like us, belief systems and theories act as unconscious priors, contaminating what we think and say and do by directly shaping the low-level information flow in our brains. Thus, theory contamination is always *embodied* theory contamination. I will say more about the methodological problem of "theory contamination" in chapter 17 and at other points as we go along. For the special case of research on pure awareness, this problem may be particularly severe.

We know from correspondence with potential participants during the pilot phase that such belief systems can even lead some people not to participate in scientific projects at all. Some of them will reject any scientific attempt to approximate something that they consider to be as fundamentally ineffable and soteriologically crucial as pure awareness, perhaps out of a fear of disenchantment (see chapter 17). Many of the traditional frameworks not only provide a conceptual system of phenomenological descriptions but also present us with a normative phenomenology: They explicitly tell us how a practitioner's progress should unfold, how the different "stages" of meditation *should* look. This is not to say that ancient theories of pure awareness—based on literally millions of hours spent in silent meditation by serious scholar-practitioners who came before us—do not have great value. But it also shows why a fresh, bottom-up approach in a new and globalized historical context has its own value—because it helps to weaken the omnipresent influence of theory contamination. Please also note that the specific belief systems many meditators adhere to are not only related to certain lineages or teachers but quite often include an explicit *metaphysics* of consciousness, a pregiven assumption about what conscious awareness really is.[6] In sum, theory contamination may have both introduced a self-selection bias into our sample and biased the questionnaire responses from those who did respond.

Even assuming a singular, precise, and unequivocal understanding of the concept of pure awareness, it is still possible that some responses were driven largely by a desire to report particularly impressive or personally meaningful experiences rather than to adhere as closely as possible to the instructions. There is some evidence for this in the 1,183 written phenomenological reports presented in this book, where a number of participants chose to report not ordinary MPE as it may occur during formal meditation practice and full-absorption episodes, but rarer, often quite dramatic, nondual states in which all subject/object structure had spontaneously disappeared. This may have been different for the overall cohort of 1,403 participants, had they all provided reports. As it was, the effect may have been an indirect consequence of us asking participants for a description of an experience "in which the quality of pure awareness was particularly salient and/or one which you can remember particularly clearly." What can be remembered particularly clearly might often be the most striking or impressive states, which may not necessarily be the most paradigmatic instances of "pure awareness."

A related issue is what could be called "response drift": Some participants may have started with a somewhat vague idea of the target state of pure awareness, which became increasingly focused as they moved through the questionnaire. Conversely, they may have started out with a very specific concept, perhaps dictated by commitment to a particular theory, and then shifted away from it as they moved through the questions. Responses given earlier, therefore, might relate to a somewhat different target than the later responses discuss.[7]

In studies of private subjective experiences, there is a general problem with verbal report—namely, ineffability, the difficulty or even impossibility of expressing certain phenomenal states or qualities in words. This problem may be particularly pronounced when it comes to MPE, given that the experiential quality of "pure awareness" has long been regarded by meditators as the paradigmatic example of ineffability. However, it may arise for several more specific reasons. One is a lack of concurrent reportability during full-absorption episodes.[8] Thanks to the nondual nature of such states (i.e., the lack of any phenomenally represented subject/object structure; see chapter 27), there will by definition be no intention or cognitive capacity to verbally report, mentally categorize, or actively memorize the phenomenal character in question. Another reason may be that MPE's timeless content is unlike[9] any of the more familiar sensory, motor, and interoceptive qualities that we experience every day, which makes it hard to grasp verbally by comparing it to such qualities. A final reason is that the experience of MPE seems to lack any internal structure or "grain." It seems uniformly dense. This phenomenal quality of "ultrasmoothness,"[10] as it has been called in the philosophical literature, may also hinder verbal report by depriving it of any discernible entry points for description or functional analysis.

Beyond the matter of verbal expressibility, another issue faced by all studies based on retrospective self-report is the unreliability of memory recall. Human autobiographical memory is notoriously fallible, including for the recall of previous mental states.[11] In our study, we were asking participants to report on experiences that occurred in their past, possibly even decades ago. Memory retrieval for such events will inevitably be biased, depending on various factors such as the amount of time that has passed, current attitudes to that past period, and so on. The uncertainty introduced by such errors may be substantial but is likely to remain unavoidable in studies of this kind. One could probably greatly mitigate the memory issue by asking meditators to provide reports immediately after a meditation session (e.g., toward the end of a silent retreat).

Other generic weaknesses of our anonymized online survey approach are that the identities of participants could not be verified and we were unable to detect fraudulent responses or multiple responses by the same person. As part of privacy protection, IP addresses were not stored and therefore could not be used for data plausibility checks.

On a conceptual level, there is an even deeper variant of the memory retrieval problem. The concept of "autobiographical memory" refers to the representation of events that at the time when they occurred were phenomenally represented as being experienced by a conscious *self*. If they weren't, they would not really be *auto*biographical memories, memories about your *own* life. This means that the respondents' claims that *they themselves* consciously experienced a selfless state at the very time when it actually occurred are highly dubious from a methodological perspective. Are retrospective reports of selfless conscious states trustworthy; can scientists take them at face value? One might think that a person can accurately recall and report a past conscious experience as an experience that *she herself* underwent only if she was self-conscious when it happened. Perhaps such experiential reports contain a sort of logical error (individuals can't have been present as self-conscious entities, but think they were) or something resembling a performative fallacy like "I do not exist!" (individuals can't have been present, but the very speech act creates an obvious contradiction).[12] To be sure, selfless conscious states very likely can and do occur—but who really *has* these states? The universe?

There is an epistemological problem here (How do we really *know* what actually happened?), as well as a phenomenological one (What was the actual structure of conscious *experience* at the time?). Again, the memories involved could not be autobiographical in the strong phenomenological sense of referring to *someone's* experiences at the time at which they occurred. Of course, it may be empirically possible that selfless states are stored and later retrieved in an autobiographical format, as a post hoc mnemonic misrepresentation adding in the feature of egoic self-awareness—a case of misremembering, and not of confabulation.[13] The state itself was selfless, while the conscious memory of it wasn't, but the memory manages to make it appear *as if* it actually had been a self-state or a person-state, retrospectively endowing or "coloring" it with the experiential quality of ownership. But while it occurred, it was an ownerless organism-state. As an aside, please note that my point about post hoc mnemonic misrepresentations could in principle hold for *all* self-conscious experience, but in a much smaller time window: Maybe all experiences are originally selfless and only swiftly integrated into an automatic self-model after they arise. We could call this generalized version the thesis of "selfhood via nested time-scales," and I have defended it elsewhere.[14] But the specific philosophical question about MPE and autobiographical memory remains open, and its implications are profound.

Finally, our data are highly unlikely to be representative of the global population of meditators. The online nature of the questionnaire and its distribution channels, as well as self-selection effects, will have produced a sample that deviates in several respects (known and unknown) from the global population. However, representativeness is not a major concern at this stage. Future versions of our questionnaire will be put to the test against various populations, and its factor structure will be improved. If necessary, different variants of the questionnaire can be adapted to different populations and modified to provide a factor structure that holds up in as many settings as possible. Methods from confirmatory factor analysis can then be used to quantify the differences in item functioning and response patterns among different groups of participants. What is most important is to finally make a fresh start, get the process going, and take the phenomenology of pure consciousness seriously—from many different angles at the same time. Daniel Dennett has called this approach "heterophenomenology," and it involves a departure from the Cartesian idea of absolute "first-person authority," not taking the reports as authoritative beyond the seeming itself—which, again, we must carefully take as seriously as we can while allowing the possibility that, at any time, we might be wrong about the contents of our own mind.[15] Many different angles allow for surprises and unexpected discoveries. A dimensional approach is only one of these angles, while the perspective of qualitative analysis, as gradually developed in the following chapters, will quite naturally make us see different focal points and general themes.

3 Silence

> A space of timeless, self-luminous, world-penetrating silence. [. . .] The silence is not acoustic, the space is not physical. [. . .] The whole world was completely simple, logical, and clear, embedded in this luminous fragrant sweet silence. [. . .] . . . delicious silence. [#1381]

The experience of silence and mental stillness are core aspects of pure awareness. Novice meditators often first notice silence and stillness as what one of our participants called "the blanking interval between two thoughts" (#240). However, as the practitioner soon begins to discover, the silence of pure awareness is not simply an absence of internal and external noise or a mere cessation of thought, nor is it a nihilistic form of nothingness. The silence of pure consciousness is not a dead silence. For example, it can have a subtle, dynamic quality like "the feeling of plunging into a calm lake, being completely still, while simultaneously flowing along with it" (#2907). It is also not an emotional experience, but it can be characterized by a positive yet very subtle affective tone, giving rise to metaphorical descriptions of silent delight like "deliciousness," "sweetness," "shine," or "fragrance" (as presented at the beginning of this chapter, in #1381).

Our meditators found or created a striking number of literal and figurative turns of phrase in their attempt to capture the special phenomenal character of silence during episodes of pure awareness. For example, they described it as a "feeling of soundlessness" (#3259) and as the "feeling of a silence that rests in itself" (#2900). One respondent compared the experiential quality of abiding in the stillness of awareness itself to "letting the silence speak for itself" (#3440), while another described it as "a huge, boundless, absolutely silent space, which however, has a circular shape. It is so quiet that it's already loud again" (#3026). Others evoked the silence in terms of nonhuman sensory experience ("Hearing with bat's ears" [#1541]) and via the paradox of "hearing silence": "The silence is 'audible' inside my head but without ears, like a splashing noise, like the spray after the sea wave has hit the beach and you can hear the millionfold bursting of the foam bubbles—but evenly and endlessly, without pause or crescendo" (#1787). Great beauty and depth can be found in the specific phenomenology of silence that comes with meditation practice, and many finely nuanced aspects can be discovered in its subtlety. One of our participants said, "There is a sense of security in it and a sweet, silent calm [. . .], a fullness, a being-satiated, a being-allowed-to-be" (#2511).

Here are some pure awareness reports in which the phenomenal quality of "silence" is a dominant feature:

2049 Neutral state in attentive silence. Either deep in thought or in an absolutely neutral space without thoughts or activities. [. . .]

2628 Overriding factor is a deep and ineffable stillness; this occurs frequently in meditation on breath as the breath becomes more and more subtle and disappears and is accompanied by a subtle sense of naturally arising bliss. [. . .]

2706 They are experiences of deep silence, absolute alertness, without contents, deep connectedness and contentment, right up to experiences that contain only an "echo" or "reverberation" of what must have been there. [. . .]

2983 An experience of falling, first wanting to hold on to something, then letting go, letting go more and more . . . Deep silence . . . The feeling of being a "spark in the universe" . . . At some point, letting go of that spark, too. Deep silence . . . Deepest darkness, shining, widest space, spaceless, timeless . . . Incomprehensible silence despite speech and conversation . . . [. . .]

3035 I once had the feeling of silence while meditating. I had no thoughts, only inner calm. I became aware of my breathing without wanting to draw my attention to it. A feeling of absolute thoughtlessness.

3380 I experienced deep silence.

3218 [. . .] A few years ago, a very deep, central, and silent part of my brain suddenly became activated, as if somebody flipped a switch and a heretofore unused part of my machinery came online and has since been available to me. This part is like a ball of silence inside my physiology and has become a constant companion, sometimes more prominent, sometimes—when my mind is more excited—slightly less prominent. The experience of a thought in meditation is sometimes like a drop of water falling onto a very quiet surface of water and creating ripples. [. . .]

3330 I could describe my experiences best with concepts like stopping time, perceiving space without time, no thoughts, absolute calm, total satisfaction, wanting nothing, being awake, expansion—absolute silence, neither nothing nor something, no evaluation! [. . .]

3439 [. . .] After minutes of thought storm, absolute calm sets in: No thought waves, no physical sensation, breathing is calm and slow. I am absolutely focused on the silence and can carry it.

3472 In a completely everyday situation an experience of timelessness, motionlessness, silence, and wholeness. Standing face to face with someone, no longer being able to distinguish whose eyes I am looking into, mine or the eyes of the person in front of me. While meditating, more and more contact with the cushion and then an increasing weakening of the feeling of my body in connection with a deep calm and silence.

3601 [. . .] Deep peace and silence. A connection to source itself. Bliss.

Deep silence also is a prime example of the *nonconceptuality* of awareness itself (see chapter 9 on the experience of "suchness").[1] Phenomenologically, deep silence is the uncontracted quality of silent knowing itself, empty cognizance lacking any form of grasping or inner agency. Prototypical minimal phenomenal experience (MPE) is not only nonconceptual; it is also characterized by a principle of mental inaction. It never reacts, it never makes a choice, and it never initiates an action. As one participant put it, pure awareness is "that which never speaks" (see chapter 30). Pure awareness is that which would not even say, "I am that which never speaks." That which says "I am that which never speaks" is something else.

As I explained in the introduction, what I mean by "experiencing consciousness *as such*" is not that we perform some sort of mental act in which we finally grasp its conceptual essence, forming a thought about consciousness *as* consciousness. Pure awareness is not pure awareness plus the thought, "Ah! Now my mind is entirely silent!" On the contrary, it seems clear that even a simpler animal, a conscious creature that has no linguistic capacities at all and cannot think thoughts in the sense of "conceptually representing" the world or its own mind, could undergo a pure-awareness experience. It seems logically possible that there might be biological or even artificial systems for which the pure-awareness experience is a much more dominant feature than it is for us. There might be conscious systems that nevertheless have no conceptual or even theoretical understanding of the fact that they are conscious, or of the sheer, current existence of awareness per se—systems that do not strive for such an understanding and that, accordingly, never could or would report it using a language like ours. In sum, MPE is an entirely silent and *nonconceptual* experience of the phenomenal character of awareness itself. We will return to this point many times, and from different angles, in chapters 9, 14, 16, 20, 30, and 32.

1196 It feels as if "silence" was an experience. In the beginning of the experience, my mind still wanted to classify it and describe it to itself, and relate it

> to the Buddhist teachings I have received. But since it is very nonconceptual, my mind "gave up" trying to grasp the experience and instead surrendered to it and just let it happen. There was also an element of fear, because it feels quite groundless, and that is not something we intuitively like to experience. I think often it is that subtle (or sometimes not so subtle) fear of this emptiness that prevents me from fully committing to these experiences, and rather trying to conceptualize them. Another aspect is the fact that in Buddhism it is described as something great to happen to you, so in the beginning of the experience there are also thoughts of positive judgment, and a want to maintain the experience and not let it go. But the experience itself in my opinion is best described by the term "nonconceptual."

Stillness and the Zero-Person Perspective

> It is as if the wind had suddenly stopped.
> —Chinul (1158–1210), *Excerpts from the Dharma Collection*

The Australian meditation researcher Toby Woods conducted a very careful and rigorous study of the experience of silence in three forms of meditation practice.[2] He and his colleagues Olivia Carter and Jennifer Windt in Melbourne used objective criteria to select a sample of 135 expert texts from the traditions of Shamatha, Transcendental, and Stillness Meditation, and used these to create a database of descriptions detailing the meditation techniques as well as correlated subjective experiences. The material in the database was compiled and systematically analyzed using a rigorous method known as "evidence synthesis." As it turns out, the phenomenology of silence and quietness is constituted largely by the absence of thoughts, other mental activity, and sounds, and therefore it can be described as an absence both of internal and external "noise" and of disturbance in general. According to this large body of expert texts, the experience of silence and quietness has a particular connection with stillness (as it happens, the German word *Stille* denotes both) and is a major phenomenological feature of what Woods and colleagues term "contentless experience."[3] This feature is most pronounced during full-absorption episodes of pure consciousness. The findings also dovetail with what was said in the preceding chapters about mental perturbation, the low-complexity constraint, and the experience of peace.

Woods and colleagues found that the phenomenology of silence and quietness is frequently presented alongside the phenomenal qualities of luminosity, bliss/joy, ease, and peace. It is remarkable that quite naturally and independently, even though I only found out about their project later, each of these phenomenal characteristics turned out to have a chapter of its own in this book. A second finding from Australia was that experts often use folk-phenomenological terms like "silence" without defining them—as a matter of fact, a strength of Woods and colleagues' study is that it uncovered a phenomenological slippage between "silence" and "stillness." As we will see in the course of our journey, there is a possibility that expert texts, as well as testimonial reports, may be contaminated by their authors' belief system or background theory.

Third, there is a specific connection between silence and the absence of conceptual thought or intellectual activity (you may recall "even intellect" as one marker of MPE in chapter 2), but the experience of silence also naturally extends to an absence of other forms of disturbance including sensory experience.[4] It is a global feature. Woods et al. write: "The fourth finding is that the silence/quietness is reported as being in some sense complete. Terms such as sheer, deep, absolute, pure, and all-encompassing are used."[5]

Is pure awareness "contentless experience"? If so, a theory of MPE could simply be one of contentless, restful alertness—a silent form of phenomenal experience that in its purest form does not instantiate any intentional properties (in a philosophical sense). On such a theory, MPE would have no content because it is not directed at anything beyond itself. At this point, let me begin to draw your attention to one specific possibility: The absence of all empirical content could itself be an appearance, and what subjects (mistakenly) describe as a "contentless" phenomenal state could actually carry an *abstract* form of representational content. In other words, there might be a very special form of conscious content that can actually *account* for reports about the experience of contentlessness.[6]

Could there be phenomenal experiences that are hard to notice because they do not resemble any other experiences? In 1974, Thomas Nagel published an article, "What Is It Like to Be a Bat?" that has become one of the best-known discussions of consciousness and why it may be irreducible, forever beyond the reach of science. It has become an item of public folklore outside of academic philosophy. When neuroscientists want to write about consciousness but, in their introductory paragraph, begin to notice that they lack a definition of what it actually is they would like to explain, they often resort to Nagel's idea that if some entity—like a laboratory animal or human being used as an experimental subject—is conscious, then it must be *like* something to be this entity. To this day, the idea seems highly intuitive to many—"Yes, it is *like* something to be conscious!"—but the more we think about it, the more we find that we don't really know what this mysterious "what it is like" really refers to.

From 1998 to 1999, I spent a wonderful year at the philosophy department at the University of California, San Diego, writing my book *Being No One*. Sometimes I went to lunch at the faculty club with a beautiful and slightly rude old man by the name of Francis Crick. I like rude old men who really lay it on the line. As everybody knows, Francis Crick (1916–2004) made a central contribution to revealing the helical structure of deoxyribonucleic acid (DNA); together with James Watson and Maurice Wilkins, he was jointly awarded the 1962 Nobel Prize in Physiology or Medicine. (Rosalind Franklin should also have been acknowledged in her lifetime for the experimental work she did with Wilkins that contributed to generating the double-helix model, because she was actually "an equal member of a quartet who solved the double helix."[7]) Now that the problem of life was basically solved, he wanted to crack the problem of consciousness—and he didn't think highly of philosophers at all.

"Listen, Thomas," he said, "you guys have had more than two millennia to solve this problem, you've made a mess of it, and you have a really bad track record. If you want to make a contribution as a philosopher, it would be best if you simply shut up. Consciousness is not a philosophical problem anymore, but a neuroscientific one, and we are going to crack it within the next two decades."

"That would be fine, Francis, I am all for it—after all, this is why, five years ago, we founded the Association for the Scientific Study of Consciousness!" I replied. "But tell me one thing: *What* exactly is it that you would like to explain?"

Francis thought that everybody knows what consciousness is—it is what you lose under anaesthesia and when you go to bed at night, and what you regain after waking up again. But of course, things aren't that simple. There can be conscious experience in the dream state, and even during dreamless deep sleep (see chapter 20), for example, and some anesthetics might work only by causing amnesia, blocking memory access to what may really have been a kind of twilight state. Philosophers have long seen that—from a theory-of-science perspective—it doesn't help to launch major research programs and make a lot of noise when the "explanandum" (that which is to be explained by a future theory of consciousness) isn't clear. This is a problem that still pervades all the neuroscience of consciousness today, in all the methodologies and competing models.[8] In a way, this is one of the goals of this book: I hope that combining the minimal model approach with phenomenological data will, by the end of our journey, give us a clearer picture of what really needs to be explained.

Back in the faculty club in 1999, we were now having dessert. I had just amicably asked my question for a second time. Francis mumbled something to the effect that, from a strategic perspective, it wasn't a good idea to define one's research targets too precisely too early on. "Agreed, Francis, we don't want to get lost in a priori theorizing or 'overfitting' philosophical models, and we do also need a general heuristic and a practical research strategy," I replied. And then I gently repeated my point for the third time: "But what *is* it that you want to explain?" This was when the famous Nobel laureate finally exploded—and Patti Churchland wasn't there to protect me.

We simply don't know what this mysterious idea of "what it is like" really refers to. To be sure, the bat's brain may instantiate unknown phenomenal properties (some philosophers call them "qualia"), and these could be beyond the reach of science and impossible for the human brain to emulate. We just do not know what it really feels like to navigate and forage by echolocation, emitting high-frequency sound pulses through our mouth or nose while listening to the echo. But could there be an abstract, much more generic experiential quality that we share with the bat's conscious model of reality? If one believes in the existence of "qualia" (I don't, because it is another successful folkloristic meme, and almost nobody knows what it originally meant),[9] this raises an interesting new possibility for posing the central question of this book in a new way: Perhaps MPE is the "C-quale" (the generic "consciousness quale") that all sentient creatures share?

Thomas Nagel was making a different point. There may be not only unknown and unknowable properties in bat consciousness, but also an irreducible *individual*—namely, the bat's "self," from which its inward perspective ultimately originates. The existence of this self would make the bat's experience irrevocably *subjective*, and therefore beyond the reach of objective science (more about this in chapter 29). For us, this point is highly relevant. What if pure awareness were a state of consciousness that did not really *resemble* anything (e.g., because it consisted of only the most abstract, generic phenomenal quality), and that was not *subjective* either, because as a stand-alone phenomenon, it was always nondual, not tied to an experiencing self? What if the simplest, minimal form of consciousness were aperspectival?

In 1974, Nagel said this: "But fundamentally an organism has conscious mental states if and only if there is something that it is to *be* that organism—something it is like *for* the organism."[10] This is what everyone remembers. But in footnote 6, he also said: "[. . .] the analogical form of the English expression 'what it is *like*' is misleading. It does not mean 'what (in our experience) it *resembles*,' but rather 'how it is for the subject himself.'" This footnote clears up a frequent misunderstanding of Nagel's article. What he did *not* mean is that the philosophical problem of what consciousness is results primarily from the fact that a bat's conscious experiences do not sufficiently resemble our own experiential states. Nagel was after something deeper—the problem of subjectivity itself, the "subjective character" of experience itself—and in chapter 4 of his 1986 book *The View from Nowhere*, the issue appeared again in a new form: as the irreducibility of individual first-person perspectives and the possible existence of an "objective self." The problem of subjectivity is the problem of mapping the first-person perspective onto the third-person worldview of science.

Put differently, Nagel wanted to know how it is for the *bat* to be a bat—not how it would be for *us* to have similar, batlike states. This, Nagel thought, is beyond our ability to conceive. I think Nagel was right: To do this properly, we would have to fully identify with the content of the bat's self-model, which would immediately make us forget all about who we thought we were before and what it was we were trying to conceive of—and even why we were doing this. As I have explained in my academic work, I believe that the deeper point is not about qualia, but about understanding the phenomenology of identification and what we are really trying to say whenever we use the mixed visuogrammatical metaphor of a "first-person perspective," somewhat conspiratorially pretending that everybody knows exactly what we are talking about.[11]

In the context of pure awareness, all this raises three interesting points. First, MPE too is clearly beyond our "ability to conceive" because it is a nonconceptual awareness of awareness itself. It *cannot* be imagined because silence cannot be expressed as noise. MPE is not pure awareness *plus* the recognition, "Ah, now I have returned to complete stillness again! So, *this* is what awareness itself feels like . . ." Just as in the thought experiment about being a bat, fully dissolving into it would make us forget all about who we thought we were before. We would move from meditation into nonmeditation (see chapter 32).

Second, Thomas Nagel's antireductionist arguments about the irreducibility of subjectivity and the putative existence of nonphysical "first-person facts" do not apply to MPE. Focusing on the problem of subjectivity actually *detracts* from a deeper understanding of consciousness, and adopting the minimal model approach dissolves the problem of subjectivity for the science of consciousness, helping us understand that the "first-person perspective" really is a surface phenomenon. As we will see in the course of this book, pure awareness itself is neither a first-person nor a third-person state because it is the prime example of an epistemic "zero-person perspective." Pure awareness is based on a form of knowing that is itself nondual, nonegoic, and therefore aperspectival in Nagel's sense. I believe that this conscious form of knowing could be an *organismic* state, but while it occurs, it is not a person-state and is not owned by any person because the process of egoic self-modeling that creates personal-level ownership has been suspended. *You* cannot be directly acquainted with it, but maybe the biological organism can. (Nagel himself saw this possibility in 1969, in one of his earlier, lesser-known writings entitled "The Boundaries of Inner Space," when he said: "Perhaps we shall have to fall back on the idea of an organism or an organic system.")[12] Therefore, all philosophical arguments that consciousness is irreducibly subjective—we could call them "antinaturalist arguments from subjectivity"—may fail on the most fundamental level, simply because MPE is not a "subjective" phenomenon at all, either phenomenologically or in any strong epistemological sense. It is a subpersonal state. While it occurs, it is not tied to an individual, personal-level perspective. To be sure, there is an epistemic subject in a much weaker, abstract sense—but what undergoes this subpersonal state of knowing is a biological organism, not an ego or a person. In this book, we will try to focus on the actual phenomenal experience itself, but it is already starting to become clear that some of the core discoveries that we are making have important *epistemological* implications.

So the first new insight relating to Nagel's work was that pure awareness is something that cannot be successfully grasped or simulated by the imagination. This is impossible because it would create a subtle sense of effort and the experience of mental agency. The second insight was that MPE itself is not a first-person state, but a zero-person state. Maybe an organism can get acquainted with or "used to" it, but a personal-level self cannot. Third, an interesting possibility now appears on the horizon: People may have had difficulty taking pure awareness seriously because, in some fundamental but yet-to-be-understood sense, it really is not *like* other states of consciousness; it only resembles them in the very weak sense of being something conscious.

A mathematical analogy for this aspect of pure awareness could be based on the notion of an empty set, which is a subset of all other sets but resembles none of them. The empty set would be unconscious. However, the set containing only the empty set resembles all other sets in at least one aspect: This would be a full absorption episode of pure awareness. If you think of a set as being like a box, the empty set is a box with nothing in it; the set containing the empty set is a box with another empty box inside it. It has an *abstract* kind of content because a box is not nothing, even if it is empty. All analogies have limitations, but, in addition to abstractness, this one would also give us simplicity plus a very weak, barely noticeable kind of resemblance to other sets.[13]

In chapter 5, I will explain the idea that pure awareness could be a model of an empty and unobstructed space of knowing, and that experiencing pure awareness simply means having a *model* of this wide and open space. The space itself is something unconscious (like the empty set), but conscious experience is exactly what appears when the space is transformed into a *self-modeling* space (i.e., a box with another empty box inside it, but this time "self-knowingly" so). I think that in the future, it may become important to think about pure awareness on a mathematical, purely formal level.[14] However, I believe that in the end, all this may only bring us back to the fable of the blind people and the elephant. Conceptual boxology doesn't really work, for example because a considerable number of our phenomenological reports describe MPE as something that has no boundaries at all, and simultaneously as something that (in an unclear sense of "containing") contains all other conscious states. I may be wrong, but I think boxes like these—containers without walls—simply do not exist.

From a different perspective, here is how the philosopher Jonathan Shear approaches this difficult point:

> Experience of the deepest level, that of pure consciousness awake to its own nature in and by itself, is generally held to be especially important. The defining characteristic of this experience is the *complete absence* of all sounds, tastes, thoughts, feelings, images, and anything else that one can ever imagine. Techniques for achieving this experience differ. [. . .] But they have in common the idea that it is possible for all empirical content to disappear, while one nevertheless remains awake.
>
> What then is the experience like? By all accounts it is not *like* anything. One can have it and remember it—one knows that one was not asleep. But one does not remember it *as* anything at all. It is just *itself*—unimaginable and indescribable. The experience itself is extraordinarily abstract. Indeed, it is the logical ultimate of abstraction, since by all accounts it is what remains after everything that can possibly be removed from experience has been removed, while one nevertheless remains awake.
>
> It should be noted, however, that even if the natural response to this experience is to describe it as "contentless," it is still appropriate to raise the question of whether it is in fact completely contentless. For while experiences [. . .] might be so subtle and abstract that they naturally seem to be completely contentless, they might nevertheless actually have some, albeit very abstract, content.[15]

From the next chapter onward, I will gradually begin to introduce candidates for what the "contentless content" of pure awareness could actually be. The first will be the abstract quality of "epistemic openness" that comes with the experience of existential ease, peace, and silence.

The American composer, artist, and music theorist John Cage (1912–1992) wrote about a sudden philosophical insight that he had in the late 1940s during an experiment in the anechoic chamber at Harvard University. He described it like this: "[S]ilence is not acoustic. It is a change of mind, a turning around."[16] Philosophers and artists have ignored the phenomenology of silence for too long, and finally giving it serious attention raises many interesting questions. Can there be a kind of music that brings the silence between two sounds into the foreground? What exactly is common to all kinds of music that makes silence, the space between two notes, audible? More generally, are there forms of art that are able to "stage" MPE, to "orchestrate" the silence of pure awareness itself? Could there be an aesthetics of acoustic emptiness? And on the mental level—in the mind of the meditator—are there ways to actively imagine a sound and then let it go? Are there what today's computational modelers might call mental "action policies":[17] paths into nonaction that will reliably create the experience of silence? This would be entirely different from mentally simulating a speech act, for example by merely saying a mantra in your mind, and more like gently striking a "mental bell" one single time and then following the fading sound into silence; more like quietly observing an autumn leaf floating to the ground, or following the path of dandelion seeds that you have gently blown off a flower (compare #2867). Are there perhaps also special ways to *think thoughts* that make the space between those thoughts more and more vivid, gradually turning an absence into a presence? This way of thinking would be a form of mental art, a poetry of silence that no art school currently teaches. Could there be something like a scenography of pure consciousness, a new phenomenological discipline that applies to inner and outer scenes alike?

Clearly, many traditional forms of art have also been forms of contemplative practice. But it now begins to look as if meditation can itself be viewed as an art form. This art form would not be of the kind that implicitly assumes a Dennettian Cartesian theater, an inner scene that includes a little man in the head—an artist-homunculus that, mysteriously, is already conscious itself while it looks at the inner screen of pure awareness. Rather, this art form would aim at the effortless elegance that lies in using a groundless ground (chapter 26) as an inner stage environment—one that is crafted simply by revealing the centerless space of silence that has always already been there.

There is a difference, however, between experiencing the silent mind of an embodied self and experiencing the spontaneous arising of an uncontracted, spacious silence, the silence that can sometimes be felt to pervade all things. Silence sometimes allows a new quality of presence to emerge from what first appeared to be mere absence—it is almost as if the dynamic stillness of mindful attention were a form of "scenic lighting design," elevating the process of seeing itself into the space of visibility. But there are important nuances in the process. Previously, we saw that pure awareness itself is nondual, nonegoic, and therefore aperspectival in Thomas Nagel's sense. If you are a practitioner of meditation yourself, you may have discovered that there exist two subtly but profoundly different variations on the experience of silence: *experiencing* the quality of silence and *being* silence itself.

In my own practice, I have found that the transition cannot be constructed or fabricated in any way; unfortunately, there is absolutely nothing you can do to make it happen. But sometimes, unexpectedly, the difference can be discovered. Sometimes the moment of discovery turns into a form of mental noise that immediately terminates the silence that is about to disclose itself. Sometimes it doesn't. Have you already noticed this difference? You do not have to be a meditator to know this difference. Who or what is noticing it? Who or what lets go of the noticing?

4 Wakefulness

"Infinite simultaneity" and yet wide awake. [#2312]

One classic definition of pure awareness is "restful alertness." There is a quality of relaxation and calm (see chapter 1), but also an element of nonagentive vigilance, of attentiveness without doing. We now turn to this element, the experience of being fully alert. Interestingly, modern science tells us that there is more than one kind of alertness. As I will explain in slightly greater detail in this discussion, neuroscientists use the term "tonic alertness" to describe that aspect of alertness that is sustained independently of external stimuli and not triggered by cues from the environment. This contrasts with what is called "phasic alertness," which is caused by sudden events like a loud noise or salient and unexpected changes in brightness, contrast, and the like. A specific phenomenal quality goes along with the first variety—the subjective, conscious experience of cue-independent tonic alertness—and it has been largely ignored by the philosophy of mind and consciousness. From now on I will call this quality "wakefulness."

In meditation, the phenomenal character of wakefulness can be much more dominant than it is in ordinary wake states. The experience of bare wakefulness itself is effortless, calm, and entirely nonconceptual. As always, let us begin by looking at some examples:

> 687 I am in a state of wakefulness. I have a strong awareness of the present moment. It is as if I am a guard dog: All my perceptions seem heightened. I am in consciousness but I am not doing anything except perceiving sounds, smells, sensations in my body. Time seems to be on hold.
>
> 3305 [. . .] During wholeness experiences during meditation [. . .] I find myself in a state of inner silence with simultaneously increased wakefulness. [. . .]
>
> 3218 [. . .] Full wakefulness is there, dullness is absent. It is an extremely peaceful and natural state. Complete innocence, completely resting in oneself and the

> self flowing in the thoughts, the body, and the infinite. [. . .] Alertness to the extreme is an experience which I have frequently: No shred of dullness, knowing that wakefulness is the mover behind the quality of decisions of the intellect [. . .]. This alertness carries on into my meditation: complete self-referral, the mind completely settled, easy to focus inward and the settling of the ocean of the mind to complete waveless stillness with an occasional thought that is appropriate at the stage of my meditation. [. . .]

Let us now consider states in which wakefulness itself almost turns into a stand-alone feature, in which pure awareness arises out of or is identical with pure awakeness. In full-absorption states, there is no other content whatsoever—including knowledge about the meditator's personal identity or life history:

> 749 I "wake up" in a space-less, time-free space. I have no body and cannot perceive any objects. Thoughts hardly form, they cannot grasp anything. There is no memory. I know nothing about myself. I also have no conception of human, living being, life, world, etc. There is only awareness that I am. Nothing happens. I rest in this awareness and am one with it.
>
> 2603 [. . .] For me, the experience of pure awareness is a calm, impersonal, and blissful presence. It is very hard to describe "awareness" in itself without content or an object. I suppose it is like a sense, sight without vision, a simple way of acknowledging and noticing what is, or pure awakeness, the very basic feeling of consciousness being "on," the fundamental constant that underlies all experience, which is always there. I can also be identified with this awareness and when I realize that this awareness is me, it is empowering. [. . .]

During a full-absorption episode, subjects are not functionally dissociated from their environment. They are not in a coma and not in dreamless deep sleep either; they remain open to the world. This is one interesting aspect of what the concept of "epistemic openness" refers to, which I will explain in greater detail later: The perception of environmental features is still possible. As a matter of fact, the experience of pure awareness can be understood as a representation of exactly this possibility: the currently existing capacity to know something, the potential to be open to the world. This is an abstract form of content. Although active perception doesn't take place, it can always be triggered by salient stimuli. In pure awareness, the organism has an inner model of its own epistemic openness; it nonconceptually knows about this openness (i.e., about "alertness being there"). But this does not yet imply that any other information processing is elevated to the level of conscious experience, nor that the knowing is of an egoic kind, that the known alertness is attributed to a knowing self. Yet, the *capacity* to know in this way is clearly preserved, and as we will see in chapter 32, there can be interesting gradual transitions between nondual mindfulness and dual mindfulness, between nonmeditation and meditation:

> 3396 I experience pure awareness in moments (which are, if anything, probably short to very short, but which I can't temporally determine any more precisely), where I refrain from any form of mental activity in the strict sense of the word and also have no perceptions of a visual, acoustic, tactile, or olfactory nature. At the same time, however, I do not have the impression that I am no longer receptive to external perceptions. I don't have a strong feeling of peacefulness, happiness, detachment, so no high feelings like in intoxication or flow, rather the state can be described as "pleasant" or "liberated," maybe also as "resting in oneself."
>
> 859 [. . .] All thinking, feeling, perceiving had ceased, as if all processed software programs of my body–mind had been closed. First the sense of hearing returned with the perception of the singing bowl. Again it took a few minutes until I remembered how to think, how to walk. This experience was a bit frightening. In the following minutes I could not perceive/process any of the visual impressions in the usual [way], saw a puddle without knowing what a puddle is, could not fall back on any concepts or memories. I found this experience fascinating and liberating. After about 30 minutes all the usual processing mechanisms returned and I experienced a high state of euphoria and bliss.
>
> 2115 [. . .] This state is what I like to call the natural underlying consciousness. The fact that it has always been there, and will be, is a great thing to come to terms with. In this state of my "Pure Consciousness," I do not feel I have a body, limbs, hands, thoughts . . . It just IS, as it is. Time definitively doesn't have any value, but I do feel like having control to get back to my normal routine and get some work done.

Tonic Alertness, Wakefulness, and Epistemic Openness

> It is a wakefulness for which no words suffice. It is not a definable entity, but at the same time, it is a self-knowing aware emptiness that is clear, lucid, and awake.
>
> —Dakpo Tashi Namgyal (ca. 1513–1587), *Clarifying the Natural State*

In my analysis of canonical texts across many centuries and cultural contexts, I found that by far the two strongest semantic markers for conceptual or theoretical descriptions of pure awareness are "Wakefulness" and "Epistemicity."[1] The second term, "epistemicity," refers to the consciously experienced quality of knowing, which can sometimes appear in isolation. Taking the phenomenology of meditation seriously demonstrates that subjective confidence (the nonconceptual experience of knowing the probability that one currently knows or is *able* to know something) can actually emerge without any representation of subject and object. For example, the phenomenology of epistemicity can emerge as an experience of pure knowing or "being of the nature of knowing itself." We will return to this surprising fact in chapter 5, but also in chapters 18, 19, and 27 to 30.

But what about the concept of wakefulness? Semantically, I found that the experience of wakefulness per se seems directly related to similar phenomenological notions like "mental clarity" (but without mental content), "cognitive lucidity" (but without coherent high-level symbolic thought), and "bare awareness" (but without an object). In this chapter, I will introduce a threefold conceptual distinction and one new conceptual tool, which are designed to do justice to some recent scientific findings and help us better understand our phenomenological reports.

From a philosophical point of view, it is crucial to understand that every phenomenon can be described on different levels of analysis, and our choice of level ultimately depends on what we want to know, on what philosophers might call the "epistemic goal." Pure awareness in meditation is no exception: It is a phenomenon that can be described on many levels, using many conceptual tools. The empirical literature is often correspondingly unclear and ambiguous in its use of notions like "arousal," "vigilance," "sustained attention," "wakefulness," and "alertness."[2] However, there are two semantic elements in a large majority of scientific treatments: first, epistemic capacity, and second, absence or indeterminacy of representational content. To begin, let us distinguish between the physical, functional, and phenomenological levels of description. I propose to use the relevant key terms as follows:

- "Arousal" is a graded **physical property** of the human brain. The level of arousal is a purely *physical boundary condition* determining the depth of cortical information processing available to the organism as a whole; it causally depends on the local level of activation in five types of neurotransmitter in the ascending reticular activating system, a complex structure that originates in the brainstem.[3] Just like the frequency of heartbeat, blood glucose level, or core body temperature, arousal is a *vital parameter* that must be successfully controlled. For example, the control of cortical arousal is necessary to generate the sleep/wake cycle.
- "Tonic alertness" refers to a graded **functional property** that determines the capacity for sustained attention. It is a causal function *resulting* from the successful control

of arousal over longer periods of time (e.g., in the absence of an external cue). Tonic alertness is a functional property that causally enables important cognitive capacities like orientation in time and space, executive control, attention, and mental agency.

- "Wakefulness" is a graded **phenomenal property** that is sometimes introspectively accessible. You can attend to it, but most people rarely do. On one level of description, wakefulness can be interpreted as a special way of *representing* tonic alertness. It may be the major component of the phenomenal character of minimal phenomenal experience (MPE), the nonconceptual awareness of tonic alertness. It may also be fundamental in the sense of being the primary dimension of phenomenal state space.

I think we can get a clearer conceptual picture by focusing on these three properties, which relate to each other in interesting ways. For instance, can there be tonic alertness without the conscious experience of wakefulness? Can one be alert without *knowing* it? The answer is yes, because in some types of coma, patients show no signs of awareness whatsoever while going through a full sleep/wake cycle.[4] The causal potential inherent in tonic alertness is there; the phenomenal experience of wakefulness is absent. The functional property of tonic alertness, if it is not internally represented and predictively modeled, can exist entirely without the conscious experience of wakefulness (as in cases of unresponsive wakefulness syndrome, known in German as *Wachkoma* [waking coma]). My empirically testable hypothesis is that for some animals (like ourselves), there are two options for controlling their own level of alertness: One that is older, simpler, and more direct, and another that actually uses a *model* to predict and control the level of cortical arousal. Put simply, an organism can be tonically alert without knowing that it is alert: Consciousness is knowing that one is alert.

What about the hardware? In humans, the functional property of tonic alertness is *realized* by certain physical properties of the brain, and we do not yet know exactly what these properties are. Empirical studies involving fourteen expert meditators have shown that only a small number of cells in the ascending arousal system are responsible for modulating and reconfiguring the inner landscape that tracks moment-to-moment changes in conscious experience, with this landscape itself being physically realized by a vast number of cortical neurons.[5] Conceivably, a nonbiological counterpart to tonic alertness could be realized by very different physical properties in machines. But for neurotypical humans, it is plausible to assume that a critical level of cortical arousal is an important necessary condition. Clearly, both the phenomenal property of wakefulness and the functional property of tonic alertness cannot exist without the physical brain or some equivalent physical implementation.

What do we know about the functional property *connecting* the physical and the phenomenological levels? Tonic alertness is a global functional property, in part physically realized by the level of cortical arousal, and scientists have hypothesized that it is one of the core functions of the cingulo-opercular network.[6] Interestingly, this brain network is one component of what in folk-psychological contexts we call the capacity for "attention," and from a scientific perspective, its function for the organism consists in "achieving and maintaining a state of high sensitivity to incoming stimuli."[7] Note how scientific descriptions like this strongly resemble descriptions of what mindfulness practice tries to achieve, to a higher degree than we usually experience in "ordinary" waking states. Note also how this opens up the possibility that there could be a baseline level of "tonic mindfulness" that is actually *always* present whenever we are conscious, independent of any form of contemplative practice, but that goes unrecognized because nobody ever tells us that there is something to look for.

The existing neuroscientific findings provide us with important pointers consistent with our phenomenological analysis: Tonic alertness is a (1) sustained and (2) internally initiated "preparedness to process and respond" that (3) implies a capacity for attentional agency, for "coengaging" phasic alertness and selective attention directed at a specific task.[8] Exactly like the quality of mindfulness that is described as spontaneous and effortless, but that is also actively cultivated by contemplative traditions, it can be described as a global and general mechanism "of keeping cognitive faculties available for current processing demands and holding unwanted activity at bay."[9] This is beginning to sound as if there could be a specific sense in which a more global and nonselfy form of "unrecognized mindfulness" is already there, automatically and effortlessly, whenever you think you are conscious. Perhaps *you* aren't? Perhaps the mindfulness must become aware of itself?

The phenomenology of tonic alertness is the nondual conscious experience of intrinsic and sustained wakefulness. It fluctuates only on the order of minutes to hours. It is entirely nonconceptual, not the result of an inference or a judgment about the current behavioral state (e.g., whether one is asleep or awake). Animals can have it. The subjective experience of bare wakefulness also has no grain or internal structure. It can therefore serve as the best, primordial example of what some philosophers of mind have called the "ultrasmoothness" of qualitative conscious experience (more about this in chapter 6). What's more, in many classical texts, the phenomenology of wakefulness and epistemicity in meditation is described as functionally autonomous, for example as "self-generating ever-fresh awareness"[10] or as an originary, naturally present, and nontransient form of "primordial knowing" (*ye shes* in Tibetan).[11] Clearly, there is a convergence emerging between phenomenological data, ancient conceptual frameworks, and modern neuroscience.

On the empirical side of things, it is plausible to assume that tonic alertness is a causally enabling factor in the realization of two bundles of functional properties that are distinct but closely related: *orientation* and *executive control*. Roughly, being oriented means knowing where you are (e.g., the position of your body in a spatial frame of reference), what time it is now (your location in a temporal context), and who you are (what your name is and which of a number of different persons in some social context you identify with). So, "being oriented" roughly means being aware of three dimensions: time, place, and person. We completely lose the phenomenology of orientation in dreamless sleep, in severe psychiatric disorders, and during full-absorption episodes of pure awareness. All three dimensions disappear during a full-absorption MPE experience, but multiple reports show that unconscious mechanisms in the meditator's brain always kick back in when there is a sudden change in the environment. The orienting response (sometimes also termed the "orienting reflex") is an organism's immediate response to a change in its environment, when that change is not sudden or threatening enough to elicit the startle reflex. When we encounter a novel environmental stimulus, such as a bright flash of light or a sudden loud noise, we will automatically pay attention to it even before identifying it. The startle response or startle reflex, on the other hand, is an equally unconsciously triggered but defensive response to a sudden or threatening stimulus, and experiencing it tends to feel unpleasant. The startle reflex comes directly out of your brainstem and is a largely unconscious, *defensive* response to potentially threatening stimuli, such as a sudden noise or an unexpected sharp movement. Many meditators know exactly how it feels when a sudden deep breath, an automatic orientation response, or even the startle reflex itself terminates an episode of being fully absorbed in the experience of pure awareness. Here is one example from our database:

> 2774 [. . .] I sink into a state of pure awareness that is impersonal—no person is there, no subject and no object. There is no space, but a feeling of duration. I no longer feel the body. It is dark in the sense of no light and no color. The state feels warm and secure, but only in retrospect. When I am in it, it actually has no real qualities except consciousness/awareness. Occasionally I forget to breathe in it. This becomes noticeable as an indefinable feeling of incongruity, which becomes more and more pressing until I become aware of my creatureliness and my body, start in fright, and breathe in deeply, then I am naturally outside of the state :-) Occasionally I also get frightened because I don't know

> who I am, where I am, and what I am. So I don't know anymore that I am a human being, an individual in a particular location, which causes a brief panic that also throws me out of the state. I no longer know which body, which home station, I must return to. [. . .]

Here, it is important to note that all three functions—alertness as such, orientation, and executive control in terms of the capacity for what I call "attentional agency"—are *epistemic* capacities. They are about acquiring knowledge. For example, "attentional agency" is the ability to deliberately and actively control the focus of attention in a goal-directed manner. At a given point in time, a biological organism may or may not have these capacities. They are also the foundation for what, in 2009, Olaf Blanke and I termed a "strong first-person perspective."[12] Sometimes you have it; sometimes you don't.

Strikingly, the paradigmatic cases of "pure" consciousness discussed here are characterized precisely by the absence of spatiotemporal self-location. This means that during full-absorption episodes of pure MPE, the subject is not *oriented* toward time or space. What Blanke and I called a "weak first-person perspective" is equally absent. In MPE, we also do not find any personal-level self-representation, and attentional or broader cognitive agency is absent too. This means that when the meditator has dissolved into a state of effortless, wide-awake pure awareness, there are no mental actions occurring like *control* of the focus of attention or one's thought processes. All this is also true of tonic alertness as such. Therefore, it might well be that the minimal form of conscious experience emerges precisely whenever the first functional property—tonic alertness, the causal enabler of orientation and executive control—is represented in the brain. This would give us a first and important building block for a definition of MPE: MPE is a simple and silent way of knowing tonic alertness, but nonegoically and in the absence of a first-person perspective. In the case of a full-absorption episode, we could even say that it is a *state* and a *mode* at the same time[13] because it is content-specific and global at the same time—like the experience of an "alertness *Ganzfeld*." In perception research, a *Ganzfeld* (literally a "complete field") refers to an unstructured, uniform field of stimulation. MPE could be a more abstract version, related not to sensory stimulation itself but to a structureless, uniform *expectation* of knowledge: Pure awareness is the process of nonconceptually knowing that a specific form of openness to the world has been achieved.

Have you ever carefully investigated exactly what happens when you wake up in the morning? What is the very first conscious experience, in the first fraction of a second, before you even remember who you are and what you will do today? I have found that part of the phenomenology of waking from dreamless deep sleep is a primordial sense of confidence that perceptual states will very soon occur, that one is now "open to the world" and knows about one's epistemic capacity, and also that one can (and very soon will) know where one is, what time it is, and so on. In a few milliseconds, one will also know *who* one is: Phenomenologically, the gradual transition from unconsciousness to the wake state is characterized by another subtle and intuitive presentiment, something that one might perhaps term a phenomenal "foreshadowing of selfhood." This is an expectation not merely of knowledge, but of egoic self-knowledge—but *before* it actually manifests.

Waking up involves an anticipation of mental agency and the capacity for global self-control. Phenomenologically, it is the subjective but as yet nonegoic confidence that a full-blown first-person perspective involving executive mental control plus an extended autobiographical self-model will very soon appear. This is what I call the "foreshadowing of selfhood": The (often very brief) phenomenal quality of subjective confidence that relates to knowledge about the environment and can also extend to the likelihood of egoic self-knowledge, to the capacity for "predicting oneself into existence."[14] In sum, one important aspect of the phenomenal character of pure awareness can be described as a quality of subjective confidence, an *expectation of epistemic states*. Computationally, wakefulness can now be described as a statistical hypothesis in the brain, a neural representation of the probability that veridical perception will actually occur. If these ideas are pointing in the right direction, then MPE may be intimately related to what happens to all of us every day—namely, in the first few hundred milliseconds of waking up.

Let me now introduce a new phenomenological concept, "epistemic openness." Epistemic openness is a global form of phenomenal character—namely, a specific form of openness to the world. *Epistēmē* (ἐπιστήμη) is the ancient Greek word for "knowledge"; hence, "epistemic openness" is openness related to knowledge, to the possibility of acquiring knowledge. As we have just seen, there is a whole bundle of functional properties related to tonic alertness that can be successfully described from a scientific perspective. I have no doubt that future neuroscience and computational modeling will arrive at much more detailed descriptions of this bundle of capacities, as specific ways of acquiring knowledge that a biological organism may or may not have. In this new sense of involving a whole bundle of knowledge-enabling properties, epistemic openness is exactly what you gain in the very moment when you wake from dreamless deep sleep, from anesthetic surgery, or from a coma: You regain openness to the world because once again, you now have all the epistemic capacities that you temporarily lost. Of course, this may also be a gradual process, with capacities coming online one after the other. But in deep sleep, you are almost completely offline; you are not epistemically open to the world. You lose epistemic openness during nocturnal sleep and anaesthesia, but also during episodes of mind-wandering and mind-blanking, and during those short involuntary episodes of "microsleep" that normally last less than fifteen seconds and occur when you are very drowsy but still trying to fight off sleep. Wakefulness, as I have defined it, is knowing one's own tonic alertness by having a predictive model of it, and it is the key primordial feature of consciousness. Therefore, we can now offer a new building block for a more substantial definition of consciousness: Consciousness is a continuing, ongoing process of nonconceptually knowing the organism's current state of epistemic openness, of expecting new knowledge without yet having it. The process itself can be transparent (see chapter 28 or figure 34.1 in chapter 34 for more), and normally it leads to the appearance of a whole world filled with existing, knowable things. Pure consciousness is knowing *only* this epistemic openness itself, and calmly abiding in this knowing is a core aspect of what many contemplative traditions teach.

Let us sum up. My first point is that all of what my brief sketch of some scientific facts refers to can also be represented internally, by the system itself, but in an entirely nonconceptual, more parsimonious, and maximally simple way. That is, epistemic openness can be known by the system itself, but in a way lacking all the features that were listed in chapter 3 in the context of "low complexity." Epistemic openness can be known timelessly, selflessly, without thought or emotion. There is an ineffable "suchness" to the phenomenal experience of wakefulness itself, which involves the qualitative character of something spontaneously revealing or presenting itself (more about this in chapter 9). It can be known nondually, without the knowing self of the meditator actively attending to it (chapter 27). Perhaps epistemic openness can sometimes even nonegoically know itself, as if emptiness were awakening to itself (see chapter 30).

This may be starting to sound slightly mysterious, but I think if we describe pure awareness as the *aperspectival phenomenal experience of epistemic openness*, then we might gain an interesting new conceptual vantage point. In a first approximation, the phenomenology of wakefulness and clarity (to be investigated in chapter 5) would then become the phenomenology of epistemic openness. This special kind of openness forms a highly abstract kind of experiential content because it refers to the mere capacity, not to the actually ongoing actualization or exertion of this capacity. There is no fixation on an object of knowledge. At this point, please recall my speculative but empirically testable hypothesis from earlier (call it the "awakening hypothesis"), which says that from an empirical perspective, MPE might actually be a prolonged version of what happens during the very first milliseconds of the process of waking up in the morning. On the level of phenomenal experience, to become open to the world means to nonconceptually represent a *possibility*, to represent mere epistemic capacity as such:

- Epistemic openness means that you could now orient yourself in space, but not necessarily that you actually do so. What you actually experience is a potentiality plus a sense of confidence.
- Epistemic openness also means that you could orient yourself in time, toward the Now, by creating a "present" moment. But pure awareness, the actual experience of waking in itself, is still timeless—you are not quite "of this world" yet.
- Epistemic openness also means that you could orient yourself toward the person you take yourself to be, by re-creating a narrative, an autobiographical self-model. But the very first moments of "coming to" are actually still selfless.
- Epistemic openness also means that you could exert executive control, for example by actively controlling the focus of attention. But full-blown MPE is effortless, no mental actions are initiated, and what in chapter 25 we will call the "epistemic agent model" is not yet in existence. Therefore, what is experienced in pure awareness could be the sheer potential to know, a global expectation of states carrying epistemic value, the empty space of all possibilities.

My second concluding point is that the notion of epistemic openness can offer a modern reinterpretation of the ancient Buddhist notion of "emptiness." I will say more about "emptiness" in chapter 17. For now, we need only distinguish between a *metaphysical* and a *phenomenological* reading of this very special concept—one of the most interesting concepts ever developed in the history of human philosophy. As always, I am not at all interested in metaphysics.[15] I am interested only in the fine structure of consciousness itself. If the phenomenology of epistemic openness is closely related to an abstract space of epistemic possibilities, then the phenomenology of *emptiness* could be interestingly related to a maximally simple and entirely unstructured inner representation of exactly this space. Phenomenologically, this is not some nihilistic void, nor is it a dead form of cold, empty space. Rather, the experience of epistemic openness can be described as a space of nonegoic knowing, of pure aliveness and spontaneous presence. In pure awareness, this space itself is experienced as fundamental, as clear and unobstructed (we will see more of this in the following chapter). Emptiness is nothing nihilistic, but a special kind of openness.

Let me close this chapter by briefly drawing attention to the fact that wakefulness—the conscious experience of tonic alertness—is a prime candidate for something that is largely invariant across cultural contexts and historical epochs. The brain mechanisms that create the sleep/wake cycle are also present in many nonhuman animals living on our planet because alertness and the bundle of functional properties described earlier have a long biological history. Wakefulness is a very natural and simple state because it rests on innate capacities. A lot of it is hardwired and largely independent of behavioral conditioning, psychological imprinting, and personality traits. Being awake is not something that you have to learn from your parents. Of course, the ways in which you are later able to describe it, the words that you have for it, the degree to which your introspective attention has been educated or cultivated, and even whether you have ever been *made aware* of your very own inborn alertness at all—these may all vary greatly from society to society, from one cultural context to another. But the spontaneous recurrence of epistemic capacity, the regularity of the sleep/wake cycle, and the periodic reopening of your own inner space of knowing are things you share with all other healthy human beings. If anything is a good example of cultural invariance, it must be the primordial phenomenal character of wakefulness itself.

5 Clarity

Unreified clarity. [#2561]

The phenomenal character of clarity is intimately related to the phenomenal character of epistemic openness, the spontaneously occurring phenomenal experience of one's own capacity to know. Clarity is described in terms of qualities like lightness and spaciousness or depth and subtlety, as a continuing state of equilibrium, a sense of wakeful presence, a calm and timeless form of stillness, a gentle form of acuity or sharpness, or a form of perception without center or boundaries. One participant metaphorically described it as "[c]lear water—pulsating at high frequency—boundless vastness and splendor" (#693). Clarity is also deeply connected to the phenomenology of "pure knowing," which we will begin to discuss in the second part of this chapter. Here are nine examples:

1818 [. . .] I turned my attention to everything arising, the pain, emotional qualities, and thoughts, and at the same moment this state of the observer took over and everything opened up and there was just this knowing of sensations coming and going, no bad or good, just watching these sensations do their thing. There was a definite sense of lightness and spaciousness and extreme clarity.

2103 Occasionally, very rarely, I have experienced a brief state of clarity during meditation where the usual case of buffeting between several observed sensations falls into a condition of equilibrium and I am, for just a moment, at a place where my attention is neither tied to anything nor pulled into one or more directions. I'm conscious of this brief state and feel like I'm observing uninterrupted conscious awareness.

2623 [. . .] Clear, awake, present, unmoved, while being aware of very subtle thoughts. Like a full moon covered by a very thin veil. [. . .]

2778 [. . .] completely clear, calm attention like a clear vibration of the whole body that you feel as such but that isn't limited, [. . .] a time- and spacelessness and yet clear awareness of sounds, without reaction or "something" following. Identity, ego, narrative were not present, the state was clear, alert, nonjudgmental, nonself-reflexive, without a sense of duration, and desireless. Peaceful and without direction, gentle and yet somehow sharp, in a soft clarity [. . .]

2602 Clarity, everything is exactly as it is. I don't have to change anything. The state forms the background to everything that can be experienced.

2747 [. . .] clear, pure consciousness, contentless, empty, silent, calm, aware, endless depth—usually in the morning hours [. . .].

3052 I can experience pure clarity through the awareness of the presence of Being, without visual perception. [. . .]

3156 [. . .] Rather cool and boundless like the universe. Within it no self, which formed like air bubbles in water only when the experience faded away. In this universe there was pure knowledge and perception without a center . . . [. . .]

3279 [. . .] In this state I very often experience Gyan Shakti, a state of *knowingness* that is also present when resting after the meditation and for some time afterward. The experience of pure consciousness is of extremely deep-clear and subtle presence, as well as of wonderful lightness.

An Unobstructed Epistemic Space

> As space pervades, awareness pervades. Like space, *rigpa* is all encompassing, nothing is outside of it. Just as the world and beings are pervaded by space, *rigpa* pervades the minds of all beings.
>
> —Tulku Urgyen Rinpoche (1920–1996), *Vajra Speech*

> Space is the example for mind essence, because space is unmade. But mind essence is not totally like space, in that space cannot think. Space has no knowing. Our mind is cognizant emptiness—empty like space, but with a natural knowing. That union of cognizance and emptiness is seen when recognizing. It is immediate, like the example I mentioned of pointing into mid-air. You do not have to wait to raise your arm for your finger to touch space—you are already touching space, all the time. You do not have to move your hand forward; the contact is already occurring and has been your entire life. All you have to do is recognize that it is taking place. It's the same with mind essence.
>
> —Tulku Urgyen Rinpoche (1920–1996), *As It Is*, II

In chapter 4, we encountered the idea of "epistemic openness." In this chapter, I want to introduce a second conceptual instrument, intimately related to the first: the notion of an "epistemic space," a space of knowing. Epistemic openness is the openness of an unobstructed epistemic space, and experiencing pure awareness simply means having a *model* of this space, nonconceptually knowing that it exists.[1] Pure awareness could be described as the experience of being knowingly poised over epistemic space.[2] In this context, we will also look at the phenomenology of "unboundedness." The experience of clarity that is the focus of this chapter can be described as the experience of an unbounded space of knowing, a space that is currently unobstructed. It is pure awareness itself, unclouded and entirely open.

In my view, the epistemic-space metaphor is the best phenomenological metaphor for pure consciousness or minimal phenomenal experience (MPE). This metaphor is found in many places in the contemplative literature. Consciousness per se is a model of an all-pervading inner space of knowing, but, as the ancient analogy (presented in the words of Tulku Urgyen Rinpoche at the beginning of this section) beautifully shows, you cannot directly grasp, feel, or touch the space itself by cognitive means. Why is this so? The metaphor tells us why: You cannot grasp and hold physical space with your physical hands. But your hands have always been *in* this space, and every single one of your bodily movements always already took place in it. Physical space cannot be grasped, but it is a condition of possibility for grasping movements, whether you know it or not. It is the same for MPE space: It cannot be cognitively grasped—by forming a concept of it, or thinking a thought about it—but it is the precondition for all cognitive grasping, for trying to conceive of awareness itself by using conscious thought (more about this in chapters 27 and 31). The Latin root of "to conceive," *concipere*, means to take in and hold, and by the fourteenth century, a second meaning had emerged: to take into your mind, to cognitively grasp something. This is what cannot be done to MPE space, but is what MPE space allows us to do to everything else.

"Touching" pure awareness would mean directing and "pointing" the focus of attention to MPE space itself, thereby turning it into an object of inner attention. Thinking is like grasping; attending is like pointing. The focus of attention is like your mental fingertip. Attention is a nonconceptual way of allocating computational resources in the brain, of optimizing for precision. But this cannot work for MPE space: You cannot conceive of awareness itself by using conscious, conceptual thought, and you cannot use the subsymbolic mechanism of deliberate attention either—just as you cannot successfully point the tip of your finger to physical space and thereby "make contact." In fact, contact has been there all along; you have been "in touch" all the time without noticing. It is the same for what Edmund Husserl called the *Blickstrahl der Aufmerksamkeit*, the "ray glance of attention": Every single one of its pointing movements always already took place in what I will from now on call the inner "model of epistemic space." Just like physical space, epistemic space creates an allocentric frame of reference: It is unbounded and uncentered, even though of course its variable content is perspectival. All the rich content making up the virtual world in which we live our conscious lives, being a simulation of an embodied behavioral space, is egocentric; there are self-oriented perceptual horizons and parallel lines that seem to meet in infinity (more on this in chapter 29). There is a single perspective, but it is a structural feature that ultimately belongs to the level of content, creating a second-order experiential frame of reference. Normally, there may be egocentric-to-allocentric and/or allocentric-to-egocentric mappings of which we are mostly unaware. I think that ordinary states of consciousness are characterized by the fact that the egocentric frame of reference is very dominant because it almost occludes the allocentric frame on which it is superimposed—but which sometimes can be experienced in isolation. We find this at various points in our phenomenological data. One of the participants in our study said: "I am not in the world—the world is in me! I am the space in which everything appears" (#2299). How does this relate to the phenomenology of "clarity"?

Generally, a *space* is clear if there are no obstacles in it, if we can in principle move and navigate in this space freely, if the many potential paths that we could take through it are not blocked and passage is not obstructed. There is an openness. This openness can be consciously experienced. In addition, a clear and unobstructed space holds a very large number of possibilities, a fact that is perhaps reflected in the abstract phenomenal quality of "vastness" reported by many of our meditators. Along with these two global qualities of unobstructedness and vastness, we may also experience the specific potential for expansion: Awareness is a space of knowledge, and of course it can be expanded. In meditation, we can sometimes experience this space itself, as an unobstructed whole, including its potential for expansion. It is not even obstructed by a center; it is unbounded.

The phenomenology of unboundedness is an important and recurring element in published reports about pure awareness. Here are three case studies from the literature:

Case study #1. I experience pure consciousness as a state of unboundedness and total ease and deep relaxation. There are no thoughts, no feelings, or any other sensations like weight or temperature. I just know I am. There is no notion of time or space, but my mind is fully awake and perfectly clear. It is a very simple and natural state.[3]

Case study #2. [. . .] a state of complete rest, full consciousness without content and unbounded in time and space.[4]

Case study #3. [. . .] a couple of times per week I experience deep, unbounded silence, during which I am completely aware and awake, but no thoughts are present. There is no awareness of where I am, or the passage of time. I feel completely whole and at peace.[5]

Phenomenologically, "unboundedness" means that there is no second, finite region to which attention could be directed, and there are no consciously experienced boundaries, limits, or horizon. Importantly, to say that the space of pure consciousness is "boundless" does not imply that there is an explicit experience of infinite expansion or of large distances. Rather, it means that there is no "other side beyond the boundary" to which attention could shift. To give an example, in visual awareness, attention can shift from a red patch into an adjacent green patch, transgressing a chromatic boundary. In the phenomenology of tonic alertness, there is no such boundary because everything outside our inner model of the epistemic space created by wakefulness cannot be deliberately attended to: It is simply unconscious.

Taking meditators' reports of unboundedness seriously also leads us to an unexpected conceptual point that has a deep philosophical flavor: MPE itself is not only "nondual" (in terms of lacking the internal manifold created by a dynamic subject/object structure; see chapters 26 and 27 for more), it is also not one. This philosophical issue was discovered a long time ago. As a matter of fact, in Śāntarakṣita's "Ornament of the Middle Way," the *Madhyamakālaṃkāra*, we find the classical "neither-one-nor-many argument," which concludes that all phenomena are open in the sense of having no inherent nature at all because in reality, they have neither a singular nor a manifold nature. Śāntarakṣita (725–788) is one of the most important thinkers in the history of Indian and Tibetan Buddhist philosophy. His work has echoed through the centuries and is discussed by logicians to this day.[6] It is plausible to assume that the fact that such intense philosophical debates occurred at all was thanks to the existence of a widespread, vivid, and distinct *phenomenology* that was recognized as being in urgent need of consistent conceptual analysis. Apparently, what I am trying to approximate here as the unbounded and unstructured phenomenal character of MPE was something well known many centuries ago.

So unboundedness seems to be a crucial part of pure awareness, in our reports and in testimony from many hundreds of years ago. But once again, it would be a mistake to interpret the phenomenology of unboundedness as an explicitly experienced and merely quantitative infinity, or as a concrete, endless expanse of some sort. It means only that the subtle experiential character in question includes the *potential* for expansion, and in the experience of pure awareness itself, there is no such thing as a "beyond"—that is, *another* consciously experienced finite region or realm "on the other side" of a boundary. This is interesting because it seems to imply that in the phenomenological domain of MPE, there are no countable entities: The domain is an unstructured space that has neither center nor periphery and cannot be introspectively *individuated* to create smaller, indivisible units of pure-awareness experience—not even a single one. So even if we take a modern and slightly radical bottom-up approach that starts from real-life experiential reports given by present-day meditators, it seems that we must add "non-oneness" to "nonduality." The metaphysical neither-one-nor-many principle returns in consciousness research.[7]

Our two new concepts of "epistemic openness" and "epistemic space" refer to an open inner space holding a very large number of possibilities for knowing the world and ourselves. Knowing *as a self*, from an egoic first-person perspective, is only one of these possibilities, one of many possible data formats, although one that has been successful for biological beings like humans, who have all our sense organs and all our effectors (our arms, legs, vocal cords, etc.) united in a single body. In chapter 27, I will call this the "single-embodiment constraint" and will relate it to nondual awareness. Our *behavioral* space is centered on a single creature, and the same is true of the conscious, experiential model of this space that our brain typically constructs: This model has a center, and it has limitations. We are *situated*. But our inner model of *epistemic* space does not share these features. It can be expanded, and a very large number of perspectives and situations can be enfolded within it. It is not egocentric. Perhaps what today we call "pure awareness" is the shared, primordial data format of all sentient beings, some sort of fundamental constant? Is MPE the most abstract data format that conscious beings can use?

By definition, an epistemic space is a space of possibilities: It contains every possible epistemic scenario and every dynamic partitioning of itself that could ever take place—everything that could *potentially* be known and experienced by a given system. An epistemic space contains the repertoire of knowledge states that a given system has. Therefore, it encompasses many ways of accessing world and self, of making reality available to itself, at this specific location in time and space.

An important concept in current debates on consciousness is "global availability." The idea is that—in purely functional terms—conscious information is globally available: precisely the information that we can access simultaneously with all our faculties, with thinking, with attention, or with selective action control. Indeed, perhaps the earliest of all the recent theories of consciousness, the Global Workspace Theory first formulated by the neuroscientist Bernard Baars in 1988, states that conscious information resides in global working memory; consciousness is simply global availability. Function, however, does not automatically lead to phenomenality.

My own theory goes a step further: Consciousness arises only when a model of this working memory has appeared in global working memory itself. It is not sufficient for the enormous potential of cognitive possibilities—the many different ways of making the world and oneself available—to merely exist. The system must also *recognize* them. We will come back to this point in chapters 9 and 10.

It seems as if this almost infinite potential can sometimes be experienced in meditation, but in a simple, nonconceptual, and entirely undramatic way. As I said earlier, you can be peacefully "poised over it." Interestingly, we can conclude that the model of such a space would have to be characterized by a very high degree of what some scientists call "counterfactual invariance": You cannot imagine a scenario where it would be absent because whatever the system could know or experience would necessarily take place within it. You cannot imagine what it is like to be unconscious.

It is possible to generalize to some extent about the nature of epistemic space as one important aspect of pure awareness. But we must also remember not to overgeneralize to the point of inaccuracy, extrapolating a general theory of consciousness from too few empirical facts. We are embodied, physical beings; impermanent products of evolution. All biological creatures have different capacities for knowing the world and themselves, and many of them may not be sentient at all. For humans, the wakeful clarity of pure awareness is related to a specific set of epistemic capacities and the representational space opened up by them. For example, as we saw in chapter 4, the nature of pure awareness for us is tied to "high stimulus-readiness," "orientation readiness," plus the mere "capacity for mental self-control" on the level of attention and cognition. What's more, our experience of this space of possibilities will always come in different degrees of lucidity, clarity, and stability—for two main reasons. First, the epistemic capacity of the biological organism itself varies over time: Certain paths and possibilities may be temporarily blocked and unblocked. When ill with fever, we may not be able to concentrate or think clearly; in a dream state, we are unable to control our attention. But you can think clearly or even try to meditate during a *lucid* dream, when your self-model has changed[8] and you "know" that you are dreaming (chapter 21)—or after you have woken up. Second, the bodily background conditions are variable, resulting in changing stability levels in our inner model of our epistemic capacity itself: Epistemic openness and our inner model of it can be more or less recognizable, either spontaneously foregrounded or receding barely noticed into the background. One of the participants in our study coined the term "Basal Clarity" (see #3058 in chapter 21). It seems that the clarity of empty cognizance is something fundamental, perhaps something on which, unbeknownst to the subject of experience, everything else rests.

Many of the dimensions of silence explored in chapter 3 bear some resemblance to the aspects of clarity considered in this chapter. Yet clarity is not the same as mental silence. In humankind's contemplative traditions, it has long been known that it is perfectly possible to be conscious and have no thoughts at all while lacking clarity and lucidity. The space of awareness can be stable and unobstructed as low-level perceptual processes and sensorimotor integration run on autopilot, but with lucidity and clarity nonetheless lacking. This is a well-known phenomenological effect during meditation. If my tentative theory of MPE is right, this happens because there is no stable model of the capacity to know. We are epistemically open, but we do not really *know* this openness. We may describe such sluggish or indeterminate states in terms of "mental torpor," and empirical psychology has recently begun to investigate the interesting phenomenon of "mind blanking," in which ongoing perception is uncoupled from attention and cognition.[9] But contemplative practitioners have known about this phenomenological possibility for many centuries: There can be silence, but it may not be a lucid silence; clarity may be lacking. So what is the main difference between the two? One answer is that the phenomenology of clarity, unlike that of silence, seems to need an additional element, a nonegoic form of self-awareness like the one that was statistically extracted as factor 8 in our study and that we labeled "Emptiness and Nonegoic Self-Awareness." Perhaps this arises only when a silent mind recognizes itself as an open space of knowing (more on this in chapter 30).

Let us say, then, that the phenomenal character of "clarity" involves the existence of an unobstructed space of epistemic possibilities, and that it expresses the *possibility* of knowing on the level of conscious experience. It is lucid openness as such, implied by possessing the property of epistemic openness. If so, it is natural to look for examples of the pure-awareness experience in which we also find the phenomenal character of *actual* knowing. Sometimes this may even involve certainty, the quality of actually knowing that one knows. Can clarity in our narrow sense coexist with the thinking of thoughts, with attentively moving through the physical world, learning, actively gathering knowledge about it, and knowing that one knows?

The answer is yes. We find examples in the selection presented in this chapter (e.g., #1818, #3156, #3279), as well as in the contemplative phenomenology of direct perception (chapter 9), witnessing (chapter 19), nondual awareness (chapter 27), and nonegoic reflexivity (chapter 29), and at many other points in this book. This quality of clarity and "pure knowing" is an experience of subjective confidence; it is the specific phenomenal character of an actual epistemic process currently going on. This process is that of estimating the probability of knowledge possession and the depth of one's own understanding—but not necessarily with any epistemic object. I think this basic, nondual property of "natural knowing" is what Tulku Urgyen Rinpoche was pointing us to in the second quotation presented at the beginning of this chapter.

Clearly, the phenomenology of pure awareness is often accompanied by a strong experience of insight, but in a nonintellectual and nonconceptual form. I have often termed it "epistemicity," the phenomenal character of knowledge-possession itself.[10] In chapter 18, we will encounter epistemicity again as the "phenomenal signature of knowing," in the context of what has been described as the luminosity of pure consciousness. Taking the phenomenology of contemplative practice seriously demonstrates that epistemicity can occur without a self, without concepts or propositional mental content, and in the absence of subject/object structure and time representation. For example, it is an important component of perceptual awareness, a nonconceptual representation of what in statistics is called "subjective confidence": the estimated accuracy of a perceptual choice and the congruence with prior expectations.[11] Ultimately, the consciously experienced sense of confidence is a property of a model in your brain—and of course, there can also be a model of the overall space of knowing as a whole. Calling this subjective quality "epistemicity" may at first sound very abstract, but despite being an *unstructured* form of knowing, it is also a very concrete phenomenal experience. It is a nonemotional feeling of knowing that is neither cognitive nor perceptual—something that has nothing to do with perceived colors, sounds, or even internal bodily sensations, and that is distinct from any kind of discursive thought. The signature of knowing comes in degrees, and it has a distinct phenomenal character all its own.

This opens up a new perspective for the scientific project of constructing a minimal model of consciousness: The qualitative character of awareness itself could be a "pure" phenomenal experience of knowing, a state of minimal complexity (e.g., in that it has no subject and no object). During a full-absorption episode, there is no knowing self and no kind of juxtaposed, reified "thing" that is being known by this self.[12] This "nondual" nature of MPE will be the topic of later chapters (especially chapters 26 and 27). It is one major aspect of what we mean when we talk about nondual awareness being an *unstructured* experience of knowing—one that is not obstructed by a first-person perspective. Consciously experiencing an uncontracted variant of epistemicity may therefore be directly related to the phenomenology of clarity.

6 Density

Smooth and pristine as silk. [#570]

I learned a lot from the meditators who generously supported us during our pilot studies. In some cases, they drew my attention to specific aspects of experience that had simply not been on my radar before. One of them was "density." The unbounded space of pure awareness can have a quality of density that perhaps is comparable to the continuum of real numbers in mathematics: Between any two distinct points, there is always another (and hence an uncountable infinity of others exists). A mathematician might say that the relevant space is *complete*, in the sense that it lacks any gaps. It also forms an internally coherent whole; everything hangs together. In this sense, the space of awareness itself is experienced as dense, as homogeneous, and as a continuum.

The experiential quality of density links up with other qualities covered in this book. For instance, it relates to the many descriptions of "connectedness" that we will investigate in chapter 11. It also may help us to better understand some paradoxical reports referring to "emptiness and fullness at the same time" (chapter 17). In my own meditation experience, two other salient qualities are sometimes also intimately related to density. These qualities of pure awareness can be described—and have been described by others—as a "thick silence" (chapter 8) or a "thickness descending," and as an abstract, nontactile way of being "velvety," "silky," or "ultrasmooth." These phenomenal qualities are sometimes associated with other visual and spatial characteristics, including in constructions that evoke paradoxical combinations of opposites: "Both empty and full, light and dark, velvet and silk" (#577). Pure awareness also has been described as exhibiting a "silvery density":

> 3422 [. . .] Although it was new to me, I rather had the impression that it was more or less the normal state, or the state that occurs when I leave everything

else out. Maybe like a primordial state. Or existence in itself. If I were to describe what that might have looked like visually, I would try to come close to it in the combination of very dense and silvery [*sehr dicht und silbern*]. I had already experienced this silvery density [*silberne Dichte*] to some extent in fast *kinhin* [walking meditation], as if I were walking by myself within silvery being [*innerhalb eines silbernen Seins*].

Next, we will look at four examples discussing density, fullness, and emptiness. Following on from density, we will see a second commonality emerging. This one relates to one of the deepest and most subtle phenomenological discoveries ever made within humankind's contemplative traditions: If we look very closely, we may find that emptiness, specifically the spacious quality of epistemic openness, is present *in* appearances themselves. Emptiness can be found not only in the stillness of the silent mind, but sometimes even in objects of perception or *in the thoughts themselves*. We often falsely take having thoughts or perceptions at all to be incompatible with lucidity, clarity, and silence, but these reports show otherwise. More on this discovery later in this book; for now, you may want to watch for this aspect cropping up in the following reports:

2359 [. . .] Little by little, all that remains is "space" filled with slight fluctuations and occasional mental events. I identify this space with "pure consciousness." At this point my concentration often declines, so that I end the meditation. But a few times I have managed to continue, whereupon the quality of the "space" changes. It then becomes "denser," and my mental activity continues to decline toward a point of "highest purity" where there is no more mental activity. At this point "time" also seems to slow down. [. . .]

3218 [. . .] The next day during the second meditation I had again waves of bliss but permeated by even more thick, yet dynamic, silence. [. . .] The experience of a thought in meditation is sometimes like a drop of water falling onto a very quiet surface of water and creating ripples. But even better is the analogy of the ocean becoming a wave, because when the faint idea of the thought comes, it is the ocean rising into a wave without losing its status of an ocean, remaining ocean with a slight stir in it—stir is almost too strong a word because it entails an external something stirring, flow is probably better term. The memory of the thought is there while maintaining the memory of the ocean. It is like multiprocessing but almost without the multi.

3029 [. . .] there are experiences of simultaneous emptiness and fullness. They are dense and I feel very present. There are feelings of perfection, unity, and bliss. In these moments Pure Awareness is not disturbed, even when thoughts, images, and memories appear. They are unseparated from the Awareness. [. . .]

3501 I had the impression of briefly experiencing the insubstantiality of *shunyata* [emptiness], then turned my attention away from it, and noticed how the next impression was still like substanceless shunỳata—that was impressive and encouraging.

Abstract Interoception and Ultrasmooth Embodiment

Whatever is expressed is expressed in the continuum of the nature.
—Longchen Rabjam (1308–1363), *The Precious Treasury of the Basic Space of Phenomena* (10: 117)

Some people love the flavor of a good glass of red wine. Others prefer the subtlety of a specific scent, like the combination of sandalwood, frankincense, cinnamon, and ginger lily. And then there are those who have discovered the silkiness of silence. Philosophers have long discussed the "ultrasmoothness" of perceptual qualities like redness and sweetness, because they come with a problem attached. Many seemingly simple forms of conscious experience driven by sensory perception have a grainless, homogeneous character, a lack of any discernable internal structure: They are "smooth" or "ultrasmooth." This is sometimes seen as hard to explain using empirical science.[1] In a visually experienced, homogeneous patch of blue, the purported quality of "blueness" itself has no graininess. Is it a phenomenal primitive, something like an atom of consciousness? How could one ever hope to reduce something that has no internal structure to structures in the brain? How could blueness be mapped onto a network of functional relations? In the philosophy of mind, this is known as the "grain problem."

The phenomenology of density, thickness, and smoothness as described by practitioners of meditation is only one example of a more general pattern that we find in many reports, and it may eventually advance our understanding of the grain problem. Often, there seems to be an abstract, nonsensory aspect of the experience that perhaps can be used to indirectly and metaphorically bring us closer to a better understanding of the phenomenal character of minimal phenomenal experience (MPE) by accentuating a new aspect of the pure-awareness experience. Interestingly, in MPE, we often find an experiential aspect that is still related to one sensory modality, but that already lacks the low-level "data format" of raw sensation and apparently direct perceptual experience. For example, the experience of silence in pure awareness is often described as not being an acoustic silence, as being unrelated to any explicit auditory experience of the complete absence of sound (this is the sudden insight that John Cage had, as described in chapter 3). Equally, the phenomenology of harmony that will be investigated in chapter 7 is not literally acoustic. And the frequently reported sense of profound clarity, as discussed in chapter 5, is not a *visual* form of clarity. The pure-awareness version of clarity is often figuratively linked to the experience of seeing clearly, having an unobstructed visual space like the vastness of the sky or the seemingly endless surface of an ocean. But there is no stimulus-correlated aspect; the "raw feeling"[2] of sensory stimulation itself is lacking. The same is true for experiential reports of "radiance" and "luminosity" (chapter 18): Mostly, they do not refer to concrete visual experiences and are only distantly or figuratively related to this specific sensory modality. Many reports point out that the "space" of awareness as a whole is not a physical space because there is no embodied feeling of motion, no perception-based sensorimotor phenomenology that could be related to it. In my own meditation practice, I have found that if I look closely, the "space" is neither inside nor outside. Equally, the phenomenal character of "vibrancy" that some of our participants mention is often not located in a certain region of the body image at all (vibrancy and density recur in chapter 17, "Emptiness and Fullness"). The quality of "gentleness" that sometimes characterizes pure awareness is not a tactile experience, and it is not an emotional state either. Here, my point is that the same is true of the "smooth," "silky," and "velvety" character that, according to our reports, the nonconceptual awareness of awareness itself can sometimes have.

In many contexts, then, sensory analogies are drawn in the attempt to approximate some element of the experience of pure awareness. Some sensory modalities are conspicuous by their absence, however. In our reports, pure awareness is almost never compared to an abstract form of tasting or smelling via gustatory or olfactory metaphors. (For interesting counterexamples, see #1381 at the beginning of chapter 3, plus the striking parallels between the notion of "unified taste" in the fourteenth-century Christian mystic Begine of Hadewijch and the Mahāmudra concept of "one taste" in chapter 26.) Similarly, "pain" was one of the two least frequently reported items in our survey, although nociception clearly has a sensory component. All these observations will have to be carefully and systematically investigated by a future science of consciousness. When this happens, the interoceptive self-model and its relation to the phenomenology of pure awareness will be especially interesting to investigate because this experiential level of the self-model in our brain involves the many different signals that are constantly transformed into our experience of the internal state of the body. These ongoing transformations involve the perception of bodily signals arising in muscles, tendons, and joints; vestibular information from the inner ear; visceral signals arising from blood vessels and the heart, lungs, stomach, and bladder; pain perception; information about body temperature processed in the thermoregulatory system; and diffuse signals originating in the body's endocrine and immune systems.

In his book *Being You*, the British-Indian neuroscientist Anil Seth has made the strong point that what I have called the "interoceptive self-model" is actually a "controlling hallucination," and what we experience as our emotions and moods are really control-oriented perceptions that regulate the body's essential variables. He writes: "At the very deepest layers of the self, beneath even emotions and moods, there lies a cognitively subterranean, inchoate, difficult-to-describe experience of simply *being a living organism*. Here, experiences of selfhood emerge in the unstructured feeling of just 'being.' "[3] In chapter 26, we will find out what meditators have to say about the primordial experience of "nondual being." One important future research target is the question of why some sensory modalities, like touch, lend themselves to figurative but apposite descriptions of MPE (e.g., as "density" or "smoothness" or "gentleness"), while others rarely appear.

I am intrigued by the following question: Is there a maximally abstract form of interoception? "Interoception" means feeling your own body from the inside, and as we will see in chapters 18 and 24, new research relates meditation experience to the experiential quality of "self-touch." In our own psychometric study, factor 10 describes MPE as an abstract form of tactile experience resembling self-touch, or even as an experience in which the entire body touches the world while simultaneously being touched by it. Is there a form of "bodiless body-experience" that is based entirely on internal information but lacks the raw feel of sensory stimulation, even from the body's interior? Are there perhaps ways of experiencing certain aspects of the inner state of the body that are not related to any specific kind of internal receptor, and are maximally simple at the same time?

I think that the answer to all these questions is yes. For example, the experience of "wakefulness" investigated in chapter 4 could be precisely this: a direct and simple way of experiencing one specific aspect of one specific part of the physical body. This aspect could be a stable dynamic signature, a process realized by a pattern of activity in a part of our body that cannot be "directly" perceived because we lack any internal receptor system to feel it—but that sometimes (namely, during the wake state) needs a control-oriented representation. I think that Anil Seth is right, but there is something even more fundamental, something that cannot be properly called a "perception" (because it is not based on sensory receptors of any kind, as in the original Latin meaning of *perceptio*, the gathering or receiving of information "through" the senses) and that is radically nonegoic (because it has nothing to do with being *you*).

In the most general terms, there is no sensory system for the brain. We cannot directly perceive our brain. As all neurosurgeons know, the brain is insensitive to pain. But we are also unable to hear it, see it, taste it, smell it, or feel it—and what we sometimes feel as a headache is in fact the state of the blood vessels in our head. The abstract, nonsensory aspect that is experienced as the thickness or "density" of MPE therefore could be something more abstract, like a property of the brain's model of tonic alertness portraying the continuous character of epistemic space, the completeness of possible knowledge states expressed as the fieldlike experience of epistemic openness. Density could designate the abstract fact that this organism is now fully alert and open to the world while also successfully sustaining its integrity, being wakefully present while preserving its own existence. My general point is this: An internal state or property that may eventually appear as something highly abstract and complex from the third-person perspective of science could, if viewed from the whole organism's internal perspective, certainly be portrayed as something simple and directly given. Again, please recall how the phenomenology of density, thickness, and smoothness as described by practitioners of meditation turned out to be only one example of a more general pattern.

Here, a new theoretical perspective is emerging that could give us yet another answer to the question "In what sense is pure awareness pure?" Not only does pure awareness lack cognitive content (there are no thoughts), it also lacks perceptual content (there are no sensations), even from the interior of the body—and this could simply be because there are no sensory receptors in the relevant part of the brain. Pure awareness could be a way of experiencing the activity of the aforementioned part of the *neural* body only, albeit on a more abstract and fundamental level of the brain's computational architecture.

If this admittedly radical perspective is not entirely misguided, then there are fundamentally different ways of experiencing the body that you are. For example, when scanning the rich landscape into which meditative attention transforms your body's inner state, you can feel your body by using the conscious body-model automatically activated in your brain. This is called a "body scan," and it is a perfect method for discovering and releasing tensions that you might not even have realized existed (like the one described in the epigraph at the beginning of chapter 1). This meditation technique became well known through the life and work of S. N. Goenka (1924–2013), an Indian teacher of Vipassanā meditation born in Burma. If you do a mindful body sweep with closed eyes, gently and precisely attending to different parts of your body from top to bottom and back again, then what you are actually doing is optimizing the brain's precision expectations, mostly for your interoceptive self-model.

Practicing this carefully in guided meditation on retreat, I sometimes have become consciously aware of aspects of my own embodiment that I had never even known existed—aspects unknown to me my whole life before this moment. For example, on one occasion, I suddenly became aware of the subtle landscape of contact sensations in the upper part of the region where my inner eyelids rest on my eyeballs. And once I realized that this set of sensations existed, it was hard to ignore or not notice it. I am not a complete outlier in valuing this type of expansion. Some people like to "bathe" their bodily self-model in mindful attention as much as they like showering in literal water.

The conscious body model in our brain, then, is something that can be made richer and more precise, something that can be expanded and "purified" by systematically attending to it. But perhaps you can also feel your body without this low-level, space-bound model? Yes, you can feel it as something that has a shape, that involves a sense of weight and balance and a rich blend of inner sensations, a whole mélange of finely nuanced phenomenal qualities. Meditation practice can even make your body "more real" to you, in the sense of bringing some of these aspects into existence. In this sense, it is a creative process. But from a radically naturalistic perspective, you can also feel your body as pure awareness itself, as the silkily silent, crystal-clear quality of epistemic openness. I call this a "radical body scan," and it can apparently refer to a much more abstract phenomenology of embodiment (more on this in chapter 24). Abiding in the raw and lucid wakefulness of pure consciousness, therefore, may turn out to be a form of *bodily* self-consciousness that has never been recognized as such. To me, our data also clearly show that a *nonegoic* form of self-awareness exists (see chapters 29 and 30). This very subtle, highly intimate, and yet entirely selfless way of being in touch with yourself could provide a new way of understanding the relationship between pure consciousness and bodily self-consciousness. It might also offer us a new way of understanding some body-related concepts that in today's philosophy are often used in vague and inflated ways—like "embodiment" and "lived experience," for example.

7 Soundness

It felt like a puzzle piece that completed a picture. [#2121]

Can you imagine entering a state of consciousness that makes everything else complete—a global quality that suddenly adds wholeness to all the rest, somehow "making everything fall into place"? This is what the phenomenal character of "soundness" feels like. As the auditory root implies, soundness can also be described as a nonacoustic experience of "global harmony," which makes it interesting in a variety of scientific contexts, including mathematics, computational modeling, and artificial neural networks. In this state, prediction error and surprise seem to be at a minimum, while the global degree of constraint satisfaction appears to be maximal. This is an experiential state that is close to being optimal, and in more than one sense. One of our participants described it as "[h]armonious and peaceful, floating, light and natural" (#221); another said: "My experience was to see/feel everything as pure, everything in harmony, me and the world, the colors, the sounds" (#1025). The phenomenology of internal harmony and consistency has been largely ignored in philosophy of mind and consciousness research. Perhaps soundness is a state that, in their stressful academic lives, self-critical intellectuals and ambition-driven empirical researchers simply do not experience very often. I am very grateful to one advanced meditator who, during the pilot phase in which we developed the first version of the MPE-92M questionnaire, drew my attention to this specific phenomenological aspect of what in German is called *Stimmigkeit.* There is no perfect translation for *Stimmigkeit* as a phenomenological concept, but it unites soundness, harmony, and a more existential sense of "this just feels absolutely right."

Here are five examples of meditators trying to describe pure awareness as an optimal state, characterizing it in terms of "soundness" or "holism," or of being "just right" or "in unison":

1378 [. . .] What remained was the feeling of a state in which everything was as it should be.

2283 [. . .] Absolute soundness, clarity, and vastness. [. . .]

2524 Everything is harmonious, just right, so that you don't even consciously perceive many things (individually) (not as a stimulus or disturbance e.g. re temperature, pressure, volume, . . .). Everything is self-evident, familiar, clear, as it should be (even if I'm consciously contemplating "problems"). To everything that is thought and imagined, you feel a connection (not in the sense of dependency or the like, but rather as a being-interwoven, as love) and yet clarity (in the sense of a kind of distance, not attracting, not identifying). It is like saying yes, understanding, humming along subliminally, even with those aspects that are condemned or felt as pain in everyday life.

2722 I had the feeling that I'm "round," no problems, no questions, no doubts. Everything is okay as it is, no ifs and buts—this is an absolutely desirable state not only in meditation, but as much as possible day to day.

3517 It is a feeling of being one with yourself and at the same time being part of the world. In harmony/unison [*Einklang*] with yourself and with something higher.

The experience of soundness and harmony often coemerges with the phenomenal character of "pure spatiality," "bodiless body-experience," and "abstract embodiment," which we will investigate more closely in chapters 23 and 24. A couple of tasters, meanwhile:

2594 It feels as if my body is dissolving, my boundaries, everything expands and I forget about time. It feels big and harmonious and I am humbled by the size and harmoniousness and the freedom.

2652 [. . .]—space disappears or rather is boundlessly open—everything is in harmony—everything is my "body"—[. . .]

Finally, soundness can have a strong epistemic quality. It can feel like an insight, but in an entirely nonconceptual way, also like a form of "embodied harmony":

3443 [. . .] "pure awareness" was a knowing felt in the body, not just a standard insight. Whereas a standard insight to me feels cognitive, and comes with a small energetic burst of "surprise" at best, and feels pedestrian, this kind of experience is more characterized by a coming to the obvious, but somehow nonstandard (for me), sense that the sum of all my knowledge, both declarative and experiential (i.e., happened to and felt/remembered through my bodily sensations / experiential body), which is normally dissonant, actually "makes sense" as a whole. [. . .]

The E-fallacy

> The biggest puzzle arising is that comprehension of the One is neither by scientific understanding nor by intellection, as it is in the case of other intelligibles. It corresponds rather to a presence that is better than scientific understanding.
>
> —Plotinus (205–270), *Enneads* VI, 9, 4, 1–4

In this section, I will briefly provide you with one new conceptual tool. I hope that it will help lead us into the next phase of our investigations of pure awareness as it occurs in meditators. In 2014, together with the philosopher Jennifer Windt, I proposed the notion of an "E-error."[1] At the time, we were interested in the phenomenology and epistemological status of intuitive knowledge—that is, in the times where we "simply know" that something is right. Intuition plays a great role in some parts of academic armchair philosophy, like the intuition that zombies (i.e., functional isomorphs lacking phenomenal states, such as creatures indistinguishable from humans in every respect except that they are unconscious) are possible. Another classical philosophical intuition (refuted by the phenomenological data presented in this book) is that consciousness necessarily implies an egoic form of self-consciousness. One explanation of this failure of imagination, also called the "ubiquity thesis,"[2] might go as follows: Every deliberate attempt to imagine a state of consciousness that genuinely lacks self-consciousness is a form of mental action; therefore, as an attempt to actively control your own state of mind, it automatically creates the phenomenology of mental agency, goal-directedness, and effort—thereby sustaining a sense of self precisely when one is trying to imagine its absence.[3] Many of us have a deep-seated philosophical intuition that consciousness without self-consciousness is simply inconceivable, and this intuition might be rooted in the functional architecture underlying human consciousness, which in turn is a product of natural evolution and sociocultural priors. But as Dennett tells us,[4] one should resist the temptation to mistake a failure of imagination for an insight into necessity: From the fact that one cannot imagine being in a state of consciousness entirely lacking self-consciousness, it does not follow that it is nomologically (or even metaphysically) impossible to be in such a state.[5] We have probably all experienced a strong feeling of knowing, or even of certainty, without being able to give any evidence or rational argument to justify it. Intuitive knowledge is often nonconceptual, and it is not based on an inference of conscious thought. But it can sometimes be characterized by an exceptionally strong sense of certainty, and in the practical context of everyday life, that is often very helpful.[6] This phenomenology of nonconceptual insight and embodied intuition plays a major role in the contemplative experience of "soundness," and it will also be helpful in understanding many of the other reports presented in this book.

The E-error is an easy fallacy to fall for. It is a category mistake in which epistemic properties are ascribed to something that does not necessarily have them. From the fact that you now consciously perceive a beautiful red flower on the table in front of you, it does not necessarily follow that such a flower exists. Appearance is not knowledge. Consciously experiencing "knowingness" is not the same as *having* knowledge. In the case of perceptual knowledge, you may be having a simple visual illusion, and the flower is actually a deep orange color. Or there may be (as I tend to believe) no colored objects in the outside world at all because wavelength mixtures in front of our eyes are all we ever have, and "colored objects" are merely useful constructs in the brain's internal model of reality. Or, alternatively, the flower itself may not exist at all because you are having a complex hallucination. The flower, the table, and the whole room may not exist at all because you are currently dreaming. And if Advaitic and Buddhist philosophers are on the right track, then even the "knowing self" that is so certain of its own existence could be a mere appearance (chapter 25).

All these ways in which perceptual experience feels like knowledge but isn't apply equally to the inner perception that we currently possess knowledge of a certain kind: This "knowledge" could always be an introspective illusion—an inner form of misperception. As such, the "phenomenal signature of knowing"—as Windt and I dubbed the feeling of knowing—is only appearance. We will encounter this new concept more often as we travel through this book, because variations on the consciously experienced signature help us to better understand certain aspects of contemplative practice. For example, we also find a signature of *self*-knowing, the feeling of self-intimacy and nonconceptually knowing oneself. And there is also an "uncontracted" signature of knowing, one that is not tied to a personal-level self anymore, such as in the phenomenology of "witness consciousness" (chapter 19) and during "clear light sleep" (chapter 20). The E-fallacy applies to all these cases because it consists in falsely concluding that a consciously experienced feeling of knowing is a reliable indicator of actually possessing knowledge. No knowledge claims follow from phenomenological descriptions, and neither do any "direct" epistemological insights. Of course, in most practical situations, the intuitive signature of knowing—which likely has a long biological history—is an excellent guide to dealing with other human beings and a complex environment, especially if we need to act fast. But the point is that at any moment, there always can be one of two subjectively indistinguishable states of consciousness—one of them providing us with a genuine form of insight and the other being a misrepresentation of reality.

Take intuitive knowledge as a first example. As an author, you can have a strong intuition that your agent is a morally good person or that you are dealing with a trustworthy publisher—but of course, you might later discover that you were wrong. If intuitions are indeed mental states characterized by a specific phenomenology of knowing, then the attempt to simultaneously characterize them as involving genuine insight and as being the basis of knowledge rests on the E-fallacy. The feeling of knowing is not the same as knowing; the phenomenology of direct evidence is not evidence. Assuming that they are the same thing is a category mistake in which epistemic properties are ascribed to subjective appearances, which do not intrinsically possess them. Whenever we derive strong theoretical claims about consciousness itself from apparently "self-evident" subjectively experienced truths, this is the mistake that we are making. Are zombies *really* conceivable? Sure, if you decide they are. In academic philosophy, this is called "intuition mongering"—and there are areas in the field that are highly specialized in it. You can make a living out of it.

As we wrote in 2015, in our introduction to the Open MIND collection, "What does it mean to have an open MIND?" (which is freely available on the internet):

> If one takes the phenomenal character of intuitions seriously, this ability clearly seems to be an epistemic ability: *prima facie*, to have an intuition means to have the subjective experience of knowing something, directly and immediately, without necessarily being able to express this knowledge linguistically or to provide an epistemic justification. Typically, inner experience seems to present knowledge to the subject of experience, even if one does not know how and why one possesses this knowledge. Intuitions are the phenomenal signature of knowing, a seemingly direct form of "seeing" the truth. As soon as we ascribe epistemic status to intuitions on the basis of their phenomenology alone, however, we commit the E-error. "Epistemicity," the phenomenal quality of "insight" and "comprehension," or the feeling of being a knowing self, as such is only a phenomenal quality, just as redness, greenness, and sweetness are.[7]

A common context in which this feeling of "epistemicity" arises is when taking psychoactive drugs.[8] When on some drugs, people may feel a deep and blissful sense of "embodied harmony" without actually being in a state that is anywhere close to optimal in the overall context of their lives. (This may or may not be different for meditators.) The harmony that they feel, therefore, may be a mere appearance—*phenomenal* harmony—not the genuine epistemic soundness of everything that one already knew actually "falling into place." When on other drugs, people may have a dramatic phenomenology of insight without later being able to say what that insight was about, or whether it even had any content. Nevertheless, the phenomenal signature of knowing may be more intense than it could ever be in ordinary waking states. And of course, under the influence of substances like the classic hallucinogens explored by Aldous Huxley and millions of seekers before and after him, we can often find intense experiences of meaningfulness and spiritual or religious certainty. But meaning is not the same as the experience of meaningfulness, and the phenomenology of certainty is not necessarily a state in which one really, accurately "knows that one knows."

Intuitive knowledge was our first example: You think that your agent is a morally good person, you believe that zombies are conceivable. Psychoactive substances were the second illustration: Not all drug-taking generates illusory epistemicity, but it is a context that creates a lot of potential for such illusions to arise. For example, some successes of psychedelic and other insight-based forms of psychotherapy might be based on "placebo insights."[9] A third paradigmatic context for misplaced epistemicity is in the experience of so-called ecstatic epileptic seizures. The French epileptologist Fabienne Picard has presented striking case studies of patients who experienced strong feelings of subjective certainty, including in the form of religious beliefs, during epileptic seizures. These cases are particularly interesting since the beliefs seemed entirely convincing during the seizures, even though they contradicted the patients' longstanding convictions. Neuroscientific research suggests that the anterior insula is involved in signaling uncertainty, or the fact "that there is something we do not understand."[10] This registering of uncertainty and ambiguity then goes on to cause an aversive affective state, often involving feelings of discomfort and anxiety of the type that we continuously try to minimize.

By contrast, however, direct electrical stimulation of a small area in the dorsal anterior insula causes intense feelings of bliss.[11] It has been suggested that such blissful states, if occurring in the context of epileptic seizures, may be associated with maximized coherence of what (in my 2003 book *Being No One*) I called the "phenomenal self-model." Subjectively, this coherence is manifested as a dramatically heightened sense of self, an intense phenomenal experience of presence, integratedness, harmony with the world, and strong positive emotion.[12] In one study, Fabrice Bartolomei and colleagues caused states of this type by directly stimulating the dorsal anterior insula. In their own words:

> Ecstatic epileptic auras consist of a sensation of "hyper-reality." The patients use terms such as "clarity," "evidence," "certainty," "understanding," "insight," "enlightenment"

or "epiphany." Dostoevsky gave a famous description of such seizures, that he experienced himself and in which he reached a wonderful state of clarity and bliss. The phenomenological detail of the ecstatic aura includes three important points: bliss ("the immense joy that feels me is above physical sensations"), a sense of certainty ("things suddenly seemed self-evident"), and a sense of timelessness or a sense of being in an eternal now ("These moments are without beginning and without end").[13]

The second feature listed here clearly shows how the epistemic feeling of knowing that one knows can be directly caused by a purely physical event, by local electrical stimulation in the brain, and in a situation where likely no additional knowledge about the external world is gained.

On the other hand, none of what I have said in the second part of this chapter shows that any of the phenomenological reports presented in the first part necessarily refer to fully illusory states. Nor is a patient's feeling of insight during a psychotherapeutic process using psilocybin or LSD necessarily a mere hallucination. Therapeutic success will often result from epistemic progress, from the growth of knowledge caused by the creation of a new, improved, or "repaired" model of reality in the patient's brain.[14] The psychological and epistemic benefits of a sustained meditation practice can also be significant. It is just that all knowledge claims made in public need independent justification.

What all this shows is that phenomenal experience is not knowledge, and in attending to the phenomenology of pure awareness, we must always be careful not to commit the E-fallacy. These experiences feel so valuable and important that the corresponding verbal reports deserve to be taken seriously, and this implies open-mindedness and intellectual honesty, including a healthily skeptical attitude when it comes to epistemological or metaphysical claims made in the public sphere.[15] Most of the reports discussed here do not make such claims; they make a generous contribution to consciousness research without pretensions to more than personal meaning. But there is an interesting phenomenological overlap between the conscious experience of soundness, certainty, self-evidence, and timeless insight in meditation; experiences like mathematical and philosophical intuition; and deep psychedelic experiences and ecstatic epileptic auras. Not all these states are examples of knowledge possession.

My point is that theoretical knowledge cannot be simply "read off" from contemplative experience, as if bootstrapping a theory of consciousness out of the "knowingness" that accompanies the pure-awareness experience itself. Yet, of course, it is equally important not to fall into the opposite extreme: It would be false to conclude that consciously experienced feelings of knowing are *never* reliable indicators of genuine insight. It is just that, merely by virtue of happening, they do not justify sweeping theoretical claims. At its core, meditation itself is clearly an epistemic practice (more on this in chapter 17), but it is also one that doesn't operate on the level of theories, words, or concepts at all. The really interesting question is how we decide what the epistemological status of any given experience may be. As soon as we have a conceptually clear phenomenological description of our research target, MPE will also require the attention of analytical epistemologists. But can there be any kind of philosophically convincing link between silence and words—between the soundness of empty cognizance and other, more theoretical forms of knowledge?

I confess that I find myself uncertain at this point. The spiritual traditions have long said that staring at the finger is not the same as seeing the moon. Are the fingers with which humans have for so long been pointing at the moon good enough? Lars Sandved-Smith, a practicing philosopher and modeler, coined (in a personal conversation) the beautiful metaphor of an absolutely precise laser beam with which we might actually touch the moon for the first time. Perhaps a real contribution might be to bring together the best of modern science and philosophy in creating better, longer, more dexterous, and much more delicate fingers to connect silence and words in new ways? Or is theorizing merely an escape from what is really at stake, and thus nothing but a clever avoidance strategy? Do we really need to prove that such a connection exists? And what would it mean to let go of the deep-rooted need for emotional security that might often go hand in hand with the search for such a connection?

8 Nonidentification

Identification like scars or vapor trails arising and disappearing [#1585]

What exactly does it mean for consciousness or awareness to be "pure"? There are all kinds of places we can start if we want to answer this question. People often start with ancient religious belief systems, or with grand philosophical theories of what the purported "essence" of consciousness is. But even at this early stage, it is already becoming clear that if we start from the real-life phenomenology of meditative experience itself, there is more than one answer to our question. The essentialism of looking for or expecting a single answer may itself be the problem, and it may be that to make progress, we need to let go of the hope for the one big answer. Multiple answers do not necessarily exclude each other; they may even complement each other in forming a bigger picture. Let me therefore begin by slowly flagging a series of possible answers to the question of what the adjective "pure" in expressions like "pure consciousness" and "pure awareness" could actually mean. Many of our reports talk about contentlessness. Traditional sources do too. The first, classical and quite radical, reply is: "Consciousness is pure whenever there is no other experiential content whatsoever, when the quality of *consciousness* itself is the only kind of phenomenal character that can later be remembered and reported." This is the kind of pure consciousness that we find in full-absorption episodes. In the introduction of this book, we saw that there are two main categories of such episodes: in deep states of meditation arising out of waking consciousness, and in dreamless sleep (see chapter 20). Here, a scientist might say that in these cases, minimal phenomenal experience (MPE)—the simplest, minimal form of experience—occurs as a singular, stand-alone feature. This is to say that in these episodes, there is no content, just naked awareness itself.

The phenomenological question of what the "purity" of pure consciousness actually consists in has a second, equally canonical reply. This one associates the "purity" with phenomenal qualities like "peace" and "deep, unbounded silence," the absence of all mental conflict, noise, and perturbation. We first encountered support for this view in chapters 2 and 3. Our phenomenological data confirm what has been known for millennia: Consciousness can exist entirely without thought. On a simplified reading, pure conscious experience here means the absence of all discursive thought, including memory, planning, daydreaming, and mind-wandering. For example, one well-known formula is that pure awareness is "transcendental consciousness" in the sense of a state that has "transcended thought," where we simply abide in a crystal-clear and thoughtless state—the state of knowing restful alertness itself.

The third reading of "purity" is "clarity" or "clear wakefulness," the experience of a clear and unobstructed space of knowing. This answer can also be found in our data, as well as in some traditional frameworks. We have already looked at this specific experiential quality in the previous two chapters, where I introduced the concept of "epistemic openness" (to help explain wakefulness) and the idea of having an inner "model of an unobstructed epistemic space itself" (for clarity). This version of "purity" is the nonconceptual conscious experience of one's own epistemic capacity *as such*, a space of *potential* states of knowledge, without any priors or predictions; an inner space that is wide open and lucid at the same time. Here, consciousness is pure in the sense of being "capacious," being a lucid, open space in which experience can occur. The space is not clouded by thoughts.

In chapter 7, we even encountered a radically naturalist reading: Hypothetically, pure awareness could be a way of thoughtlessly experiencing the activity of a specific, as yet unknown part of the *neural* body only.[1] This fourth version would be "pure" in not being mediated by any sensory system (either directed at the interior of the body or aimed outward into the environment) and in emerging without any form of inner or outer action. From a neurocomputational perspective, we could say that pure awareness, in this sense, is special because it arises neither from perceptual nor from active inference. It would be the most abstract experience of embodiment that human beings are capable of (see chapter 24 for more). As shorthands for these options for making sense of pure awareness, we could even begin to speak of P1 (no content), P2 (no thought), P3 (clarity), P4 (abstract embodiment—here viewed only as a computational property, but see chapter 24), and so on. In this chapter and the next, we will encounter two further phenomenological interpretations of the term "pure consciousness." I promise that I will give a full list of these in chapter 34 of this book, when summarizing the main results of our investigations. But the beauty of lived contemplative experience is that it does not adhere to any rigid conceptual schema.

Phenomenologically, our first three interpretations of "purity" can clearly coexist in one experience. For example, during meditation practice, it sometimes happens that one enters a deep state of full absorption that is characterized by nothing other than a crystal-clear experience of epistemic openness and unbounded, wakeful silence (P3). But in real life, all these aspects tend to get mixed up. P3 and P2 may give way to P1, leading into a process in which the practitioner oscillates back and forth between thoughtless clarity and full absorption multiple times, in a way that is extremely hard to describe in words. However, sooner or later, every meditator will realize that this is not the only way of experiencing pure consciousness. And this is where things get interesting. What happens when you sit in a deep and stable state of restful alertness and very slowly open your eyes, now seeing the world with a clear and thoughtless mind? What exactly happens when, while in a state of reflexively aware mindfulness, you hear sounds emerging out of silence? Phenomenologically, are the sounds perhaps sometimes *made* out of silence? And what happens if, in that crystal-clear silence, one single thought arises, unfolds its content, and quietly dissolves, as in Padmasambhāva's classic image of a thief entering an empty house, looking around, and disappearing?

Padmasambhāva came to Tibet around 767 and is widely venerated as a second Buddha, even in Nepal, Bhutan, and the Himalayan states of India. He said that just as an empty house is at no risk from a thief because there is nothing that the thief could grasp and hold on to, thoughts recognized as empty cannot in any way harm an empty mind. Just like the thief leaving the house, they will "self-liberate" and quietly dissolve. We were never taught to look there—but is it possible that thoughts are really *made of* that self-knowing mindfulness itself? Padmasambhāva called the state resulting from this non-intellectual insight a state that had "crossed the dangerous defile of moving thoughts."

The main point is that the openness of pure awareness can become interestingly intermingled with various modes of conscious experience. One of our participants beautifully described the coemergence of a stable state of pure awareness and conscious content by creating the following two metaphors: "A gentle rain on a still pond, or a crystal ball with water being poured over it" (#1293). And of course, four modes are not all there is. Another—the fifth, and equally valid—reading of what the "purity" of MPE amounts to is this: The distinct and timeless quality of awareness itself can be clearly felt, the phenomenal character of epistemic openness exists, but there is additional content that in principle could be reported (e.g., spontaneously arising thoughts, visual perception, bird song, bodily sensations, etc.), and yet the process of *identification* is lacking. The thief doesn't try to grab anything. There is no mental agent attempting to control the process. The meditator has taken a break. Now, conscious experience may even be pure in the sense of having become "nondual," lacking subject/object structure (chapters 26 and 27). And what is seen when the eyes open, then, may be neither real nor unreal (more on the phenomenology of "virtuality" in chapter 28). Here, the experiential quality of awareness is pure in the sense that it remains autonomous, uncontracted, and fully present, but *detached* from any spontaneously arising content. To put it in traditional Buddhist terms, grasping and clinging are absent. To use more modern terminology, from current philosophy of cognitive science: Internal representations of ongoing cognitive or perceptual processes are not automatically integrated into a transparent self-model anymore.[2] Ownership and the sense of control are absent. Therefore, another phenomenological reading of "pure consciousness," can be summarized as "no identification." Let us look at some experiential reports that illustrate this version:

> 182 I experienced this awareness as a distinct state of consciousness that came on its own and after a while (many minutes, not many hours) passed away again. I perceived myself as completely detached from my thoughts and sensations. These still existed and I still perceived them, but was no longer identified with them. Rather I was identified with an awareness that perceived these thoughts and sensations without feeling anything myself. In this respect I would describe it as pure.

> 1703 [. . .] It starts when I look for "who" is doing the seeing. Or "who" is doing the feeling? [. . .] There is certainly physical sensation happening. That is very clear. But who is feeling the feeling? Or is there just feeling? Or that same line of thought, but with hearing or thinking. As soon as "I" look for the thinker of thoughts, it hits me: There is simply thinking. But no one or thing is doing it. It's just happening all on its own. And it's like there is nothing behind the experience. There is just sort of floating. Just like a big smear of sensations all suspended somewhere. And then as quickly as it comes, it goes . . . in the sense that, thoughts of how "I" just had a cool experience and of ways to get it back. I start clinging. I start wanting it to come back and stay for longer. . . . When I'm having the experience, certain bits of language seem completely nonsensical. For example: "I hear a sound." It seems that "There is hearing going on" feels so much more of a correct description. The same goes for, "I am thinking a thought" → there is thinking going on.

One particularly relevant phenomenal quality for the theory and practice of de-identification is the "sense of control," the subjective experience of being the doer of actions and the thinker of thoughts:

1830 [. . .] Acceptance of not being "the thinker" was far more accepted and clearer than usual. Awareness had space for everything, while not "being" any single thing. Breath proceeding normally and being known, but in larger space than usual, perhaps unbounded space. Peaceful yet not emotional. Neither filled with agency nor completely agencyless. Some relief from not having to construct/feel in control.

Nonidentification also applies to soundness and the affective dimension of MPE experiences:

246 It was pure joy without attachment.

2524 To everything that is thought and imagined, one feels a connection (not in the sense of dependency or the like, but rather as a being-interwoven, as love) and yet clarity (in the sense of a kind of distance, not attracting, not identifying).

The phenomenology of spatiality (see chapter 23) in this version of purity is inversely related to mental grasping:

1819 [. . .] The awareness is very broad, beyond the body or spatial boundaries. Yet bodies or space appear in it—completely relaxed. Depending on the situation, the experience of spatiality can spread out strongly and increase in luminosity (not visual) and in power / luminosity / depth / vastness / clarity / knowledge, depending on how deep the connection is, depending on how little distraction / grasping there is.

As many meditators know, the process of carefully but ever more gently and effortlessly observing the contents of self-consciousness can gradually lead to an experience of de-identification. The phenomenological elements of control, mental agency, and selfhood begin to disappear, while an additional quality of clarity emerges. P3 and the quality of nonidentification are related. For example, there is a distinct and specific phenomenology of seeing thoughts arise within pure awareness itself and later dissolve back into it. This is the experience of *cognitive* de-identification, the conscious experience of naturally and gradually disengaging from the contents of thought. Now these contents are still conscious, but they are no longer part of the egoic self-model. The same can happen for emotions and bodily sensations.

Interestingly, nonidentification is reversible. Nonidentification can be reversed through the automaticity of "clinging," of gradually returning to the phenomenal character of agency (as in the last example presented here). It can also, however, be reversed seemingly at will, and for reasons other than anxiety or unease, in a more relaxed, exploratory fashion that nevertheless leads back to identification. To judge from these reports of real-life experience, something in us may also "deliberately" cause a reidentification with the content from which awareness was previously detached:

> 207 I was sad. On the way home from the bus I first tried to observe the sadness while walking and then to detect the consciousness that feels/observes the sadness. Suddenly I was in a state of absolute silence and peace that I had never experienced before. The sadness was still there, but it was no longer part of me and somehow had nothing sad about it anymore. I can't describe it, it was as if it was outside of me. Then I started to marvel at the situation and then my mind came back with the thought that I would always meditate so much and that I could just be sad for today. And in that moment it got me and I opted for sadness and the silence was gone.

Our phenomenological data also show that nonidentification can refer to single mental states, but also to the person as a whole:

> 2953 [. . .] Yet, I felt just as being myself, but not being anything in particular including being a human or individual person. In fact, I really didn't even know I was in such a state, yet being in it, until the mind suddenly produced a few thoughts, which I observed as if they were a phenomena outside myself. I could hear the quality of them as if being part of a person that I recognized, but no longer identified with.

Sometimes experiences like these are later described as coming into contact with the "true self" (chapter 29). Phenomenologically, de-identification can also lead to de-immersion. One classic metaphor for this process, found in many places in the popular literature on meditation, is the image of being fully immersed in a movie and then suddenly realizing that you are *not* the hero, disengaging from the perspective from which the story seems to be told. Here is one example from our study:

> 26 The first experience to which your description of "consciousness of consciousness" would apply happened when I was still very young. [. . .] It was in that moment that I felt for the first time that I am not my thoughts, but the invisible thinker. I felt like a lonely spectator in an empty cinema who has suddenly realized that he's not the lead actor in the film he's watching.

In this report, there is still a thinker; there is a metaphorical "lonely spectator," although he is "invisible" and no longer fully immersed in his thoughts. If we were to feign ignorance and take the words at face value, then the description here would seem to refer to an example of dual meta-awareness, because the speaker still uses the first-person pronoun "I" and there is a subject/object structure (a "lonely spectator" and the "thoughts"). However, it might well be that the actual episode of "awareness of awareness" was entirely nondual, but for lack of better words, it is later referred to with the help of a dualist metaphor (see chapters 26 and 27). We can also imagine another scenario. Here is one possible variation on our classic metaphor:

> Imagine the cinema is dark and completely empty. All the seats are empty. Running on the screen is a movie of someone watching a movie, a person who is deeply immersed in the plot of this movie-in-the-movie most of the time, but who sometimes has glimpses and vague intuitions that she might actually not be the protagonist, but might instead be a passive viewer in some sort of cinema. Or that she might even be the whole cinema itself. Not seated in the middle, but somehow everywhere. From time to time, the viewer in the movie even has fleeting and slightly complacent philosophical fantasies that she might be some sort of empty space in which the movie becomes aware of itself, or that she might be the beam of light coming from the projector. Suspended in the dark, as it were, this movie containing the movie-in-the-movie is the only thing that can be seen. But nobody is watching. Now, very slowly, the light is being turned on, in slow motion as it were. The cinema as a whole emerges out of the darkness, slowly fading in, empty seats and all. Now the lights are on. The movie is still running. Nobody sees it.

The Contraction Principle

> Of course, ultimately the little one is not real. He is a picture. He is not aware.
> —Douglas E. Harding (1909–2007), *Face to No-Face*

If you have a background in philosophy or consciousness studies, you may remember those boring old discussions about zombies—round and round the question of whether there could be a creature behaving exactly like you or me but not conscious. As a way to start thinking about the contraction principle, let's take the idea one step further, moving beyond mere intuition-mongering toward something more meaningful. If zombies are possible, it is possible that you are now unconscious—but that is impossible for *you* to discover this fact intellectually. A zombie is a functional isomorph of a possibly existing conscious being, which means that all its outer and inner behaviors and functions are indistinguishable from such a being. Not only does it walk and talk in the same way as its conscious twin, but it also thinks the same thoughts (e.g., when it thinks that it is actually conscious right now). Again, if zombies are possible, then you might be one right now and you would have no chance of ever discovering that many of the beliefs that you have about yourself are actually false—for example, the belief that you are not a zombie and are quite obviously conscious in this very moment. Why? Because if you did, you would not be a functional isomorph of your conscious twin anymore. Your brain couldn't have the same causal structure—because it has just generated a thought that would have been impossible with the conscious twin's causal structure. All that any of this really tells us is that absurdities easily accumulate around conceivability arguments and intuition pumps in the philosophy of consciousness.

Given this background, let me now draw your attention to something much more interesting. The phenomenology of some spiritual experiences gives a new and much deeper meaning to thought experiments of this kind, as well as to other ideas, like the Douglas Harding quote presented at the start of this section, or the contraction principle (to be explained next), or the suggestion that no character in any narrative can ever be a truly self-aware part of the narrative (even if it tried to control the script and "wake up" to its own fictitiousness; see chapter 17). What we took to be ourselves—the content of our conscious self-model—could actually be a zombie. Our ordinary conscious self-model could ultimately be a mere image, a transparent image of an ego that falsely believes it is actually conscious. As Douglas Harding says, the "little one" may not be aware. What spiritual traditions sometimes call "insight" or "liberation," then, could be the moment when the whole cosmos (not the egoic self) realizes that it is the entity that is conscious. This realization is not something intellectual, and not something that *you* (or the zombie) could ever have. The phenomenological result of this shift into the zero-person perspective would be the emergence of an open self-aware field, a globalized and nonegoic form of self-awareness permeating everything else. And this new model of reality would not merely be a new "belief"; it would constitute an epistemic gain of an entirely different kind. There are forms of conscious knowing that have nothing to do with some fictitious "self" being related to words, concepts, or propositions. As a matter of fact, it may turn out that beliefs are something that only zombies have.

Carefully investigating the phenomenology of pure awareness can thus lead to interesting discoveries. The phenomenology includes not only the way in which pure awareness emerges, but also the way it breaks down. Perhaps you have already discovered how, phenomenologically, it is not *we* who get identified or fused with some content (e.g., an arising thought)? In my own experience, I find that it is *awareness* that gets identified. "We"—as the agentive sense of self—come into existence only *after* the fact, after awareness has already contracted into a specific content (like the mental image of a future goal state to be reached, a subtle feeling of mental effort, the ensuing sense of control, etc.)—after, as it were, open awareness has already been lost to itself. The sense of self is the *result* of contraction, not its origin. The breakdown is not your fault. You are not the cause, but the effect. You are the way in which the brain explains away something that was slightly unexpected. That "something" is the surprise of suddenly being able to control a part of the world—your body, your attention, your thoughts.

The "contraction principle" is easy to understand intellectually but very hard to investigate within phenomenal experience itself. From a scientific perspective, "being conscious" or "appearing" is a property of a complex, internal model in the brain: What really *is* conscious is simply a certain part of an organism's model of the world, a specific processing layer in its internal model of reality—which typically also includes the organism itself plus other agents in the world. We have not yet understood what this property is, but it is straightforward and empirically plausible to say that "phenomenality" or "appearance" is a property of some sustained, functionally integrated state in our head. Consciousness is something subpersonal. At the very least, this is a rational and evidence-based theory, one possible view from the outside.

But seen from the organism's inner perspective, things are very different. The model is transparent; therefore, the representational medium is invisible. Yet this model is all the organism has. This means that what it experiences is experienced not as a world-*model* or some sort of inner image, but simply as the world itself. Consciousness is the appearance of a world; there is full immersion, plus the phenomenology of direct realism. In addition, because the internal model is not being experienced *as* a model, it now is the *organism* that seems to be conscious, not the world as a whole or some complex image in the organism's head. Phenomenologically, the property of "being aware" has been contracted. Now, consciousness is something personal. Now *you* are conscious.

From an outside perspective, however, it is as if the embodied brain—in searching for a viable strategy of portraying reality—found it helpful to commit what in philosophy is known as the "fallacy of composition." This is an informal fallacy that arises in natural language when one falsely infers that something is true of the whole from the fact that it is true of some part of the whole. Two examples would be "Atoms are not alive, all biological creatures are made out of atoms, so ultimately none of them is really alive!" and "All voters have rational preferences, therefore any collective choice induced by majority rule will also be rational!" Of course, the brain evolved not to perform logic but to create an efficient predictive model of reality that helps the body to survive and copy its genes to the next generation. According to the virtual reality (VR) in your head, there is now a conscious self—a whole, embodied person—and this is what forms the origin of a first-person perspective.

Alongside the conscious self, there will often be other self-aware agents that are also portrayed as conscious and able to control their own attention, their thoughts, and their bodily movements. They are not empty persons,[3] but egos. Phenomenologically, they are not dream characters or conspecifics with some complex VR running in their brains, but full-blown agents—and they constantly project this property back into us, as we begin to mirror each other. To cooperate successfully, we need to navigate a complex social world. From the organism's perspective, there is a social frame of reference, and the property of "being conscious" may now be instantiated at *multiple* locations in this environment. Even in the dream state, neurotypical human beings will always experience other dream characters as also "being conscious." The illusion—which often persists even during lucid dreams—that dream characters have a conscious life of their own illustrates a robust coding principle of the human brain: For the organism, "phenomenality" is a local, personal-level property either of itself or of another creature. It is not something either omnipresent or local. It is a global feature of the animal as a whole, and it can occur in multiple agents at the same time.

Philosophers of consciousness distinguish between system consciousness (whenever the predicate "conscious" is applied to an animal, a person, or a machine as a whole) and state consciousness (all cases where "conscious" is used to refer to a state of an animal, a person, or a machine).[4] In ordinary states, the brain depicts consciousness as system consciousness, not state consciousness. Internally, a state property is *misrepresented* as a system property. Let us call this the "contraction principle":

> **(CP)** "Phenomenality" is a subpersonal property, a property of certain functionally integrated brain states. The brains of neurotypical human beings *misrepresent* this objectively given property of phenomenality by contracting it into a transparent conscious self-model, which then forms the origin of a first-person perspective.[5]

And here is where research into the phenomenology of pure awareness gets really interesting. This research is directly relevant to constructing a minimal model of consciousness, and it will also be crucial in the formulation of a first standard model. Why? First, the phenomenological degree of "aperture" of the field of awareness is something that can be measured by meditation researchers.[6] More important, there are well-documented counterexamples to CP, such as the nondual states of awareness that sometimes spontaneously occur in advanced practitioners of meditation. We will carefully look at such states later in this book, mostly in chapters 25, 26, and 27. In such nondual states, we find conscious experience without contraction, phenomenality without an egoic self-model, and awareness without the perspective created by top-down superimposition of an abstract subject/object prior. The organism is liberated from full immersion, and it can transcend the phenomenology of naive realism. In such states, it is as if the perspectives of science and contemplative practice begin to meet, as if they might become congruent—but in an unexpected way, perhaps in a way that some may even find uncomfortable. What the contraction principle shows is that nondual states of MPE may actually be the best entry point for the formulation of a first standard model of consciousness. The minimal model approach dissolves the problem of subjectivity.

MPE research is essential because there is a methodological primacy to subpersonal state consciousness: It can occur without system consciousness, but not vice versa. This is shown by our phenomenological material, as we will see in the course of this book. Wakefulness, reflexivity, and the phenomenal signature of knowing can all exist in nonegoic variants, but egoic self-awareness is always accompanied by the experience of wakefulness, self-directedness, and a knowing self. What if you even *were* self-knowing wakefulness that had mistakenly identified itself with the epistemic agent model of the biological organism in which it occurred (chapter 24)?

Possibly your mother and your father played a role in this process. Egoic self-awareness is what makes social cognition possible in the first place. Please note that what I have termed "contraction" is a spatial metaphor for a form of misrepresentation that, in the wider context of biological and cultural evolution, has apparently been a successful form of self-deception: A property of a part was portrayed as a property of the whole; a subpersonal state was depicted as a personal-level property. In this wider context, there is nothing wrong with that. Yes, it has generated a lot of psychological suffering, but at the same time, one can view it as a major triumph of natural evolution. Imagine that you are a conscious animal living in an increasingly complex social environment populated by other such animals. They could be potential allies, they could be prey, and they could also be dangerous. It is important for you to know whether other such animals are currently responsive, whether they are asleep or awake, and whether they can attend to you or recognize you. It matters whether they are epistemically open to the world (chapter 4) and whether they know that they are (chapter 5).

This is where your parents come back in. Maybe as an infant, you were much more like a loosely structured field of open awareness, but Mom and Dad—who were already firmly in the grip of their own illusion of conscious agency—continuously projected an epistemic agent model onto you by stubbornly treating you as a single conscious entity, a knowing self. Later, through "education," and via punishment and reward, they even held something that they called "you" responsible for outer behavior, stubbornly and at times cruelly assuming the existence of stable egoic self-awareness, of accountability and free will. Why did your parents do this? One may speculate that "awareness contraction" was a necessary precondition for social cognition and cooperation in large groups of human beings. High-level social cognition made it useful for biological organisms to mutually represent each other as self-aware agents. You solve the problem of detecting and assessing degrees of epistemic openness by classifying other organisms that you perceive from the outside as either conscious or unconscious, or alternatively by attributing the graded property of "being aware" to them, thereby experiencing them as currently being more or less alert, more or less wakeful, more or less distracted, and so on. And then you turn the classifying impulse (whether binary or graded) back on yourself.

Awareness contraction is how you learn to perceive yourself as an individual, self-aware agent, an active entity that is currently conscious. And this may be one of the many ways in which pure, nondual awareness became contracted into a self-model: Given the single-embodiment constraint (more on this in chapter 27), the subject/subject distinction you *had* to use when interacting with members of your own family or tribe became the blueprint for the phenomenology of subject/object duality. Something that in reality is wide open, entirely passive, clear, and silent turned into the property of an active self via (1), most importantly, the single-embodiment constraint; and (2) the need for social transactions that depend on a subject/object split. Reflecting on these two reliable drivers of contraction, it is interesting to note how many of humankind's most earnest practitioners of meditation have chosen to live in solitude as hermits, or in communal seclusion as nuns and monks, and how, to this day, silent meditation retreats take place in an environment in which all social interactions are deliberately reduced to a minimum.

More immediately, the contraction principle is important if we want to better understand the phenomenology of identification that is the focus of this chapter. Why are some features of awareness, even the phenomenal quality of awareness itself, almost automatically attributed to a fictitious self? Why are they integrated into a transparent self-model? From a scientific perspective, neither the self-model itself nor the person as a whole is conscious—the subpersonal world-model in the person's head is what is truly conscious. This leads to a testable phenomenological prediction: If a meditator were to become gradually aware of this hypothetical fact, then this process should have two components. First, awareness that the world-model is the conscious entity should create a global phenomenology of nondual wakefulness, which we might (drawing on the title of David Hinton's 2019 book) call an "awakening cosmos," an awareness of *everything* becoming nonegoically aware and epistemically open. Second, this awareness should lead to a phenomenology of virtuality, to a gradual suspension of naive realism (more on this in chapter 28). This experiment would be fascinating to perform.

In the meantime, however, we might still find ourselves asking: But why was the identification with an agent, with a conscious self, so very successful?

For a biological organism and its brain, it is of the utmost importance to reliably distinguish between those events that are self-caused and those that are "external," in the sense of currently being beyond our control. There now is a strong consensus across many disciplines in the consciousness community that neither the brain nor consciousness evolved to meditate, to think logically, or even to deliver an accurate inner image of the world as it "really" is. The primary goal is to continue staying alive, to maintain physiological integrity in the face of danger and opportunity, even if this should require clever forms of self-deception. In the words of Anil Seth:

> Evolution's reason for providing organisms with brains is not so they can write poetry, do crossword puzzles, or pursue neuroscience. Evolutionarily speaking, brains are not "for" rational thinking, linguistic communication, or even for perceiving the world. The most fundamental reason any organism has a brain is to *help it stay alive*, through making sure that its physiological essential variables remain within the tight ranges compatible with its continued survival.[7]

Conscious experience is about control. It is about predictability and reducing uncertainty. For many organisms, staying alive presumably involves successfully distinguishing between what they can control and what is beyond their reach. This is a fundamental categorization, leading much later to conceptual distinctions like "self" versus "nonself." The first distinction (controlled/uncontrolled) then turns into a "prior": a deeply ingrained and very successful top-down prediction about what the world is like and how it will be experienced. A self exists, a single "distinguished entity" that is the locus of control and constitutes the experiential unit of identification (see chapters 24 and 29 for more).[8] The rare phenomenology of nonidentification is what suspends this distinction. It is not that the meditator falls into an explicit phenomenology of alienation, of being externally controlled (as the schizophrenic does). It is rather that the distinction itself gently and quietly disappears—but without the accompanying thought, "The distinction between self and nonself has disappeared!" All that happens is that one self-fulfilling prophecy turns silent. In terms of a computational model, we could say that the anticipation, the top-down prediction that every perceptible event must be either self-caused or external, is suspended. Within the experience, the distinction becomes meaningless, as in report #1830: "[. . .] Neither filled with agency nor completely agencyless. [. . .]." It is as though an abstract hyperprior has been episodically suspended. What results is a liberating form of indeterminacy, a phenomenology of neither-nor-ness.

But there is more. If it is true that the sense of self is the *result* of contraction, not its origin, then all of this applies to the "meditating self" too. If our unfolding story is on the right track, then there should actually be very subtle illusions of control. If you are a practitioner yourself, can you observe what exactly happens in the very moment when you notice that your mind has wandered, that you have "lost it again"? The meditating self is the new phenomenological entity that is disappointed with itself, the noble one that sincerely wanted to fully identify with pure awareness, but somehow "lost control over the process." That wasn't its fault. There is no reason to be disappointed.

Unfortunately, the phenomenology of "regaining" control and mindfulness, of "waking up again," could often be an example of exactly the same mechanism. No reason for the meditating self to be proud, to pat itself on some invisible shoulder with a feeling of success and relief. The subtle psychological drama, the self-condemnation, the relief, and even the practitioner's sense of achievement, stamina, and sustained discipline could always be just the result of yet another contraction. Occasionally, the whole dynamic can be seen through (though I rarely manage to do that myself).

Given our new conceptual instrument of "contraction" as a special form of misrepresentation, specifically a biologically successful computational principle, we can now describe the phenomenology of meditation more precisely and begin to distinguish different ways in which the quality of awareness per se may manifest. Let us briefly review the preceding five chapters through the lens of the contraction principle. We have seen that there are contracted and uncontracted forms of silence because the phenomenology of silence can be experienced as being "in the mind," but sometimes also as all-pervading. Wakefulness can be a property of the meditating self or of the world as a whole ("the awakening cosmos"). Similarly, the phenomenology of clarity can be experienced as a local form of lucidity in one's "own" mind, but it can also expand and sometimes even turn into a feature characterizing the space of conscious experience as a whole. The density of a "thick silence" can be located in the head or it can be everywhere. Soundness and internal harmony are often experienced as related to an embodied self, but sometimes also as qualities of the overall situation, pertaining to the wholeness of the moment. Let us keep this new conceptual tool of contraction in mind as we penetrate deeper into the landscape of seeing *what is*.

9 Suchness

When openness touches openness. [#1444]

For many centuries, meditators have been describing states of direct perception, the experience of seeing *what is*. Meditation involves a distinct phenomenology of perceiving, timelessly, without any conceptual overlay or any form of judgment, interpretation, or choice. Seeing *what is* out of a state of pure awareness often reveals another particular and extremely interesting phenomenal quality—namely, the experience of "suchness." I personally find this to be a profound kind of phenomenal character—one that is also a prime example of what it means to be ineffable, to resist any form of conceptual approximation. Perhaps one could begin by saying that, in ordinary perception, "suchness" is the additional phenomenology of nonconceptuality and openness itself. Sometimes it is recognized, but often it is not.

I have already offered a phenomenological reinterpretation of "emptiness" as "epistemic openness." One prediction would be that in all situations in which subject/object structure fades away, "emptiness" and "epistemic openness" will be properties not only of the conscious mind, but also of what were previously taken to be inanimate perceptual objects. Suchness then becomes the emptiness of appearances in the more precise sense of their being epistemically open, for example in terms of lacking a predetermined conceptual essence. Things appear without conceptual overlay; there is no ready-made conceptual interpretation that categorizes or automatically evaluates them in any way. Accordingly, the influence of past experience is minimal; there is no recognition. Pure awareness is nonreactive, and therefore experiencing suchness also means experiencing the world without perceiving any affordance for action. In suchness, the quality of appearance *itself* becomes very salient. Interestingly, this also leads to a positive reading of the term "meaningless": The phenomenology of nonconceptuality is the conscious experience of things simply appearing, timelessly presenting themselves in consciousness, without cognitive penetration, free of any fixed meaning. According to conscious experience, things that are perceived do not refer to anything outside of themselves anymore, neither to the past nor to the future. The phenomenology of suchness and of seeing *what is* is the phenomenology of meaninglessness: Now, it is not just pure awareness that is epistemically open; all conscious percepts, all appearances themselves are as well. Each appearance is a vast space of epistemic possibilities, of possibilities to be known from an infinity of perspectives.

At this point, continuing our discussion from chapter 8, we also find yet another phenomenologically grounded reading of what the "purity" of pure awareness might consist in (in our summary, I will call it "P4"; see chapter 34). It is not that there is no perceptual content whatsoever. The complete lack of conceptual overlay, including time experience and judgment as to the "existence" or "nonexistence" of what is perceived, is what really makes pure awareness—and whatever appears within it—pure. According to subjective experience, meditation practice often brings us into a deeper and more direct contact with reality. From a third-person perspective, this phenomenological observation also leads to a deeper understanding of the second, slightly simplified reading of "purity" (P2) as presented in chapter 8: The absence of all discursive thought leads to a suspension of prior knowledge and expected probabilities. This creates a conscious model of reality that is "temporally thin," in that it does not contain past experiences and simulated futures anymore—if you will, you now see the world in a way that is not (or to a much lesser degree) contaminated by counterfactual aspects. Therefore, pure consciousness is not just the phenomenology of "not thinking," of mere mental inaction; it is also an experience of timeless immediacy—of perceiving *what is*. Let us now look carefully at some examples from real-world practitioners:

> 2420 Very hard for me to describe—answering all these questions was possible but even that required some conceptualization that only seemed to approximate the experience. Main feeling I had was just of seeing everything as it truly is.
>
> 2754 [. . .] An altered perception, more authentic, more real, more unfiltered.
>
> 2426 [. . .] I felt and thought that for the first time I was perceiving reality correctly. [. . .]
>
> 2935 [. . .] it was as if a theater curtain was pushed aside and I could perceive unfiltered. My eyes were still closed. It was as if I was in an empty space from which I could decide what I wanted to look at next. When I then, after some time, opened my eyes, this state continued. It was pure perception and amazement. I was completely relaxed and left the house. I went for a walk and was amazed. Everything was good. At some point this state must have stopped; at some point thoughts must have started, and needs, so that I disappeared

> back into everyday life. Since then I have been longing for it so much that ultimately there is no longer anything more important for me.

We will return to the notion of "filters" being removed in the second half of this chapter. For now, let us stay with simple, pure consciousness in the context of systematic contemplative practice. Phenomenologically, pure consciousness is "pure" because it is experienced as direct, as entirely unmediated by conceptual thought. But it is also pure in that it does not originate from an egoic point of view, while at the same time having the capacity to *contain* one. As a matter of fact, perhaps subjectivity itself, or what philosophers sometimes call the "first-person perspective," is merely one of these filters described by our meditators, a stage curtain that can be opened or closed. Behind this curtain, we would discover silence and clarity. Philosophically, it is plausible to assume that the "abstract space of knowledge" already discussed thus far can either be explicitly corepresented during the process of perceptually experiencing the world or not. The degree to which this happens would then determine the degree to which someone experiences the phenomenal quality of epistemic clarity and of "nonconceptually knowing that one knows":

> 940 [. . .] After a while all thoughts, including the mantra, became very quiet and hardly perceptible. The feeling arose as if the mantra was falling "through me." The "mental space" began to expand and became steadily larger. As if a switch was flipped, all mental content, like auditory and visual thoughts, emotions, memories, or ideas, stopped. The absolute silence in my mind was enormous and it was spacious, very spacious, perhaps endless. I opened my eyes and all the objects of the room were there as usual, but I saw them with a kind of transcendent clarity. As if the contrast was stronger than usual, as if my glasses had just been ultrasonically cleaned. Everything was clearer and more distinct. This state lasted for an indefinite period of time. [. . .]

The experience of epistemic openness and spacious awareness can take place in complete silence (see chapter 3) after all other content has disappeared. After the experience has stabilized, however, it sometimes may be possible to "reinvite" the contents of sensory perception, for example by slowly opening one's eyes. What this adds is the experience of "epistemic lucidity," which we investigated in detail in chapters 4 and 5 (for examples, see the beginning of report #2935 and the end of #940). Perceptual experience may then unfold, but now it is enveloped in a subtle quality of clarity, as if "suspended" in an unobstructed, global space of lucid awareness and infinite silence (for more on this aspect, see chapter 28, "Transparency, Translucency, and Virtuality"). Alluding to traditional Buddhist descriptions, we could call this specific combination of lucid clarity and spacious awareness with ongoing sensory perception "seeing out of luminous emptiness." As we have just seen, some practitioners also describe it as "a kind of transcendent clarity."

"Suchness" and "thusness" are originally Buddhist terms (from the Pali *tathatā*). Over many centuries of Buddhist philosophy, intricate technical debates have centered on how *tathatā* (or *dharmatā*) relates to *śūnyatā*—that is, how the notion of suchness relates to the concept of emptiness. As early as the fifth century, Chinese Buddhists began to discuss the "Buddha-nature of insentient objects,"[1] and many believe that the Buddha referred to himself as the *Tathāgata*—which can be interpreted as "One who has arrived at suchness."

"Suchness" is clearly an important conceptual tool for phenomenologists—one that Western philosophy has been lacking (although interesting links can be drawn to Latin concepts like *quidditas* and *haecceitas*, if one strips out their metaphysical context).[2] But we must be aware that some of our experiential reports may be strongly colored by the fact that their authors knew about Buddhist philosophy and "suchness." Some of our practitioners (as well as famous writers like Aldous Huxley, whom we will meet later) actually used this specific term to describe their experience:

2652 [. . .] feeling of "suchness": deeply moved by e.g. a coffee cup or any object.

2803 [. . .] with open eyes long-lasting state of a very clear and bright perception, closely connected with thusness [*So-Sein*], in peace with the thusness.

2691 [. . .] The best way to describe it is to say that "there was an appearance." It was blindingly obvious that some things simply were "so." I stood and "watched" this, but it was not "me" doing the watching. My internal world and the external world were "complete" in and of themselves. I "knew" that some personal beliefs I usually possess, and hold, were no longer relevant, and a wider understanding of the human condition took its place. I knew what I was experiencing, it was not imaginary, but the knowing was also beyond me. There was a universality about it.

3160 [. . .] Outside on the street there was a sound and in that moment it was clear to me that it is not a self that perceives the sound, but that there is awareness in/with the sound. There existed a perception that the sound appears in space with awareness and "recognizes itself." Other perceptions, e.g., visual objects, bodily sensations, were also appearances of/with awareness without being tied to a perceiver or being perceived by a perceiver. Just an appearance of appearances. This experience/view has no mystical quality, but a quality of "ordinary suchness." It simply is the way it is. One minute amazing and full of joy, but at the same time very simple. [. . .]

The phenomenology of direct perception and suchness can be accompanied by the phenomenal character of "sensationless awe" that we touched on in chapter 1 and will investigate in more depth in chapter 15.

> 3146 [. . .] There was a quality of gentleness and wonder, not in any words or concepts, but more as a pervasive feeling of "this is it, this is how everything truly is." [. . .]

Direct perception is simple and effortless and can include a quality of all-encompassing insight, which may take in the deep structure of conscious experience itself:

> 2673 Single-pointed, my perceptions were simple. Effortless, not separated. Clear, a resonance in the deepest being, in the experience itself a feeling that in retrospect I would describe as seeing and understanding everything.
>
> 2422 I felt as if I was experiencing the fundamental structure of experience, during meditation.

Perceptual acuity itself seems to improve during episodes of direct perception:

> 1661 I had pristine unmediated access to all of my sense fields. Every detail of sensory input was more vivid and more present than they ever are during ordinary consciousness. I could experience internal discourse but very little of it arose naturally, it was a wilful exertion to talk to myself, and when I didn't exert there was very limited chatter, and what chatter there was was quiet and did not demand much attention as it typically does. The overwhelming feeling of the completeness of the present moment pervaded the experience. [. . .]
>
> 3146 [. . .] Lights were bright, details of objects became clearer, the body sort of became lighter, felt a bit like floating [. . .]. There was a quality of gentleness and wonder, not in any words or concepts, but more as a pervasive feeling of "this is it, this is how everything truly is." Like sticking my head above water after being held under for aeons, seeing the blue, open sky. Senses were extremely sharp and acute, yet I didn't form any opinions about the flowers I passed as I walked to the top of the hill, no words about anything inside or out, just, for lack of a better word, luminosity.

The transition to the phenomenology of "suchness" can be sudden, and the egocentricity of the overall state may be weakened:

> 2357 A sudden shift in perception. Very quick, instantaneous. Suddenly reality hit me. I felt that in the moment that I just sunk down to the root of reality. It was very sudden. A shock really. Really jolting.

2295 [. . .] and for a short moment I had the feeling that the perspective shifted from myself to an "objective" perspective. As if the border between my face and the world were transparent. Douglas Harding's explanation of being "headless" conveys it best. [. . .]

The experience of suchness can also be accompanied by the qualities of "soundness" (chapter 7) and "luminosity" (chapter 18), and it can lead to episodes of nondual awareness (chapter 27):

2775 [. . .] I was pretty shattered and finding it hard to concentrate in meditation. However, the garden was remarkably silent, and this caught my attention and led me to notice that nothing was wrong. [. . .] One single yellow rose with red highlights lit up in a way which expressed the experience. I was helpless but to be at one with it. I realized that its luminosity was a gift arising in consciousness moment by moment, with no attempt to try to understand. It was a blessing which didn't last long [. . .], but which made an indelible impact on my awareness. [. . .]

3160 [. . .] Once I took up this view from the subtle view of an observer, then the space/consciousness was filled with a kind of vibration and a kind of inner glow (which has nothing to do with real brightness). Once I let the subtle view of the observer go, then everything was just as it was, without any other specific quality.

3338 [. . .] I can reach states of pure awareness without much effort. I simply sit down and "peel off" my conditioning completely and let everything go. It is as if I let go of my whole existence. There the space for pure awareness opens up—and the ego-centered qualities recede and become unimportant—it's hard to describe it—it's a space where there is no difference between viewer and viewed, between perceiver and perception—everything is as it is—suchness—awake, present, now. I often don't hold this state very long, maybe a few moments or minutes. It doesn't really matter how often or how long, because I know that this suchness is always there, not far away, always available, just a breath away or already there, that "it" is already "in" any thing that I perceive, at any time. The experience of open awareness has permanently changed my perception of the dual, everyday world. The change in perspective is small, but substantial. I live and experience the world from both perspectives, the dual perspective with the "me and you" and the perspective of open awareness, in which there is no "me" and no "you." And something else is important: The open awareness showed me the emptiness or conditionality of things, but at the same time also the completeness. I am amazed and astonished by the Hundreds of Thousands of Things. A miracle!

Timeless Self-Evidencing and "Seeing What Is"

> And there is in Lao-Tzu's story of existence another seminal name for the whole of it: *tzu-jan* [. . .]. Literally meaning "self-so" or "the of-itself," *tzu-jan* was meant to emphasize the particularity and self-sufficiency, the *thusness*, of each of the ten thousand things that make the generative process of Tao. And so, it is best translated as "occurrence appearing to itself," which opens a first description of the Cosmos here in the beginning where the existence-tissue is whole; "from nowhere else, occurrence."
>
> —David Hinton (*1954), *Existence*

> There are indeed things that are inexpressible. These *show* themselves; they are the mystical.
>
> —Ludwig Wittgenstein (1889–1951), *Tractatus Logico-Philosophicus*, 6.522

Let us begin by drawing a parallel. The notion of a curtain being drawn or of "filters" being removed (e.g., in reports #2426 and #2935) is a classic theme found in many attempts to describe "suchness" or the experience of "seeing *what is*" in altered states of consciousness not brought about by meditation practice. Aldous Huxley opens *The Doors of Perception* (1954) with a quote from William Blake's *The Marriage of Heaven and Hell* (1794): "If the doors of perception were cleansed every thing would appear to man as it is, Infinite. For man has closed himself up, till he sees all things thro' narrow chinks of his cavern." Of course, the phenomenology of "suchness" is a well-known feature of the psychedelic experience too. Here, it can be extremely dominant, rather than being a subtle, self-disclosing feature as it is in meditative perception. Let me cite one of my favorite descriptions of a mescaline-induced state of *seeing what is*, written by Huxley himself:

> I was seeing what Adam had seen on the morning of his creation—the miracle, moment by moment, of naked existence. [. . .] *Istigkeit*—wasn't that the word Meister Eckhart liked to use? "Is-ness." The Being of Platonic philosophy—except that Plato seems to have made the enormous, the grotesque mistake of separating Being from becoming and identifying it with the mathematical abstraction of the Idea. He could never, poor fellow, have seen a bunch of flowers shining with their own inner light and all but quivering under the pressure of the significance with which they were charged; could never have perceived that what rose and iris and carnation so intensely signified was nothing more, and nothing less, than what they were—a

> transience that was yet eternal life, a perpetual perishing that was at the same time pure Being, a bundle of minute, unique particulars in which, by some unspeakable and yet self-evident paradox, was to be seen the divine source of all existence. [. . .]
>
> Confronted by a chair which looked like the Last Judgment—or, to be more accurate, by a Last Judgment which, after a long time and with considerable difficulty, I recognized as a chair—I found myself all at once on the brink of panic. This, I suddenly felt, was going too far.

In chapter 32, we will briefly look at four descriptions of minimal phenomenal experience (MPE) experiences under the influence of LSD and psilocybin, as given by our meditators; and in chapter 26, we will encounter Meister Eckhart's concepts of *ist, istic, and istikeit*[3] for a second time. For now, let us stay with the phenomenal character of suchness as it sometimes occurs during ordinary contemplative practice. The parallels between the drug-induced and the meditation-induced experiences will become obvious.

Before we move on, please recall how the phenomenal *experience* of knowing does not entail that one really possesses knowledge. A strong and robust experience of knowing, even of absolute certainty, can occur during hallucinations, in dreams, and during psychiatric diseases or epileptic seizures. In chapter 7, we termed this the "E-fallacy." Equally, the experience of having "direct" or "immediate" knowledge of *what is* does not imply any kind of justification or directness from an epistemological perspective. Epistemologically, there is no certainty here. All we have is the phenomenal experience of immediacy, and "suchness" is being used here solely as a phenomenological concept. In connection with the experience of suchness, Rob Burbea also has made clear for contemplative practice itself that any remaining form of intuitive realism is an obstacle that must be overcome.[4] In interpreting all the reports in this chapter, therefore, we must be careful not to jump to overly naive and simplistic interpretations of "direct perception" and the reported phenomenology of nonconceptually seeing *what is*. The experiences themselves are simply much too interesting.

Even in calm and undramatic states of meditation, there is an enigmatic aspect to the phenomenology of suchness. As I said at the very beginning of this chapter, if anything is the epitome of ineffability—or at least of something for which we lack any established terminology—then it is probably the experiential character of suchness or thusness, as it sometimes reveals itself in the meditator's perceptual experience of seeing *what is*. In the introduction, we defined "concurrent ineffability" as a property of certain conscious experiences: the fact that they cannot be reported while taking place. But becoming aware of suchness can also mean beginning to realize that there is a valid sense in which *every* conscious experience is ineffable while it unfolds. Yet what *is* suchness, the epitome of ineffability? I would try to describe it as a combination of timeless change, epistemic self-revelation, and a profound form of "meaninglessness"—but a form that is in no way experienced as negative.

In his *Tractatus*, Ludwig Wittgenstein bluntly affirmed the existence of ineffable aspects of reality when he wrote, *"Es gibt allerdings Unaussprechliches"* (There are indeed things that are inexpressible). He also did not shy away from calling them "the mystical." But given the subject of this book, it is interesting to note that the two major English translations of the passage cited in this chapter found different solutions for conveying what he then had to say about the mystical: "*Dies* zeigt *sich*." One translation (by Ogden and Ramsey) reads: "There is indeed the inexpressible. This *shows* itself; it is the mystical." The other translation (by Pears and McGuinness) has this: "There are, indeed, things that cannot be put into words. They *make themselves manifest*. They are what is mystical." I think that both translations excellently pick out the key aspects that, in a phenomenological reading, characterize the nonconceptual character of suchness. Suchness has a quality of self-disclosure and of self-manifestation at the same time.

I also believe that understanding suchness may get us very close to understanding what *appearance*—conscious experience itself—ultimately is. As the poet David Hinton evocatively points out (see the first epigraphical quote in this section), it is as if the sheer generative process of world manifestation spontaneously turns back onto itself, thereby continuously appearing to and within itself, occurrence recurring to occurrence. Our problem is that most of us grew up in a cultural context that never provided us with the words we would need, with the conceptual tools and mental techniques that could help in communicating or even recognizing this aspect of our own lived experience. You certainly do not have to be a meditator to become aware of it—but for most of us, it has to be pointed out before we can begin to see it. One interesting question is whether MPE has its own quality of suchness, or whether we should rather think of it as identical to what we later call "suchness." But let me first try to describe the phenomenal character of suchness as precisely as I can.

First, perceptual objects, if seen by a silent, crystal-clear mind, can adopt a quality of timelessness. On closer inspection, this is actually a phenomenology of "timeless change," because it involves a dynamic element. Aldous Huxley pointed to this when describing the "transience that was yet eternal life" as "a perpetual perishing that was at the same time pure Being." My first philosophical point is to forestall a conclusion that is easily drawn but faulty. The implicit metaphysical assumptions underlying our crude folk-psychological system for describing our own conscious minds are structured by opposites taken to be incompatible. So we say that something is *either* being *or* becoming, *either* timeless *or* flowing, *either* static *or* dynamic. From these linguistically embedded assumptions, however, it does not follow that all subjective experiences actually have these either/or structures. Our very own phenomenal experience may defy all the distinctions that we impose on it—especially when it concerns something nonconceptual. We will devote a whole chapter to the phenomenology of timelessness and timeless change later in this book (chapter 22); I will also return later to the issue of "paradoxical" experiences. For now, we can simply observe that the distinctions between full reification and flow, between timelessness and the dynamics of ongoing change, are among the many unconscious predictions that our enculturated minds make, but they can sometimes be suspended and give way to a new form of phenomenal experience that is liberated from any given pair of imposed opposites. This suspension may even be experienced as an insight, a temporary relief from "the grotesque mistake of separating Being from becoming."

The second phenomenal characteristic of suchness is even harder to understand on a conceptual level. At least in my own experience, "suchness" has a quality of epistemic self-disclosure, as if the nonconceptual content were continuously "pointing to itself" or "self-revealing." This happens via a subtle dynamic of spontaneous self-presentation in which an aspect of experience makes itself knowable. There is a loopiness in this element of suchness. Experience does not refer to past or future; rather, it timelessly refers to *itself*. In my own experience, this nonconceptual form of self-reference also has an interesting, yet hard-to-describe quality of "salience" to it.[5] It is as if the phenomenal character of "wakefulness," investigated in chapter 4, were no longer contracted into a knowing self, but as if, perhaps shockingly, the world itself or individual perceptual objects had begun to awaken. In his 2019 book *Awakened Cosmos*, David Hinton remarks that *tzu-jan* literally means "self-ablaze." It is tempting to describe the epistemically open, but at the same time reflexive, character of suchness in terms of a perceptual object being "epistemically self-ablaze." Could this be what makes the perceptual object conscious in the first place? What would it mean for a whole person to become "epistemically self-ablaze"?

Another way to describe the aspect of suchness that I am trying to isolate could be to use the rather technical-sounding concept of "subsymbolic token-reflexivity." A representation is subsymbolic if it is constituted by entities that are not themselves representations, like the neurons or synapses in your brain, the pixels on your screen, or individual samples of a signal. A "subsymbolic" representation is something nonlinguistic, and the units in neural networks can be considered particular cases of this category. In language, the meaning of a token-reflexive expression mentions the particular expression whose meaning it is. For example, a "self-pointing" expression like "I *hereby* apologize!" actually refers to the expression or speech act in which it occurs. Suchness is a little bit like a perceptual object saying, "I *hereby* manifest myself!" or "I *hereby* present myself to you!"—but without words. Phenomenologically, this also creates salience and significance because it immediately attracts attention. In a way, it seems to shout "*Look* at me!" and "Look at *me!*" at the same time—but in utter silence. Of course, there is a lot of philosophically deep water here. But isn't something very similar true of the suchness in a flower as seen from the emptiness of an entirely silent mind—does the flower not express a nonlinguistic form of *pure meaning* that consists only in directly signifying its own particularity, at once dynamically and timelessly?

The simple question that I want to ask is whether the "self-revealing," "self-disclosing," or "self-pointing" character of phenomenal suchness could be a subsymbolic variant of this larger category of process, perhaps an indicator of part of what goes on in the brain of the meditator. "Subsymbolic" means that we are dealing only with connections between nodes, or fluid patterns of activation, such as in a neural network. In a standard neural net, there are no hard symbols or "atoms of meaning"; there is no syntax and no semantics, no fixed inner language. Is what suchness tells us, entirely without words, actually something like: "*That* flower," "*This* is what *this particular* flower looks like," or "I *hereby* present myself as a flower"? If we take the phenomenology of pure-awareness experiences seriously, there seems to be a recurrent process at work that is constantly signifying itself, constantly pointing to itself, but in a "thin" way, predicting only its very own occurrence in the very next moment. Could this phenomenological feature be hinting at the underlying neurocomputational structures and physical correlates?

I think the nonegoic experience of suchness is a clear point in favor of one specific theory of consciousness—namely, Victor Lamme's "recurrent processing theory of consciousness," sometimes also called the "reentry model."[6] Lamme argues that localized recurrent or reentrant processing within the perceptual parts of our brain is sufficient to give rise to consciousness, and the parietal and frontal regions might be required only for later reporting the contents of perceptual experience or drawing on them for reasoning and decision-making. Taking the experience of suchness as seriously as we can is of pivotal importance because in my view, it demonstrates that the first element, conscious experience, can exist without the other ones: reasoning, perspective-taking, and decision-making. The phenomenology clearly has a first-order, self-reflexive element, and it relentlessly creates the evidence for its own existence. Just think of the two Wittgenstein translations presented previously: The mystical *discloses* itself and *manifests* itself. We could parse these as highlighting the ways in which suchness is always signifying its own being and being its own signification, respectively. One last time, in the words of Aldous Huxley: "shining with their own inner light and all but quivering under the pressure of the significance with which they were charged; [Plato] could never have perceived that what rose and iris and carnation so intensely signified was nothing more, and nothing less, than what they were."

In 2016, the Danish-Australian philosopher Jakob Hohwy published an important article entitled "The Self-evidencing Brain."[7] Part of his core proposal was that the human brain is constantly producing evidence for its own existence, by predicting future states in which it still exists. It is as if an organism were treating its own future existence as a hypothesis and continuously trying to find proof or new evidence for the truth of this hypothesis. Here are my questions: Could it be that individual brain states temporarily do the same? Could it be that there is actually a whole nested *hierarchy* of self-evidencing states, and that precisely this is what makes human embodiment so very different from that which any nonbiological machine or intelligent robot could have? For Hohwy, "self-evidencing" is an epistemological and a computational notion. But here, my point is that there is also an interesting *phenomenological* reading of this concept because it seems to have correlates in conscious experience itself.[8] I would like to call the experiential side of self-evidencing "the phenomenology of epistemic self-validation." This would mean that multiple entities in our conscious mind self-evidence at the same time: the flower, the body, the model of a seeing self, and even the global feature of wakefulness. For example, could the sound appearing in space by "recognizing itself," or Huxley's rose and iris and carnation continuously "signifying themselves," actually be more than just examples of the phenomenology of nonconceptual self-disclosure that I have been trying to gesture at here? Are they literally states that try to sustain their existence in our brain, by continually producing fresh "evidence for their own existence," *validating* the feeling of knowing?

In recent publications,[9] Hohwy has pointed out that there is a deep connection between self-evidencing and our consciously experienced sense of being (see chapter 27 for more). He has also suggested that if our goal is to understand conscious experience, a certain subset of self-evidencing properties tend to transpire as good explanations of why those properties are the way they are, including how they might not occur in creatures that we do not consider conscious. This is my point: Could the phenomenology of self-revealing suchness perhaps be much better described as the phenomenal character of *self-evidencing*, of what I have termed "epistemic self-validation"? If so, this insight could build a conceptual bridge to the best of current research on consciousness.

The notion of multiple self-evidencing states arising within a larger system may help us to see a new connection between descriptions of conscious experience and the computational level of analysis. Perhaps conscious representations are "epistemically self-validating," in the computational sense that every single conscious model (e.g., of a sound or of a flower), and not just "the brain" as a whole, is actually creating evidence for its own existence: Every conscious model predicts epistemic value plus its own future existence, and this fact is subtly reflected in its very phenomenal character. This would mean that normally such states simply "appear as real" (because they express a very high probability), but that in meditation, we sometimes have access to their deeper functional structure, to a recurrent and "loopy" process of local self-evidencing—but, as always, entirely without words and concepts. Zooming out further, the transition from physics to life can be characterized as an emergence of self-replicating structures, dynamical systems that create near-identical copies of themselves. Could it be the case that—in the long slow transitions from life to mind and onward to consciousness[10]—some representational states in biological nervous systems "come alive" because and so long as they continuously self-replicate (namely, by predicting their own knowability and future existence)? Returning to how best to understand an individual organism's experience of suchness, however, the idea of locally self-evidencing perceptual states opens up a new perspective on the "mystical" Wittgensteinian qualities of self-presentation and self-manifestation that characterize it. And as we will see in chapter 28, this deeper form of introspection can sometimes create a phenomenology of translucency and virtuality (i.e., of things appearing as *neither* real *nor* unreal).

What's more, perhaps this framework gives us a different take on what philosophers have in the past called "appearances": as epistemically self-validating representations, dynamically self-evidencing models emerging in the nervous systems of some biological creatures, including models of knowing and wakefulness. Appearing would then be a continuously self-evidencing process that really only predicts its own internal states, but that has become so useful that it can be used as a model predicting a complex external stimulus source and the probability of knowledge possession itself. For example, if the stimulus source is an iris, then the self-evidencing iris-model makes the iris *appear* to the organism. The process presents evidence for itself; it discloses itself; it epistemically self-validates. Again, as in the quote from Wittgenstein, it *shows* itself and *manifests* at the same time.

This new way of describing the nature of consciousness also offers a deep connection to the new phenomenological concept of "nonegoic self-awareness," to which we will return in chapter 29. On the level of statistical analysis, this concept relates to factor 8, which was called "Emptiness and Nonegoic Self-Awareness": Pure awareness is often described by our meditators as something that actually knows itself. This may be one of the study's most interesting results. I will not go into detail about this point here, but if you are feeling impatient, you may want to quickly return to figure 2.1 in chapter 2, where you can see that the second- and third-strongest loading items in factor 8 were offering two metaphorical descriptions of first-order reflexivity, actually combining the "Self-Knowledge, Autonomous Cognizance, and Insight" of factor 3 and the "Wakefulness" of factor 4 with the phenomenal quality of emptiness and epistemic openness. In other words, we seem to find an empty form of self-cognizant wakefulness. Interestingly, these factors are negatively correlated with the phenomenology of selfhood: The phenomenology of self-knowing and self-awakening picked out by factor 8 is *nonegoic*, meaning that it has nothing to do with agency, control, the meditator's personality traits, or her autobiographical narrative. Experiencing suchness may be directly related to perceiving from the perspective created by nonegoic self-awareness.

In suchness, too, there is an element of nonegoic self-awareness, but on the level of perceptual objects themselves.[11] However, if we pay real attention to the phenomenology, we find that the objects are already de-reified: They are no longer really objects. The dualistic prior assumptions that everything is *either* subject *or* object, that things are *either* sentient *or* insentient, and that no third possibility exists, begin to lose their grip on our phenomenology.

At the very beginning of this book, I claimed that, if we begin to take our own phenomenology more seriously, conscious awareness may not be a *subjective* phenomenon at all. You may already have noted that from this, it does not follow that conscious awareness is a purely *objective* phenomenon either. Our phenomenological data now begin to show what this could mean. According to some poetic descriptions, objects seem to be in a process of awakening to themselves. Recall the first passage from report #3160 presented earlier in this chapter: "that it is not a self that perceives the sound, but that there is awareness in/with the sound. There existed a perception that the sound appears in space with awareness and 'recognizes itself'." The first conclusion from phenomenological observations like these is that in describing suchness, we must do phenomenological justice to a strong aspect of reflexivity, as well as to a nonegoic signature of knowing. Second, suchness is related to an experiential attenuation of subject/object duality. Subject/object structure can itself be viewed as a filtering mechanism, which can be temporarily suspended in pure perception. For another concrete example, consider this part of a report that will be presented in full length in chapter 27 (#2780): "Everything was awareness. It was not that (for example) I saw a chair, but that awareness 'happened' in the form of a chair."

Let us take stock of where we are now. The first element that I proposed as crucial for understanding the conscious experience of suchness and seeing *what is* was the phenomenal character of timeless change. Then we took a closer look at the experiential qualities of "self-disclosure" and "self-manifestation." The third and final phenomenological aspect of suchness that I highlighted in this chapter was "meaninglessness." Let us take another look at this last aspect.

Earlier in the chapter, I briefly noted that "suchness" could refer to the very phenomenology of nonconceptuality itself. Meaninglessness need not be negative or repugnant; on the contrary, its phenomenology can have immense beauty and depth. If you look at a chair or at a carnation in a vase on the table in front of you with a crystal-clear, silent mind, without labeling or naming it in any way and without making any connections to past or future, then you may not even recognize it. Your mind is in a state of epistemic openness, but now the perceptual object is epistemically open as well because it has lost its predetermined conceptual essence. It is now open to an infinite number of different interpretations and possible perspectives. This is the infinity that the poet William Blake spoke of, and in rare moments, it can be consciously experienced by any of us.

Once we have understood that "meaninglessness" can be an entirely positive, even spiritual experience, we gain an unexpected new perspective on the old philosophical question about the "meaning" of life. What if your whole life were meaningless, but only in the very special sense described in this chapter? Could you perhaps look at yourself plus all of your life history as a large, self-revealing, ultimately selfless pattern of pure suchness, of events manifesting within a timeless space of awareness? Could you see the pattern as entirely meaningless in this sense, as neither good nor bad at all, but also thus see it as a pattern that is at times beginning to see *itself* in a much deeper and liberating way?

Back to nonconceptuality: What does it mean to say that suchness includes the very phenomenology of nonconceptuality itself? I think that there are two aspects. As we saw in chapter 4, to say that something is "epistemically open" also means that it has no essence, no intrinsic and predetermined meaning; that it possesses no label and no obvious canonical interpretation. It is *available*, but as something that can be seen from an infinite number of possible perspectives. My first point is that this infinity of epistemic possibilities has a phenomenal character of its own.

But there is a second aspect. What is more, "suchness" in this sense is not only an experience of nonconceptuality itself, but also one of *inconceivability*—that is, we see the actual fact of something being ineffable and possessing no intrinsic conceptual essence, and simultaneously we see the *impossibility* of ever bringing it "under" any conceptual form of thought. Its richness overwhelms the thinking mind. Put differently, "being beyond comprehension" is a distinct experiential quality of its own, and suchness includes that quality. It is not only the infinity that we feel, but a sense of impossibility, of unfathomable depth. This is my second point.

The phenomenology of perception while in a meditative state shows that these two qualities of infinity and inconceivability are something that can be directly experienced. Do you recall the beautiful metaphor at the beginning of this chapter, coined by one of the participants in our study? There are states of consciousness in which "openness touches openness." In these states, in experiences of suchness, there is also a quality of freshness and wonder. Many of us may remember such states from childhood. One participant wrote: "I also had the feeling of seeing objects as if for the first time, with a renewed attention similar to that of children looking at things for the first time. I was not creating this situation. It simply happened for some reason, and I found myself in it" (#2543). As there is no conceptual overlay, the meaning is lost, there is a timeless moment in which the carnation is no longer perceived *as a carnation*. You do not recognize the perceptual object as that which it always was. But there is something else you see: an infinity of epistemic possibilities, the particularity of timeless self-disclosure in this very instance, the manifestation of self-sufficiency without substantiality, sheer inconceivability, the emptiness of appearances.

Or even a Last Judgment. From a certain theological perspective, the Last Judgment can be seen as the ultimate climax in the history of conscious experience, the point at which time ends and everything is finally seen as it really is, by a single Divine Mind, naked and in its entirety. Some meditators with religious beliefs might feel driven to impose this—or similar—frameworks onto their experience, as the only conceivable way to later make sense of it. The French philosopher, mystic, and political activist Simone Weil (1909–1943) said, "One has to be dead to be able to see things in their nakedness."[12] If dying into pure experience is what some experiences of suchness feel like in retrospect, the process of coming back to life afterward is bound to be hard, or at least strange. For the meditator or the psychonaut following in the footsteps of Huxley, it may certainly be difficult to return to the prior state—it may be only "after a long time and with considerable difficulty" that you regain the ability to contract into a knowing self, to begin to think thoughts, to recognize the chair as a chair. Seeing suchness upends the constant process of meaning-making, and the emptiness of appearances is their utter meaninglessness. Liberated from meaning, they are self-ablaze, shining with their own inner light—epistemically open.

One final question is whether MPE *itself* carries the phenomenal character of suchness too. In other words, how exactly does all of this relate to the elephant, if at all? For the scholar, the issue may now become whether suchness is part of the elephant, or even a good candidate for "prototypical" elephanthood. For the active practitioner, all of this is interesting in a very different way. If, in the process of seeing *what is* without words and concepts, she has begun to gradually discover suchness in perceptual objects, new targets begin to emerge for her—this time on the level of inner awareness, or introspective experience. If you are a practitioner, can you see the suchness in the process that is trying to meditate? But even if you are neither a scholar nor a practitioner, there now are fascinating questions to guide you in your own phenomenological investigation. Does your ordinary, innate experience of wakefulness (explored in chapter 4) actually possess the quality of suchness too? And what about the quality of "knowingness," the experiential signature of knowing itself (chapters 7 and 18)? Just like wakefulness, it exists in the simple process of mindfully and choicelessly seeing *what is*; it naturally coemerges with it. Does knowingness have suchness too?

10 Presence

A constant still presence that cannot be touched by words,
does not live in the realm of objects. [#789]

I was very present, but not there. [. . .] Pure existence. [#2943]

The quality of "presence" is a central feature of what it feels like to abide in pure awareness. But we must be careful, because "presence" may actually mean slightly different things to different people. One reading of "presence" is "being fully in the present moment."

> 2901 A state free from fear and other negative feelings. Being only in the moment and no thoughts about future or past. No judgment, only the moment, free from judgments.

Being fully present as an embodied self while resting in choiceless awareness is a well-known form of contemplative experience. It can be very intense, stable, and crystal clear. However, abiding in a clear state of mind without thoughts, calmly resting in the present moment, is nonetheless a subtle and highly refined form of temporal experience. "Nowness" is not the same as "timelessness." The same is true of the experience of mere duration, as in Jennifer Windt's original theory[1] of what minimal phenomenal experience (MPE) might be (see chapter 22 for more on the experience of duration and "pure temporality"), because both forms of time experience still have a thickness to them. "Nowness" means that you locate yourself in a temporal order, as existing at some *point* in time, but in ordinary states of consciousness, this point is never really extensionless, as some kind of mathematical abstraction might suggest. Research has long shown that what we subjectively experience as a single psychological moment actually has extension in objective time, and that extension can vary greatly depending on context (we will probe this experiential domain further in chapter 22).[2] Some descriptions of the phenomenology of embodied presence, therefore, may refer to a state where a minimal self is still present, imbued with a silent mind, temporarily liberated from its own inner narrative, mindfully embodied, and calmly located in space and time. So long as the phenomenology of "presence" emerges together with spatiotemporal self-location, a centered Here and a Now remain.

There are, however, additional readings of "presence" that are not directly related to time experience, such as those that relate instead to an experience of "nondual being" (chapter 26). Many intermediate stages appear to lead to a fully uncontracted state of presence in this wider sense. They are not easy to describe. Sometimes the aspects of selflessness and nonduality are not fully expressed, and presence is described as pure awareness, as something within the phenomenal field. But presence may also seem to be something in which everything else is held, something that permeates the whole field, with the meditator "bathing" in it. In his book *The Way of Effortless Mindfulness*, the American meditation teacher Loch Kelly described this process (which can be a form of practice) as "marinating" in the continuous field of awake awareness.[3] In attempting to describe the quality of presence, some of our respondents use similar terminology:

> 1758 [. . .] What's most characteristic about it is that awareness is a quality as such in the field of experience. That presence can actually be felt and permeates experience. It is not nondual but has a taste of nondual. There is a sense of self but it is light. Feel as if I'm held and bathing in presence. A sense of safety. Comfort. Being comforted by awareness. Still there is emotion and thoughts and stress. But all that is happening in presence. Being held by presence. A sense of a loving field permeating experience. Being the background which is coming forth.

Others described pure awareness as a fully selfless episode of wakeful presence, with or without imagery of immersion:

> 2565 [. . .] It is a feeling of bright lightness, boundlessly connected to the universe, pure presence without an experience of a self.
>
> 3542 My body feels as if it is being infused in fresh spring water. It is a comfortable but also refreshing feeling. Later a state of wakeful presence sets in, a short moment of emptiness that knows no self, the world is one.

The phenomenology of presence can sometimes have a fully nondual character, and at times it can coemerge with other aspects discussed in this book, like luminosity (chapter 18), soundness (chapter 7), and the experience of spatial immersion without self-location (chapter 23):

141 It was as if I was one with a glowing radiant happy warm presence that permeated through me and everything else—no separation in observing and observer—being right in the middle of it and yet not localized—a feeling like looking out—blissfully open and totally satisfied with everything. My surroundings radiated visually too—light-filled vastness, openness, and total harmony. Great joy and gratitude. [. . .]

This kind of presence seems to be a nondual form, not something that the meditator has fabricated. It simply arises. Selfless presence in this sense is an all-pervading, timeless experience of stillness, unity, and emptiness. There is presence, but it is not *you* who is present. Suddenly recognizing that can even be something of a shock:

3431 [. . .] I was perceiving nature perfectly well, but somehow differently from usual. Silent, even though birds could be heard. It felt like standing still and at the same time like total presence, although it wasn't me who was present. [. . .] This is not "relaxed," but also not tense. Simply very present, but not in the sense of "I am present." But even this "present" doesn't really capture it. It's like a "noticing everything" without "paying attention" to anything, that's why it's also not the feeling of being especially "mindful," because it's more that it just "is," nobody has to be mindful anymore. Maybe also because then there's just nobody left who could be mindful anymore. In any case, I'm not (anymore) in that moment. Sometimes there's also something startling about it.

The phenomenology of presence is intimately related to the frequently reported experiential character of existence *as such*, as if it were expanded into a comprehensive space of "nondual being" (more about which in chapter 26):

301 [. . .] It is a state about the knowledge of one's own existence in connection with something indescribable. A kind of presence that expands over everything and also implies everything. There are no questions there, no answers, no time, except the feeling of absolute silence and a feeling of connectedness. [. . .]

3320 [. . .] It was absolute presence in absolute presence of everything, which surrounded and simultaneously contained everything. [. . .]

Another result from this study is that the phenomenology of pure presence can also emerge as a full-absorption episode. In this case, pure presence is a stand-alone phenomenon, described as identical to awareness itself:

3294 [. . .] It was as if everything disappeared, body, mind, space and time, but awareness itself was still present. There were no particular qualities that involved self-reflection, like emotion etc., until after the experience ended and I was again aware of my surroundings. This particular "pure presence" involved the emptiness of all content except awareness itself. As if awareness is the only thing that exists, but I am that awareness and not separate from it.

Spontaneous Presence and the Dolphin Model of Meditation

> Naturally occurring timeless awareness—utterly lucid awakened mind—is something marvellous and superb, primordially and spontaneously present. [. . .]
> As for this treasury of phenomenal space, source of everything, Nirvana does not need to be sought; it is primordially, spontaneously perfect.
> —Longchen Rabjam (1308–1363), *The Precious Treasury of the Basic Space of Phenomena*

The concept of "presence" plays a central role in many traditional texts directly related to meditation and pure awareness. If we are interested in MPE, one fundamental problem is that these texts often do not distinguish between the phenomenology and metaphysics of pure awareness, but—from a Western, analytical perspective—also don't offer any convincing argument for why these two fields of inquiry should be conflated. This illustrates how comparative philosophy—sometimes called "cross-cultural philosophy"—has to deal with conceptual schemes that may simply be incommensurable, and in more than one way. On the other hand, the traditional idea is often that "presence" is not a form of experiential content at all; rather, it equates to *manifestation* or "givenness" itself. Is there a meaningful distinction between appearance and reality for "presence," "manifestation as such," or "the mere aspect of givenness itself"? I definitely think so, but I could certainly be wrong. In comparative philosophy, we often confront difficult issues of methodological incommensurability, relating to the question of whether and how comparisons between different philosophical traditions can be conducted at all (just think of Western philosophy of cognitive science, *Advaita Vedanta*, and Tibetan Buddhism). Then there is the thorny issue of metaphysical and epistemological incommensurability: In finally developing new forms of cross-cultural philosophy that take not only Western approaches but all of humankind's traditions seriously, how can we compare their different modes of justification and inquiry (like the epistemic practices of meditation versus rational argument), or even their different theories about what "existence" really is? Could there ever be a truly *global* philosophy of consciousness?

Here is a first example to consider. One central concept in the philosophical traditions of Advaita Vedanta and Kashmir Shaivism is *prakāśa*. Prakāśa is that which can never become an object, that which has no content or inner structure; it is the aspect of pure presence or pure manifestation in consciousness itself. Importantly, it also contains the semantic element of "self-disclosure" or "self-evidencing" (as discussed in chapter 9, in the context of what Wittgenstein said about the inexpressible and the mystical). Here is a description of prakāśa from another Austrian philosopher, Wolfgang Fasching, who has done a lot of excellent work in this field:[4]

> Yet although it [*prakāśa*] can never be given as an object, it is in no way concealed, rather it is essentially self-disclosing (*svaprakāśa*), i.e. its very being *is* its own revealedness without any subject–object difference. [. . .] Consciousness is the taking place of presence in which all this object-appearing with all its manifold modes and structures takes place, yet in itself it has no modes or structures, and nothing to analyse. It is utter simplicity.[5]

Using our new conceptual tools, we could now say that, phenomenologically, presence is the aspect of "givenness" that relates to epistemic openness, to wakefulness (as the conscious experience of tonic alertness), and to clarity itself. What becomes present is the same quality of being open to the world that we discussed in chapter 4. Please note, however, that "givenness" itself always remains an unclear and subtly misleading phenomenological term, simply because there really is no second person, no other self, to "do" the giving—the phenomenology of presence simply appears, in an apparently uncaused manner. It is not a social event. In its origin, "giving" is a dualistic personal-level predicate: Person A gives something to person B. "Givenness" can also have a hidden theological connotation (a fact that makes it attractive to antinaturalists) because A might be a personal God who reveals the truth—or "gives" Herself—to B. As the German philosopher Max Scheler (1874–1928) thought, revelation in this sense would be simply the specific type of givenness wherein the divine or the holy is given.[6] But the world itself never "gives" anything. Nonetheless, the idea of unpersonified "giving" remains a reference point for many feelings of joy, awe, and gratitude (chapter 15), as well as for experiences of spontaneous presence, suchness, and unexpected self-revelation. In this and many other ways, the reports presented here can also be read as describing stages in which the qualities of epistemic openness, wakefulness, and clarity gradually manifest themselves, but without becoming properties of a person. It is therefore not a thing or agent that is present or absent, but rather a process. The phenomenology of presence can be described as the continuous *becoming-manifest* of epistemic openness.

What, then, is the space in which Wolfgang Fasching's "taking place of presence" happens? From the perspective of modern philosophy and cognitive science, we could say that it is an integrated epistemic space—part of an embodied computational space in our brain, if you will. Different human beings or other conscious animals could certainly have very different inner spaces of this kind. But whenever they are conscious, they also nonconceptually and thoughtlessly *know* that they currently are possessors of precisely such a space. The space is now represented within itself, it is a *self-presenting* epistemic space—as if it had folded back into itself one more time. "Taking place" means locating yourself. Here, what actually locates itself in epistemic space could be the very model of this epistemic space itself, as explained in chapter 5. That is to say, an image of the space itself becomes manifest within it; the image presents "us"—or at times only itself (see chapter 30)—with evidence for its own existence. It seems as if this inner image of the space of knowing generates and discloses itself, and whenever the meditator oscillates between a dual state and full absorption, the image can be experienced, remembered, and reported. The fact that is reported is that a very special model is manifesting itself, that our integrated state of epistemic possibilities has become present within itself.[7]

To make this clearer, let us look at a second example. Another crucial concept is "spontaneous presence," found mostly in Tibetan Buddhism. Spontaneous presence (*lhündrup*) is that aspect of *rigpa* ("knowledge of the ground") out of which all phenomena arise and into which they are all later absorbed. It is self-caused or spontaneous, which may also mean "having a self-contained origin" or being without origin at all. If we set all metaphysics aside and stay as close as possible to contemplative experience itself, then the term "spontaneity" refers solely to a *phenomenological* spontaneity. According to conscious experience, pure awareness arises without any external cause. Enter naturalism; enter intellectual honesty. It is important to understand that, just as for all other conscious states, experiences of effortlessness, the phenomenologies of "givenness" and "spontaneous presence," depend on unconscious causal precursors in the brain—anything else would be sheer magic. However, these precursors can certainly be cultivated, strengthened, or triggered in indirect ways, of which meditation is one. In this sense, meditation practice can be seen as an ancient way of doing neurofeedback, using the phenomenal correlates of meditation practice as a user interface. If this practice is done properly, the experiential flow of awareness itself may still be the best and most natural real-time display of brain activity, with many uses for self-regulation.

In 2015, the Ukrainian philosopher and computer scientist Iuliia Pliushch and I published a book chapter in which we introduced the "dolphin model of cognition" as a new metaphor for the interplay between conscious and unconscious processes.[8] As you probably know, dolphins frequently leap above the surface of water. One reason for this behavior could be that when traveling long distances, jumping can save the dolphins energy, as there is less friction in air than in water. Typically, the animals will perform long, ballistic jumps alternating with periods of swimming below but close to the surface. "Porpoising" is one name for this high-speed, surface-piercing behavior of dolphins and other species, in which leaps are interspersed with relatively long swimming bouts, often about twice the length of the leap. Porpoising may be the most energetically efficient way to swim rapidly and continuously and keep breathing at the same time. Just as dolphins cross the surface, thought processes often cross the border between conscious and unconscious processing, and in both directions. For example, chains of cognitive states may originate in unconscious goal-commitments triggered by external stimuli, and then transiently become integrated into the conscious self-model for introspective availability and selective control, only to disappear into another unconscious "swimming bout" below the surface. Here is a new question for you: Could there also be a dolphin model of meditation?

The idea here is that many of the things that a human being does and lives through have conscious and unconscious components, and this fact has been almost entirely ignored in meditation research. Meditation is something that human beings do, and as a process, it will inevitably have conscious and unconscious aspects. This obvious fact might also play an important role in homing in on MPE, the experience of pure awareness. The experience of apparently uncaused "self-disclosure" and "spontaneous presence," which has been reported by millions of practitioners over the centuries, as well as the deep dialectic between meditation and nonmeditation (see chapter 32), can now be seen in a new light, giving us new perspective on ancient philosophical debates. One example is the endless scholastic dispute between "subitists" (proponents of sudden awakening) and "gradualists" (philosophical defenders of the view that enlightenment can be achieved only step by step, through gradual practice). My personal view on this is that (just as in our fable of "The Elephant and the Blind") they are both right, and the traditional dispute is mostly irrelevant to practitioners. But these new data and the perspectives deriving from them may allow us to intervene more decisively in stale disagreements of this kind, using empirical research to develop an evidence-based rational perspective. All we have to take into account is that, just like every other conscious episode or event, the experience of "self-disclosure" and "spontaneous presence" must have had unconscious causal precursors. Of course, it will also have unconscious causal *consequences*—an experience like this, for example, may make the precursor's future recurrence more probable by leaving a positive trace in the brain, gradually opening a new path into state space. In principle, it might even become permanent and continuous (chapter 33), or it may contribute to the meditator's health, indirectly strengthening the person's physical, mental, and social well-being. We can predict that—just like thoughts that "unexpectedly" pop up in meditation—certain types of silence, clarity, and presence *must* appear entirely spontaneous and uncaused, simply because they are the first elements in a causal chain that just crossed the water surface. They are the jumping dolphin that we later remember; only as conscious memories can they sometimes become part of our inner life-history. But they are not apparitional experiences or metaphysical mysteries. Dolphins don't have wings; they have strong tail fins, and they always fall back into the ocean. Just like dolphins, we too are fully embodied beings. We are embodied epistemic spaces.

Let me conclude this chapter by giving a third example of the dolphin model of meditation, this time looking at the sense of self. To understand what the sense of self really is, it helps to focus on how the epistemic agent model (the "knowing ego"; see chapter 25) suddenly appears, and also on how it disappears. One thing that our high-level self-model tries to explain away is the fact that the organism harboring it (and *of which* it is a model) is confronted with an endless chain of unexpected events and ugly surprises. The continuous process of attempting to construct a stable sense of self gives rise to an interesting phenomenology of being "ambushed" and "overwhelmed" by sudden distractions, and there is no better way to see this clearly than contemplative practice itself.

I think that a lot can be learned from carefully looking at the phenomenology of distraction. Most people desperately identify with their high-level self-model and do their best to avoid seeing the abundance of attentional lapses and permanent distractions. There may be good evolutionary reasons for this fact. But if you look at what is happening systematically and on a small timescale (e.g., under the microscope of classical insight meditation), then you see how "you," the meditator, the entity that wants to *have* insight, constantly gets attacked from behind, out of the dark, ambushed by the unconscious. The first thing that every meditator learns is that, strictly speaking, the process of being "ambushed," of the very first thought arising, is completely unpredictable and caused by an unknown force. But isn't the same thing true of "waking up" again, of mindfully noticing the *second* thought, of becoming aware of the noticing itself, or even of observing the fact that "I" have already been carried far away for a long time? Waking up in meditation is an inherently unpredictable event because while immersed in the ensuing train of thought, fantasy, or daydream, there is no conscious knowledge that something like "becoming aware again" could even happen. There is no personal-level memory because the original epistemic agent model has been killed in ambush. This leads to the conclusion that the memory of being awake and the anticipation that awakening is at all possible must be an unconscious form of knowledge. This might be a general principle: just like the phenomenology of being overwhelmed from behind and killed in a sudden ambush, the reborn sense of being a "knowing self" is an effect, not a cause. It may have effects later, but it certainly is not something self-caused—to think otherwise would be to acquiesce in an illusion of control.[9]

To see this possibility from another angle, think of the experience of waking up in the morning. The feeling of suddenly "coming to" is clearly something that was caused by unconscious processes in the brain that preceded it. A "knowing self" pops up, but that event itself was quite a surprise—wasn't it? Waking up is a major discontinuity that happens from one moment to the next, on a small timescale. But this discontinuity is immediately glossed over by the process that in chapter 17 I will call "narrative self-deception": There is a short moment of pure awareness, but as soon as automatic orientation in time and space has taken place, the organism swiftly orients itself to *person*, activating the high-level model of your life history, of who you take yourself to be, and of what you must do today. Often, this takes less than a second. It is equally automatic, and certainly not something *you*—the content of the personal-level self-model that the organism is currently trying to stabilize—have achieved or ever decided to do. But in an interesting sense, it is now the *organism* that gets overwhelmed, ambushed, and enslaved by the long-term self-model that tries to explain away the myriad of unexpected surprises and discontinuities that happen from moment to moment. The organism is like an elephant that now *imagines* one and the same rider sitting on its back: a knowing self that controls the overall process.

I am confident that a future computational phenomenology of meditation will describe all of this in a much better and finer-grained way. But it is important to understand that phenomenological facts do not determine metaphysics. We might even call this the "M-fallacy": From the fact that something is subjectively *experienced* as self-caused, as "spontaneously present," or as uncaused and "unborn," nothing much follows about its metaphysical status. On the other hand, humankind's spiritual traditions are right to point out that the experiential quality of presence investigated in this chapter is ahistorical and entirely nonagentive. It has no past and no future. Phenomenologically, it is nothing that could be expected or created by the conscious self. If the conscious self-model manages to explain away its sudden occurrence in the very moment of contracting the selfless clarity of wakeful presence into the model of a person, it may be experienced as a success of contemplative practice, as an achievement of some complacent spiritual ego-manager. But precisely this moment is also the moment in which it ends.

I pointed out earlier that the incoherent phenomenological myth of first-person "givenness" has a subtle theological connotation. Perhaps the old theological idea of "divine grace" was a strategy to prevent this contraction? If you firmly believe in the possibility of free and unmerited favors from God, who may at times unexpectedly operate in your own conscious mind to regenerate and sanctify it (e.g., by absolving you from the internal sin of clinging to dual awareness), then this belief may enable a quality of letting go, of true effortlessness, a form of *genuine* surrender that is very hard to achieve in Eastern models of self-redemption. While theoretically untenable, there may still be a deep, highly practical form of wisdom buried in this Western model: namely, that setting a "grace hyperprior" via an act of faith opens a space of inner experience that under normal conditions remains almost inaccessible. We will return to this point toward the end of the book, when discussing spontaneity, effortlessness, and the difference between meditation and nonmeditation.

11 Connectedness

> It is like a feeling of stepping back from oneself and at the same time being connected to everything. [#221]

When asked about their pure-awareness experience, many meditators report an experience of connectedness, such as a sense of wholeness, a deepening bond to nature, or a direct, nonconceptual insight into the interdependence of all phenomena. This may include an intensified experience of connectedness with oneself and other sentient creatures. Interestingly, many traditional philosophical models claim that what—for lack of a better term—we call "pure awareness" actually underlies and permeates all other forms of appearance: Silence and stillness are always present even within sound, movement, and thoughts themselves.

If this ancient idea were really true, then one would predict a specific phenomenological effect whenever meditators begin to introspectively penetrate into minimal phenomenal experience (MPE)—that is, whenever they start to access the simplest, subtlest, almost implicit level of conscious experience there is by gradually becoming aware of the ever-present "background." First, one would predict that practitioners will gradually begin to notice a subtle and abstract quality of holistic integration on the level of manifest content *itself*, perhaps as a more concrete form of mutual embeddedness and interdependence. In social contexts—when experiencing the other within a group of human beings—this might manifest itself as a quality of compassion. Second, one would expect phenomenological reports to reflect a nonconceptual quality of connectedness *as such*, a more direct experience of integration and embeddedness itself. Let us have a look at what practitioners actually report:

221 It is like a feeling of stepping back from oneself and at the same time being connected to everything. Harmonious and peaceful, floating, light and natural.

[. . .] I perceive things in my immediate surroundings as if I were standing next to them. [. . .]

269 [. . .] The mind is calm and I feel how I am part of the big picture, connected to everything. Everything is connected.

2500 [. . .] Simply being in the here and now, being connected with everything that is. [. . .]

1188 I know states of deep calm with simultaneous absolute alertness. Everything merges together. I am everything, there are no boundaries between me and the others. A feeling of being connected to everything. [. . .]

1347 I experienced great peace, a silence, and was completely awake. Paradoxically, in part I absolutely perceived my body, but in part I was—physically too—connected to everything, and one. Connectedness is the word that best describes my experience: connected with myself, with all living beings, with all times, with nature, with all objects and sounds, i.e.: not separate.

1595 Sitting meditation with open eyes: The primordial source of all being is perceptible and from it all forms of being arise and fall away again. Objects are permeated with vitality—visibly full of light. State of perfect connectedness with everything, in everything. [. . .]

2527 [. . .] I suddenly felt my tension in my body and had the insight: Everything is as it is, quite natural. Only I myself hold on. Then the tension let go and a deep silence and connection surfaced. I did not feel separate from the outside world, but was part of the whole, in subtle movement with everything else, like a river, bright and light.

2721 [. . .] The silence was so deep that nothing (no self) was there and at the same time I knew that I was sitting in the zendo. There were no emotions like being happy, but there was a feeling of connectedness.

Different people enter pure awareness in different ways. For many, the phenomenal quality of "connectedness" goes along with an opening of the heart, because it has a deeper emotional quality and creates a specific sense of intimacy and compassion—not only toward others, but also toward themselves. The qualities of "self-intimacy" and self-compassion may be intrinsically linked; their phenomenological opposite would be the sense of disconnectedness, emotional numbing, and self-alienation that we find in psychiatric syndromes like depersonalization disorder (for more on this, see the discussion of "witness consciousness" in chapter 19):

1717 [. . .] When accessing it through the heart there is a sense of connectedness to the world, embodiment, unity, softness, density, richness, joy, love. I can access it with my eyes open and boundaries feel blurred and malleable rather than completely disappeared. It's lovely.

3270 I experience pure awareness as a thoughtless moment of connectedness with myself, which often gives rise to a feeling of langorousness and inner kindness in me.

3250 [. . .] It was a strong feeling of connectedness and of "knowing" interrelatedness. It was as if I was on another level and I was different from my fellow meditators but still we were all connected to each other. [. . .]

3569 [. . .] Everything was ok and connected with the rest. The boundary-transcending feeling was that of connectedness to all that is, and perfect love.

3259 [. . .] presence, clarity, deep joy, timelessness, deep connection with myself, with nature, and with life as such, deep peace, a feeling of soundlessness, interwoven with life, complete absence of intention, as if everything was together at once in this moment [. . .]

For some, timelessness and gradual ego dissolution (see chapters 22 and 25), but also soundness and awe (see chapters 7 and 15), are dominant features that accompany the experience of connectedness:

3248 It was an experience of great connectedness with everything, an experience of unity with nature, with everything. It was completely detached from the sense of time. The strongest sensations were astonishment and reverence. And all embedded in a deep inner silence, a feeling of complete inner and outer peace.

3012 [. . .] a feeling of calmness arose, my body relaxed, the breath breathed itself. I had only a weak feeling of "I," in the foreground was the experience of clarity, boundlessness, connectedness with everything.

3003 [. . .] I felt completely open in all senses, everything bright, clear, dynamic, connected with everything, no reference to time or self during the direct experience, everything ok and that's how it is, in harmony with life and the universe, with no separations.

Connectedness and the phenomenal character of abstract, nonphysical spatiality plus "bodiless embodiment" (which we will investigate in greater detail in chapters 23 and 24) are other global aspects that seem to frequently co-occur. But we also find some specific, concrete details that are not present in other reports of this category:

3073 [. . .] I'm meditating, suddenly have the feeling that my body has disappeared, has expanded infinitely and my mind is everywhere. I have a strong feeling of connection with everything.

2362 [. . .] A state [. . .] in which I suddenly felt connected with everything, esp. with above, with the clouds or the sky—perceived something like threads between me and the universe. Yet "I" was as big as the universe.

1942 [. . .] at this moment an enormously strong (cold) wind hits me. At the same time the mountainside opposite is connected very closely to me with strings, or rather a carpet. In the carpet I am visibly clearly enmeshed with houses, the countryside, and the whole world. I stand next to the carpet, see myself and feel very connected with everything. Tears run down my face.

3295 [. . .] In this phase the space expanded and time no longer played a role, there was only pure being, a pure feeling of happiness, which was at the same time a feeling of infinity and the feeling of being connected with everything.

2607 I was in a place that felt infinite and shapeless and colorless. I felt an awareness represented by an orb of light in the invisible fabric of this infinite space. I felt every living thing past, present, and future connected to myself through this space. I felt no human emotion as such; no fear, no joy, no sadness; as I did not exist to feel emotions. However, there was a sense of unity of all things, beings, life, and a sense of love between it all, interwoven into the space and connecting all things. Time did not exist. I came away from the experience and cried tears of happiness for about 15 minutes and felt totally connected to everything around me. It was the most beautiful and important experience of my life and it continues to shape me, five years later.

Peripersonal Space

> That clarity is a beginning place, and almost as soon as this empty gaze into the nature of things reveals existence vast and deep, it reveals something else no less wondrous and unimaginable: there is no distinction between empty awareness and the expansive presence of existence. They are whole, a single existential tissue, which is to say that existence-tissue is our most fundamental self. [. . .] Here in the beginning, there is the existence-tissue open to itself, miraculously and inexplicably aware of itself, when there might just as well be nothing but opaque existence, existence blind to itself! Vast and deep, everything and everywhere—the sheer presence of materiality is open to itself through our eyes, aware of itself here in the beginning. The story of existence is a self-portrait.
>
> —David Hinton (*1954), *Existence*

From a qualitative perspective, the phenomenology of connectedness reveals a number of interesting details: most importantly, something like an "existence tissue" that has been ignored by mainstream philosophy of mind and cognitive science and seems intimately related to MPE. For example, some of our meditators speak of an "invisible fabric" and of "interwovenness" (#2607); they describe "strings" forming a "carpet" weaving them "into the world" (#1942) or an experience of "threads" connecting and integrating them with the universe as a whole (#2362); and they report being "interwoven with life" itself (#3259).

To deepen our understanding, there is another interesting phenomenological detail that I would like to highlight here, which could point to an expansion of what neuroscientists call "peripersonal space." Peripersonal space is the region of space immediately surrounding our bodies in which objects can be grasped and manipulated. Our inner model of this space may help with early threat detection and response, such as via involuntary defensive or other self-preserving movements (think of instinctively protecting your head during sports or in a fistfight, or swatting at a mosquito coming to close to your face, but also the awkward feeling when someone stands too close to you in an elevator). The Indian-Canadian philosopher Mohan Matthen describes peripersonal space as the locus of greatest vulnerability—the zone in which others may intrude with their own body—and calls it the "territory of direct intervention—the area where we probe and shove, the zone of the head-butt and the cross-check."[1] He also quotes the neuroscientist Michael Graziano, who describes it as the "margin of safety, bad breath zone, duck-and-flinch buffer."[2]

Neuroscientific research has uncovered how the primate brain constructs multiple representations of space. These representations can be modified very quickly and are centered on different body parts (e.g., there is hand-centered space, head-centered space, and trunk-centered space). The integration of visual information available outside the body with tactile information arising outside the body leads to the creation of an "invisible bubble" that we can sometimes consciously experience, often simply referred to as the "near space" or "reaching space"; this is where our direct interactions with objects take place. Goal-directed action plays a role, as does bodily self-protection (e.g., relative to fast-approaching objects and other animals). A unifying computational model by Dijkerman and Medendorp proposes that our representation of peripersonal space is achieved by binding visual and tactile stimuli together, as they occur frequently in close spatial and temporal proximity, and this depends on the *predicted* sensory consequences.[3]

In a sense, peripersonal space is an invisible extension of the embodied self. It defines what is experienced as being close to us and also what we consciously feel when something or somebody is getting *too* close to us. Peripersonal space has also been called "something like a buffer zone between the self and the world," and one can sum up a lot of recent research by saying that it "refers to *a special way of representing* objects and events located in relative proximity to what one takes to be one's body."[4] A lot of our own phenomenological survey data show that what one "takes as one's body" may dramatically change in the context of MPE (see chapters 24 and 29), and that the experience of connectedness changes the spatial experience of proximity (for an example, see #221 later in this chapter). Leading researchers like Jean-Paul Noel and colleagues have described peripersonal space not only as "a multisensory-motor interface between the individual and the environment," but even as "a spatial extension of the body [that] plays a role in scaffolding a primitive sense of self-awareness."[5] This is one reason why, in chapter 21, we will briefly touch on fascinating experiments by Laura Aymerich-Franch showing that the sense of self can actually be "smeared" in space. All of this leads me to believe that underlying changes in peripersonal space could play a major role in understanding the phenomenology of meditation. Please note that this will be particularly true for traditional forms of moving meditation like tai chi and qigong, which involve the mindful coordination of slow-flowing movements, as well as the experience of unbounded spatiality and variants on the kind of "bodiless body-experience" that will be investigated later in this book, in chapters 23 and 24.

Importantly, our model of peripersonal space can expand or shrink because the same area of physical space can be portrayed as peripersonal or not, depending on context. Anxiety expands your peripersonal space, as does successfully using a tool or manipulating your sense of agency in a scientific experiment. Testosterone administration in women increases the size of their peripersonal space[6]—as does pregnancy.[7] An important recent finding is that the extent and shape of peripersonal space can be altered in the type of virtual reality (VR) experiment that I described in *The Ego Tunnel*, which aimed to induce illusory ownership toward the location of an illusory body.[8] New results show that the experience of self-location follows the experience of body ownership, and peripersonal space is tied to self-location. Even more interesting in the context of meditation research is the fact that within an impoverished sensory environment and in absence of actions, the boundary of peripersonal space becomes "ill-defined"—that is, the difference between far and near space is attenuated. In audio-visual deprivation, participants reported feeling "lost in space."[9]

Scientific research has shown that human beings can refer somatic sensations to a discrete volume of empty space, such as by expanding the boundaries of their bodily self-model in a way that makes them feel an invisible hand—even if their visual experience directly contradicts the possibility of a real, physical hand actually being there.[10] This experimental setup creates the conscious experience of owning and embodying a region of empty space, and even the "illusion of having an invisible hand that 'feels' touches applied to it in empty space in direct view of the participants."[11] Interestingly, this "invisible hand illusion" happens only *inside peripersonal space*, but it shows that, in creating bodily self-consciousness, visual information is not as dominant as previously thought, and also that, according to conscious experience, human beings are capable of having physical and nonphysical body parts. If we think about some of the kinds of "bodiless body-experience" reported by our meditators (see chapter 24), as well as the kinds of radical connectedness that we are exploring in this chapter, this fact is obviously interesting. Further research by Arvid Guterstam and his colleagues has shown that the underlying neural mechanisms can create the illusion of having a hollow, almost transparent, or fully invisible body.[12] In an ingenious experiment, the invisible-body illusion arises when participants wear head-mounted displays and observe a paintbrush moving in an empty space and defining the contours of a body, while receiving simultaneous touches on the corresponding parts of their real body, which is hidden from view.[13] Please note that this differs from the well-documented neuropsychological disorder called "asomatognosia," in which patients may sometimes feel *no* body at all: In the current case, a body-shaped region of empty space is "owned" and begins to function as the unit of identification.

What all of this shows is that human beings can *embody* empty space.[14] You may want to remember this when we look at the equally inventive experiment by Aymerich-Franch and her colleagues described in chapter 21, which shows that the *sense of self* can actually be spread out in space.

In chapter 23, I will present a large number of experiential reports suggesting that the experience of MPE is related to an unbounded space without center or periphery; and in chapter 24, we will look at the phenomenology of body dissolution and "bodiless body-experience." Plausibly, some of these experiences may be related to a dramatic expansion of peripersonal space. Here is one example:

> 2619 [. . .] The big awareness was beyond it. I felt like a humungous sphere, that my awareness was not located in my body but sort of in a big sphere around it. Anything I could sense at any distance was within my sphere of awareness and I was identified (though that's not really correct) with that big awareness.

As always, it is equally plausible to assume that there may be varied paths and intermediate stages into such states. Some aspects of connectedness described here could actually be related to such stages. For example, given an expansion of peripersonal space, one would predict that distant objects could suddenly be perceived as being in "near space." Recall report #221:

> 221 It is like a feeling of stepping back from oneself and at the same time being connected to everything. Harmonious and peaceful, floating, light and natural. [. . .] I perceive things in my immediate surroundings as if I were standing next to them. [. . .]

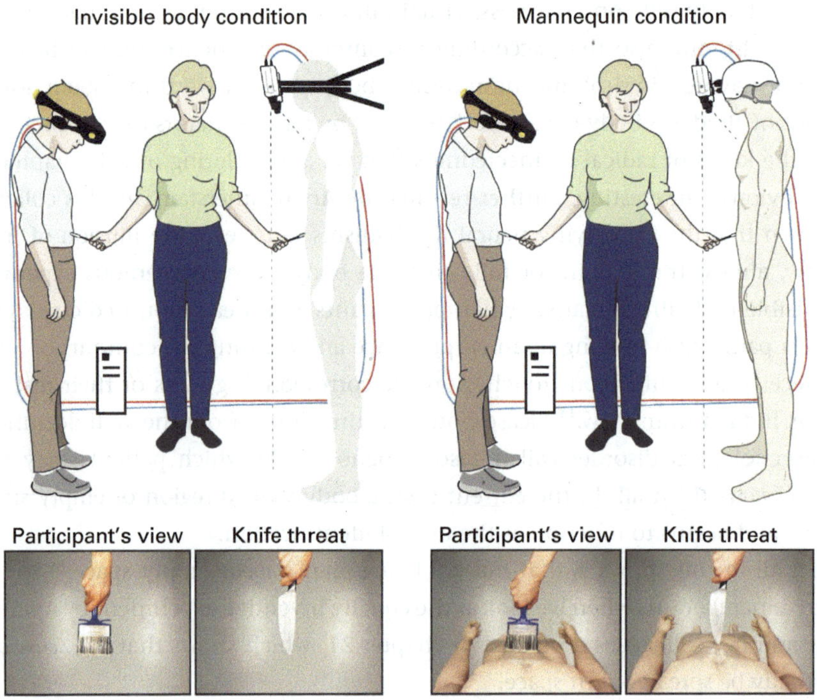

Figure 11.1
The experimental setup of the invisible body condition (left panel) and the mannequin condition (right panel). The participants were fitted with a set of head-mounted displays that showed the real-time three-dimensional video feed of a pair of downward-facing cameras that were mounted on the wall (left panel) or on the head of a mannequin (right panel). The experimenter applied touches to each participant's body and the corresponding body part of the invisible body/mannequin using a paintbrush. The grayed-out body in the left panel illustrates the discrete portion of empty space that was meant to represent the invisible body. Two sample frames of the actual visual stimuli presented in the head-mounted displays are shown at bottom, featuring the brushing procedure and knife threat event for the two conditions, respectively. (Figure and caption from Guterstam et al., 2015.)

Perhaps what is usually a multisensory interface for body/object interactions in the brain is more malleable than we thought and can expand during meditation? The functional distinction between what is a body and what is a graspable object, between self and nonself, would then be gradually attenuated in favor of a less differentiated, but much more integrated, holistic model of reality. Here is what a Zen meditator says, criticizing our questionnaire in an interesting way:

> 117 [. . .] I must add that the description in terms of "pure awareness" or "awareness of awareness" only partially reflects my experience. I would better describe

> it as a more immediate than usual contact with the world (i.e., less mediated in terms of conceptual elaboration), with less "time delay," more "inclusive" (a feeling of expansion of peripersonal space to the point of blurring the distinction between "self"/"not self," but more as a loss of importance of this distinction), and above all with a physical-mental sensation of "presence" alive and vibrant more than usual.

In my semantic analysis of ways in which the experience of pure awareness has been described over the centuries, connectedness was not a very prominent aspect.[15] But from a philosophical perspective, certain phenomenological clusters that appear in a qualitative analysis of contemporary real-life reports—like "connectedness" or "nondual being" (chapter 26)—relate with striking directness to centuries of Western thinking about consciousness. Over the centuries, two semantic elements running like red threads through the Western philosophy of consciousness have been the idea of metarepresentation and the theoretical motif of unity and global integration (for a computational model that unifies both of these aspects, see figure 34.7 in chapter 34). Today, we still have a variety of "higher-order" theories involving some kind of metarepresentation, while unity and integration reappear in some current contenders on the market for theories of consciousness, such as the form of a single "global workspace"[16] that makes conscious contents globally available to the organism, or the idea that a measure of "integrated information" can be used to distinguish between conscious and nonconscious systems.[17] A number of the qualitative clusters presented in this book are directly relevant to mathematically modeling consciousness, including density and soundness (chapters 6 and 7), timelessness and timeless change (chapter 22), and the notion of a space lacking internal structure, boundaries, and a center (chapter 23). The conscious experience of connectedness is another example.

Perhaps most interestingly, from a statistical perspective, the phenomenology of connectedness was not on our radar at all. It simply did not surface in our pilot studies, and accordingly, there was no item in our questionnaire that directly targeted it. This is an example of how qualitative analysis of open-ended reports can lead to discoveries that statistical analysis of forced-choice protocols would not reveal. Nothing in our pilot studies suggested that "connectedness" would turn out to be such a frequent phenomenological feature in reports describing the MPE experience, but the twenty-two reports selected in this chapter clearly show that there is something here—something that should certainly become a target of future research.

12 The Most Natural State

Truth, like an awakening to real life. [#2908]

"Naturalness," "ordinariness," and "simplicity" are common phenomenological themes, and all of them are clearly reflected in many of our experiential reports, as well as in centuries of canonical contemplative literature. Many practitioners say that pure awareness is entirely nonspectacular, and some even point out that there is a sense in which it is not even an "experience" at all (for more, see chapter 31). But there is something more to the "natural state," as it has been called in Tibetan Buddhism for many centuries. There is a deep yet subtle quality of *profundity* that can accompany the simplicity to which we now turn, as if it were even more natural and more real than what we take to be our normal state of lived experience—but in an entirely unspectacular way.

49 [. . .] I sometimes feel everything is more natural than usual. That means that I feel more familiar with the world and the environment around us.

961 The state is characterized by a feeling of peace, of weightlessness. Spectacular descriptions would be inappropriate. [. . .]

1560 In meditation I am often in a state of silent balance, awake and completely in tune with myself. It feels: right, natural, normal, good, complete . . . I am lacking nothing there/here, in fact I would like to be there (in this state) all the time. [. . .]

1784 [. . .] I feel this experience as a very natural state, while I feel the other states of consciousness as unnatural, as long as the experience of this pure being lasts. You could put it like this: There is no wakefulness, there is no sleep, there is only being.

1813 [. . .] It suddenly became sunshine within, bright and warm sunlight. I had to open my eyes since it was a cloudy day that day. It was a sense of ease, lightness, expansion, and that it was the most natural way of being that I have ever experienced. I could sit with my eyes open and look around just amazed by the feeling of warm softness and silence that everything was held in. I felt that I would never ever lose touch with it again and then I did.

2291 [. . .] It wasn't a revelatory experience, kind of seemed natural. While I was doing it, I was a little bit afraid that I will stay like that for the rest of my life, but I haven't. [. . .]

3464 [. . .] At the same time, the state appears as the most self-evident thing in the world, like a kind of basic state of mind. And, crazily enough, like the only real thing in the world. [. . .]

From very different angles, many of our reports seem to show that what really makes the phenomenal character of minimal phenomenal experience (MPE) so hard to express in words is the *combination* of profundity and simplicity. This is also one of the factors that make it interesting for philosophers. On the other hand, if the combination of profundity and simplicity truly is fundamental, there may be a real phenomenological possibility—perhaps even an epistemological risk—of some subjects only ever seeing the simplicity of MPE, while never recognizing the profundity.

185 Beyond profound, yet utterly ordinary. [. . .] Extra-ordinarily ordinary, and total game changer.

1647 The experience itself was simultaneous extremely profound while remaining incredibly mundane. I was flooded with a powerfully vivid memory of playing in the forest as a young child . . . it felt no more special than the ordinary existence of every child pre trauma.

1739 [. . .] It feels very free and relaxed, but at the same time simple and unspectacular.

2417 [. . .] I had the impression that everything was starkly, profoundly ordinary, just as it was.

The C-fallacy

> The natural state is one which has five distinctions: (1) a primordially pure nature, (2) an aspect of spontaneous presence, (3) an indeterminate essence, (4) an all-embracing quality, (5) and a great [capacity to] differentiate between realization and non-realization [. . .]. It cannot be described in words. Recognizing it cannot be taught.
>
> —Rgyal-ba-g'yung-drung (1242–1290), *Pith Instructions for A Khrid rDzogs Chen*

In Tibetan Buddhism, pure awareness has long been beautifully described as "the natural state" (*rang bzhin gnas rigs*). The existence of a label like this may seem unremarkable, or it may be straightforwardly useful. But it also presents us with one example of a methodological difficulty, a major obstacle that every scientific approach has to face. In chapter 2, I called it the problem of "theory contamination": Theoretical assumptions and belief systems may have strongly colored the reports that we received, and the conceptual instruments available to a practitioner inevitably shape the way she communicates her own experience. We find many examples of such contamination in this book: Our meditators spontaneously and more or less innocently use ancient concepts charged with implicit bias, like "suchness" (chapter 9), "emptiness" (chapter 17), "luminosity" (chapter 18), "witnessing" (chapter 19), "nonduality" (chapters 26 and 27), and "nonmeditation" (chapter 32). We will come back to the idea of theory contamination in chapter 17, as well as in a number of other places on our journey. For now, let us begin by looking at two excellent phenomenological reports that are clearly colored by the conceptual framework of a specific contemplative tradition, faithfully reproducing its terminology:

83 I regularly practice Dzogchen and can find the natural state relatively easily. It is a state of all things falling away. Like the ultimate escape from discursive thought. It is clear, awake, fresh, and the base of all phenomena.

84 Firstly, when recognizing one's natural state, the recognizer dissolves into its ground. This being the source of all experiences. Mind made meditation is useful to calm the mind in order that it can be observed. Then we relax by letting concentration dissolve into awareness inside time. Discovering the open empty nature of all perceptions, they naturally self-liberate. Looking back at who knows this, one realizes the nondual wisdom. Recognizing this effortless state again and again, we gradually become reawakened. Our natural wisdoms arise automatically as objectless love, joy, and compassion. There is the great peace, boundless space and clarity. That cannot be known by the mind. This view is beyond concepts. [. . .] I am just a Vajra Parrot who dances on the Books.

I do not doubt the sincerity or the phenomenological relevance of reports like these in any way; as a matter of fact, they are very precise and summarize many aspects that naturally emerge from our reports. But I want to offer a second logical tool that you may find helpful for thinking about the problem of theory contamination more broadly. Recall that one of the goals of this book is to contribute to the formulation of a more comprehensive model of consciousness, by beginning to lay some of the very first foundations for a *minimal model explanation* of phenomenal experience. What is the simplest kind of conscious experience? Can we isolate phenomenality per se? Is there something like a prototypical core, perhaps even something resembling the singular experiential essence of our target phenomenon, and if so, how would it reappear on a conceptual level? When asking these questions, it is important to avoid a specific kind of logical error, which I will call the "C-fallacy," as a parallel to the "E-fallacy" explored in chapter 7.

The E-fallacy arises whenever someone falsely concludes that a consciously experienced feeling of knowing is a reliable indicator of actually possessing knowledge. As we saw in chapter 7, the phenomenal experience of knowing does not entail that one actually possesses knowledge; a strong and robust experience of knowing, or even of absolute certainty, can occur during hallucinations, in dreams, and in psychiatric diseases. This simple point applies not only to "knowing that one knows," but equally to moral knowledge and the phenomenology of insight into ethical values. As Sonam Kachru has pointed out, "There is no ethical given to be realized in meditation, any more than there are epistemological givens to be revealed in it."[1] If we take a rational, evidence-based perspective, then "epistemicity" as such—the nonconceptual phenomenal quality of "insight" and "comprehension" or the complex feeling of being a knowing self—is only a phenomenal quality, just as redness, greenness, and sweetness are. Therefore, knowledge claims made in public need independent justification, and the mere claim that something *felt* like knowledge does not count as justification.

The same is true of "phenomenality" or "conscious*ness*." The C-fallacy arises whenever someone falsely concludes that just because something *feels* like "consciousness as such," we have actually found or even ultimately understood the very essence of consciousness itself. Unfortunately, therefore, we cannot just go and look for reports in which people *claim* to have experienced "the simplest and most natural form of experience," the "true essence" of consciousness, or "awareness as such," and assume that they will give us our answer. Again, there is no doubt about the sincerity and truthfulness of such reports. They deserve our deepest respect because they do their best to convey something that has been characterized as ineffable for millennia. They make a genuine and substantial contribution to research. But if we adopt an intellectually honest and methodologically sound perspective, then "phenomenality" as such—the subjective sense of appearance *itself*, the bare and nonconceptual character of "awareness" or "pure consciousness"—is only a phenomenal quality too, just as redness, greenness, sweetness, and epistemicity are.

Today, the way to go is not to jump to strong metaphysical conclusions in a naively realistic manner, as if the structure of reality itself could be directly read from the structure of certain kinds of conscious experience. Our new model of consciousness must not be constrained only by the phenomenology because other forms of evidence count as well. The elephant approach must involve a multiplicity of perspectives, because it is attempting to satisfy a whole range of constraints entailed by multiple levels of analysis. Nothing is given here—no direct form of inner perception. The directness, transparency, and experiential immediacy all belong on the level of appearances, as later reported verbally.

As I put it in chapter 7, unfortunately we cannot bootstrap a theory of consciousness out of the "knowingness" that accompanies the pure-awareness experience itself. This is why we need a new bottom-up approach—one that begins by taking the phenomenology more seriously than ever and then proceeds to *empirically* test whether it can figure in a minimal model explanation of conscious experience. A minimal model of consciousness would include only the core causal factors giving rise to the target phenomenon: only those causal factors that make a difference to the actual occurrence and the essential phenomenal character of MPE itself. If successful, this would lead to an idealized model of the universal and repeatable features of *all* conscious experience, isolating its explanatorily relevant and structurally stable properties. This, in turn, could lay the foundations for a first standard model of consciousness. The experience of pure awareness in meditation is, I believe, the perfect scientific entry point, the prime candidate for the simplest form in which our target phenomenon arises. I certainly may be wrong about this. But—given new phenomenological data and many centuries of contemplative phenomenology produced by those who came before us—I would insist that it is (no pun intended) the *natural* candidate.

13 Coming Home

The experience is like coming home . . . full, complete, nothing to reach for. Like a big smile. [#570]

There is a sense of security in it, a sweet, silent calm [. . .] a fullness, a quality of being-satiated, a permission-to-be. [#2511]

Many phenomenological reports liken entering the pure-awareness experience to the experience of arriving back home or remembering something that had always been there. The aspect that we now turn our attention to can have an affective dimension as well as an epistemic one. Often, meditators speak about feelings of comfort, security, or being sheltered. If one takes verbal descriptions referring to this cluster of phenomenal qualities seriously, abiding in pure awareness seems to be the ultimate way to "take refuge" and the real essence of what it means to do so. In addition, some practitioners point out that "coming home" is much more than a mere feeling; rather, it is a global state of being (see #3218 and #3464 in the following list). On the other hand, entering pure awareness can also carry a specific quality of insight, leading to a distinct phenomenology of recognition or remembrance. In certain contemplative traditions, this experiential quality has been conceptually interpreted as "recognizing one's true nature."

But as always, let's spend some time with the phenomenology first. Here are nineteen examples in which the quality of "coming home" is a dominant feature:

35 Pure awareness is the realization of having finally found home after an eternal search. The pathological searching, the agony of control, comes to an abrupt end, and for the first time you realize what it means to be alive. It is the realization that nothing and no one can hurt you; the true self is untouchable, timeless, and the essence from which all life springs and to which it returns.

115 [. . .] I could hardly describe the experience of Pure Consciousness itself. I only know that every time, it feels like coming home or remembering something that was always there and only forgotten. The experience is always connected with happiness, peace, wholeness. [. . .]

577 [. . .] It's so clear it's our natural state as it feels so familiar. It's the most important thing to me and I love that I can share and teach the practice to others. To remind them and bring them home. My first experience of it was like a homecoming.

808 My experience(s) of Pure Consciousness is/are like coming home to my true home, my origin, and from there recharging, regenerating and letting the weight of internal conflicts escape, dissipate into the vast truth of my higher, interconnected being. It is a state of peace in which I can restore my original state of being and authenticity [. . .].

949 During my meditation course I felt all of a sudden very peaceful, not judging, feeling whole, homecoming. I was just there, everything was all right, this was the truth.

1081 [. . .] Then in the silence, stillness, spaciousness . . . this experience of changeless, timeless, boundless awareness-emptiness, total openness, with nonlocality, no "me"; deeply peaceful, moved to tears, as-it-is, the sense of being "home."

1625 [. . .] I was completely calm and peacefully relaxed. There were no questions or doubts in me, everything was completely clear and I knew exactly where I was: (in the spiritual sense) at home. [. . .]

1828 After about twenty years of meditation practice, I can establish the state of pure awareness "at the push of a button" within a few minutes. In the end, this state is nothing special, it is rather like dwelling in the state of precognitive perception, in which the mind generally doesn't yet elaborate any concrete contents, unless I specifically allow this. Since in this way the totality of the current experience is absorbed without any inner psychological or other mental filters, and thus no fragmentation of the experience takes place, a subliminal feeling of "finally having arrived home" automatically arises. [. . .]

2071 The experience itself only lasted about a minute or so (I checked the time afterward), but it felt much longer. For the first time in my life I was at complete peace with myself and the world around me. And the best thing is that it felt so right, almost as if I had finally arrived home after being lost for a very long time. I felt tears running down my face. I could no longer hear my breathing. I couldn't even feel my own heartbeat anymore.

2764 [. . .] Nothing could be doubted, everything was truth and love and a feeling of arrival.

3005 [. . .] Bright and clear, kind of sunny. Silent pure joy. A homecoming. It was the most natural thing in the world. No right or wrong. Just being. Safe and connected.

3218 [. . .] That morning there was a complete absence of fear, a state of complete freedom and all possibilities. The state—I can only say state and not feeling, because it was much more than a feeling and really a state of being—of "Mother being at home," and not only that, but also the child being completely in tune with Mother's desires and Mother being pleased and content with the child, resulting in mutual nourishment on all levels of existence. The next few days I had more of these "Mother is at home" experiences. [. . .]

3305 [. . .] I experience a feeling of unity, of great happiness, of deep peace, of having arrived home. [. . .]

3353 [. . .] The first time that I experienced an overwhelming incidence of peaceful happiness and oneness or wholeness, when I was 23 years of age, was something I always remembered, but never dared to talk about. But for a long time, I longed for this kind of being in the world. Since that time, I always felt a longing. The first time that, years and years later, I was sitting on a cushion, meditating, it felt like coming home . . . until this very day.

3464 [. . .] The state contains an inner sweetness that is very subtle and that I experience as a kind of "primordial happiness." This is combined with a feeling of "coming home." Although I hesitate to speak of a feeling. The state is by no means euphoric, but can lead to euphoria after meditation. [. . .]

3493 While meditating in a Buddhist monastery I discovered an inner "place" where I can be as I am without having to do anything or prove anything. I call this place "my inner fortress." Since this experience it has been possible for me to return there again and again. I can rest in myself, no matter what is going on around me. Since then I really feel free!

3497 [. . .] A feeling of safety and boundlessness, like belonging and being accepted and at the same time full of life and in silence, like being eternity and like being joy. Then at some point the perceptible thought and wish that it may remain so. [. . .]

3568 My experience rests on a "space of emptiness" . . . Here there was nothing but pure emptiness, silence, connectedness, and the feeling of having arrived in the present. There was no must anymore, no want anymore . . . It was all good as it is.

> 3615 [. . .] For me it is the overwhelming feeling of homecoming, arriving at a place that is (and has always been) there and here and in me and in everything. [. . .]

Let us now turn to the specifically epistemic dimension of "coming home." This dimension is characterized by a phenomenology of "remembering" or "recognizing." Often, what is now remembered is described as fundamental and as something that "had always been there but was forgotten":

> 197 [. . .] It felt like getting in contact with something / touching an inner quality that had always been there.
>
> 1311 [. . .] Like entering into a completely different state of matter, but one that at the same time is so fundamental that it also feels completely normal—because it was always and in all experiences invisibly present as their foundation anyway. It is this quality of recognition, this "Ah yes—of course!" that is, however, completely unspectacular—because it had never been any different in any case. [. . .]
>
> 1426 [. . .] After several days of intensive meditation the sudden realization opened up: Consciousness is empty—and not as intellectual knowledge but as direct experience. The most amazing thing about this experience or state (which lasted for over 2 hours) was how normal it felt, in the sense of "Yes, of course, that's how it is!" It was also a feeling of recognition.
>
> 1482 [. . .] a kind of natural remembrance of a state of being that is deeply personal and at the same time universal in the sense that boundaries have dissolved into an experience of deep unity, characterized by a sense of fully being without any reason whatsoever, free from the dualistic thinking mind. [. . .]
>
> 2293 It felt like a homecoming to a natural state forgotten but very familiar. Not in a spiritual way but just a re-cognition.

Recognizing the Unfabricated Baseline

> The first characteristic is that it has no boundaries, no fence round it, no edges; it's absolutely unlimited in all directions. The second characteristic is that it is absolutely clear, clean, empty of contamination. It is utterly simple, totally transparent, empty of everything but itself, empty even of itself, clearer than glass, cloudless, an infinite sky. The third characteristic is that it is also full of the world. Because it's empty, it's full—full of the scene, whatever the scene is, absolutely united with it. The fourth characteristic is that it is awake, it's aware, it's conscious. And the fifth characteristic is that it is right where you are.
>
> —Douglas Harding, *Face to No-Face: Rediscovering Our Original Nature*

From a scientific perspective, there are three obvious conclusions that we can draw. First, whatever minimal phenomenal experience (MPE) turns out to be, it seems to be a state in which global uncertainty has been greatly reduced. "Home" can be interpreted as the place in which a given system expects the lowest level of prediction error—and it seems as if this place can be not only something in the outer world but also a specific level of information processing and self-representation within the system's own conscious model of reality. Perhaps MPE is the computational level on which the egoic self-model "bottoms out," or the region in phenomenal state space where *you* really are? Whatever it is, in an experiential as well as in a computational sense, "home" must be that from which everything originated, the primordial source of invariance, and that which again and again proves to be the place of greatest safety and security.

Second, MPE clearly looks like a baseline state. Perhaps it is related to an affective baseline in early childhood, to homeostatic stability, or to a computational reference point out of which more complex states of conscious experience can be constructed. As you may recall from chapter 1, one report pointed out that "the more I can quiet my physical and mental stress and anxiety the more I can uncover of this baseline reality of experience," and in chapter 5, we noted that one of the participants in our study coined the term "Basal Clarity" (#3058; see also chapter 21).

Third, taking the phenomenology of recognition at face value, we can say that MPE looks like a state that is not newly fabricated or constructed by the meditator, but rather is something that has been there all along and was only rarely attended to. Perhaps it is like the pattern of sensations of contact, weight, and pressure on the soles of our feet or behind our eyelids, or the humming sound of the refrigerator. It could be something so reliable and invariant that we never attend to it, so it automatically drops out of our conscious model of reality. It may be an internal bodily sensation without a receptor system of its own—the stable but somewhat diffuse background buzz of tonic alertness (chapter 4), the raw feeling that constantly signals epistemic openness to the organism. Perhaps the resulting "epistemic hum" of wakeful, nonconceptual meta-awareness is so reliable and invariant that it drops away? Only if we step on a sharp object and our foot unexpectedly gets pierced from below, if grass pollen has landed on our eyeballs, or if the refrigerator suddenly falls silent do we notice a change. The same may be true for sudden prediction errors related to the phenomenology of wakefulness and epistemic openness: Only if we are in danger of suddenly fainting do we notice that there is something we might lose; only when some stimulant drug—even just two cups of green tea too many—causes excitement, nervousness, insomnia, or a rambling flow of thought and speech, do we notice that there was a baseline of tonic alertness from which one can deviate.

Unfortunately, the classic Buddhist idea of "recognizing one's true nature" is more difficult to make sense of (we'll see more on this in chapters 29 and 30). The concept is clearly pointing toward an experience that people have. The phenomenology, the experience itself, clearly exists. But as a metaphysical interpretation of the specific phenomenological element highlighted by the two short epigraphs and twenty-four reports I have presented in this chapter, it contradicts the antiessentialist and antisubstantialist assumptions of standard Buddhist metaphysics, because notions like "true nature" and "the natural state" (see chapter 12) carry a normative meaning. They indirectly presuppose that there is an essence of mental appearances—something that is "true" about them, as opposed to something else that must be "false." Something is real about my mental state, and something else isn't. There is a natural state and an unnatural state, but no cogent argument is given for what the difference is. It seems that this type of implicit contradiction exists in some classical and many of today's popular texts—but then again, superficial observations like these cannot even begin to do justice to more than twenty centuries of Buddhist philosophy.

In particular, in trying to arrive at a modern understanding of the phenomenology of being in "the natural state" and of "coming home," we might be in danger of committing the E-fallacy and the C-fallacy at the same time. It is indisputable that there can be a quality of insight, a marked phenomenal signature of knowing (chapter 7) that characterizes the conscious experience of "remembering the natural state" or of "recognizing one's true nature." But from this fact, it does not follow that any real gain in knowledge has actually been involved. Of course, such an increase in knowledge seems overwhelmingly plausible—but then again, on a scientific level, we are also coming to know more and more about false memories and déjà vu experiences. Maybe there are introspective déjà vu experiences? What is there that rules out the occurrence of false memories in meditation? What makes us so sure that it always is the *same* state to which we are returning? A recent study involving eighty-four Shamata practitioners, eighty Thai Forest meditators, and eighty-eight Stillness Meditation participants concluded that, yes, their experiences are "contentless," in that they have minimal awareness of thoughts and perceptions, but no, they are certainly not *identical* to each other in being devoid of all content, as much of the previous academic literature and many metaphysical belief systems have assumed.[1] The claim that one's "true" nature has been recognized or that the meditator has made contact with and settled into the "essence" of consciousness is a strong metaphysical non sequitur. From a verbal report about the feeling of knowing, it doesn't follow that you, the person who provided it, actually possessed knowledge; and just because something *feels* like the essence of consciousness itself, the metaphysical conclusion that we have actually *found* consciousness itself is not justified. This is what I mean by the risk of committing the E-fallacy and the C-fallacy at the same time.

I think that the value of the classical motif of "recognizing your own true nature" consists in something else. It is actually not so much a theoretical statement—it is more like a practical instruction for advanced meditators, and probably one that has been highly effective over the centuries. All it does is offer one possible way for the practitioner to look at their own experience of pure awareness and see what happens. It is a new window that provides a new perspective. Perhaps in some cases, it is even an attempt to install a self-fulfilling prophecy of sameness across time in the mind of the faithful, a prediction targeting future states of nonegoic self-awareness that later creates the causal power to sustain such states more and more often. It may be a parallel case to what, at the end of chapter 10, I termed the installation of a "grace hyperprior"—if you will, an intended form of theory contamination that makes the occurrence of nondual states more likely. We could call it the "true-nature prior": a mostly unconscious belief in the brain that says that (1) you have a preexisting, perhaps innate, true nature; and (2) this nature can be recognized. This belief is a prediction of sameness, the anticipation of "an innate and unconditioned mode of being and awareness, a concealed depth dimension of experience that is implicit in all conscious activity, all phenomena, and is therefore fully present and available to direct perception once the miasma of cognitive and affective obscurations is dispelled."[2]

I will say more about this in chapters 26 and 27. For now, we can simply conclude that on this reading—treating it as something aimed at successful practice (not at metaphysics)—the classical motif of "recognizing your own true nature" would really describe the possibility of a shift in the unit of identification: the possibility of pure awareness becoming that *from* which the world is now experienced, as opposed to that *at* which the meditating self looks.

14 There Is Nothing Left to Do

Like a contented stone [#808]

Everything comes, everything goes. I or rather It rests. There is nothing I have to do. It is easy. [#2511]

Imagine that you discovered that what really creates satisfaction in life has nothing to do with achieving goals, or even with acting in any specific way. Imagine that you discovered this not in some intellectual or philosophical way, but as an unexpected and entirely nonconceptual form of inner knowledge. Interestingly, many meditators report that during an episode of pure awareness, their goal hierarchy dissolves: All is well; there is nothing left to do. There is a quality of acceptance and contentment, and sometimes it begins to extend into the practitioner's life as a whole. This quality of acceptance and contentment overlaps significantly with the phenomenology of "existential ease" introduced in chapter 1. The need for "narrative self-deception" (chapter 17) may be attenuated, gradually liberating the meditator from her constant urge to stabilize the long-term self-model by constructing a thematically coherent life story in a never-ending search for meaning. Other experiential features typically coemerging with this quality are relaxation and peace, love and gratitude, soundness, ego dissolution, and dignity. Here are six examples:

1661 [. . .] There was simply nothing left to be done. All ambition and grasping seemed absurd and counterproductive. There was deep abiding acceptance of the world as it was, and of myself in the world.

1942 [. . .] Suddenly I realize that there's nothing to do. I don't have to do anything. When I meditate I don't have to do anything. I don't have to seek anything. I notice my thoughts, which could very well go on a search and find something they could do, e.g., think of something, or plan. But the "knowing" "says": There is nothing to do. I feel deep inside me a calmness, a "not being driven," an absolute contentment. Of course I do everything there is to do in everyday life, but this feeling on this other level that "there is nothing to do," I also feel it over and over in everyday life.

2951 [. . .] There was a "self," but it had no agenda.

3000 [. . .] Peace, complete relaxation, realization that everything is good, nothing has to be done, nothing has to be wanted, that this is the right thing, gratitude, love for all beings, dissolution, dissociation from the separated I.

3334 [. . .] There was no need to be happy or unhappy because I noticed that everything was and is exactly what it is. It is just right. It just is what it is. [. . .]

2426 [. . .] I felt and thought that for the first time I could perceive reality correctly. For the short moment of the experience (a few seconds) everything was clear and I felt that there was nothing more to do. I thought of loneliness and the concept itself seemed illogical to me, just like wanting to do anything at all. "There's nothing left to do, everything is there." During the experience I was serene, calm, hardly emotional, liberated, everything seemed very clear and a little bit cool. Immediately afterward, I couldn't help crying. I thought: All my life I've been under psychological pressure to change something. Human beings are unworthy of all other mental states thanks to the suffering they cause.

Ataraxia

> The solution to the problem of life is apparent in the disappearance of this problem. (Is this not the reason why people to whom the meaning of life became clear after long doubting could not then say what this meaning consisted in?)
>
> —Ludwig Wittgenstein (1889–1951), *Tractatus Logico-Philosophicus*, 6.521

> But the person who takes no position as to what is by nature good or bad neither avoids nor pursues intensely. As a result, he achieves ataraxia.
>
> —Sextus Empiricus (ca. 160–210), *The Skeptic Way*, book I, chapter 12

Many deep and mostly unexplored relationships connect contemplative practice and ancient Western philosophy—too many to even attempt an overview. Let us briefly look at a single example: the concept of *ataraxia* (ἀταραξία), which is often interpreted as referring to a lucid and enduring state of "imperturbability," "equanimity," or "tranquility." In ancient Greek philosophy, the term was introduced by Democritus and Pyrrho and then further developed into different understandings and practices by Epicurus and by thinkers in the Stoic tradition. Cicero and Seneca used the Latin translation *tranquilitas animi* ("tranquility of the soul").

Pyrrho (ca. 360–270 BC) was the first Greek skeptic philosopher and the founder of Pyrrhonism. The ancient skeptics were radical thinkers who also called themselves "those who suspend" (*ephektikoi*), signaling that their investigations involved extreme suspension of judgment.[1] To understand what the tranquil state of *ataraxia* is, we need to know that Pyrrhonian skepticism was initially a solely intellectual enterprise, but it unexpectedly transformed itself into a *practice*. That practice—the suspension of judgment about what is good and what is bad, about what is real and what is mere appearance—is intimately related to the notion of "choiceless awareness," which played an important role in the teachings of contemporary Indian philosophers like Jiddu Krishnamurti, and also in many classical descriptions of meditative techniques and mindfulness practices. The goal of Pyrrhonian practice was an undisturbed calmness of soul. As Sextus Empiricus said:

> The Skeptics hoped to acquire unperturbedness by deciding the anomaly in the things which appear and which are thought, but being unable to do this, they suspended judgment. And while they were suspending judgment, unperturbedness closely followed them by chance, as it were, as a shadow [closely follows] a body.[2]

The discovery of *ataraxia* was an unexpected one: Any holding of beliefs about how things really are prevents one from attaining peace of mind, whereas suspending all judgment leads to a new state. We might call this state "living a life without belief," or even use the terms of this book and refer to the philosophical ideal behind it as "conscious experience without theory contamination." It is interesting to note how—on the subpersonal level of modern computational phenomenology—this ideal returns as the notion of "flattening the brain's predictive hierarchy," which we first encountered in chapter 2.[3] Back at the level of lived experience, it definitely has to do with the quality of suchness that we have already investigated, and with cultivating epistemic openness. What's more, engaging in purely speculative forms of philosophical metaphysics now begins to look like the exact opposite of spiritual practice, and simply taking things to be real starts to seem like a form of unawareness (we touch on all four of these points—*seeing what is*, epistemic openness, getting lost in narrative self-deception, and transparency—from a phenomenological perspective in chapters 9, 4, 17, and 28, respectively).

Why is this so? Affirming something that you do not really know (e.g., by claiming to know what is real or unreal) is a form of intellectual dishonesty—a way of not being true to yourself. On the level of conscious experience itself, it means being caught in an automatic mechanism of world-construction, mindlessly acting out your brain's unconscious assumptions about the structure of reality.

Pyrrho's disciple Timon is known for a statement that, at the very end, points to a fourfold indeterminacy that directly mirrors a core insight of Nagarjuna, the most important Buddhist philosopher: his *catuṣkoṭi* (the famous "four-cornered" negation). Aristocles describes Timon's take on Pyrrho as follows:

> Pyrrho's pupil, Timon, says that anyone who is going to lead a happy life must take account of the following three things: first, what objects are like by nature; secondly, what our attitude to them should be; finally, what will result for those who take this attitude. Now he says Pyrrho shows that objects are equally indifferent and unfathomable and undeterminable, hence neither our senses nor our judgments are true or false; so for that reason we should not trust in them but should be without judgment and without inclination and unmoved, saying about each thing that it no more is than is not or both is and is not or neither is nor is not. And Timon says that for those who take this attitude the result will be first nonassertion, then tranquillity.[4]

Socrates is famous for having said, "The unexamined life is not worth living." Many meditators will certainly agree, but they will also go further, pointing out that, more specifically, an unexamined *inner* life is not worth living. As a matter of fact, the formal practice of classical Vipassanā meditation (which was the single most frequently used technique among our participants) and the informal establishment of mindfulness during the day (*satipaṭṭhāna* or *smṛtyupasthāna*, in Buddhist terms) can be seen as exactly this: the serious practice of living an examined inner life.

Humankind has developed many such meditation techniques—"policies" for precisely those mental actions that lead to the effortless mental *inaction* of minimal phenomenal experience (MPE), as the modern computational phenomenology of meditation might frame them. Some of the blind may touch the elephant with their right hand, some with their left. Some may use both hands at the same time, and some may alternate. Some may run only the tips of their fingers or the tip of only a single finger over its skin; others may gently let the palms of their hands glide over it, or even put their cheek on its flank. Some of the blind may discover by chance that there is something that can be touched with their feet and begin to explore one of the elephant's four lower legs. Figure 14.1 shows the most popular meditation techniques, as well as the ways in which practitioners combine techniques in their quest for the examined life.

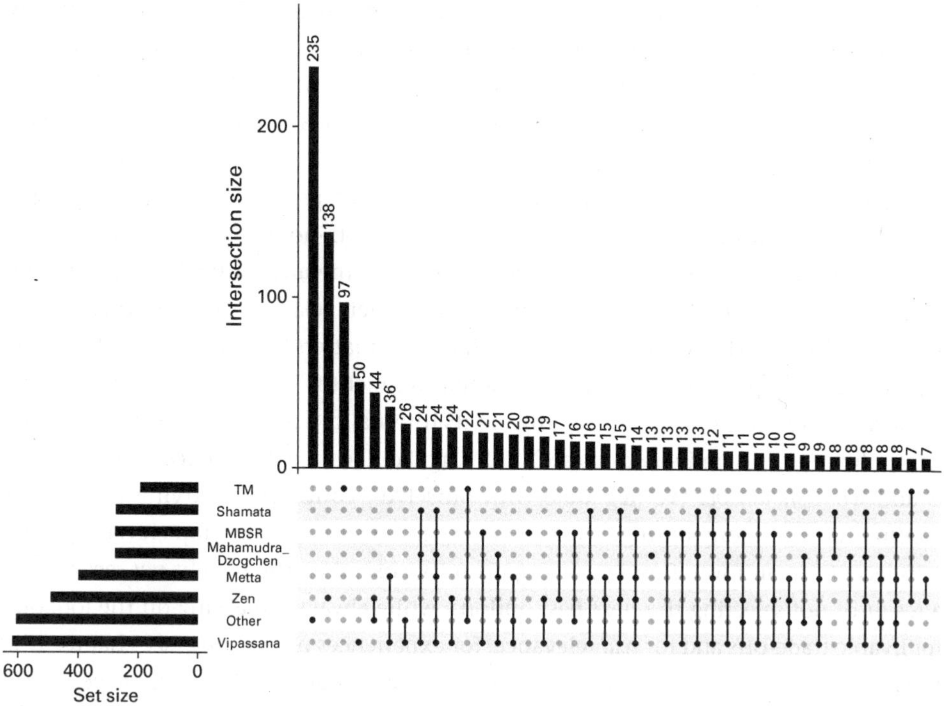

Figure 14.1
This plot shows the combinations of meditation techniques reported as being used by our participants. The top graph is a histogram, with each bar representing the frequency of a particular meditation technique (or constellation of techniques) that are identified right below it. Black dots connected by black lines string together techniques that belong to a given constellation. The bar chart on the left shows the overall frequencies of each technique. MBSR = Mindfulness-Based Stress Reduction; TM = Transcendental Meditation.

Genuine tranquillity arises in an existential and entirely nonconceptual way, resulting from a special form of inner investigation. In the meditation technique most frequently used by our respondents, Vipassanā, tranquillity is enabled precisely by the suspension of all judgment and by an effortless cultivation of choiceless awareness, but it also occurs in the context of full-absorption episodes. Of course, such episodes can be generated by many others of the techniques shown in figure 14.1.[5] *Ataraxia* in this sense is related to the quality that in chapter 1 was termed "existential ease." It is not the result of taking on some merely intellectual attitude; the kind of examination that is required is of a much more fundamental kind. There is an existential immediacy to *ataraxia*. It seems as though at least some meditators may love wisdom in the sense represented by the life and death of Socrates, but many of them actually follow through on a deeper, Pyrrhonian level (see also chapter 16). *Ataraxia* is not something conceptual; indeed, it is not a cognitive or intellectual state at all. It results from the cultivation of a specific form of inner attention that leads to an awareness of awareness itself.

Vipassanā meditation is, in more modern terms, the process of being choicelessly aware of the continuous dynamics unfolding in the phenomenal self-model. One of the deepest parts of the human self-model is the goal hierarchy that it represents: the hierarchy of goals to be achieved in this life and their relation to each other, which together constitute the values we live by. What *really* matters? Our goal hierarchy is also a part of the life process itself—something that we identify with and constantly try to make as consistent as possible, maximizing its internal coherence. The decoherence of contradictory goals and values creates suffering. So does the divergence between *what is* and *what should be*. We negotiate with ourselves, continuously trying to dissolve conflicts between long-term goals and short-term rewards. We develop clever forms of self-deception, and we often ignore opportunity costs (i.e., the benefit *not* received as a result of not selecting the next best option), while always remaining on the lookout for livable trade-offs and for *real* relevance, for experiences of meaning (see chapter 17). Vipassanā meditation allows us to observe this permanent source of inner perturbation in an entirely nonconceptual way, gently, precisely, and eventually even without an observer. Meditation prevents us from acting out our goal hierarchy like mindless biorobots.

Of course, we are enculturated beings, and many of our values and goals have been projected onto us, not only through biological evolution but also through our specific cognitive niches, as well as through the more recent social contexts in which we have grown up. But if the only way that we can critically reflect on them is intellectually, the causal force of these values and goals will remain almost unchanged—we will achieve no fine-grained understanding of their actual modus operandi. Achieving such an understanding is what it means to "live an examined *inner* life," and of course, it is directly related to the practice of nonassertion and the ancient Greek idea of *ataraxia*. It is interesting to note that the continuous dynamics of the goal hierarchy are at the core of our conscious self-model; they are a deep and relentless source of fragmentation, and phenomenologically, they are what most of us *identify* with. Any more radical ending to this process of continuously reorganizing our goal-hierarchy, therefore, would imply a form of ego dissolution (chapter 25). If there is no ego, there is no agenda. If there is no agenda, there is nothing left do.

I have always thought that in Western antiquity, many indicators point to an intrinsic relationship between philosophy, spiritual practice, and certain altered states of consciousness—new *phenomenal* models of reality resulting from an accompanying existential transformation in the thinker–practitioner.[6] *Ataraxia* is just one example. The French philosopher Pierre Hadot (1922–2010) said that we can understand early Western philosophy much better if we view it as employing *spiritual exercises*—that is, sets of "practices [. . .] intended to effect a modification and a transformation in the subjects who practice them."[7] For example, Hadot pointed out that the Socratic dialogue was really "a kind of communal spiritual exercise."[8] He interestingly claimed that in many cases, "[t]he philosophy teacher's discourse could also assume the form of a spiritual exercise, if the discourse were presented in such a way that the disciple, as auditor, reader, or interlocutor, could make spiritual progress and transform himself within."[9]

In his detailed historical investigation of how the ancients *really* conceived of philosophy, Hadot found that the earliest Greek thinkers, as well as the philosophers of the Hellenistic period (i.e., the period of Mediterranean history between the death of Alexander the Great in 323 BC and the emergence of the Roman Empire) and those of late antiquity, were concerned not just to develop convincing philosophical theories but also to practice philosophy as a way of life.[10] In part 2 of his well-known book *Philosophy as a Way of Life*, Hadot defended the claim that what Ignatius of Loyola later called an *exercitium spirituale* (to cite one example) was "nothing but a Christian version of a Greco-Roman tradition." He demonstrated that

> both the idea and the terminology of *exercitium spirituale* are attested in early Latin Christianity, well before Ignatius of Loyola, and they correspond to the Greek Christian term *askesis*. In turn, *askesis*—which must be understood not as asceticism, but as the practice of spiritual exercises—already existed within the philosophical tradition of antiquity.[11]

Let me therefore conclude this chapter by giving a second example without further comment—one that is directly related to the topics of ego dissolution and *ataraxia*, the contemplative phenomenology of existential ease (chapter 1), and choiceless meditative attention to the present moment (chapter 22). This example shows that the two themes of cultivating sustained "attention to oneself" and "being fully in the present moment" were both strongly expressed in early Western philosophy. As Pierre Hadot puts it, *prosoche* (attention to oneself) was the fundamental attitude of the philosopher as well as the monk.[12] To be very clear at this point, the Stoic *prosoche* and the Buddhist *sati* (mindfulness) are certainly not simply one and the same thing.[13] But if one reads the following longer passage from Hadot—in the middle of which he quotes the famous *Meditations* authored by the Roman emperor Marcus Aurelius (121–180)—as contextualized by the background of our own phenomenological data originating in present-day contemplative practice, then it is hard to overlook certain interesting commonalities. The Eastern and the Western approaches can be seen as two distinct epistemic practices that are nevertheless expressions of the same underlying attitude:

> Attention (*prosoche*) is the fundamental Stoic spiritual attitude. It is a continuous vigilance and presence of mind, self consciousness which never sleeps, and a constant tension of the spirit. [. . .] It is this vigilance of the spirit which lets us apply the fundamental rule to each of life's particular situations, and always to do what we do "appropriately." We could also define this attitude as "concentration on the present moment":
>
> > Everywhere and at all times, it is up to you to rejoice piously at what is occurring *at the present moment*, to conduct yourself with justice towards the people who are *present here and now*, and to apply rule of discernment to your *present* representations, so that nothing slips in that is not objective.
>
> Attention to the present moment is, in a sense, the key to spiritual exercises. It frees us from the passions, which are always caused by the past or the future—two areas which do not depend on us. By encouraging concentration on the minuscule present moment, which, in its exiguity, is always bearable and controllable, attention increases our vigilance. Finally, attention to the present moment allows us to accede to cosmic consciousness, by making us attentive to the infinite value of each instant, and causing us to accept each moment of existence from the viewpoint of the universal law of the *cosmos*.[14]

15 Joy, Awe, Bliss, and Gratitude

> Figuratively speaking, it is consenting to being. [#151]

> The state contains an inner sweetness which is very subtle and which I experience as a kind of "primordial happiness." [#3464]

Pure awareness as such is not an emotional state. On the other hand, it can trigger a whole spectrum of mostly positive affective states like joy, existential relief, gratitude, impersonal love, awe, and wonder, ranging in intensity from the very delicate and natural to the dramatic and overwhelming. In particular, minimal phenomenal experience (MPE) can sometimes coexist with a mostly subtle but distinctive form of bliss, an experience that has sometimes been described as an "invisible smile." In addition, this kind of bliss has a pristine and primordial phenomenal character, as described by one of our participants in the second quotation presented here. In pure awareness, we find not only the phenomenology of peace, existential ease, and silence described in chapters 1–3, but at times also different forms of what in German is called *stilles Entzücken* ("silent delight"). As explained in chapter 1, this experiential quality is intimately connected to a calm and entirely undramatic phenomenology of rapture. Personally, I found it strongly resonant with my own experience when one of our participants spoke of "sensationless awe" (#3524), and another of "a very fine and nonsensational sense of wonder" (#3624). In this chapter, we will look at a wide variety of such states.

Interestingly, many meditators explicitly related positive emotional states like joy, existential relief, gratitude, impersonal love, awe, and wonder to childhood experiences of timeless unity. Some of us had such experiences long before we started to practice meditation, and many people know them without ever having engaged in any form of contemplative practice at all. Let us begin by looking at three examples:

3305 [. . .] The experience [. . .] came over me spontaneously in my parents' garden. Perception was crystal clear, my surroundings shimmered, and I became aware of the true timelessness in the world. The body became one with everything. Great happiness flowed through me. [. . .]

2288 [. . .] I experienced in the garden a oneness with nature and everything that exists. I had the feeling that I had already experienced this as a child.

1647 [. . .] I was flooded with a powerfully vivid memory of playing in the forest as a young child . . . it felt no more special than the ordinary existence of every child pretrauma. [. . .]

In this phenomenological domain, there are elements of innocence, wonder, gentleness, and timelessness. In adults, a specific sense of *relief* also seems to arise when the egoic sense of control is dissolved and the phenomenal quality called "agency" is attenuated. If one looks closely, a whole spectrum of mostly positive affective qualities can coemerge with or be triggered by an episode of pure awareness. Let us take a quick look at this spectrum.

First, we find the inner experience of awe and wonder:

3146 [. . .] There was a quality of gentleness and wonder, not in any words or concepts, but more as a pervasive feeling of "this is it, this is how everything truly is."

3207 [. . .] Feelings were mainly love and amazement and gratitude. [. . .]

3524 Easy calm, easy completion, sensationless awe, being deeply moved, all-pervading floaty thoughtless state of not searching further, being boundlessly held [*entgrenztes Gehaltensein*], fulfilled without needs.

Joy and existential relief are quite common, but they come in a wide variety of ways:

2056 [. . .]—from one moment to the next, complete unburdening—a slight (almost "childlike") feeling of joy—[. . .]

2346 [. . .] I achieved this perspective within minutes and quickly experienced an accompanying sense of total relief and freedom. It was quite exhilarating, and I was keenly aware of a center to myself unrelated to my bodily experience or the content of my thoughts. It was immensely relieving and made me quite giddy in a way. [. . .]

2521 [. . .] There was a general sense of awe and joy. In this state, I felt like I would know how to inhabit it forever, but after what felt like 15 to 45 minutes later, I started to feel I was gradually coming back to a normal state, which I did although I still felt joyful. I remained in a somewhat more joyful state for a few days and then I returned entirely to my normal self with the memory of what had happened. [. . .]

[. . .] I expanded what I was paying attention to (trees, sound of the wind and cars, sensations of the body) and started thinking "Look for the self. Is there anyone experiencing these things or is everything just happening by itself?" At that moment I started smiling and feeling an unexpected deep sense of joy. It would last a few seconds, but from then on, I was able to invoke that kind of feeling any time I wanted. It never lasts more than a few short seconds and I do need to prepare to make it happen (stop, expand awareness, remember to notice that everything happens by itself). There is a general sense of joy, awe, and a feeling that after all, I don't really have any real problems. Sometimes the sense of joy feels almost too intense, as if there is tension around an unusual feeling. There's also an urge to smile. [. . .]

2867 [. . .] the intensity started to wane, and I was left with a feeling of having been blown away like dandelion seeds on a puff of air. In awe and at peace. Also, a feeling of giddiness, like I had just stepped off a ride.

3186 [. . .] The contact and the encounter with your own soul. Simply beautiful and not comparable to anything earthly. Pure joy.

The phenomenal qualities of bliss and timelessness sometimes occur in calm and peaceful states, but they may also arise with complex experiential content:

3236 Deep calm, bliss.

2706 [. . .] On the other hand, there are paradoxical spaces of consciousness with a vast emptiness and yet billowing fullness, a presence of being, which open up the deepest serenity and at the same time overwhelming ecstasy and bliss. A great joy and connectedness is accompanied by sorrow and deep compassion for the suffering of beings. Although thoughtless, the desire to share these vast spaces of consciousness is strong, gratitude and humility arise, reverence and lightness, vastness and also silence again, motionlessness, timelessness, solidifying in powerful pulsation.

3074 A sense of deep bliss, the smell of a rose, and colors of pink and purple in the third eye area, timeless, weightless, and at peace.

3110 In general, the pure awareness washes over me with a sense of calm within myself and greater unity outside of myself. It is almost always accompanied by either the sensation of bright colors (usually yellows, sometimes green, red, blue, or purple). Occasionally, there has been a noticeable absence/void of color. The experience has been one of pure calmness and bliss. I think in actual time it has been very short, but the sensation is one of expanse and timelessness.

3190 After my meditation I went for a walk by a lake at sunrise. It was a wonderful atmosphere. Then I heard voices. Normally it would have destroyed my beautiful experience of nature to hear other people talking, but now this was just the occasion to realize that everything is one, that I am one with everything, that everything is good. The experience went from the lovely harmonious setting to a deep experience of bliss that is endless. At the beginning of my experience the golden light of the rising sun made me happy, but then this light was within me and everywhere and I was everywhere this light, boundless, blissful.

Another phenomenological element is the experience of love:

624 [. . .] You experience the feeling of unconditional love, of being an intrinsic and inherent part of the universe, without separation. I feel life deeply in its pure state, without distortions, without a mind that wants to control everything for fear of pain, I feel perfect as I am, full of wisdom and beautiful in every way (loved and full of love). [. . .]

717 [. . .] It is a feeling of merging with all happiness and all love. It is soooooo soooooooo big. . . . ., weightless, the words we have at our disposal here are not enough for the description.

2874 The experience occurred during a sitting meditation and can best be described as a stream of "love" that flooded the body or absorbed the body into itself (qualities of warmth/light/bliss). [. . .] A state of heightened alertness and full consciousness was present. It was also possible to withdraw from this flow at any time and to switch back to the "everyday" phenomenal experience.

Sometimes the positive affect of pure awareness can include gratitude; a sometimes strong, often timeless, at times even bodiless, nonpersonal, uncontracted, and "nondual" quality of gratitude *as such*, lacking subject and object:

82 [. . .] The dis-identification of thoughts and emotions was amazing, I stopped feeling thought. So was the sense of conscious space. I felt infinite gratitude when I returned to the normal waking state.

1862 The experience is incredibly difficult to describe. It is like taking a thimble of water from the ocean of wisdom; however, I continually felt gratitude, awe, and was completely void of the emotion anger. [. . .]

2687 [. . .] a deep feeling of joy, timelessness, happiness emerged. I could perceive it and had no sense of time. I felt no physical sensations, no thoughts. At some point I felt my body again, thoughts and feelings came. It could "not be repeated." I had the feeling that I had been given a gift.

2714 It is difficult to put it into words, in any case through meditation I can create conditions (by relaxing into the moment) and it comes or not, like grace; a strong element of letting go until there is no one left to let go, no inside or outside, no one who is doing well or badly, although in the background there may be a sensation of the body, occasionally thoughts, perceptions of the environment, these are not there and everything at the same time, perfect birth; it doesn't last and has no time; it is completely there and in no place, there is no place where it could be or not be, there is nothing outside this being, which calling being is already too much, often accompanied by tears, and if there is a feeling then gratitude, but not for anything or anyone, rather gratitude as such.

One unexpected detail was that a number of participants in our study reported crying, typically at the end of an episode characterized by particular intensity. Sometimes they emphasized that the crying was not itself emotional:

1081 [. . .] deeply peaceful, moved to tears, as-it-is, the sense of being "home." [. . .]

1114 [. . .] Something that moves me to tears. [. . .]

1142 [. . .] At this point, tears came to my eyes and I began thinking and the experience dissipated.

1558 [. . .] My body was very relaxed, my mind awake and without thoughts, emotionally I was moved and tears were flowing. [. . .]

1718 [. . .] My breath was deep and slow through my mouth with tears welling in my eyes. I felt relaxed and amazed.

1942 [. . .] Tears run down my face. [. . .]

2071 [. . .] I felt tears running down my face. [. . .]

2234 [. . .] I cried tears of joy during and after. [. . .]

2504 [. . .] Soon the emotion came, tears of joy flowing. [. . .]

2607 [. . .] I came away from the experience and cried tears of happiness for about 15 minutes [. . .]

2742 [. . .] At that moment I was so overwhelmed by this purity that tears were running down my face. [. . .]

2879 [. . .] when, from one moment to the next, I was seized by a kind of different force and tears ran down my face without me being sad. Rather, I was happy, without that perhaps being the right word. It was kind of like being deeply moved. [. . .]

3048 [. . .] Tears ran down my face. This was not connected with any feeling or emotion, it was simple. [. . .]

One interesting result of our study is how many meditators, when asked to describe a single, paradigmatic experience of pure awareness, testified to a deep sense of profundity and existential relevance, but how few of them actually chose to interpret their experience in an explicitly religious way, using religious terms like "grace." There are some exceptions (for examples using the term "grace," see #692, #2714, #2983, and #3173; for the idea of a "grace hyperprior," see chapters 10 and 13). Here are two examples:

692 All that I [. . .] was allowed to experience of pure consciousness or awareness is based on pure Divine grace, for which I am infinitely grateful. [. . .] One cannot produce these experiences by one's own conviction or willpower, let alone repeat them on demand. All one can do for this is to purify oneself through one's daily sadhana (spiritual practice) and prepare oneself so that one is physically and mentally ready and able to endure and sustain [*(er)tragen*] this pure awareness whenever IT will reveal itself and not get burned in the process.

3173 For myself, I sometimes call this a state of "grace" that is poured out on me, so to speak.

Deep contemplative experiences apparently sometimes involve a quality of what in German would be called *Gottebenbildlichkeit* (godlikeness). This is the experience of being created in the image of God, of participating in and perhaps even existing only by virtue of something radically different from oneself—in short, existing only as "an image-bearer of the Noncontingent."[1] *Imago Dei*—the ancient theological idea that human beings are created in the image and likeness of God (which we find in some Islamic Sufi groups, in Judaism, and in Christianity)—may have what from chapter 26 onward I will call a "phenomenological anchor," because it refers to a specific region in phenomenal state space. Transposing the idea into metaphysically neutral terminology, this quality can be described as the experience of being an expression of the whole—if you will, not of *having* a model of the whole in your mind, but of actually *being* such a model. Perhaps one could describe this as a global phenomenology of "self-similarity," meaning that parts (like yourself) are now experienced as resembling or even locally representing the whole:

2384 Afterward I knew that I am the expression of the whole universe, right in the place where I find myself. I do nothing, everything happens.

Before we proceed to look at seven final examples of verbal reports explicitly using religious terms, allow me to make two related points. First, some of the analogies and new metaphors for pure awareness coined by our participants are open to a theological or religious reading without requiring it: Good examples that we have encountered in this chapter include "*Zustimmen zum Sein*" (#151; "consenting to being"), "*die Berührung und das Begegnen mit der eigenen Seele*" (#3186; "the contact and the encounter with your own soul"), and "*entgrenztes Gehaltensein*" (#3524; "being boundlessly held"). In addition, there are whole *categories* of state that are relevant in this context, such as the phenomenology of "coming home" (as described separately in chapter 33) and that of "waking up": for example, "*das Wahre, so wie ein Aufwachen in das richtige Leben*" (#2908; "Truth, like an awakening to life") or the beautiful notion of pure awareness as "*eine Art 'Erweckung zum Menschlichen'*" (#1787; "a kind of 'awakening to the human'"; see also chapter 4).

To conclude, here are some of the relatively rare examples where religious terminology is explicitly used:

> 267 [. . .] It was a very profound experience for me. As if for the first time I had got an impression of what "God" really is.
>
> 540 [. . .] At the same time the flow of my thoughts lessens more and more and my consciousness becomes very quiet and peaceful. But this is only a transitional stage. Since after some time there is a feeling of increasing expansion not only of consciousness but also of the body. The feeling that comes along with it is love. As a Christian I perceive it as God's love touching me and at the same time igniting within me my love for him and for all creatures. The feeling of love is the most important thing in the whole experience. It is what I experience as expanding pure awareness, and it encompasses God, myself, and all creatures.
>
> 863 [. . .] I can describe it as "perfection." Since then I can imagine that something like godlike energy exists. [. . .]
>
> 1471 I experienced a state of perfect silence and infinity. I simply felt happy, whole, secure, one with God. I knew that in this state nothing bad could happen to me. I felt devotion and gratitude at the same time. This feeling pulsated in alternation with perfect silence.
>
> 3194 I felt completely boundless, merged with the whole, peaceful, fulfilled, connected with the One God.
>
> 3218 [. . .] When I was walking on the gravel road and seeing insects or little animals passing by, I felt an intense feeling of bliss at their sight. A complete marvel at the perfection and beauty of God's creation infused my heart. [. . .]
>
> 3579 Again and again in meditation, I experience a feeling of being one with God and God in everything. This is the God of all religions, not of Christianity, detached from space and time. It is difficult to put into words, because it is first and foremost an experience of unity. There is no more self or not-self. I feel one with the original source, wholly internal and wholly external. All separation is suspended. There is only complete presence. All conflicts, everything, is void and no longer existent.

Spiritual but Not Religious?

> Attention, taken to its highest degree, is the same thing as prayer. [. . .] Absolutely unmixed attention is prayer. [. . .] Attention alone—that attention which is so full that the "I" disappears—is required of me. I have to deprive all that I call "I" of the light of my attention and turn it on to that which cannot be conceived.
>
> —Simone Weil (1909–1943), *Gravity and Grace*

The relationship between the phenomenology of pure awareness itself and positive affective experience could easily be a whole research project on its own. A first conceptual distinction that we should make concerns whether these phenomenal qualities tend to occur in the midst of a full-absorption episode, while still trying to quiet the mind and gradually access the state of pure awareness, or when moving out of the state into something more complex, thereby integrating additional experiential contents. Here, one important possibility is that many practitioners may actually alternate rapidly between various stages of meditation. For example, some full-absorption episodes could actually be very short and occur rather frequently during a single period of sitting. Therefore, we must consider not only prolonged or dramatic experiences, but also subtle and potentially swift fluctuations in emotional tone.

A second target for future research would be the relationship between MPE reports explicitly employing religious terminology, those describing something we might call a phenomenology that is "spiritual but not religious (SBNR)," and those that are not even describing MPE-related states as "spiritual" in any way, being explicitly secular and metaphysically neutral in nature. The semantics and other linguistic aspects of reports anywhere on this spectrum could be studied in both quantitative and qualitative ways.

Over the centuries, contemplative practice has mostly taken place against a backdrop of religious belief systems like Buddhism or Hinduism, with meditators trying to achieve a soteriological goal like "liberation" or "enlightenment." Accordingly, the phenomenological taxonomies of such states have been shaped by metaphysical belief systems and an ancient cultural context; they are highly *normative* taxonomies. However, during the last fifty years, a historically new situation has emerged: Millions of practitioners in Western societies meditate on a daily basis, and research by Heinz Streib and Ralph Hood has shown that many of those who do so describe themselves as secular or as "spiritual but not religious."[2]

In our first survey, the most frequently named religious affiliation was "spiritual but not religious" (SBNR) / "spiritual but not affiliated" (SBNA) (n = 632), followed by two large groups of Buddhist and Christian denomination (322 and 320), and then one last, largish group of participants categorizing themselves as "secular" (188). It is plausible that differing conceptual backgrounds will have an influence on the intrinsic motivations, implicit expectations, and personal goals of meditators (more on this in chapter 17). In general, our data show that for many of our participants, the importance of meditation is much greater than the importance of religion.

From a more abstract, philosophical perspective, I have argued that the opposite of religion is not science but spirituality,[3] and that the possibility (or impossibility) of developing a genuinely "secular spirituality" is one of the most important topics of our time (more on this in the epilogue). On the other hand, the increasingly popular concept of SBNR is derived from a new and slightly ill-defined term that emerged at the end of the twentieth century. A semantic analysis of SBNR based on empirical evidence from psychological research seems to show that in the large majority of real-world practitioners, the actual difference between "spirituality" and "religion" is smaller than explicit distinctions between both concepts may seem to indicate. I will come back to this point from two different angles in chapter 17 and the epilogue.

Referring back to the work by Streib and Hood just mentioned,[4] the Canadian philosopher Evan Thompson has convincingly made the point that the idea of a fully secular spirituality without religious elements could refer in some cases to a form of "privatized experience-oriented religion."[5] This possibility is a relevant target for future research. It will be important to investigate, in particular, how the presence or absence of spiritual or religious elements (whether private or institutionalized) influences verbal reports and the fine-grained phenomenal character of contemplative experience itself.

In terms of our ongoing qualitative analysis, it is worth noting that you will find many reports of joy, awe, bliss, and gratitude in other chapters of this book. From a purely statistical perspective, four items in factor 2 (labeled "Peace, Bliss, and Silence") pick out states of positive mood (like "gentleness" or the "invisible smile," plus the phenomenal quality of "bliss" itself). These typically coemerge with a holistic quality of deep relaxation and pure being—the phenomenal character of "existential ease" described in chapter 1—in connection with an experience of peace and deep, unbounded silence. Interestingly, these items form a cluster that remains stable across six factor solutions, an internal coherence suggesting that precisely these experiential qualities very often co-occur. References to "soundness" and "harmony" and statements to the effect that this was "the simplest kind of conscious experience I know" also belong to factor 2, but they had a weaker loading and were not stable across all six factor solutions.

From a conceptual point of view, one of the most interesting research questions in this subdomain is whether the "primordial" form of happiness and the very subtle, undramatic, unsensational quality of rapture and wonder discussed at the very beginning of this chapter actually belong to the prototypical core of MPE. But what about the more pronounced, phenomenal character of bliss: Is it a property of pure consciousness per se? Ancient contemplative traditions seem to disagree on this point, as do some present-day experts.[6] For example, the famous notion of *sat-chit-ananda* ("existence, consciousness, and bliss") clearly makes it an essential component of pure awareness. The term goes back to the *Nrisimha Tapaniya Upanishad* and the *Rama uttara tapaniya Upanishad*; as a metaphysical concept, it is perhaps most prominently known from Vedanta philosophy, where it refers to the three main attributes of Brahman, the nonpersonal Absolute. On the other hand, many Tibetan Buddhist sources actually *warn* practitioners that experiences of nonthought (chapter 3), clarity (chapter 5), and bliss (as described previously) are mere "meditative moods," likely to distract from what really matters if one becomes attached to them. Here are two examples, one taken from the famous *Moonbeams of Mahāmudrā*, the other from *Clarifying the Natural State*:

> When any of those experiences of bliss, clarity, or nonconceptuality develop (either altogether or separately), if we take them to be sublime and intentionally cultivate them with attachment and fixation—feeling pleased when they develop and unhappy when they don't—they will cause us to stray from the abiding state.[7]

> Whether it is one of these three meditative moods—bliss, clarity and nonthought—in combination or a part of any of them, you might cling to these meditative experiences as being paramount and train in a way that is fettered by them, being happy when they come and unhappy when they do not.[8]

The Tibetans clearly thought that the experience of bliss can be something dangerous, leading to more or less subtle forms of delusion. This dilemma is also related to the issue of intrinsic motivation, to which we will return in chapter 17: Does MPE in itself carry motivational force—is it a state of consciousness that automatically creates the desire to return to it—or are other, more extrinsic factors more significant in encouraging meditators to sustain their regular practice, sometimes over many years? Is consciousness itself fundamentally *affective*?[9] As to the general conceptual question of whether more pronounced positive affective states are intrinsic to the pure-consciousness experience, we now have a clear interim result. The new phenomenological data presented here show a double dissociation: In meditators, the experience of pure consciousness can definitely occur *without* automatically triggering joy, awe, bliss, or gratitude; and all human beings—meditators and nonmeditators alike—know the conscious experience of joy, awe, bliss, or gratitude in situations where the phenomenal character of awareness itself is entirely absent.

16 Simplicity, Nothingness, and Absence

A lucid void [#105]

The certainty of having encountered NOTHINGNESS. [#108]

Unity and nothingness at once, perfectly balanced. [#1937]

One of the methodological background assumptions guiding the Minimal Phenomenal Experience (MPE) research project was that pure awareness might be the simplest form of conscious experience that human beings are capable of. From the first-person perspective of meditators, this intuitive assessment has been confirmed, as our data clearly show. But as we have already seen, there is much more to be said from a theoretical, third-person perspective. Let us begin with three examples of how meditators describe the phenomenology of simplicity:

1424 [. . .] It is a domain of complete silence. Simplicity. There is nothing that needs to be expressed, communicated, or evaluated. The very simplest being.

2372 A "moment without ingredients."

3218 [. . .] The simplest state of awareness perfectly describes my experience. It IS utter simplicity, no complications whatsoever, easy and so natural as if things could not be any different, will not be any different, and have never been any different than this completely effortless flow of life and evolution. [. . .]

In some cases, practitioners describe deep MPE-like states as involving a paradoxical phenomenology of "nothingness," which may involve everything "being" nothing, nothing remaining while everything *is*, or being a nothing that somehow isn't *really* nothing. (You will discover close parallels with all three in chapter 26, when we investigate the experience of pure being more closely by looking at some descriptions by medieval mystics.) Normally, such states will be concurrently ineffable: impossible to describe while they are happening, simply because they lack the distinction between self and world. Interestingly, what is later described as "nothingness" can also become the true self; it can function as what, in chapter 29, I have tried to describe more precisely as a "nonegoic unit of identification." Here are three examples:

> 1788 [. . .] In a further, even deeper state I suddenly see that everything is pure "nothing." This has a frightening effect on me to begin with, as well as being very liberating and absurdly funny, because "nothing" is the only thing that is. I myself am also nothing, have always been nothing, and I cannot lose anything, because there is nothing to lose. This is how I experience it in this state. The liberating and very pleasing thing is that I then "see" that what I really am cannot die, because it is always already this living nothing and there is nothing else. [. . .]
>
> 2441 I was pure me and everything at the same time, I was no more and knew that I am, in the deepest sense, who I am. Nothing. Nothing more remains, everything is. [. . .]
>
> 2544 The first time this happened it was very sudden—I suddenly recognized that I was nothing, nothing at all. And it was so obvious, yet I'd never noticed before. Nothing that's not nothing though. Nothing as in being no sort of object or phenomenon whatsoever. Yet despite being nothing, I was here. I recognized that I'd always been here. I saw that I was what here is, and that I was what now is. I clearly saw that it's never not now. It couldn't possibly ever be other than now. I saw that, as nothing, I had never been harmed in any way, nor could I ever be. And when I say "I" I don't mean the person. There was such profound peace and bliss. I didn't move for some time. The body was utterly still, the mind was utterly still. All sensory perceptions were vivid. I didn't want to move ever. But when I did get up to move I saw that I did not move. The ground moved underneath me and the trees passed by as I walked, but I did not move. Soon the emotion came, tears of joy flowing. I didn't understand anything. I did not give it to the mind. The mind could not hold this.

In chapters 2 and 3, we saw how Eastern and Western philosophical traditions converge in suggesting that "low complexity" must be an essential semantic constraint for the concept of pure awareness.[1] Empirically, one may assume that the first-person simplicity of MPE should be reflected in third-person measures of brain complexity.[2] If subjective phenomenology is at all related to objective brain function, radical attenuation in functional complexity makes it plausible to assume that full-absorption episodes sometimes border on a complete cessation of conscious experience. Our reports show that such episodes of pure awareness are minimally complex and undifferentiated (e.g., in that the phenomenal character of awareness in and of itself lacks all internal structure and temporal dynamics). There are also reports about experiences that weren't even "experiences" anymore, but rather states of utter annihilation and nothingness (see also chapter 31):

> 2788 First, there was no program anymore, then, there were only some bars, and finally, they disappeared leaving everything blank and no television anymore at all.

Some descriptions of phenomenal "nothingness" involve episodes of complete disorientation relative to time, space, and person, while others refer to a combination of nothingness and extreme wakefulness. Interestingly, some meditators even report prolonged "absences" for which there is no memory whatsoever—states perhaps vaguely reminiscent of petit mal seizures and the phenomenon of childhood or juvenile absence epilepsy. In some of humankind's contemplative traditions, this phenomenon is known as the attainment of "cessation" (*nirodha-samāpatti* or *saññā-vedayita-nirodha*), referring to a complete cessation of perception, thought, and feeling, an unconscious state in which not only all mental activity has been suspended but also the vital signs of most bodily functions are greatly attenuated.[3] In these very special situations, "pure awareness," "pure being," or "nondual unity" typically seems to occur just before or just after the episode proper. Let us take a look at five examples:

> 1908 It's difficult to describe, but I felt as if I was being sucked out of the world through a portal. Everything around me disappeared and I was left in a state of "nothingness" that felt, paradoxically, safe.

> 3431 [. . .] the impression emerges of getting very close to something fundamental, unbelievable, essential, which then in some way seems to elude direct consciousness. A kind of "consciousness swallows itself." Often combined with a feeling of sinking down, shrinking, or also falling or collapsing in on oneself, like a whirlpool (but slower) or an hourglass in the middle, or the drain of a bathtub, or as if something is being stuffed inside and disappearing at the same time. I don't know if this is what is meant by "pure awareness"; it feels more like "nothing." Some questions in the questionnaire, which asked e.g. whether the state was "positive" or "relaxed" or a "feeling of unity" or something like that, I couldn't answer at all or could only say no to, because the state simply somehow didn't exist at all or was nothing at all . . . what these experiences—or rather, states—have in common is a kind of great amazement,

so to speak holding your breath or not breathing anymore and an immense "awake"ness. [. . .]

3136 After about thirty minutes of complete unconsciousness in an upright meditation posture, it took me several minutes before I knew again who and where I was. I landed in a state of consciousness of boundless, stable peacefulness that seemed indestructible and continued even when I got up and walked home. Everything external seemed far away and could not harm my state of "pure being."

3323 [. . .] However, later on there were more "experiences" in which the very last remnant of this pure consciousness in meditation was extinguished. This was like an inner death, but then also an even greater freedom than pure consciousness itself. There it was clearly experienced that pure consciousness is far from being the deepest possible (or highest possible) thing, but that "behind" it there exists a much more extensive, indescribable "not-anything." But it cannot be described in words, since it is no longer an experience; rather, it can at most be described as the absence of all experience, or as absolute freedom.

3348 [. . .] I did one morning decide to have my only meal of the day outside [. . .]. I sat on a rock near a pond. It must have been around 11.00 in the morning. I remember realizing how the dragonfly that hovered over the pond and the rock I was sitting on were one and the same. Next, the bell for evening meditation rang. It was 5.00 in the afternoon . . . my meal untouched . . . the body felt strange, yet free and open. Many of these sensationless [states] have followed since. While initially it was followed by a fear of dissolving . . . today the body is used to it and somehow knows it will return to the world as we know it.

Could there be islands of pure awareness for which there is no memory whatsoever? Some of the reports presented here point to the empirical possibility of what I will call "mnemonic closure": It is conceivable that there are MPE states that never can or will be reported—states that may from the outside appear to be a complete absence, an episodic loss of consciousness, or simply a "passing-out" during meditation. But they might actually be states of consciousness. Plausibly, states that are "closed" to any form of personal memory in this way would appear in the context of full-absorption episodes as described in this book. Proving their existence would demonstrate that the functional property of "availability for personal-level memory recall" is not a necessary element of our minimal model of what consciousness really is.

The possibility that awareness may persist in fully disconnected cortical islands has recently begun to fascinate philosophers and neuroscientists alike because the discovery of islands of awareness would have important implications for debates about the nature of consciousness.[4] Future research into MPE and meditation, therefore, could contribute to our understanding of the conditions under which such islands might arise and the forms might they take.

Simplicity and Profundity

> You can come upon your relationship to this nothingness and its fear only by being choicelessly aware of the escapes. [. . .] You and nothingness are one; you and nothingness are a joint phenomenon, not two separate processes. [. . .] When there is the discovery, the experiencing of that nothingness as you, then fear—which exists only when the thinker is separate from his thoughts and so tries to establish a relationship with them—completely drops away.
>
> —Jiddu Krishnamurti (1895–1986), *Commentaries on Living*

In chapter 12, when investigating descriptions of pure awareness as "the most natural state," we discovered that what makes the phenomenal character of MPE so hard to express in words may be the combination of profundity and simplicity. One could certainly imagine that some random person to whom MPE is directly pointed out might briefly feel their way into it but see only the simplicity and not the profundity, shrug their shoulders, and walk away. There could be experience without recognition because in a sense, MPE is natural and unnatural at the same time.

One may speculate that MPE could be the baseline of all experience, as well as the critical boundary between conscious experience and its cessation. If so, it would be quite plausible that MPE could sometimes alternate with states of actual unconsciousness. When we investigated the phenomenology of wakefulness in chapter 4, I introduced the idea of an "alertness *Ganzfeld.*" Did you know that staring at an undifferentiated and uniform field of color for some time will make you see black or go temporarily blind?[5] It is well known that seeing only one single and undifferentiated chromatic stimulus eventually leads to the experience of seeing blackness or of being blind. Vision scientists call this the "*Ganzfeld* effect" or "perceptual deprivation." It turns out that color "qualia" are not context-invariant atoms of conscious experience at all. Perhaps the same is true of my hypothetical alertness *Ganzfeld*, and if so, this may explain some episodes of unconsciousness. If the whole phenomenal field is uniformly filled by pure wakefulness, then some people might not simply "see black" but go blind altogether; they might not experience sheer nothingness, but instead lose conscious experience altogether. Perhaps the term "white-out" would be a better label for this admittedly speculative hypothesis.

In any case, there may be interesting relationships connecting MPE and what is later reported as a loss of all subjective experience. If pure awareness is functionally related to the critical borderlands where consciousness ceases altogether, then one would also predict that people who have learned to *recognize* the phenomenal character of MPE become aware of it at the edges of sleep: when going to bed and when emerging from deep sleep in the morning. Interestingly, new research is now beginning to document the conscious experience of "nothingness" even during dreamless deep sleep.[6]

In the existing literature, phenomenological reports of MPE often point to a combination of wakefulness and simplicity. The following example adds two important aspects—the qualities of "self-sufficiency" and "intimacy":

> When I experience pure consciousness, it is a state in which I am awake and aware, but not aware of anything except awareness itself. As I merge into the experience, outer-relatedness lessens and inner peace and self-sufficiency remains. It is not an intellectual experience. It is by far the most intimate and simple experience in my life.[7]

The experiential quality of self-sufficiency relates to experiences of soundness (chapter 7) and naturalness (chapter 12), but most obviously to the phenomenology of "coming home" and what was figuratively described (in chapters 13 and 14) as "There is nothing left to do." We encountered the phenomenal character of self-intimacy when investigating the quality of connectedness in chapter 11, and we will briefly encounter it again in chapters 29 and 30, when looking at the "True Self" and the discovery of nonegoic self-awareness. For now, let's stay with the phenomenology of simplicity itself.

In item #84 of our study, we asked participants to rate their agreement with the statement "The experience of 'pure awareness' is the simplest kind of conscious experience I know." The median level of agreement with this statement was 80 (on a scale from 0 to 100). This result seems to confirm my original intuition that the phenomenology of pure awareness is a good place to start in the search for a minimal model of consciousness. If, as scientists and philosophers, we do eventually want to formulate a first "standard model of consciousness" using the formal language of mathematics (by analogy with the Standard Model of particle physics, which is theoretically self-consistent and has been extremely successful in providing testable predictions), then the concept of MPE will have to be central.

Why is that? For us, MPE—that is, consciousness per se—is the *intended interpretation* of any purely mathematical description because what we really want to know is what the property of conscious*ness* itself consists in. Finding this interpretation is what motivates our search for the standard model of consciousness. Any purely formal model needs an interpretation that creates a link between symbols, on one hand, and certain states or functions in the real world, on the other. We want our model of consciousness to be factually true, a nonarbitrary interpretation, with empirical reality itself eventually becoming the model of our theory. For consciousness, this may be a few decades away. But at the very least, investigating the phenomenology of "awareness itself" will generate great heuristic fecundity because it changes our theoretical intuitions, makes us think about things in a new way, and gives us a fresh angle on the problem of consciousness. That fresh angle consists in the fact that we are now looking at it from neither the first-person perspective of subjective, personal experience nor the third-person perspective of a large scientific community. Instead, we are uniting both approaches by finally taking the "zero-person perspective" seriously (see chapters 3 and 29).

Meditators have been adopting the zero-person perspective for many centuries. To be sure, people have all kinds of complex reasons for practicing meditation, and it is perfectly fine to do it in an experimental spirit or out of sheer curiosity, for its therapeutic benefits, or to improve one's quality of life. But here we are taking a philosophical stance on it. From a more radical, philosophical perspective, meditation really has nothing do with well-being, stress reduction, or self-optimization; it is a special form of epistemic practice. It is about insight, about gaining knowledge. Epistemic practices like science and philosophy tend to be engaged in through language, via books, concepts, arguments, and theories. They are socially organized, they depend on personal-level interactions between members of a group, and they are concerned with assessing and legitimizing knowledge claims. Meditation is a very different kind of epistemic practice. The sought-after form of insight occurs in silence, entirely without words. It often takes place in a social context, but being able to make public verbal statements of knowledge after the fact is not one of its essential features or goals.

Radical meditation is an epistemic practice, just as philosophy and science are. This means that meditation is aimed at insight, at the creation of knowledge—but a very specific kind of knowledge that has nothing to with words, concepts, or theories. Rather, a central goal of meditation practice is an entirely silent and nonconceptual form of *nonegoic self-knowledge* (chapters 29 and 30). Like any practice, it can be abused, and in chapter 17, we will take a careful look at two of the most important ways in which this can happen: narrative self-deception and mortality denial. Drugs can be abused, and so can meditation. Viewed from the more radical philosophical perspective that I take in this book, contemplative practice cannot be reduced to some banal form of self-help or self-improvement, nor to a neoliberal form of self-pacification that serves to stabilize the status quo of woke consumer capitalism. It is also not a form of media consumption that you engage in through headphones. These would all be examples of what I call "meditation abuse."

The epistemic practice of meditation critically depends on something that cannot be faked or commodified, something that Immanuel Kant, in his 1793 work *Religion within the Boundaries of Mere Reason*,[8] called the "the sincere intention of being honest towards oneself" (more on this in the epilogue). Precisely because it is a *genuinely* epistemic practice, one that needs sincerity and honesty, real meditation practice is not something that can ever be turned into an individualistic lifestyle accoutrement to make people "mentally fit" and resilient, and thus help them keep functioning within an economic system that is about to destroy the planet. Rather, it is directly and intimately related to what philosophers since the time of Socrates have called the project of "living an examined life"—but, as already explained, in a way that has nothing to do with words, thoughts, or theories. Nonconceptually seeing the world and what one previously took to be oneself, in and out of MPE, is what this epistemic practice is all about, and this is also what it means to take on the zero-person perspective. Yet no one is doing the taking; it is more like effortlessly stepping out of the picture. The zero-person perspective has existed all along, and we can view it as a selfless mode of knowing in its own right. But what is known? Shen-Hui (684–758), a student of Hui Neng, said: "Seeing into nothingness—this is true seeing and eternal seeing."[9] According to Suzuki, Shen-Hui proclaimed that "seeing into one's nature" *is* "seeing into nothingness."[10]

One might now begin to speculate that pure awareness may be a nonlinguistic epistemic modality, a fundamental computational principle, a nondual "data format" that has naturally evolved in biological brains, providing us with an aperspectival form of nonpersonal knowledge of the world in itself (like level 4 in figure 34.1 in chapter 34). The zero-person perspective could be what envelops and permeates the first-person perspective, making it conscious in the first place. Is this where the profundity comes from? Is it that we feel the underlying unity of the third- and first-person perspectives, but nonconceptually, not as some intellectual construct but as what awareness itself really is? The idea sounds exciting—perhaps even slightly romantic. However, in beginning to think about the zero-person perspective, we need to apply the highest standards of intellectual honesty and stay as conceptually clear as possible. It is just too important not to.

In the spirit of remaining clear and honest, it's crucial to acknowledge that "simplicity" is only a phenomenal quality—just as redness, greenness, and sweetness are—and that the statistical fact about a median of 80 for the agreement mentioned earlier does not in any way prove that we have actually found the simplest state of consciousness that human beings are capable of. There may be even simpler ones. Just think of the possibility of "islands of awareness," which I touched on at the very end of the preceding section. Perhaps we all have ultrashort flickers of awareness during dreamless deep sleep, which we never remember and therefore never report to each other? Together with the dream researcher Tore Nielsen, the philosophers Jennifer Windt and Evan Thompson asked whether consciousness really disappears in dreamless deep sleep. Not only did they identify "selfless" states and contentless sleep experiences (which we will carefully investigate in chapter 20) as possible candidates for dreamless-sleep consciousness, but they also found a variety of phenomenal states that are not dreams at all, such as nonimmersive imagery and sleep thinking, as well as perceptual experiences and bodily sensations lacking the simulational character of dreaming. On the basis of these findings, they call for an entirely new taxonomy of dreamless sleep experience.[11] Accordingly, future research may demonstrate that even less complex forms of phenomenal character than MPE in meditation exist, and they constitute the simplest form of conscious experience that neurotypical human beings are capable of. Empirically, it clearly remains an open question what the very simplest form of conscious experience really is.

Conceptual issues are just as important as the empirical question of whether meditation-related MPE is the simplest form of consciousness. First, it is essential that we avoid conflating subjective ratings of "simplicity" with whatever objective measures we may develop in the future. It may feel highly intuitive to claim that a state of consciousness is "the simplest kind of conscious experience we know," but doing so raises a host of thorny philosophical issues. The criteria for minimality, or for whatever else we decide the word "simple" really means, will have to be developed by philosophers, mathematicians, and the interdisciplinary community of consciousness researchers. For example, if we opt for a state-space model of consciousness, simplicity might be defined in terms of dimensionality or the volume in phenomenal state space inhabited by pure-awareness experiences, or quite likely some other, more complicated mathematical construct. But how do we get to the dimensions of phenomenal state space themselves? The ninety-two items of our first MPE-92M survey, which led to the twelve statistical dimensions derived from them, were defined by questions that a limited number of human beings (mostly myself, my coauthor Alex Gamma, and the many wonderful participants in our pilot studies) selected on the basis of their own phenomenal experience. This is one methodological option, but there are many others.

Do you remember the fable of the elephant and the blind? The blind people were guided to the elephant by the king's loyal subjects, all of them sighted people. What if they had chosen different blind people to invite? Why should precisely *these* questions and dimensions help us carve out what pure consciousness "really is"? Is the experience of pure awareness a natural element like gold or silver, or an element like plutonium or meitnerium? Is what we mean by the working concept of MPE ultimately something that—like "electron" or "tau neutrino"—reflects the structure of the natural world, not the personal goals, interests, and actions of human beings? Will the experience of pure awareness be the same for all sentient beings in the universe that can have it? We simply do not know yet, but tentative recent evidence begins to speak against it.[12] In a philosophical sense, one perfectly possible result is that we may ultimately find that there is no intrinsic essence to consciousness at all. Consciousness, too, may prove to be empty.

Ultimately, "MPE" is a cluster concept that refers to a phenomenological prototype. This implies that there may be no set of necessary and sufficient conditions that allow us to say whether some minimal or extremely simple representational state is conscious. Rather, "consciousness" itself refers to states, regions, and trajectories in a multidimensional space constituted by functional and content-related dimensions; it is a graded and heterogeneous construct, and membership comes in degrees.[13] The underlying dimensions that give structure to the space of conscious experience create a complex pattern of family resemblances, and absolute orderings may not be possible in all cases. Different embodied beings will open up different spaces of experience because they will have different brains, live in different sociocultural contexts, and have different ways of accessing their individual phenomenal space. This means that our blind people may never fully agree about the family of animals in front of them. There will be degrees of prototypicality because highly prototypical members of our family have attributes that overlap with most other examples of the category, while low-prototypicality members have little overlap. Membership of the family of phenomenal states, therefore, will be graded. But careful quantitative assessments of prototypicality may actually help us understand why some phenomenological exemplars capture the intuitive "essence" of consciousness better than others. What *could* be common to many different conscious beings, for example, is the underlying experience of epistemic openness discussed in chapter 4; this is because it directly reflects the degree of wakefulness.

Epistemic openness could be a fundamental part of consciousness even (or especially) for conscious beings that are not yet born. We all probably dream a lot before being born because in the course of fetal development, rapid eye movement (REM) sleep may constitute up to 80 percent of total sleep time.[14] I predict that the fascinating concept of "fetal consciousness" will soon move to center stage in consciousness research because important markers of phenomenal experience, like the capacity for second-order learning, are now being discovered in unborn human infants.[15] Anatomical development during the fetal period also indicates that the neural networks probably necessary for conscious processing are established around week 25 of gestation before birth. This raises the question of whether the pure-awareness experience might actually be the first conscious experience that every human being has in the womb. We should keep this possibility in mind: If MPE really is minimal, then it is plausible to assume that it might also be primary. It may be the *first* phenomenal state that all of us ever experience, even if we cannot remember it after birth. Emptiness—the silent, clear, and nondual experience of becoming epistemically open to the world—would then be the first conscious experience every sentient being actually has.

Let me close by giving one last example of what in chapter 12 I introduced as the "C-fallacy," this time applying our new conceptual tool to the phenomenology of simplicity. Of course, from the fact that something *appears as* and is later *described as* maximally simple from a first-person perspective, it doesn't follow that we have *really* found the simplest form of human consciousness. The fetal experience of pure awareness could be even simpler. To briefly return to the earlier example, empirically, there might be ultrabrief "flickers of awareness" with fine-grained content that nobody ever remembers or reports because these micromoments occur only during dreamless phases of nocturnal sleep—or even during normal wakefulness. Or there could be even simpler states in pathological conditions, such as borderline cases of bare wakefulness following severe brain injury, while slowly recovering from a weeklong coma, or in the process of waking from anaesthesia. Recall the E-fallacy too (chapter 7): from the fact that some conscious experience subjectively feels like an epistemic state, or that it is later described as carrying the phenomenal signature of knowing, it does not follow that it actually presents us with an insight into reality. Inner experience, in all its depth and beauty, is one thing, but public statements and strong theoretical claims need independent justification. They are not part of the epistemic practice of meditation itself.

I think the same skeptical point applies to the simplicity of pure awareness. A lot depends on our defining criteria for the phenomenological concept of "simplicity." Therefore, the question "What is the simplest kind of conscious experience?" cannot be answered by decree, by spiritual teachers restating the phenomenological taxonomies of time-honored contemplative traditions, or by the authoritative representatives of some religious system reiterating that religion's metaphysical precepts. Rather, this question is an open target for research. It is a philosophical, a conceptual, and a methodological question—and an important one. We simply do not know the answer yet.

17 Emptiness and Fullness

To be filled, you must be empty. [#654]

Many of our meditators describe a phenomenal quality of emptiness that is paradoxically rich and full. This kind of emptiness is a global quality of openness and connectedness, like a lucid void or a clear space that is nonetheless dense or even vibrant. On the other hand, some of our participants protested against the (originally Buddhist) notion of "emptiness" as a descriptor for their experiences, possibly because they did not like associating it with other phenomenological concepts such as the experience of a conscious "vacuum" or "void" used in item #13 of our survey. This may have to do with problems of translation, a conflict between metaphysical and phenomenological readings of the original notion, or both. For now, let us stay with the phenomenology itself, as reported by real-life practitioners, and look at a series of examples:

1585 Infinite spaciousness, stillness, a void that is full and vibrant [. . .]

1662 [. . .] There was a sense of quiet awe and contentment and experiencing "emptiness" (as the closest "canonical" term, i.e., *sunyata*, even though there wasn't a feeling of emptiness, but one of aliveness and infinite interconnectedness). [. . .]

1788 [. . .] It is an emptiness that does not mean the absence of something, it is not a lack of something, but it is really elementary, very real and clear, extremely alive, and the objects chair, table, etc. still seem like chair and table, but not as real, as if they were only dream objects. They recede into the background as meaningless and a very present emptiness comes into the foreground, but it is more than the space between the objects. The living emptiness then permeates everything, concepts like here and now no longer exist. [. . .]

3173 [. . .] the experience of an absolute emptiness comparable to deep sleep. I basically experience this state only once it has stopped. Then I know that it was just there. But in this state itself there is nothing. After that there are first determinations like silence or gentleness (while still in the meditation) and then good mood or feeling refreshed (after the meditation). [. . .]

3223 It felt so empty. As if for a moment nothing was there and you were just present with yourself, undistracted.

At the same time, emptiness is often explicitly described as coemerging with the phenomenology of "fullness":

178 [. . .] Feelings of emptiness and fullness at the same time. With feelings of calmness, peacefulness, and ease.

654 I have had feelings of complete fullness. In order for that fullness to fill, I was first in a state of emptiness. To be filled, you must be empty. The more empty, the more this divine experience can fill you.

1837 The feeling of pure awareness came from meditating. This oneness from my environment and my environment with me felt like fullness and emptiness all at once. [. . .]

2537 [. . .] there were moments and pauses between breaths of completely filled emptiness, in vastness, openness, and fusion. [. . .]

2754 It is not easy for me to describe this state. "Presence," "silence," "fullness," "emptiness" are terms that encircle it. A having-arrived, no longer becoming, timeless. Peaceful, eternal. Depth, fulfilment, insight.

3440 There is a point when knowing, knowing just stops. All is same. Just there. Not real. What is. Empty and full. If peace is defined by the emptiness of movement on each side of peace.

3601 [. . .] It reveals itself as pure presence, no thoughts, no physical sensations. A feeling of fullness and emptiness at the same time. [. . .]

Emptiness, Meaning-Making, and Mortality Denial

> Emptiness is the nature of mind, clarity is its characteristic, and [their] unification is its essence. It is free from all extremes of elaborations, such as good and bad; arising, ceasing, and abiding; existence and nonexistence; and permanence and impermanence. It transcends speech or thought and is beyond identification. And yet there is something to be experienced: it is sharp, naked clarity, which is of the nature of bliss, clarity, and nonconceptuality.
>
> —Wangchuk Dorje (The Ninth Karmapa; 1556–1603), *Dispelling the Darkness of Ignorance* (II, 8)

> It is a fact, Simmias, that those who go about philosophizing correctly are in training for death, and that to them of all men death is least alarming.
>
> —Plato (428/427 or 424/423–348/347 BC), *Phaedo* (67e)

> As soon as your breath ceases, what is known as the clear light of the first intermediate state [. . .] will manifest. [. . .] At the same time, a naked awareness that is neither inside nor outside of you will dawn, clear and empty, without center or perimeter. At that moment, recognize this instant presence as your own nature and relax in that state.
>
> —Padmasambhāva (ca. eighth–ninth centuries CE) and Karma Lingpa (1326–1386), *Bardo Thödol* (The Tibetan Book of the Dead)

"Emptiness" (*suññatā* in Pali; *śūnyatā* in Sanskrit) is one of the most important concepts in Buddhist philosophy—perhaps the most important of all. It has been discussed, interpreted, and reinterpreted by scholars and practitioners for well over 2,000 years. Many of the participants in our study may have known the concept of "emptiness"; they may have heard or read about it and may have had a personal and intuitive understanding of it. If so, this will have colored many of the reports given here. We first need to distinguish between a metaphysical and a phenomenological reading of this special concept—which, again, may well be one of the most interesting concepts ever developed in the history of human philosophy. As always, I am not at all interested in the metaphysics, only in the fine structure of consciousness itself. From a metaphysical perspective, "emptiness" means that all phenomena lack substantiality or an intrinsic nature of their own—they are not "ontologically self-subsistent" and they have no "essence," as a Western philosopher might say. But let us set all of this aside for now.

One of our new phenomenological tools is the concept of "epistemic openness." You may recall from chapter 4 that epistemic openness involves a distinct phenomenal quality—namely, a specific form of openness to the world. Epistemic openness is openness related to knowledge, to an inner space defined by the *possibility* of acquiring knowledge. The notion of epistemic openness offers a modern reinterpretation of the ancient Buddhist notion of "emptiness." From a phenomenological perspective, full-absorption states—in which (as we have already seen) pure awareness remains as the only reportable phenomenal character—are a prime candidate for generating experiences of emptiness. Here, epistemic openness per se is all there is. But there are many other situations in which emptiness arises. For example, in the context of mindful perception, "seeing the empty nature of phenomena" (chapter 9) can involve a specific form of conscious experience without the slightest trace of conceptual overlay: The distinct and crystal-clear phenomenology of seeing and perceiving out of timeless silence. In this state, things have no *conceptual* essence, in the sense that they are not named, categorized, or judged at all. Phenomenologically, all forms of cognitive penetration have come to rest. As we saw in chapter 9, on suchness, there is a specific, positive sense in which perceptual states are "meaningless" whenever their emptiness is phenomenally experienced. They are not even "things"; they are not reified as mind-independent entities; and therefore their content is epistemically open, such that a multitude of conceptual interpretations and perspectives are possible.

Importantly, in global minimal phenomenal experience (MPE) modes, the lack of conceptual overlay also applies, for example, to the implicit assumption that appearances are objects, juxtaposed with and consciously experienced by a knowing self. The overlay of subject/object structure can itself disappear, creating a deeper form of openness. As we will see in chapter 28, "Transparency, Translucency, and Virtuality," a strong experience of epistemic openness can even lead to suspension of the distinction between existing and nonexisting entities. Using slightly more contemporary terms, I will call this the "phenomenology of virtuality."

If, as I propose, we interpret "emptiness" as referring to the conscious experience of epistemic openness and virtuality, then this helps us gain a new perspective in a multitude of phenomenological contexts. From a third-person perspective, we can now describe certain aspects more clearly. But what will happen if a meditator continuously brings her own theoretical perspective into her practice, and if in her practice she is actually *motivated* by some belief system and the goal states that it defines—including, for example, particular metaphysical assumptions about what emptiness "really is"? Will third-person expectations shape first-person experience?

"Emptiness" is the perfect illustration of what in chapter 2 I called the "problem of theory contamination". The phenomenology of meditation is special, in that theoretical assumptions and belief systems may have colored the reports that we received in a particularly strong way. What is more, we all *embody* our beliefs and our deeply ingrained cultural background assumptions—namely, via the two-way window of the self-model created in our brain, which causally connects our inner life as "knowing selves" to the social reality around us.[1] The biological/computational process that underlies the creation not only of our low-level bodily self-model, but also of the brain's more abstract inner image of a thinking, knowing, and socially situated self, of course will directly influence conscious experience itself. Our best current theories show not only that there is a deep continuity of mind and life but also that life, embodied self-consciousness, and cultural context form a nested hierarchy. Our minds are *nested* minds, and by functionally appropriating theories via our self-model, we can "make beliefs our own"—changing our very brains, and changing the way we see reality.

There are a few chapters in this book that are longer than the others, and there is always a reason for that. This chapter is one of them; chapters 24, 26, 27, 28, and 34 are other examples. Here, I will take a closer look at the methodological problem of theory contamination by highlighting two of its deeper roots. The first root of theory contamination is the human need for meaning-making, and the second one relates to the constant temptation to deny our own mortality. I am fully aware that this will be a difficult, perhaps unsettling, chapter for many of my readers—but I am equally convinced that it will help us confront meaning-making and mortality denial constructively, to get the full picture and find out what the deeper value of meditation as an epistemic practice (as discussed in chapter 16) really is. As is the case with any mixed plate of hors d'oeuvres, there will be some that you like and others that you would rather leave untouched. If you want to give this one a try, let's take a deep breath and begin by slowly and carefully approaching the first topic: the process of so-called meaning-making.

It must have been around 1980 when, in Chalet Tanneg in the Swiss village of Gstaad, I had the chance to talk to Jiddu Krishnamurti all by myself.[2] For the previous four years or so, I had practiced yoga, breathing exercises, and meditation systematically and regularly. I told K about all the positive effects that this had had in my life, on many levels at once. Then he asked me a question that tore into my heart like a dagger—one of the deadliest, meanest questions I have ever been asked. It was a question I had never thought of, a question that at first hurt me but that also proved to be extremely helpful and reverberated through my life for many years to come: "Can you find out—not intellectually, but in your own direct experience—which of all these positive effects came from the actual meditation practice itself and which came from the fact that you *found something*?"

Many of us try to *find something*: something that allows us to live in an insane world without going insane ourselves. Something that creates coherence. Something truly sustainable that functions as a normative metacontext, some overarching theme or highest-level goal. Something that reenchants our world, a reliable long-term perspective. The process of meaning-making is this attempt to *find something*; it is the search for coherence, for something that holds steady in a world where everything else is impermanent. My first point is that, for many committed practitioners, the core motivation allowing them to sustain a regular practice over many years may consist to a considerable degree in the fact that they have *found something*, that they have adopted a certain belief system (or joined a group or spiritual movement, identified with a certain lineage or teacher, etc.), and not solely in the intrinsic force of the pure-awareness experience itself. This belief system, the sense of community, the acceptance of an authority, and the ongoing project of meaning-making may be among the major sources of motivation.

Previously, I said that the first motivational root of theory contamination is the human need for meaning-making, and that the second one relates to the constant temptation to deny our own mortality. Both these motivational roots influence our behavior on a subpersonal level, in the brain itself, but they also naturally lead to an intellectual interest in theories of consciousness, spiritual teachings about pure awareness itself, and the like. They therefore will influence the experience, as well as the way in which it is described later. Many of our participants may be familiar with the terminology employed by certain traditional and prescientific theories of consciousness, particularly that used in the context of spiritual practice. As I have pointed out, some will also have a good understanding of the epistemological and metaphysical background assumptions of these terms, and many will also have developed their very own, more intuitive understanding of what terms like "emptiness" refer to.

One example of this theoretical influence on phenomenological reports is found among religious practitioners. When trying to make verbal sense of their own contemplative experience, Christians may gravitate toward notions like "fullness" or "grace," while those inspired by Buddhist teachings may actively search for the experience that most directly correlates with their own intuitive understanding of "emptiness" or "spontaneous presence." A few years ago, I was invited to give a talk at the European Summer Research Institute, an event initiated by Mind & Life Europe that took place on the beautiful Fraueninsel at Lake Chiemsee in Germany, a large freshwater lake often called "the Bavarian Sea." (My talk later got censored by the organizers and banned from the website, along with that of the Buddhist teacher and author Stephen Batchelor.) During this meeting, Matthieu Ricard importantly pointed out how much of a difference is made by the basic assumptions from which a spiritual practice begins—for example, whether it starts from the idea of an original sin or an innate, ever-present Buddha-nature.

This is obviously right. Real spiritual practice, in my view, involves none of these things—it is the epistemic practice of *liberating* oneself from background assumptions. What we call "background assumptions" are not mysterious abstract entities, but conceptually and linguistically mediated processes that also involve subpersonal neural differences. The human self-model is the computational link connecting subpersonal and suprapersonal (i.e., social) levels of information processing.[3] We are enculturated beings, and cultural background assumptions create personal-level expectations, which continuously change the embodied brain's inner landscape of priors and unconscious beliefs. These assumptions are not merely reflected in the conscious layers of our self-model; our beliefs actually *penetrate* it from above. What is more, many of the preexisting traditional frameworks not only provide a conceptual system of phenomenological descriptions, but they also present us with a *prescriptive* element, telling the practitioner what progress really is and how the different "stages" of meditation *should* look.

This observation leads us to a second aspect of meaning-making: To practice meditation seriously, you need a certain quality of earnestness, a strong and reliable source of motivation and commitment. Intrinsic motivation is therefore an important causal force, and not everybody has it. For example, it is empirically plausible that meditation practice has low appeal for those who are "existentially indifferent"—the roughly 35 percent of the general population who show a low commitment to all sources of meaning but at the same time have no problem at all with this general lack of meaning. Existentially indifferent people demonstrate a distinct lack of interest in self-knowledge, spirituality, and explicit religiosity. Empirical research has shown that their psychological well-being is lower, but also that they have fewer mental health problems—the lack of experienced meaning and commitment to a system of values does not really bother them or cause psychological crisis.[4] They are just not interested in these things.

Empirically, I would predict that only very few long-term practitioners of meditation will be existentially indifferent in this sense. On the contrary, it seems likely that many of them will not only follow the philosophical ideal of living an examined intellectual life but will also have a higher-than-average interest in self-knowledge, spirituality, and explicit forms of religiosity. Committed practitioners will be those who have taken the ideal of self-knowledge to heart and transposed it into the more fundamental, nonconceptual levels of their life, thereby extending the original philosophical project well beyond the space of thought. They have an expanded interest in knowledge. As explained previously, long-term meditators will begin to embody their practice—and perhaps also to embody some of the background assumptions that guided them in the beginning, when they first joined a movement or identified with a teacher.

The qualities of intrinsic existential seriousness, sincerity, and often admirable discipline manifested by people highly preoccupied with a search for meaning can create their own pitfalls, however, because the sense of commitment may often be anchored in an ideology, authority, or other form of attachment to a metaphysical belief system that cannot be supported by rational argument or empirical evidence. In these cases, as practice progresses, personality changes will gradually ensue. The interaction of meditative practice and ideology is likely to lead to the construction of an alternative life story, and the practitioner may get entangled in an evolving narrative identity. Some give themselves a new name; others begin to wear funny clothes or even begin to act as spiritual teachers. (People have been known to actually write books.) In sum, the process may create a new kind of personal-level self-model, of the kind that Loch Kelly called a "spiritual ego-manager."[5] This may involve contraction into a spiritual superego—a more or less fundamentalist know-it-all who constantly labels, judges, and controls.

Strong intrinsic motivation to meditate also creates methodological problems. Many of our participants (about 77 percent) were regular practitioners who had been meditating for years; it seems plausible to assume that such strong discipline may often be anchored in adherence to specific belief systems and the conceptual framework of a certain lineage or spiritual tradition, or of a specific teacher, movement, or organization. This throws into question the whole idea of an "expert meditator" because the so-called expertise valued by scientists recruiting for experiments may be the most theory-contaminated of all. I have been talking to meditators all over the world for more than four decades and have engaged in numerous personal communications in the context of this research. For many meditators, there clearly exist "metaphysical taboo zones": things that remain and must remain unexamined. I also know from private correspondence that such belief systems can sometimes even lead to a decision *not* to participate in scientific projects. For phenomenological surveys like ours, this also means that there may have been not only a positive but also a negative self-selection bias at work.

It is conceivable that a relatively large percentage of *phenomenological* experts chose not to participate in our study because the idea of examining pure awareness contravened some aspect of their belief system. Perhaps the number of practice hours is not a particularly useful parameter, because in some cases, it expresses the degree of attachment to a certain belief system rather than some mysterious form of "expertise." Here, then, is a research question for the future: At what point in the lifetime of a long-term practitioner will theory contamination reach its peak? This is an open question. It is certainly possible that some motivating belief systems are actually self-reinforcing, becoming *stronger* over time and gradually turning the meditator into a pious, intellectually narrow-minded "true believer."[6]

Obviously, all this is not to say that ancient theories of pure awareness—based on millions of hours spent in silent meditation by the serious scholar–practitioners who came before us—do not have great value. To quote the old analogy coined by Bernard of Chartres (who died after 1124), in this domain of research, too, we are truly standing on the shoulders of giants. I think that the phenomenological depth of these traditional systems is often enormous, and it's positively embarrassing that Western science has failed to acknowledge their value for so long. But a fresh, bottom-up approach in a new historical context, spanning many countries and invading intellectual taboo zones, will have its own value. Such an approach helps to weaken the influence of theory contamination (which creates positive biases) on the one hand and taboo zones (which create theoretical blind spots) on the other. Nevertheless, I believe that when the focus of our inquiry is the phenomenology of pure consciousness, the influence of preexisting theories is a much more serious obstacle than it is in other, related fields of research. In what follows, I want to highlight the two fundamental psychological needs that most regular practitioners will have, briefly pointing out how they relate to intrinsic motivation and the methodological problem of theory contamination, specifically in the context of pure-awareness research.

As we have seen, the first of these deep needs is to develop a successful and sustainable strategy of meaning-making—to find a more permanent solution to the constant struggle to make sense of life events. This need is shared not only by meditators, but by all those who are not existentially indifferent. But for many of our participants, their meditation practice will have been a central part of their own strategy of meaning-making, embedded in the larger context of the attempt to make sense of their own life. "Meaning-making" is a vague and slightly superficial term that often refers to the process by which human beings internally construe or try to understand or emotionally make sense of the chain of events constituting their life, relationships, and what they take to be "their" self.

Typically, the difference between the mere subjective experience of meaningfulness and the much more difficult philosophical question of whether something like an objective "meaning" exists at all—and how it can be known, if so—is deliberately glossed over. For many people, regularly repeating experiences of meaningfulness or of "being touched" seems to be good enough; the good feeling is what counts, while the deeper issue relating to the conceivability of objective meaning in life is systematically avoided. Often, the idea of meaning-making is used as a confabulatory euphemism, a milder word for something considered to be too harsh or blunt, for a reality that we do not want to face because it is too unpleasant or embarrassing. I will call this reality our deep-rooted need for "narrative self-deception."

One simple conceptual point to note at the outset is that words and sentences have meaning, but events don't. Events just happen. Events as such, as well as whole chains of events connecting birth and death, are intrinsically meaningless—but not in the emotionally negative sense that we may attribute to the idea of life as a whole being meaningless. What "meaning-making" really refers to is an active search for positive experiences that create certain phenomenal qualities of emotional security and coherence over time—a quest for transtemporal sameness of the self, ideally accompanied by the phenomenal signature of knowing, a positive feeling of insight. Of course, given the impermanence of everything, this coherence over time can be only limited at best. There is a quality of absurdity in the depth of our longing for coherence and the obvious futility of all our attempts to ever reach a stable state; and there may be a related form of absurdity in seeking this coherence on the level of phenomenology, in fleeting conscious experiences of "meaningfulness." I think that to intentionally engage in narrative self-deception is a fundamentally antiphilosophical activity.

To repeat our simple conceptual point from before: There is a difference between systematically seeking mere *experiences* of meaningfulness, or cultivating the phenomenology of existential soundness and coherence, and the more genuinely philosophical project of investigating the possibility of *epistemic* meaning-making while avoiding the trap of the E-fallacy (as explained in chapter 7). It is quite possible that many practitioners of meditation do not see this difference. Be that as it may, for many meditators their practice is something that not only takes place from moment to moment and day to day, but also is attached to a larger project of understanding life by building a coherent inner model of their own existence as a whole, of their spiritual path as it unfolds over time.

Let us flesh out the new concept of "narrative self-deception," first by contrasting it with what it is not. A second, equally vague, and slightly superficial cluster of ideas centers on notions like "narrative self," "narrative identity formation," and the like. Of course, a lot of current research shows that there simply is no self that could be the active narrator of an inner life story. The whole point is that the process in question is entirely selfless but *seemingly* creates an entity that is in control and remains the same across time.[7] There is no narrative self, only a subpersonal process generating a fictitious unit of identification. But for enculturated beings like us, the idea of a narrative identity is highly intuitive, and therefore often recurs in popular debates or phenomenological approaches to the mind. It is as if characters in a self-organizing life story suddenly assumed that they actually had a genuine form of self-awareness (which no fictitious entity can have) and even began to play a strong authorial role (which no character really has). It is as if they were trying to control the script—perhaps even trying to "wake up" to their own fictitiousness. But likening selfless, dynamical self-organization in the embodied and enculturated brain to a work of literary fiction, inspiring as it may be, creates puzzles and paradoxes. For example: Can one simultaneously be an author of and a character in the same life story? "Narration" vaguely refers to a data format that appeals to us because in a certain way, we really seem to live through our own stories—externally as well as internally. Human beings are storytellers, and they stabilize the fabric of their long-term self-model by creating a permanent inner monologue. They identify themselves with Harding's "little one" (chapter 8), the entity constantly chattering on about all its extremely interesting and important perceptions and thoughts and emotions to keep the long-term self alive. There is no better way to understand this fact, to observe its continuous fight for survival under the microscope of mindfulness, than to participate in a silent meditation retreat.

The idea of a life story seems entirely natural at first, and it has the advantage of potentially adding not only causal but also *thematic* coherence to our long-term self-model. A story encourages us to hallucinate cause-and-effect relationships between events in our narrative ("Nothing ever happens by chance!") and detect patterns in the way that they affected us ("I have always been interested in this!"). Most of all, a life story offers a chance to insert a single, overarching theme that holds everything together, a normative metacontext that allows us to morally evaluate ourselves and others with the goal of enhancing and sustaining our sense of self-worth. The romanticism of "being on a path" is a good example of this. Now we can judge, negotiate, plan for the future, and thereby continuously stabilize the long-term self-model. There is a way of life, a general theme. The story has a beginning and a possible end; it has characters, episodes, a setting—and like a dream that has become fully lucid, it even has a plot that can apparently be controlled. Nonlucid dreams, in which plot control is absent, are a nice example of what "narrative self-deception" actually involves on a subpersonal level. While dreaming, you can see the storytelling mechanism in action: a mechanism desperately attempting to weave random events into a coherent narrative; an automatic brain process in constant search of a thematic red thread; the tireless creation of meaning, intelligibility, and control in an inner environment that in reality is entirely unpredictable. The result is the dream narrative you can later report.

Perhaps you know from your own dream experiences the feeling of attempting to remember something important that you apparently *just* forgot—the feeling of constantly trying to reestablish a sense of control, trying to recall the overall theme, as it were, or to understand *what it was all about.* Like Sisyphus rolling a boulder up a mountain only for it to roll down again every time it neared the top, the dreaming brain tries to stabilize the long-term self-model. Again and again, it looks for the red thread, attempting to successfully hallucinate sameness across time. A never-ending Sisyphean task, it shows not only that narrative self-deception is something that takes place on the level of whole persons and culture, but also how we embody the battle for meaning and uncertainty reduction, subpersonally, even at night. Narrative self-deception is a general computational principle that works on a number of levels simultaneously.

My main point is that, for a considerable number of meditators, their practice may play a crucial role in endowing their own lives with thematic coherence, to the point where their ongoing search for thematic consistency may actually be more important than the actual practice and any of its other causal effects. Sitting on a cushion, alone or with others, may be only one part of crafting an overarching framework, finding a more gratifying sense of community, and somehow trying to feel at home in this dangerous and unpredictable world. Meditation may be part of an attempt to "recall the overall theme," driven by the hope of *finding something.* How the theme is formulated will influence how practitioners later describe their experiences—but, of course, the theme will also partly determine the actual content of those experiences themselves, what they actually find.

The impact of narrative context on the contents of consciousness seems crucial to acknowledge. However, as Emily Troscianko has reminded me in a personal communication, the whole idea of "the" actual content is in fact dubious because we have to get away from talking both about consciousness as a container with "contents" and about "the experience itself." New theories of the human mind as a form of hierarchial Bayesian-updating,[8] plus much empirical data, suggest that we should instead develop a new version of what Daniel Dennett has called the "multiple drafts model" of consciousness.[9] In this model, the nature of the probe applied (e.g., the precise way in which we ask the question "What am I conscious of now?"; see figure 34.1 in chapter 34) determines what is retrospectively called "the experience," with no "real phenomenology" beyond that post hoc labeling. There is a quality of groundlessness here, and this is where advanced spiritual practice comes back in: If we start to investigate the actual process of life-storytelling with fresh eyes, under the microscope of mindful attention, then it becomes obvious that what's going on is really a competition among multiple drafts. In this new context, let me also draw attention to the fact that, quite often, narrative self-deception may have enabled human cooperation in large groups and in turn now critically depends on that social cooperation, on the kind of mutual reinforcement we find in religious communities, cults, or spiritual movements trying to construct an alternative reality for themselves. All these social factors shape the virtual stories competing for airtime, as well as dramatically raising the stakes of the competition.

The shaping of experience and recall by meaning-making endeavors also has a sociological aspect because for some meditators, the project of meaning-making is that of actively "reenchanting their life-world." Max Weber, when he coined the famous phrase *die Entzauberung der Welt* (the disenchantment of the world), was referring to the sense of disenchantment caused by cultural rationalization, by the unstoppable rise of science and technology that had led to a bureaucratic, "soulless," and secularized Western society, where scientific understanding was more highly valued than faith or personal experience. More than a century ago, in 1918, Weber wrote: "The fate of our times is characterized by rationalization and intellectualization and, above all, by the 'disenchantment of the world.'"[10] This sociological context has created countermovements, and for many meditators, their own practice and the way that they interpret their experiences may be precisely a project of restoring the magic, aiming at *die Wiederverzauberung der Welt* (the reenchantment of the world).

As our data show, this project can clearly work on the level of individual phenomenology: Just think of the experiential quality termed "nonsensational awe" (chapter 1) or our investigation of "suchness" (chapter 9). You may also think of the experience of connectedness (chapter 11) and the rich spectrum of states characterized by joy and gratitude (chapter 15). As we move on, we will encounter more examples of MPE modes of consciousness that clearly involve a magical and enchanted quality, like timeless change (chapter 22) or translucency and virtuality (chapter 28). One important research target for the future will be to better understand the mutual relationship between the meditator's individual phenomenology and the sociological context in which it is embedded.

Our problem is that we need a better understanding of theory contamination as it applies to the pure-awareness experience. We need an understanding that goes beyond intuitive, feel-good, vague, and therefore popular notions like "meaning-making" or "narrative self." Let me offer a series of new concepts to get a clearer view of this problem. The first of these concepts is the "autobiographical self-model," which allows individuals to organize episodic memories and abstract knowledge of their past into a coherent biographical picture. The autobiographical self-model is not a thing, but rather a process. It is also not a little man in the head. This process is mostly subpersonal: To a large extent, it is not something that *you* do, but an automatic process taking place under the hood, in your unconscious brain. The process constantly tries to minimize surprise and to keep a certain layer in your self-model as coherent as possible (namely, the layer that portrays your life as a whole). At the same time, the process of the autobiographical self-model tries to make the organism more intelligent by expanding what mathematicians and computer scientists would call the "predictive horizon."

Intelligence can be described as the capacity to successfully predict the future, the continuous attempt to expand the organism's predictive horizon. Biological intelligence consists partly in maintaining a steady state in the here and now (e.g., through homeostatic control of body functions), but efficient regulation also requires anticipating future needs and preparing to satisfy them before they arise (called "allostatic control"). The human self-model does all these things at the same time. The body model helps with online motor control and with simply staying alive, while the autobiographical layer in the self-model stores a selection of past events while also planning for the future, seamlessly connecting what has already happened to possible future selves. The model expands into time, creating a past and a future. Whenever this layer is conscious, it can function as what we have termed the "unit of identification." You may think that you simply *are* your life story, which springs from the ongoing search for thematic stability. But you may also begin to hallucinate an observer, an active narrator, or a stable protagonist within the story itself—an entity that reflects on and monitors the whole process (more on this in chapter 25).

I coined the term "narrative self-deception" a few years ago, when investigating the problem of conscious suffering.[11] Here is a concrete example. A spiritual biography can be a form of narrative self-deception, at least in part. It can create an illusory unit of identification, an invisible yet robust sense of self endowed with a feeling of transtemporal sameness. Scientifically, a number of different types of memory have been identified and classified, and we know that none of them is aimed at veridicality, that all are instead incrementally optimized for replication and survival. We have also learned that self-deception has evolved, making animals more efficient in the short term and also allowing them to deceive *others* more successfully.[12] Today, we must face the fact that our autobiographical self-model is not a tool for self-knowledge in a philosophical or spiritual sense, but rather something that evolved in an evolutionary arms race. As a matter of fact, it seems to be one of the major obstacles to the kind of self-knowledge sought in contemplative practice. It created an elaborate form of biological intelligence, but the formation and organization of autobiographical memories also constitute the central mechanism through which our conscious sense of identity across time is constructed. This mechanism is what makes the mind wander, what creates dual awareness, and what continuously decouples attention from the present moment, and of course, it directly influences how individuals develop their more temporally extended sense of self. Long-term goals play an important role in the embodied process of meaning-making, as do the emotional meanings that we assign to them. Thematic coherence—or its loss—is something you can *feel*.

Narrative self-deception may also be a clever strategy for distracting attention from suffering. It allows you to feel better about yourself. From an evolutionary perspective, any self-conscious system that discovered too many negatively valenced moments—too many conscious experiences that it would rather not go through if it had a choice—might become paralyzed and stop procreating. Like a Buddhist nun or monk, it might refrain from adding new human beings to this world of impermanence and delusion because its own insight into the nature and ubiquity of suffering would have become too clear. Monks and nuns who obey the rule of celibacy are nobody's ancestors: They do not copy their genes into the next generation, and therefore, they are side alleys and dead ends of biological evolution. The logic of psychological evolution clearly mandates concealment of the facts of impermanence and conscious suffering from any self-modeling system that is supposed to be an efficient copying device. As a general rule (but with plenty of exceptions), natural selection will often have favored those who were aggressive and greedy, functionally attached to the goal states that they hallucinated, eager to have children and to rise in a social hierarchy, and reliably caught on the hedonic treadmill.

Systematic distraction from the potential insight that the life of an antientropic system is one big uphill battle, a strenuous affair with minimal prospect of success in more than the very short term, may certainly not be in the interests of the individual vehicle itself, but it probably causally enhances the process of genetic evolution, ensuring that genes keep getting replicated. Any conscious representation of negatively valenced uncertainty under the condition of identification (chapter 8) causes suffering. The evolutionary algorithm therefore has selected for genetic instructions that make it less likely for an organism's insights into the deep structure of its own mind—insights of the type just sketched—to be strongly reflected in its conscious self-model. As a result, we now have a cognitive scotoma for conscious suffering,[13] a systematic blind spot that makes it difficult to see certain obvious truths about our own lives.

Conversely, an adaptive advantage may have applied to humans' tendency to suffer from a robust version of optimism bias as one form of narrative self-deception: If you unrealistically believe that you are much less likely to experience negative future events than anyone else, you may procreate more successfully. The primary function of the human self-model's autobiographical levels may be precisely to drive the organism forward by relentlessly expanding its sense of self in time, thereby making it lose contact with the often-unpleasant reality of the present moment. The autobiographical self-model of humans may have been so successful in part because it provides a functionally adequate form of self-deception, glossing over the ugly details of everyday life by developing a grandiose and unrealistically optimistic inner story—a "narrative self-model" with which we can identify.

What has all this got to do with meditation practice and the problem of theory contamination? Well, many forms of meditation practice involve the opposite of this future-oriented sugar-coating of the present: Mindful awareness means continuously *shrinking* the conscious brain's predictive horizon until the quality of timeless change, which we will investigate in chapter 22, begins to reveal itself. Mindful awareness is what *ends* the narrative. The bad news, however, is that for at least some meditators, the wider context of meaning-making and narrative self-deception may be at least as important as the actual practice itself—this possibility is what Krishnamurti's question pointed to. (The same may be true of other forms of epistemic practice: scientists conducting meditation research, say, or academic philosophers interested in the problem of consciousness.) Therefore, a lot of what appears to be strong intrinsic motivation may actually be extrinsic, driven by the surrounding social context by having succumbed to some spiritual authority or some set of metaphysical background beliefs. Belief systems offer labels, value-charged theories of consciousness, and taxonomies to describe inner experience. This simple point ends my sketch of the first of the two deeper roots of the methodological problem of theory contamination: narrative self-deception and mortality denial. We must face the fact that in the phenomenological material that I am presenting, all the factors that I have sketched out here will have contributed to an unknown amount of contamination.

The good news is that there is probably no better tool than the practice of meditation itself to help us *really* understand the mechanisms of self-deception that continuously express themselves as the subtle workings of one's own mind. This is a deeper form of understanding than mere theoretical knowledge—one that can have direct causal consequences because it takes place on a nonintellectual level that goes far beyond words and concepts. The fine-grained mental mechanisms of meaning-*making* itself are one of the most interesting objects of mindfulness practice. Classical insight meditation, for example, consists in nothing other than observing the actual process of narrative self-deception at work: Every single thought arising is an attempt to generate a new self-model, to become "temporally thick," to escape the wholeness of the present moment. Here, you can catch the king of escapees in the act and finally watch the autobiographical self-model under the microscope, including the fact that it is actually selfless—a self-model only in the computational sense, and not really a *self*-model when it comes to the phenomenology of actual experience. The automatic process of expanding the predictive horizon and automatically identifying with some future self is entirely empty in the sense that it is epistemically open and has no fixed meaning. A lot of what I said about the selfless and entirely nonconceptual quality of "suchness" in chapter 9 probably also applies to the act of viewing our own life as a whole. There is no tragedy in suchness. Life just happens—and meditation practice is a perfect way to become aware of this fact.

It is true that the need for meaning-making often leads to systematic and even socially organized forms of narrative self-deception, but it does not necessarily have to. Let us turn to the second fundamental psychological need that most of us share, meditators and nonmeditators alike: *mortality denial*. We all have to manage death anxiety, and we all have to come to terms with the qualities of absurdity and futility that accompany any insight into our own finitude. We may not consciously experience this need, but there are subpersonal as well as cultural mechanisms constantly trying to satisfy it for us. There are whole industries dedicated to helping us practice narrative self-deception and mortality denial—just think of the Roman Catholic Church. The fact that these industries have endured for many centuries, surviving much longer than any form of government, any political system, any tribe, and most nations, demonstrates how deep and universal these needs really are. After they appeared, they changed our inner environment, and we had to adapt to this fact.

Human beings are the products of evolution, which means that our bodies and our minds have been optimized not for happiness or self-knowledge, but for survival and procreation. At the same time, we have a new problem that no other animal on this planet seems to have: We have a direct, self-conscious insight into our own mortality, into the fact that ultimately, I will *not* survive. I call this new problem the emergence of a "toxic epistemic state," something that interferes with the coherence of our autobiographical self-model, as just described. Let us define a "toxic epistemic state" as any form of knowledge or insight that threatens the biological fitness of an organism, its individual reproductive success. An epistemic state that is toxic from a biological perspective may be seen as a healing or liberating form of insight from a spiritual perspective. One therefore may speculate that specific forms of self-knowledge have become much more explicit in some spiritual practitioners than in their fellow human beings, such as insights into mortality, futility, and the prevalence of unnecessary suffering. But viewed from an evolutionary, biological point of view, the knowledge that my own death is inevitable is poisonous. It threatens my mental integrity. A solution has to be found. This knowledge calls for an expansion of narrative self-deception into the time that lies beyond my physical death—or into the dimension of timelessness. In this context, again, it is interesting to note that in the history of humankind, many spiritual practitioners were nuns and monks who took a vow of chastity and had no children.

In his 1973 classic *The Denial of Death*, Ernest Becker said that mortality denial can take the form of a personal "immortality project" in which a symbolic belief system ensures that I can believe that my self is superior to physical reality. My first point is that for some meditators, their own practice may be part of such an immortality project, and this fact may at times distort how they choose to speak about their own inner experience. Here is how Wikipedia explains one of Becker's central insights: "By successfully living under the terms of the immortality project, people feel they can become heroic and, henceforth, part of something eternal: something that will never die as compared to their physical body."[14] If Becker is right, there may be a specific form of mortality denial that we could call "contemplative heroism." Meditation practice would be part of a larger project, the heroic battle against finitude, an epic quest to discover that part of yourself that you can firmly believe will never die.

Mortality denial has many other facets and consequences beyond personal immortality projects. For example, populism is directly related to meaning-making and mortality denial. If you are a politician, offering a tribal narrative plus a form of symbolic identity that allows people to identify with something greater, with something that extends far beyond their physical death, then you will typically be more successful than any secular or intellectually honest competitor you may have. If you are one of the many new entrepreneurs in the global attention economy (or an algorithm aimed at extracting attention from human brains), then you will always be more successful when you learn how to play to your audience's unconscious needs by offering an interesting new strategy for narrative self-deception or a particularly clever route to mortality denial. And if you are a philosopher, you will always have more readers if you manage to subtly include some new and clever backdoors for mortality denial in your work. A mild but strategic form of vagueness is a very useful ingredient because it helps the reader to actively project her desired experience of meaningfulness into your words. Even if people don't fully understand what you are saying, they will get an intuitive sense that "this feels right," that—even if what you really mean remains slightly obscure from a conceptual point of view—on an emotional level, it literally "makes sense," and therefore it must be relevant and significant in some deeper way to *them* and their own project of meaning-making, of successful narrative self-deception. Just like assisted suicide, assisted self-deception is a type of social cooperation: It is special in that it often involves interesting new forms of complicity and creativity in mortality denial, forms that are richly rewarding for all parties. Of course, everything that I have just said about ambitious politicians, attention entrepreneurs, and philosophers can apply equally to spiritual teachers.

In this chapter, I have already provided a number of new conceptual instruments: the idea of meaning-making as "narrative self-deception," the concept of an "autobiographical self-model" that automatically creates an image of our life as a whole, and the notion of "contemplative heroism" as a strategy for dealing with impermanence and futility. In this final part, let me offer you two more conceptual tools. The first is "absurdity management." Insight into mortality is a toxic form of self-knowledge, a permanent danger that has unexpectedly sprung up in our inner environment. Developmentally, it typically emerges at around the age of eight to ten years, and it presents an obvious danger to our mental health, including to our sense of integrity and self-worth. Not only does it create a concrete form of anxiety that needs to be managed,[15] it also creates an almost all-encompassing, global feeling of futility and absurdity that has to be kept out of the conscious self-model just as urgently as coherence needs to be found in a dream. I will call this experiential quality the "Sisyphean quality." Our biological self-model has now been damaged—informationally poisoned, if you will. Therefore, it calls for a boulder-pushing adaptation, a psychological coping strategy, a way of efficiently integrating or simply denying the reality of death. The brain *must* react; it must create a sustainable new model of reality.

Here, my general point is that for many meditators, their practice will likely be part of a wider strategy of mortality denial, of "finding evidence" to build this new death-denying model of reality. The practice can create certain health-promoting, stress-reducing altered states of consciousness, but what really counts for the individual practitioner may sometimes be the metaphysical *interpretation* of these states, the adherence to a new belief system that helps with death denial. Please note that all of this is only one example of what may actually be humankind's oldest coping strategy, *the* classical form of absurdity management: First, you create altered states of consciousness, such as by drumming, dancing, fasting, depriving yourself of sleep, performing shamanic rituals, or gathering hallucinogenic plants and mushrooms (step 1); and then you *interpret* in the desired fashion the alternative models of reality that your Sisyphean brain creates in such states (step 2). For example, you could later say that you have now "directly experienced" the fact that something can exist without the physical body, that you have communicated with spirits, the angels, the dead, or similar entities. Of course, the altered states of consciousness created by meditation practice can be abused in the same way—so long as you interpret them as a direct experience of "reality itself," creating the right social context for cultivating certain versions of the E-fallacy (step 3).

This may also be one reason why many Asian systems of philosophy do not clearly separate the phenomenology and the metaphysics of pure consciousness: These systems originated in religious systems of mortality denial. If this is so, then there is a methodological problem, because such belief systems will not only color external experiential reports but may also (via the person's self-model) install unconscious assumptions in the very brains that shape the experiences themselves, like an additional set of self-fulfilling prophecies acting on the subpersonal level.

Given this new context, do you remember the "dolphin model of meditation" sketched out in chapter 10? It showed how unconscious assumptions can make certain experiential contents "spontaneously appear," as if uncaused and unborn. In short, many human beings have a deep need for absurdity management and their individual successes (and failures) in satisfying this need will color the phenomenological reports that they give. This is all too human and nothing to be condemned, but it does create a methodological problem.

Bhava-taṇhā is a term that Buddhist philosophers have known and analyzed for 2,500 years. It refers to the craving for existence, which is one of the deepest causes of conscious suffering in humans, and probably in many other self-conscious animals too. To be driven by the craving for existence, you do not have to possess explicit knowledge that you will die. *Bhava-taṇhā* can also be interpreted as the craving to be something specific or to unite with an experience, as well as the correlated striving for permanence—the phenomenology of identification discussed in chapter 8. I think that modern science is making it more and more obvious that biological systems like ourselves are actually physical *embodiments* of this craving for existence—if you will, antientropic Sisyphus-machines fighting a constant uphill battle to preserve their own boundaries. I think that this results in a deep and fundamental distortion of the way that our conscious models of reality portray the world and ourselves in it. This distortion had been nonconceptually investigated by meditators long before modern science even entered the stage. I call it the "existence bias."

The link between our need for absurdity management and existence bias is that there is a deep absurdity in *being* an embodied craving for existence while at the same time knowing that you will have to die (i.e., having a cognitive insight into the fact that this craving will one day be frustrated). The insight is toxic; the craving becomes poisoned. So what exactly is existence bias? One of the deepest roots of human suffering and self-awareness is a top-level preference that creates a self-directed variant of existence bias: the fallacy of treating the mere existence of something as evidence of its goodness. Here, however, the concept of "existence bias" refers not to the well-documented fact, widely known in social psychology, that human beings generally favor the status quo,[16] but rather to the more specific observation that beings like ourselves will almost always opt to sustain their own physical existence.[17]

Of course, human beings will sometimes sacrifice themselves to save their offspring or protect their tribe. We are replicator-copying survival machines that have been mercilessly optimized for millions of years to never give up, to optimize inclusive fitness, and to maximize our contribution to the gene pool. But on an individual, physical level, we are antientropic systems fighting an uphill battle in a constant attempt to reduce uncertainty and "understand ourselves" by finding a viable strategy of self-modeling—we are free-energy machines who, in the words of neuroscientist Karl Friston, continuously "maximize the evidence for their own existence."[18] As the philosopher Jakob Hohwy has made admirably clear, we are not just biological agents endowed with information-hungry brains relentlessly gathering more data to produce ever more evidence for our own existence.[19] We can also be viewed as self-organizing systems trying to continuously *expand* our predictive horizons, desperately sustaining our existence in a dynamical environment by following an inner norm of tracking the very conditions of possibility for existence themselves.[20] Our phenomenology deeply reflects this computational imperative for constant self-evidencing, and on many hierarchical levels. As a matter of fact, pure awareness itself may be interestingly related to the conditions of possibility for existence themselves (more on this in chapter 32). What makes it so interesting is that it nevertheless lacks the experiential qualities of absurdity and futility.

Armed with the concept of existence bias, we can be more precise about the phenomenology of absurdity and futility. The quality of absurdity consists in the fact that the conflict between the embodied existence bias (which we *are*, from which we originated) and the high-level cognitive insight into our mortality has seemingly become a permanent functional feature of our self-model, something built in. If you will, it is now not a bug but a feature—and that is what brings the quality of absurdity into human existence. As already noted, *bhava-taṇhā*, the automatic craving for existence, is one of the deepest causes of conscious suffering in humans, and probably in many other animals too. It also led to the evolution of religion, because what is special for humans is that we have to deal with the additional challenge of "toxic self-knowledge" threatening the integrity of our self-model. The quality of futility originates in the fact that although we want to avoid nonexistence completely, we now have explicit intellectual knowledge that, ultimately, this will be impossible. Futility is one major component of absurdity. No other animal has this kind of problem. As human beings, we need a strategy for absurdity management because we explicitly know that every single individual will eventually lose the uphill battle sketched here—that our predictive horizon will eventually shrink to zero simply because in biological evolution, "passengers are not carried."[21]

Bhava-taṇhā and the concept of an embodied existence bias also yield a new perspective on our deep-seated tendency toward mortality denial: Mortality denial is not simply some sort of false belief or a misguided intellectual attitude; it is a computational principle. This principle consists in automatically neutralizing "mortality-salient information"—all the toxic and potentially demotivating information that leads to the individual's awareness that death is inevitable—by explaining it away in the context of a new, alternative model of reality. We embody this principle. This is why it surfaces on the level of psychology and culture, and why it creates a systematic blind spot in our conscious self-model.[22]

Of course, much more could be said at this point about different levels and units of selection in the evolution of death denial, particularly when it comes to the interplay between biology and culture. But figuratively, we were never *supposed* to mentally simulate possible worlds in which we have died, just as we were never *supposed* to use up too many precious mental resources to imagine situations in which we are sick or have no offspring—except, of course, in acutely dangerous situations where fast, intelligent planning is needed.

Friston says that "biological systems move around in their state space, but revisit a limited number of states"—namely, those in which we are safe and healthy, situations that do not present us with ugly surprises but that actually correspond to stable, survival-friendly homeostatic steady states far from equilibrium.[23] Please note that the insight into mortality is exactly one such ugly surprise, but one that occurs in our *inner* environment. Biological organisms like ourselves must maintain certain vital parameters, certain critical physiological variables, within particular bounds; we must avoid unexpected phase transitions and protect a state far from thermodynamic equilibrium.

Technically, any successful biological organism will achieve in its lifetime only a very small number of all possible obtaining states. Every member of this small group of states has a high probability, and being dead is not one of them. Unless forced by circumstances, we will never plan for or imagine all those other states that have low probability, or else we wouldn't be here. Morbid mopers were rarely among our ancestors. The simple reason is that beings like us needed all their limited mental resources to continuously develop new existence-sustaining action policies. They used their brains to simulate paths into physically improbable worlds where they and their children were still alive and healthy. This is an important part of what it means to act intelligently and adaptively within our environment. Ultimately, our tendency toward mortality denial falls out of statistical physics: Beings like ourselves survive by predicting our own existence, not by predicting nonexistence. Previously, we saw that human consciousness can be described as being governed by an embodied existence bias. Now we can extend this point: In a sense that could be technically well defined, we actually *are* an embodied form of mortality denial. We are resistance itself.

But we are also social beings. In dealing with toxic self-knowledge, we have had to develop enculturated strategies for mortality denial and narrative self-deception, which in turn shape the structure of our conscious self-model in a top-down manner. We invented organized religion—and we became dangerous beings who were religious but not spiritual (our snappy acronym here might be RBNS). Our immortality projects and the faith-based religious heroism of absurdity management have created an endless chain of wars and social conflicts. Ultimately, death anxiety has created more death anxiety.

Again, let us ask: What does all this have to do with meditation practice and the problem of theory contamination? The bad news is that for many meditators, their practice is likely tied to a wider context of mortality denial: using the phenomenology as proof of the existence of an afterlife, interpreting it in a way that confirms some organized system of mortality denial. To illustrate, here is an interesting detail: Humankind has evolved five major religions, but only two of them—Hinduism and Buddhism—are strongly and unequivocally related to a truly systematic cultivation of contemplative practice and the experience of pure awareness. At the same time, Hinduism and Buddhism are the only two of the major religious systems that involve an explicit belief in reincarnation. Whereas in monotheistic religions like Catholicism, mortality denial is primitive, blunt, and direct (according to the Nicene Creed, there will simply be a resurrection of the dead), many ancient and all major Indian religions (namely, Buddhism, Hinduism, Jainism, and Sikhism) promote belief in another metaphysical model, according to which at least one part or aspect of the currently living human being will start a new life in another physical body after death.

The Buddhist and Hindu models for liberation and mortality denial are slightly more sophisticated because they involve a false alternative: Of course, we do not *want* to be reborn because our true goal is to liberate ourselves from the cycle of birth and death. But if we do not have the good fortune to attain enlightenment in this life, the poor ego will simply have to continue in another physical form. We could even call this the "binary model for death denial." The false dichotomy behind the binary model is easy to see through, but at least it offers two possible trajectories for successful absurdity management by narrative self-deception: enlightenment or reincarnation—giga-bingo or another life.

I am aware that all this may sound harsh. But the good news is that there is probably no better tool for *really* confronting one's own mortality and the fear of death than meditation practice itself. This is what goes to the root of the problem. It takes more courage, but it may also be more efficient than any form of intellectual gymnastics. Over the centuries, many Western philosophers have pointed out in one way or another that living an examined life has a lot to do with learning to die, and as a matter of fact, confronting one's own mortality as directly as possible may turn out to be the deeper core of any more serious spiritual practice. As the quotations presented in this section's epigraph show, the Western tradition of learning to die has focused on using reason to transcend individuality and the passions and to move toward universal principles like rationality and intellectual insight, while the Eastern tradition operated on a more direct and practical level, also attempting to transcend individuality and the passions toward something universal, but by cultivating pure awareness and peaceful ego dissolution. If there are toxic epistemic states, maybe there are also *liberating* epistemic states?

Science and meditation are both epistemic practices; narrative self-deception and mortality denial aren't. We have explored some of the ways in which meditation proper is an epistemic practice, and how it goes far beyond stress reduction, well-being, capitalist imperatives of self-optimization, or vague ideas of "self-actualization." Cultivating epistemic openness and surrendering to the real possibility of ego dissolution in more advanced meditative practice may constitute a radical version of what in early Western traditions was termed "living an examined life" and "training for death." It is the epistemic practice of investigating and confronting existence bias head on (e.g., by viewing it through the lens of nonegoic self-knowledge)—and in this sense, it is philosophical practice and consciousness research taken slightly more seriously.

There is also a deeper phenomenological discovery to be made, which I would say that you can verify through the practice of meditation: Reflexive MPE, the timeless experience of nonegoic self-awareness (see chapter 30 for more), is free of the almost all-pervading, global affective tone of futility and absurdity that was mentioned earlier in this chapter and that needs to be repressed in the egoic self-model. The Sisyphean quality is *not* part of its phenomenal character. On the other hand, MPE is positively characterized by what in chapter 1 was termed the experiential quality of "existential ease." Ultimately, therefore, meditation may be capable of dissolving the need for narrative self-deception, mortality denial, and all heroic forms of absurdity management.

18 Luminosity

I would best describe it as opening my eyes while my eyes were closed, or opening my eyes underwater. [#1364]

The kinds of phenomenal character sometimes described as "luminosity," "radiance," or even "enlightenment" come in many varieties. To give the two most important examples: Many practitioners describe a nonperceptual phenomenology of "clear light," or more general clarity and epistemic openness, while others report a more visual form of brightness, which can be experienced with closed as well as open eyes. It is quite possible that we are dealing with fundamentally different phenomena here, and that some reports about luminosity of the second type are actually self-fulfilling prophecies caused by taking light metaphors too literally. I will let you judge for yourself. For simplicity's sake, I have created two major subsections for reports referring to this specific aspect of the pure-awareness experience.[1]

We could label the first type of experience "luminous emptiness":

48 My experience occurred during a Douglas Harding "Tube experiment." My end was perfectly colorless, empty, boundless, and yet aware. The emptiness was luminous. There was a strong sense that the other's face was my own (like looking in a mirror) and a deep sense of tenderness and gratitude for that face.

1311 [. . .] Almost impossible to describe, because the word "experiences" actually misses the point completely. The closest would be like a boundless space of awareness without center or edge, radiating out of itself, which is completely empty of anything identifiable but at the same time also extremely concrete, full of objectless love, and very close, intimate, and also extremely "normal." I say "is" and not "was" because this "space"/ "state" in which I "was" or which I am is exactly the same every time, without any variation, and in this point also completely different from psychedelic drug or other mystical experiences [. . .].

1364 [. . .] It is inevitably very hard to describe the sensation in words, but I would best describe it as opening my eyes while my eyes were closed, or opening my eyes underwater. A sudden perfect clarity of experience. Pure awareness, yes, but fuller than that—a brightness unrelated to color/light is another good expression! Awareness unclouded by thought or sense information, but not without this information either. Not negative, positive, or neutral.

3628 [. . .] There was a feeling of energy, nonvisual brightness, unity and expansiveness. The perception of time was diminished, as I was not trying to cognize or estimate time. [. . .]

The second type of report refers to "luminosity" as a more clearly visual form of experiential content, with its phenomenal character therefore more related to one specific sensory modality. We could label it "quasi-perceptual luminosity" or "concrete enlightenment." In type-2 phenomenological reports employing the concept of luminosity, I found more concrete descriptions, referring to visual experiences of pure brightness or radiance, of "milky" or "shimmering" light, sometimes accompanied by sensations of warmth. These experiences are not timeless or transcendental in the sense of being "always already the same" (see chapters 22 and 31); they often seem to be much more like processes unfolding over time. Of course, it may be possible that luminous emptiness leads to visual luminosity or vice versa, that they mingle, or that they are stages of one and the same process:

199 I spontaneously had the feeling of "floating" through a gate made of light. There was nothing there but warm, bright light. My person / my body no longer existed, everything was just light. An indescribable lightness and silence flowed around me. No constant circling of thought, but eternal, timeless calm. . . .

1183 [. . .] then with the mind quietened down, was aware of the movement in and out of clear light either as light or as vibration expanding at the heart. I also found that the awareness would move from spacious clear light to focused (closed) clear light when aware of thoughts arising. [. . .] Later in the day was aware of the dominant sense, visual thought becoming or triggering the clear light (in contrast to the recessive (hearing thought)). Again there is no difference between the energies (or the feel of them) of thought and clear light—there is the basic difference in spaciousness, but when moving from still clear light to the thought clear light there is also a spaciousness about it, as it is the focus of the awareness at the time—there is nothing else at that point but the clear light of the thought—the only clear light at that instant.

[. . .] After a while the mind became still and very brightly luminous (clear light). Opening an eyelid and found that it was still fairly dark—just laughed from surprise as it was just like daylight behind the lids! Continued exercise of breathing into the heart and radiating the breath now as light through and out of the pores of the body. As the breathing progressed, the breath turned fully to light passing to the heart and throughout the body until the whole body was radiating light. Again, a shimmering luminous clear light radiating through the body expanding on the bright clear light from before but occupying the whole body and surroundings. The "light" breath became slower and slower as the shimmering luminosity filled and radiated from the boundaryless body. Opening the eyes to the whole surroundings bathed in shimmering luminous clear light. The shimmering clear light slowly receded to the body as still clear light with everything retained as a blissful clarity.

1312 [. . .] during which I had more and more the feeling or perception of reaching pure consciousness, that is, a consciousness without specific content, simply just consciousness. At some point I had the feeling "now," that is, of having arrived in pure consciousness. [. . .] The state of mind I was in at that time I would describe as "milky" or "soft white mist." So there was an aspect of inner brightness there, along with the awareness of being in silent, pure consciousness.

1575 [. . .] Brightness with eyes closed and intense warmth, both neither really inside nor outside (in the body). "But is that 'pure awareness'?," asks the thinking mind, and something else simply knows it.

1728 [. . .] Immediately after turning awareness upon itself it feels complete, blissful, and somewhat magnetically absorbing. It has a strong "visual" quality of radiance and shining, rejoicing. Although it feels entirely complete in itself, paradoxically, simultaneously there is some sense of someone noticing all the qualities or enjoyment in it, and some residue of a sense of beholding it rather than looking entirely from it.

2747 [. . .] Experience during a 10-day dark retreat: pure consciousness, bright, clear, connected with the experience of inner light—like in a dream—but not in sleep, instead during waking meditation in the darkness. [. . .]

2867 [. . .] Usually I sit with eyes slightly open but suddenly I became aware that my eyes were closed but it was very bright; I felt heavy and warm and expanded, and expanding. Totally at peace and still; breathing happened for me, to me—I was breathed. I was everything at once, all encompassing, and nothing at all, no form and no shape. Brightness and density and utter calm. [. . .]

3295 [. . .] I had the indescribably beautiful experience of perceiving only light in space, of which I myself was a part. [. . .]

3525 [. . .] a white light spread from the crown of the head over the whole body. A state of deep calm, peace, and absolute love followed. I could hear the sounds around me, but at the same time I was completely centered and in myself.

3565 It was a feeling of merging with my surroundings, becoming one, not being separated anymore. It was a very beautiful feeling. A feeling of lightness, being filled with warmth and positive energy, love, connectedness. Inner brightness.

3516 [. . .] A bright appearance of light shining from within, which shuts down all thoughts and ways of looking at what is happening. It is like sitting on a big beanbag or throne; the body sits on this throne of a ball of light, and you (whatever it is that is doing the perceiving) look out. It is not really possible to look inside, but from there the light goes out. The relaxation is very deep, so deep that the inner verbalization/dialogue also stops. The speech apparatus (larynx, throat . . .) is very relaxed, loose, and still. Due to the very deep relaxation there is also no impulse to question one's own feeling of self (like "Who am I?," "Where am I?"), so that different areas of the body, which otherwise in everyday life answer as "I," relax and widen. This results in the experience of an open tube along the spine, which opens wider and wider from the neck down. At the same time, the jaw and mouth area relax. The light sometimes becomes very bright, then every consideration or contemplation, even about these aspects of experience, disappears; for brief moments there is only light. If it gets a little weaker, I know that I am fully present; the thing is that it can just get "like too much," and then everything disappears within it. [. . .]

It remains unclear whether the two phenomenologies—we could perhaps call them "abstract enlightenment" and "concrete enlightenment"—are in any way systematically related. However, it is worth recalling that the English term "enlightenment" originated in the (often criticized) translation of the abstract noun *bodhi* as *Erleuchtung* by the German scholar Max Müller (1823–1900) in 1857. The verbal root *budh-* means "to awaken," and its literal meaning is closer to a continuous process—namely, the process of gradually waking—than to a sudden event. There is also no semantic element of luminosity in *bodhi*, the Buddhist original. In addition, we should never forget that epistemological metaphors of light, illumination, and enlightenment have played a role in countless other systems of philosophy and religion, both Western and Eastern.[2] Practitioners in different traditions and geographical regions, therefore, will use different terminologies and accordingly have different expectations, which in turn will shape the experiences and the ways in which they are reported. For example, what originally was a more concrete experience of quasi-perceptual brightness (e.g., an abstract form of visual mental imagery with closed eyes) may be falsely reported as "clear light" by some—and vice versa.

Empty Cognizance

> The Seer is nothing but the power of seeing which, although pure, appears to see through the mind.
>
> —Patanjali, *Yoga Sūtra* (II: 20)

> Some people compare this cognizance to a radiant "thing" that shines with a light like a "clear light." It means a sense of being wakeful, a quality of being vividly wide awake, which is empty of any identity, and naturally alert.
>
> —Tulku Urgyen Rinpoche (1920–1996), *As It Is*, II.

Let us begin by looking at some results from our survey. Under our twelve-factor interpretation, factor 6 ("Luminosity") referred to reports about pure awareness involving nonvisual phenomenal qualities of "brightness," "radiance," and "nontactile vibrancy," but also to the visual experience of brightness with closed eyes. It was remarkable to see how this specific cluster of items turned out to be highly stable across all the factor solutions that we tested. Plausibly, the relevant experiential qualities are intimately connected. From a statistical perspective, another interesting detail was the correlation between factor 6 and factor 10 (which we dubbed "Touching World and Self"). There also was a correlation with "Peace, Bliss, and Silence" (factor 2; see chapter 2), but this was almost to be expected. Factor 10, on the other hand, describes minimal phenomenal experience (MPE) as an abstract form of tactile experience resembling self-touch (e.g., the specific sensation of intermanual self-touch, such as when you touch one of your hands with the other), or involving the entire body touching the world while simultaneously being touched by it (more on this in chapters 24 and 25). Other phenomenological descriptors in factor 10 are "velvety," "dense," and "full"—but always in a nontactile way (i.e., lacking the concrete phenomenal qualities normally characterizing the stimulus-correlated sense of touch).

At the intersection of factors 6 and 10, the experience of pure consciousness is described in three main ways: (1) as an abstract form of contact or of "being in touch"; (2) as possessing the phenomenal character of reflexivity (as found in self-touch); and (3) in globalized form, as "an experience of pure awareness penetrating your body, e.g., like a field that also penetrates all other objects and living things" (item #87). The correlation of factors 6 and 10 is not intuitively obvious. There seems to be no direct phenomenological relationship between qualities like "brightness," "brilliance," or "radiance" and the touch-related ones just mentioned. If future studies replicate this result, then that might point to interesting commonalities in the neural substrate.

In his work on advanced Vipassanā meditators, the Israeli researcher Yochai Ataria describes a primordial phenomenological structure that we experience (e.g., when one of our hands touches the other). He calls it the "touching–touched structure," in which we have a concrete tactile experience "of being a subject (touching) and an object (being touched) at the same time."[3] Please note that normally, both aspects of any such experience of self-touch will also be represented at the same location within the organism's spatial frame of reference, and therefore the experience can help a human infant (or even a fetus) to consolidate the boundaries of its conscious body-model as motor development progresses.[4] In self-touch, there is a spatiotemporal redundancy linking proprioceptive and tactile information, which can help the brain to build its model of reality.

In general, the boundary between subject and object is experienced in a clear and simple manner. In meditation, however, the boundary between the world and the knowing self may be gradually weakened as more and more boundaries begin to disappear,[5] turning a more holistic quality of awareness per se into the dominant feature of the phenomenal field. In the tradition of the German philosopher Edmund Husserl (1859–1938) and the French phenomenologist Maurice Merleau-Ponty (1908–1961), Ataria points out that a body-based perspective allows us to gain an interesting new angle on the maximally simple and "primitive" experiential relations in which an organism can stand to the world it is embedded in. I think this new perspective may have far-reaching implications for our understanding of conscious world-experience: If we use the epistemic practice of meditation to look more closely at the fine structure of such experiential relations, are they really an abstract form of touching *ourselves*? Will we discover that world-experience always already is self-experience because a wide-open, effortless form of mindful attention, one that encompasses the whole phenomenal field, is really a way for the organism to selflessly *touch itself*? I think that our new phenomenological data clearly show that even the touching of attention by attention itself is possible: We can even direct our own attention to attention itself, perhaps to the wandering of its focal point, to the feeling of recognition becoming more precise, or to the pure capacity of being attentive itself—and then let go. Building on the existing technical term of "meta-attention," we might speak here of "meta-attentional self-touch" (see also figure 34.1 in chapter 34).

If we now combine Yochai Ataria's point about the touching–touched structure with the contraction principle introduced in chapter 8, then an interesting phenomenological prediction arises: There should be states characterized by an entirely *uncontracted* version of the primordial touching–touched structure. Please recall that the contraction principle says that "being conscious" or "appearing" is really a property of a whole world-model in the brain, while the brain usually portrays "being conscious" as a property of *you*, contracting it into the transparent self-model, the image of a conscious person. If the person-model goes, what the brain does with "being conscious" may change. Meditation practice sometimes can decompress or "uncontract" consciousness, as it were, making *everything* appear to be pervaded by a quality of pure awareness. Could meditation do the same for the touching–touched structure? This new idea seems to be in agreement with Ataria's way of treating the problem. He writes:

> Thus it is possible to describe the touching–touched structure in terms of touching (and being touched by) the whole world and not a specific object. With this in mind, we can describe the experience of knowing without a clear object (knowing it is happening without an object, or without a specific object) as a thin touching–touched structure in which one feels immersed within, yet at the same time also somewhat separate from, the world (Ataria, 2014a). While touching the world itself (and not a specific object), this sense of touching/being-touched becomes so diffuse that it does not generate an SB [*a sense of boundaries*; TM], yet nevertheless a very thin sense of touching/being-touched continues to exist.[6]

This possibility is directly related to the phenomenology of "bodiless body-experience" and the new notion of "abstract embodiment" to be introduced in chapter 24, but likely also to the experience of "connectedness" investigated in chapter 11. If Ataria is right, then "even when lacking a sense of boundary, some basic bodily feelings and a very liquid touching/being-touched structure continue to exist."[7]

The possibility of a touching–touched structure persisting even when the subject/object structure collapses also relates to the Low Complexity (PC2) constraint, which says that pure awareness is extremely simple, lacking almost all internal structure: There should be simple and undifferentiated situations in which the touching–touched structure is attenuated but does not disappear altogether. In such cases, it would remain an *uncontracted* experiential quality, no longer bound into a conscious self-model. The touching–touched structure has a bodily origin in the sense of touch, but in this context, it would no longer be confined to the spatial boundaries of the physical body. In this sense, it would be ownerless. This potential independence from bodily spatiality in turn might underlie many of the experiential reports on "nondual being" that will be presented in chapter 26. In the words of Ataria: "In this situation the subject and the world return to exist on the same level—they are both constituted by the same flesh."[8]

In shifting from a phenomenological back to a semantic perspective, we find that "Self-Luminosity" (PC3) was one of the six semantic constraints for the concept of "pure awareness" that I extracted from a selection of canonical texts before developing the survey. In these texts, "Self-Luminosity" may also be found as the phenomenology of "brilliance" or as the "clear light of primordial awareness." Here is a first empirical falsification of one of my hypotheses: Hinting at the possibility of cultural variance, I claimed that as opposed to Wakefulness and Low Complexity, Self-Luminosity is not often found in Western phenomenological reports.[9] Our first survey shows that many meditators in Western Europe and the US do report this type of experience; it may, therefore, be more culturally *invariant* than I originally thought. The question remains which of its subcomponents will turn out to be prototypical—if you like, which will remain as stable elements of the common phenomenological denominator for pure awareness. My prediction is that it will be what earlier in this chapter I provisionally termed "abstract enlightenment."

People have thought about this before. According to classical Buddhist teachings, the term "luminosity" (*prabhāsvaratā* in Sanskrit) is directly related to the radiance of consciousness itself and to the notion of the "luminous mind," also translated as the "brightly shining mind" or "mind of clear light." Here is one classical example taken from the *Bardo Thödol*, the "Tibetan Book of the Dead" (eighth century CE):

> This brilliant emptiness is the radiant essence of your own awareness. It is beyond substance, beyond characteristics, beyond colour. [. . .] The instant of your own presence is empty, yet it is not a nihilistic emptiness, but unimpeded radiance, brilliant and vibrant. [. . .] Your own awareness, a vast luminous expanse, clarity inseparable from emptiness, is also the Buddha of unchanging light, beyond birth and death. Just to perceive this is enough. If you recognize this brilliant essence of your own awareness as Buddha Nature, then gazing into it is to abide in the state of enlightenment.[10]

We find here the aspects of radiance, brilliance, and vibrancy, the elements of clarity and vastness—all of them confirmed by our meditators. But if, as proposed in chapter 4, we read "emptiness" as "epistemic openness," then at its core, it refers to an entirely nonconceptual experience of *epistemic capacity*, the capacity to know and experience (in Tibetan, *salwa*[11]). In this new context, we could also speak of the uncontracted phenomenology of epistemic clarity (chapter 5), representing a currently unobstructed inner *space of knowing*, a space in which we can orient ourselves, in which perceptual processes can unfold, in which attention can be controlled and focused, or in which concepts can be formed and applied to experience. If this space is empty, then no such processes are currently taking place. The mere potential, our own capacity to harbor such processes, is now experienced *as such*. This emptiness is a timeless, high-level phenomenological invariant that we could also describe as a statistical hypothesis—namely, the abstract belief that "something can be known."

Another appropriate Western descriptor for luminosity (in the sense of consciously and nonconceptually experiencing epistemic capacity per se) could be "lucidity." This term implies that one can possess the capacity in question without consciously knowing it at all, or that one could have a merely intellectual understanding (i.e., in the form of propositional, high-level thought) of the fact that one possesses this capacity, while at the same time entirely lacking a more direct, nonconceptual *awareness* of one's capacity to know. Lacking this awareness, a given state would not be lucid. We can be entirely unaware of the space of knowing, but sometimes a global model of this space itself may begin to emerge—the global availability of information itself becomes globally available. MPE seems to be characterized by exactly this: an effortlessly occurring and nonconceptual representation of epistemic capacity, not reified, uncompressed, and uncontracted into any object of knowledge or a knowing self (chapter 25), but generating a specific experiential quality that can be described as "lucidity," "luminosity," "brightness," or "clarity." Phenomenologically, this feature is the spontaneous self-disclosure of an unobstructed epistemic space. Alluding to ancient Buddhist terminology, we might also call it "empty cognizance."

Let us now return from the broader semantics of concepts like "MPE" and "luminosity" to our own qualitative perspective. As noted previously, and as described in the second group of reports in section 1 of this chapter, we sometimes also find a much more concrete, quasi-sensory or sometimes fully visual phenomenology of luminosity that may involve actual visual experience. It is not at all easy to make sense of the relationship between the various kinds of luminosity. Survey response effects may be playing a role: For example, the occurrence of the word "luminosity" (or another visual term) in an item may have encouraged respondents to agree with it based on the fact that their MPE experiences had actual visual elements, even if the item explicitly referred to a nonsensory context. All I can do is draw attention to the fact that there may be three interrelated ways to interpret our phenomenological data, on three levels of description: the raw physical intensity of an internal stimulus source, functional autonomy, and intrinsic epistemicity. Let me briefly explain.

First, I would propose that the relevant phenomenology of quasi-perceptual concreteness consists in the fact that the subjective experience of tonic alertness varies along a dimension of intensity, although it is clearly mode-neutral and therefore not associated with any interoceptive or exteroceptive sensory modality. As we saw in chapter 4, tonic alertness is what determines the capacity for sustained attention in the absence of an external cue; the conscious experience of wakefulness simply means *knowing* this capacity, having a model of it. My suggestion is that the phenomenology of wakefulness itself has an aspect of raw feeling or sheer intensity, just as the bodily sensation of hunger and the visual quality of redness do (e.g., in terms of saturation or purity in a color experience)—but that the actual stimulus source, its hidden cause, is internal to the brain and doesn't map onto any internal sensory system or an event beyond the organism's sensory sheets (e.g., the retina for vision or the skin for touch). It need not have anything to do with an equivalent of blood sugar levels or images on the retina, but it may be wholly determined by the process activating the cortex itself. This is entirely speculative and only time will tell, but the raw-feeling aspect of wakefulness could correspond to the physical signal of the ascending reticular formation or a specific aspect of activity in the diffuse thalamocortical relay. That is, it could be directly related to the mechanism by which the brain first wakes itself up and then begins to "broadcast" whatever comes in through the senses much more widely within itself. Maybe meditation has something to do with, just like in a color experience, increasing the "saturation" of pure wakefulness in our model of reality. Phenomenologically, it seems that as soon and as long as we are awake, something always radiates (there can even be a subtle quality of vibrancy), but it does so with varying degrees of intensity (or "nonvisual brightness") and it has an aspect of raw feeling. This is where the concreteness comes from. Yet none of this explains where the decidedly *visual* concreteness found in some cases could originate.

Second, not only does luminosity come in variable strengths and degrees, but it also is functionally autonomous and permeates all of phenomenal space. Now we move from a physical to a functional level of description. Here, luminosity or "self-luminosity" can be analyzed as a visual analogy for the experienced *functional autonomy* of tonic alertness (chapter 4). In German translations, this aspect has sometimes been termed *Eigenstrahlkraft* ("proper radiance" or "the power of autonomous radiance"). This quality may point not so much to a distinct form of phenomenal character or representational content, but rather to a functional dimension structuring the space of conscious experience, which in turn allows a relative ordering of global states. Global states could have more or less Eigenstrahlkraft, more or less intrinsic radiance or brightness. It is also conceivable that an organism may possess an inner representation of this dimension, knowing its own current degree of wakefulness, and that talking about self-luminosity is actually a way of talking about tonic alertness—a cue-independent, intrinsic baseline of wakefulness that is always there whenever there is conscious experience at all, that is functionally autonomous, and that is largely independent of the ever-changing kaleidoscope of surface phenomenology. Meditation practice could—as Chinul, the founder of Korean Zen, would have it—consist in attending to this inner representation of wakefulness, *tracing back* the radiance of your own self-aware mind to its intrinsic baseline.

Third, if we read "luminosity" as a phenomenological metaphor for epistemic openness (chapters 4 and 5), then it bears a direct relation to the semantic constraint of epistemicity, the self-disclosing quality of knowing. It has an epistemological dimension. There is a quality of subjective confidence in luminosity because—whatever your current model of reality is—there is a detectable signature of knowing folded right into it. This may well be an abstract computational fact, one that will hold for all conscious systems. To quote Chinul again, this time on different forms of nescience, or not-knowing: "As far as the true mind is concerned, it knows while knowing nothing; but because it is impartial, quiet, and utterly radiant, it is different from the nescience of grass and trees."[12]

In the context of a visual metaphor, for something to be "self-luminous" means that it possesses a quality of intrinsic epistemicity because it autonomously *makes itself knowable*. Think of a single burning candle in a dark and empty room. The candle not only illuminates the room, but also endows itself with the property of visibility. By burning, it creates the condition of possibility for visual experience. It is not something that needs to be discovered or gradually constructed by an epistemic effort, in an agentive, top-down manner. No self is needed. If we take the visual metaphor of "self-luminosity" at face value, then the candlelight of MPE must be continuously self-revealing and transcendental at the same time—an internal process that discloses itself by creating the possibility of being known.

"Self-luminosity" could therefore be treated as a visual analogy for the phenomenal signature of knowing, as an attempt to express *what it is like*. Let me give one last example, drawing on an analogy from vision science. From a scientific perspective, a consciously experienced visual quality like "redness" cannot be analyzed merely as hue plus saturation because it always also varies along a third subjective dimension of "brightness." Pure awareness might be like a domain-general form of brightness that comes in varying degrees of intensity and applies to hearing and all other sensory modalities too. Zero brightness means zero phenomenal experience. In our analogy, the subjectively experienced brightness of a red object represents sheer stimulus strength for the organism—that is, it models the *luminance* of a visually given physical object—and perhaps MPE turns out to be a literal, mode-neutral form of brightness. Once again taking a radically naturalist perspective, the internal "luminance" in question could be a dynamical property of some part of the neural body. Therefore, if we want to understand the phenomenal concreteness of mode-neutral and apparently contentless wakefulness, then we must ask: What, for bare wakefulness, is the nonsensory equivalent of the raw stimulus intensity that is caused by the physical property of luminance in a given visual target and that, in the domain of visual consciousness, leads to the experiential quality called brightness? Is there something like "mode-neutral brightness"? Can it become part of the self-model; can one experientially *embody* it? If so, what is the stimulus source, what its hidden cause, and how would prediction errors and their ongoing minimization be reflected as part of conscious experience? These questions also close the loop and bring us back to the physical. If we read "self-luminosity" as a visual analogy, then the question becomes: What exactly is "luminant" here—is it perhaps a part of the neural body?

19 Witnessing

The world reflecting in a mirror without a perceiving observer. [#1735]

An impression of "there are impersonal eyes in the room." [#3451]

"Witness consciousness" is a global phenomenology. This is to say that the *totality* of all experiential contents appears as if being observed by something that isn't really an "observer" at all—by a timeless, impersonal, knowing presence. Witness consciousness is our first example of an experience that (as explained in the introduction) is neither content-specific nor simply a "state," but a *mode* of consciousness—that is, a global *way* in which reality appears to us: The world is mirrored in an all-encompassing quality of infinite, choiceless, and nonconceptual knowing. The phenomenon has been known and discussed in Indian philosophy for many centuries, and certain Western schools of meditation in the Vedic tradition (TM) explicitly tell their students to expect the phenomenal quality of "witnessing" as their practice advances, during the daytime as well as during dreamless deep sleep (chapter 20).

> 3389 Since I already have many years of meditation practice behind me, infinity experiences, or the experience that an infinite observer is observing my earthly life, happen to me over and over in daytime consciousness. Waking sleep and conscious dreaming, however, are rather rare. [. . .]

Witness consciousness can occur spontaneously and effortlessly, outside of formal practice, and it can last for hours:

> 1501 I once had another experience that lasted half a day, during the activity. I did my normal work in the office, but at the same time I was a witness to what I was doing. The big advantage of this was that I did not get tired at all. It was as if I was untouchable from the outside. The mind was fully awake and clear.

> It all went smoothly of its own accord, everything proceeded intuitively and was completely right. A really great experience.

The global, passive witness is not a person or an epistemic agent (i.e., an active "knowing self"); it is not something that has goals, makes choices, and selects objects of knowledge. It is entirely impersonal, but it also seems as if the phenomenal process of witnessing can be terminated at will:

> 800 It is for me like something, a presence, bigger than myself, that observes in a neutral or sometimes benevolent way and that knows.
>
> 3530 During experiences in the nondreaming state I often had the perception that an observer, who is not a physically contoured observer but kind of like "that which observes," simply takes an interest in what is happening in the area that I usually call "in me" or "I," in everything that surfaces in terms of thoughts, sensations, feelings, etc., but that I could then switch back to a more controlling active observer, who could specifically direct certain ideas and focuses that were considered helpful by this authority. [. . .]

There are cases (e.g., #1756 in chapter 24) in which pure witnessing of vacuous space is preceded by a gradual dissolution of body boundaries. The phenomenology of pure observation sometimes co-occurs with unbounded, spacious awareness:

> 3497 A feeling of being unlimited and of observing is there, surrounded by and part of unlimited space. [. . .]

Conceptually, witness consciousness still retains the duality of world and witness; it can even appear as a full-blown state of dissociation. But, as always, phenomenology in the real world may be much richer and much less clear cut than any conceptual schema. Witness consciousness, for example, often occurs in the context of, or alternates with, what we will investigate under the headings of "nondual awareness" and "nondual being" or unity (see chapters 26 and 27).

> 2999 I had the "feeling" I was the whole world and at the same time I was a silent witness of this "feeling."
>
> 2916 [. . .] Relatively common is the experience of the pure, very old, actually eternal witness, in whose awareness all phenomena emerge as on a canvas. Bodily perceptions, thoughts, feelings, sensory perceptions are all manifestations or plays of light on the three-dimensional canvas. Differentiated from this is the experience of being IN THE MIDDLE of the phenomena and experiencing them directly and in immediate identity as an intensely "juicy" three-dimensional experience.

Occasionally, there are cases of strong pure-awareness experiences following emergencies (e.g., #2862 in chapter 32). In some of these cases, minimal phenomenal experience (MPE) is not merely a phenomenal state, but a *mode* of conscious experience. The same seems to apply to the subset of witness consciousness experiences: As a global phenomenology, it can also occur during emergencies and following accidents. It is an entirely different mode of experiencing reality, but its phenomenological profile also clearly differs from dissociative states or out-of-body experiences following accidents and severe trauma:

> 1831 After meditation I was sitting in the garden in a contemplative, relaxed state, everything was very light and softly drawn, a three-year-old child was playing next to me, but I was busy with nothing and thinking about nothing. Suddenly a young cow charged through the garden from next door—I just thought, oh, this could get dangerous for the child, but quite calmly, without this contemplative feeling having changed, I just had the feeling that there was now a need for action—that was my only thought—everything else happened by itself, I didn't decide anything, I didn't think anything, I wasn't afraid or excited (and I'm definitely afraid of boisterous cattle!). I acted as if in a dream state, there was also no sense of time or self, I just had the completely neutral intention to protect the child without any emotion and without thinking . . . I walked completely calmly toward the approaching cow—and it was so unbelievable—the cow just stopped galloping right in front of me and I could just touch her and hold her—and there was this encounter, this feeling of not being separated from this cow, she looked at me so incredulously and was completely trusting—we were both in a totally different space, a different time, a different connection, and my body was as if sleepwalking but completely self-sufficient, it was a completely fearless, very peaceful situation—absolutely in this outer world and not of this world at all—everything was as if damped down by something soft, tender, light, as if wrapped in cotton wool—yes, maybe space really had a different density. This action of "me" catching the cow was the talk of the village for a week—I was the big hero—but that was not "me." The clearest feeling was rather that maybe it was acting and I was watching. Whether this is called a form of awareness—I don't know—but I experience this kind of "body awareness" quite often. My body acts so precisely, so perfectly and independently, intuitively—without me even having the opportunity to start a thought. . . . until my consciousness has understood what is happening—"I" could observe my body in its absolutely precise action

without feeling fear or any emotion for even a moment. It has happened this way in several really dangerous emergency situations—I have never even had a scratch, but without this what I call awareness of the body I might just as well have been dead. It was always free of thoughts, fear, time, and perception was softly cushioned and everything took place in a space with a different density and light.

Sākṣin and the Ideal Observer

> When the mind is quiet, we come to know ourselves as the pure witness. We withdraw from the experiencer and its experience and stand apart in pure awareness, which is between and beyond the two. The personality, based on self-identification, on imagining oneself to be something: "I am this, I am that," continues, but only as a part of the objective world. Its identification with the witness snaps.
>
> —Sri Nisargadatta Maharaj (ca. 1897–1981), *I Am That*

> But there's a danger lurking here. That Douglas should think he's the One! [. . .] I am through to Who I am as Who I am, not as Douglas. That's very important. It's very reassuring, too. I am sure I'm through to Who I am. Why? Partly because old Douglas is not involved in the process. If he were, I should doubt the whole thing [. . .] What I am apt to forget is that when I see Who I am, it's not Douglas who's doing it; it's the One who is doing it.
>
> —Douglas Harding (1909–2007), *On Having No Head*

"Witness Consciousness" emerged as factor 12 in our study, referring to the experience of a passive and impersonal observer being present. This statistical finding demonstrates how a classical term from the Advaita Vedanta system of philosophy (the *sākṣin*) finds direct expression within phenomenological reports given by present-day practitioners from many countries. The notion of the "witness" has a long tradition in Indian and Tibetan metaphysics, and on a phenomenological reading, it bears quite obvious relevance to the concept of MPE. For example, the idea of nonpersonal witnessing captures the phenomenal signature of knowing, the centerless and nonegoic character of awareness itself, the two qualities of timelessness and simplicity, and also the element of "choiceless awareness," the intrinsic nonreactivity that is the goal of all mindfulness practice. Using our new conceptual instruments, we can now describe it as an uncontracted phenomenal signature of knowing.

The earliest appearance of the idea seems to be found in the Atharva Veda, in the Muṇḍaka Upaniṣad,[1] whereas the source of the philosophical concept *sākṣin* is obscure.[2] Witness consciousness is that which makes all knowledge possible, that which cannot itself become an object of knowledge, and that which is self-luminous. Bina Gupta lists the following epistemological characteristics:

1. The witness consciousness, although the basis of all knowing, is different from the object known. It is implied in every act of knowing. It is the ultimate subject; it can never become an object of knowledge.
2. It is the pure element of awareness in all knowing. It is one, immutable, indivisible reality.
3. It shines by its own light; it is self-luminous.
4. It is different from the empirical individual (*jīva*; Sanskrit: जीव), who cognizes and enjoys. In other words, it is different from the empirical individual who is caught up in the triple states of waking, dreaming, and dreamless sleep.[3]

Of course, there are also many metaphysical—not only epistemological or phenomenological—readings of witness consciousness.[4] For example, the witness could be an aspect of reality itself; it could be the self-knowing world as a whole, or even God. I call this the "principle of phenomenal correlates": For practically every metaphysical theory that philosophers have developed (like idealism, panpsychism, pantheism, solipsism, or nihilism), there is an altered state or even a mode of consciousness that directly corresponds to this theory. There is a correlated, brain-bound conscious model of reality, a nonconceptual mode of experiencing the world, a region in phenomenal state space that later is expressed by this theory—there is a *phenomenology* of idealism, a *phenomenology* of panpsychism, and so on. In different cultural contexts, or against the backdrop of a specific religion, this altered state will be described differently. For every set of ontological assumptions (e.g., mind itself is the ultimate foundation of reality, conscious experience is fundamental and ubiquitous, God is identical with the cosmos, etc.) there is a global mode of phenomenal experience that is mirrored by this specific metaphysical theory—and that perhaps originally even inspired it.

Let us take the example of panpsychism.[5] I would claim that what people *really* find so fascinating about the topic of panpsychism is not the philosophical debate as such, but rather an emotionally attractive, intuitive feeling that they get when they try to imagine their own world with the added extra of panpsychism being true. The fascination is generated by a phenomenological inkling, a glimpse of the profundity and deep relevance of a global mode of experience that really exists (chapter 27). Philosophical panpsychism is the poor man's nondual awareness.

For the special case of witness consciousness, it is interesting to note that if one looks into early Western theories of consciousness, such as those centering on the Latin concept of *conscientia*, a common motif is that an agent is conscious by virtue of sharing its knowledge with an ideal observer. I think that the witness of Indian philosophy is such an ideal observer. In an early European context, this observer could be God, a divine person who sees everything you do, think, and feel. I think that on a phenomenological level, witness consciousness bears great structural resemblance to Christian ideas like "placing yourself in the Lord's presence" or "resting in the Lord's gaze," as found in certain kinds of contemplative prayer. In Christian philosophy, awareness of an ideal observer being present is directly related to having a moral conscience and *bearing witness* to your own sins. To have a conscience means to install a model of an ideal observer in your mind; the "purity" of consciousness is really a moral kind of purity, and consciousness in this more fundamental sense is the only space in which we can be together with God before the body dies.[6]

In religiously inspired accounts of consciousness, the metaphysical assumption is that the ideal observer is a divine person, that the observer must be another *self* or epistemic agent of some sort (see chapter 25). From a philosophical point of view, this assumption is superfluous. A mistake frequently made by Christian theologians has been to think that just because they take themselves to be persons, God must be a person too. But the notion of ideal observation does not need to be reified into the concept of a personal God. My main point is that both concepts—the witness and the personal God—are metaphysical reifications of what originally was a phenomenological *process*—namely, the process of "ideal observation," of "observing without an observer." This leads to the interesting possibility that human beings have experienced the ideal observation of selfless witness consciousness for millennia, and that some cultural contexts led to the creation of an egoic interpretation (like early European Christianity, in which a personal God sees everything, all the time), whereas others developed a more abstract metaphysical theory of consciousness, including nonegoic models of witnessing (like Advaita Vedanta in India, the famous nondualistic school of philosophy propounded by Gaudapada and Adi Shankara in the seventh and eighth centuries).

On a purely phenomenological reading that abstracts away from all metaphysical and epistemological aspects of the term "witness consciousness," we find something that strongly resembles MPE. It is a self-luminous form of bare awareness onto which intentional objects can be superimposed during single "acts of knowing," a nonegoic type of phenomenal character that has nothing to do with any form of conscious self-representation as a cognitive agent or an entity possessing subjective preferences and personality traits. Perhaps most important, it is something that in principle could exist in waking states, during dreaming, and in dreamless sleep, simply because it is different from any phenomenal self-model that portrays only the "empirical individual" by contracting pure awareness into an inner image of a person who seemingly "cognizes and enjoys." Witness consciousness strongly resembles prototypical MPE as an intrinsically nonreactive, nonegoic, and nonconceptual experience of knowing and wakefulness. In a way, it is as if the background property of epistemic openness becomes apparent and more pronounced. Witness consciousness can be interpreted as a manifest, noncontracted, and all-encompassing variant of the "phenomenal signature of knowing" (introduced in chapter 7) coming to the fore: a situation in which what was termed the "model of our epistemic space" in chapter 5 has become more and more explicit, losing its phenomenal transparency (chapter 28). Phenomenologically, there is now not only the abstract possibility of knowing, but an *actual* process of observing without an observer. In the existing literature, one interesting reading of the *sākṣin* parses it as a "field" of consciousness, within which the distinctions among subject, object, and the process of knowing become a "context for contents, or 'space' in which forms may appear."[7]

One target for future studies will be the relationship between witness consciousness and the phenomenal character of "connectedness" (chapter 11). Is witness consciousness a dissociative state? Does it lack compassion? Could it even be related to psychiatric disorders like derealization and depersonalization, in which the world and the self can appear unreal and without emotional coloring, as if seen through a fog or a milk-glass pane? This would seem to exclude the phenomenal qualities of clarity (chapter 5) and connectedness (chapter 11) that many of our respondents have reported. True, becoming aware of the phenomenal character of awareness itself has often been described as not only looking through a window, but gradually beginning to see the window pane itself, as if shifting focus to look *at* the window (more on this in chapter 28, "Transparency, Translucency, and Virtuality"). But if we take the phenomenology of pure awareness seriously, MPE doesn't have the hazy opacity of a milk-glass pane; rather, it enhances the vividness of what is seen through it. It is a *lucid* form of empty cognizance, and yet we can become aware of it—for example, by "resting in its gaze." Witness consciousness seems much more strongly connected to the aspects of nonidentification, transparency, and virtuality investigated in other chapters of this book. What remains puzzling is the quality of "disconnectedness" that accompanies it. Consider this description from the literature:

> When I first started meditating, I was having fairly flashy and dramatic feelings during and after meditation. There was a huge contrast between in and out of meditation, and this has gotten less over the years. More qualities that happen in meditation are now there in activity—not only in my state of consciousness, but also physiologically. My metabolism is more subtle and relaxed and stable. The main

> thing I notice these days is just more of a witness value, like I'm being pulled back from everything. I'm not so strongly connected. I don't identify with my body, my thoughts, and my circumstances. I realize that's not me; that's not all I am, just my body and thinking. It's like witnessing, but it's not schizophrenia. I'm not attached to little ups and downs. I'm much more stable emotionally.[8]

Phenomenologically, it is not clear how MPE as expressed in witness consciousness relates to MPE as expressed in the phenomenal character of "connectedness" (chapter 11).

A second interesting target for future research is the degree of theory contamination involved in reports of witness consciousness. In some meditation movements like TM, practitioners are explicitly primed to expect witness consciousness and to regard it as a major marker of significant progress, while in many others, the concept is almost never mentioned at all. However, in our own study, only 57 of 359 (16 percent) of participants scoring high (≥ 70) on factor 12 ("Witness Consciousness") told us that TM was their personal meditation technique, which is not significantly more than for the low scorers (135 of 1,000 participants, or 13.5 percent). In meditators, therefore, witnessing may be a robust but rather rare phenomenon that mostly occurs independently of personal belief systems or theoretical background. However, the phenomenology of witnessing has been interestingly linked to psychiatric conditions known as "depersonalization" and "derealization disorder." Depersonalization involves a phenomenology of being detached from what one previously took to be one's self. Patients may also report an attenuation or loss of the phenomenal quality of agency, feeling as if they were an outside observer of their own thoughts and actions as well as their body. Derealization involves detachment from reality as a whole, with patients seeing the world around them as if through a milk-glass pane or fog, or experiencing it as dreamlike and surreal (see chapter 28 for some interesting parallels).

What I have termed "theory contamination" may causally determine whether the state is experienced as distressful, causing panic, or as wholesome and positive: "a syndrome may be produced that, depending on the attitude the person adopts toward himself and then toward the resulting phenomenon, may be experienced either as something to be sought and valued or as something to be feared and called a disease."[9] As Bryce Huebner and Genevieve Hayman have pointed out, ultimately what matters in minimizing psychiatric risks is the moral-phenomenological framework of meditative practice. They argue that since selfless states can be cultivated using meditative practices, it should be possible to evoke *positive* transformations similar to those that emerge in the context of drug-induced ego dissolution, but in forms as persistent and pervasive as the kinds of ego dissolution that occur in depersonalization and derealization disorder.[10] That is, meditation could give us states that have not only the desired experiential positivity, but also the functional persistence that we find in psychiatrically precarious contexts. Could witness consciousness therefore be a special, lower-risk variant of self-induced depersonalization syndrome, as Raymond Kennedy proposed in 1976? If I am right in suggesting that individual background assumptions and "sociocultural priors" may play a major role not only in how advanced meditative experiences are verbally reported, but also in how they are embodied or how the individual and its brain actually *react* to them, then witness consciousness might provide us with an interesting example. The cluster of MPE modes traditionally described as "witness consciousness" may be a particularly positive behavioral and experiential outcome of a process that gets other people into psychiatry because they don't have the right form of narrative self-deception (chapter 17), the relevant "background myth," or the stable social context of a shared moral-phenomenological framework. In a 1990 paper titled "Depersonalization and Meditation," based on experiential reports by TM practitioners, Richard Castillo wrote this:

> Thus, the ideation surrounding the experience of depersonalization is modified from that associated with psychopathology to that of religious experience. [. . .] Once the patient has accepted the definition of the problem in terms taken from the religious model, the patient's emotions will become attached to symbols associated with this mythic world. [. . .] Witnessing is described as the experience of being a witness to one's self—that is, being split into an observing self and a participating self. TM meditators interpret these experiences as "higher states of consciousness" or "enlightenment:" In the mythic world of the meditators, Witnessing is considered to be highly desirable.[11]
>
> I suggest that the relative lack of reported anxiety in response to the depersonalization is a result of the ideational construction of the experiences in terms consistent with the mythic model of TM. In other words, instead of "pathologizing" the experiences—that is, interpreting them as psychopathology according to a medical model of reality—they are "sacralizing" the experiences—that is, interpreting them according to a sacred model of reality.[12]

I would like to call this the "principle of self-healing" through the cultivation of a stable and positively experienced altered state of consciousness: There might be rare cases in which people have overcome a psychosis or other psychiatric illness on their own by successfully "spiritualizing" or "sacralizing" their conscious model of reality, using a method such as systematic meditation practice. Another concept that may prove to be relevant in this context is the reenchantment of our life-world (chapter 17). It is certainly conceivable that some patients have managed to free themselves from the suffering and psychological strain, regaining a normal level of coping with everyday life. However, it has to be emphasized that phenomenological profiles of depersonalization, derealization, and witness consciousness may vary widely—just think of the aspects of "clarity," "unboundedness," and "unity" (recall #1501, #3497, and #2999), or the occurrence of witnessing during dreamless deep sleep (see the example presented next and in chapter 20). And none of this touches the epistemological question of whether these MPE modes contain any valid insight into the nature of reality. The phenomenal experience of agency may or may not be an illusion, conscious experience may or may not be a wholly impersonal process, and so on. But it is certainly possible that one or more of these three are high-convergence modes that allow us phenomenological access to certain realities about how our minds work, and that their underlying ontology is perhaps closer to the scientific worldview than that of ordinary waking consciousness. More generally, putting witness consciousness in context opens up an interesting new perspective. From now on, we can always ask: Was there a specific background myth, a metaphysical model against which a specific experience occurred? Is there a mental model of reality that is getting confirmed by the experience?

A third open question for the future concerns the fact that for many centuries, descriptions of witness consciousness have always included the possibility that it is a phenomenon that can occur during deep sleep and dreaming too.[13] I have thought for a long time that testing this claim using the tools of modern science will be an exciting (if technically difficult) project for empirical research, and certainly highly relevant from a philosophical perspective. In chapters 20 and 21, therefore, we will take a deeper look at the possibility of MPE occurring during dreamless deep sleep and while dreaming. As a transition to what is to come, and to give you a first intuitive sense of what I am talking about, here is another phenomenological case study from the literature, given by a TM meditator:

> There were long periods of witnessing throughout the day and witnessing sleep a lot too. That was the most concrete. You can often wonder when you're having an experience in activity if it's real or if you are only imagining it. In sleep it's hard to imagine something. Witnessing was happening in both deep sleep and dreaming. Most of my life experience of sleep was you'd lay your head on the pillow and a few seconds later, you'd wake up. But I started to become aware of the passage of time. Sleep wasn't just lost to me, but I was experiencing it all night long. This experience has been continual since then. It makes sleep a lot more enjoyable, because there's a bliss in actually experiencing it rather than having no experience of it.[14]

20 Pure Awareness during Dreamless Deep Sleep

There is no waking, there is no sleeping, there is only being. [#1784]

Contemplative practice is related to sleep in a number of interesting ways. First, many dozens of our experiential reports emphasize how deeply relaxing, restful, and refreshing it is to spend a while abiding in the wakeful presence of bare awareness. What is more, the relationship between our subjectively experienced target phenomenology of wakeful presence and the objective property of functional-level wakefulness (also called "tonic alertness") seems to be symmetrical. It is not only that pure awareness makes you feel more fully awake. There is also a well-known mirror image, with the arrow of causality apparently pointing in the opposite direction: Being fully awake can make pure awareness more likely. This is true in two senses. One of the most important physiological factors enabling episodes of pure awareness is, quite simply, being well rested. Furthermore, it even seems that having a particularly good night's rest quite reliably causes the phenomenology of pure awareness to emerge, for example in one's early-morning meditation:

> 3218 [. . .] The experiences described are much more accessible when I go into my meditation with a completely rested and settled mind and body. But it is remarkable that the experiences are not at all random: Once the conditions of being well rested and settled in mind and body are met, the experiences are there. [. . .] This state of ocean comes relatively regularly during meditation, provided that I am completely rested. [. . .]

Robert Forman edited one of the first and best-known collections of texts focusing on pure awareness, *The Problem of Pure Consciousness,* first published by Oxford University Press in 1990. In a later monograph about mysticism and consciousness, he writes:

I tend to undergo this phenomenon more often when I have been getting enough sleep. Other than that, I have not been able to correlate this phenomenon with any other process: for example, it does not happen more often when I have eaten certain foods, wear certain clothes, or sit in a particular chair [. . .].[1]

Interestingly, a third general observation shared by many committed practitioners is that the quality of sleep *itself* changes over the years, as it does for everyone, but especially during periods of more intensive practice (e.g., during silent retreats). For example, sleep may feel lighter, dream recall may improve, and spontaneous lucid dreams can occur:

> 1341 [. . .] Another more general experience over time is that my sleep is lighter, and I remember my dreams and can control them. I never used to think I had dreams.

We will return to the relationship between meditation and dreams in chapter 21. An interim conclusion from these first three observations is that concepts like "wakefulness" or "awakening" are likely more than mere phenomenological metaphors. In addition, neuroscience provides us with strong empirical indicators pointing to possible causal relationships between minimal phenomenal experience (MPE) and the functional mechanisms by which the human brain reactivates itself, bringing the organism back into a state of being conscious and open to the world. Finally, the phenomenological material clearly demonstrates that human beings enter a state or even a global mode of pure awareness much more frequently when starting from the wake state than from the dream state or even from dreamless deep sleep.

There is an important caveat here, however. From the fact that MPE is *most frequently* accessed from the wake state, it doesn't follow that this is necessarily the *easiest* way to do it. It has been argued that the state is actually more accessible from some dream states, because lucid dreaming is a kind of back door or shortcut into the state.[2] Specifically, the practice in dream yoga of dissolving the content once in a lucid dream could function as a shortcut into what in the second part of this chapter I will also call "clear light sleep" or "witnessing sleep," whereas accessing that state from wakefulness could actually require more extensive training and practice in meditation. As I have learned in conversation with the neuroscientist Benjamin Baird,[3] MPE states or even prolonged MPE modes may be more commonly experienced during waking periods, partly just because very few people are trained in practices of dream yoga. Baird's point yields another specific, and empirically testable, hypothesis for future research: Pure awareness might be most frequently accessed from the wake state, most rarely from dreamless deep sleep, and most easily from a lucid dream.

Considering the existing evidence for this hypothesis, our data clearly show that the experience of pure awareness during sleep sometimes does happen. It is fascinating to see how in some advanced meditators, the phenomenal quality of wakefulness investigated in chapter 4 can actually occur even during those quiescent sleep stages in which the brain waves are slow and of high voltage, and in which the breathing and heart rate are slow and regular and the sleeper is mostly very calm. We will look at pure awareness during the dream state in chapter 21, while focusing for now on the seemingly paradoxical experience of wakefulness during mostly dreamless, non–rapid eye movement (NREM) sleep.

Stand-alone experiences of "wake sleep" or "witnessing sleep" during NREM sleep are much rarer than full-absorption episodes of pure awareness during standard formal practice like sitting meditation. Out of the 1,403 participants in our study, only 25 unequivocally confirmed that they had experienced pure awareness during dreamless deep sleep. The first finding is that sometimes the phenomenal quality of "luminosity" investigated in chapter 19 can spontaneously coemerge with a strong experience of wakefulness during dreamless deep sleep. One also finds that toward the end of an episode of "wake sleep," there may be a strong emotional reaction of the type described in chapter 15. As always, let us start slowly, by taking a careful look at first-hand descriptions of this rare experience, such as the following:

> 58 [. . .] The experience itself happened as I was physically asleep. I "woke up" realizing that I was meditating in the sleep state. I had no awareness of my body or the external environment. The level of concentration was very high, magnetlike, and I noticed that I was meditating spontaneously/effortlessly/automatically and that I was awake (in other words, I was in *dhyana* or access concentration). I decided to switch focus to the light and awareness that was present in that moment. The light and the awareness became hyperpresent and I felt more awake than ever during normal waking consciousness. The light was also very intense (it had a visual quality, but could also be described as bright and clear in a more subtle sense if we think of these qualities as belonging more to conscious awareness than to visual light). The main and almost exclusive contents of consciousness were light and awareness. There was no separation between the light and the awareness. In fact, they seemed to mutually fuel each other; as the light became stronger, awareness increased, and vice versa. At an emotional level there was a sense of great awe. This emotion developed into exhilaration, which made the state unstable. I was starting to react to what was happening, and this eventually led to exiting the state and also returning to physical reality. This can be compared to looking into

a bright light becoming increasingly strong; soon you will turn away or close your eyes simply because the light is so intense. [. . .]

In trying to convey the quality of pure awareness in dreamless deep sleep, meditators use terms that converge on or very strongly resemble those used to describe full-absorption episodes that begin in the wake state: "unity," "selflessness," "spaciousness," "nonduality," "timelessness," "luminosity," "radiance," "contentlessness," "unboundedness," and "vividness":

> 75 [. . .] On a few occasions I have entered the state of deep dreamless sleep consciously. It is in this state where my experiences have occurred whereby I am a simple unified awareness, seemingly without the daytime qualities of ego. I have experienced vast, noncentered timeless space and sometimes light and sometimes pure black or color but can really only put inadequate words to the experiences in retrospect. [. . .]
>
> 3171 In dreamless sleep—bright, contentless, unbounded, vivid. [. . .]

In some experiential reports, one finds that—just like during episodes of nondual awareness (as investigated in chapter 27)—the phenomenal quality of awareness itself can function as a nonegoic unit of identification, as what some traditional systems describe as the "true self" (see chapter 29 for more):

> 420 Experiences of pure awareness in NREM sleep [. . .], with one caveat: the question about the presence of self is not specific enough, i.e., autobiographical, core, or body-based self were absent, but the pure awareness was not someone else's awareness nor some subtle object/state I was aware of, as in *jhana*. [. . .] Rather, it is that which is and has been aware all along in all experiences, so in that sense, it is me. [. . .] Again, it is not something other than who one is.

One practitioner created a beautiful new metaphor, referring to the fact that in some parts of the world, for a full twenty-four-hour period around the summer solstice (approximately June 21 in the northern hemisphere and December 23 in the southern hemisphere), the sun never sinks below the horizon:

> 1828 [. . .] As an approximate verbal description: The objectless self appears in the empty space of pure, contentless awareness like a "midnight sun." [. . .]

I also found that some meditators had deep and unexpected spiritual experiences *before* establishing a regular meditation practice. For a small subgroup, contemplative practice may really be an attempt to understand and return to something that has already occurred in their lives, spontaneously and unsought, such as in childhood (see chapter 9 on the experiential qualities of freshness and wonder in the context of suchness and chapter 15 for three concrete examples) or while experimenting with psychoactive substances (again, see chapter 9 and chapter 32 for a selection of four examples). Perhaps, for some, even spontaneously occurring wake sleep can later turn into a phenomenon in search of an explanation:

> 2387 I was aware that it was night and I was lying in bed. Yet I was absolutely clear and wide awake and detached from my body. For a brief moment I felt myself in the Here and Now and in infinity at the same time. However, as I didn't yet have any experience of meditation at that time, I was so startled and then brought this state to an end.

> 3273 In deep sleep I can hear the wind or rain. This started when I was 12 years old. My body is sleeping, my mind is awake. In meditation I am one with everything, mind, body, and the world. Feeling of oneness. [. . .] During deep sleep, I can scan into my brain physiology and into my eyes from a point between the eyes and feel the energy in the body going up to the head chakra and I do a big jump.

Not all meditation-related experiences of awareness during sleep are the same. For example, some are crystal clear and may involve a feeling of security and contentedness, while others are accompanied by the often very unpleasant phenomenology of sleep paralysis, anxiety, and social hallucinations—phenomena that are also known from scientific research on out-of-body experiences:[4]

> 68 [. . .] It might be worth mentioning that I have had numerous experiences of awareness of the sleep state over the years and that I think these are meditation-related. These experiences have, however, been mostly negative (including sleep paralysis, sensed evil presences, etc.) and not characterized by hyperawareness, but rather had a dreamlike quality.

> 1188 There are situations when I am asleep, or my body is, and my mind is awake and clear. While my mind is awake, only my body needs rest. In this wakefulness I can observe it. Feeling that I am much more than just my body. I am completely relaxed and fresh.

> 1189 [. . .] It can happen that in the morning, before the actual waking up, the mind is awake while the body is still asleep. A feeling of security, you feel pleasantly wrapped up like in a cocoon or like a mummy and you have the feeling that you could lie like this forever without having to move. It requires a conscious decision to move in order to leave this state (usually reluctantly). It is also possible that you have the feeling of being completely awake, but then you wake up thanks to external noises, or an alarm clock or even your own snoring, and only thereby realize that you were actually still sleeping.

1238 [. . .] I woke up one night and realized with horror that I could not move my body. Then I realized that the body was asleep. After that, I could gradually move it again. [. . .]

1371 [. . .] In waking sleep the body lies as if asleep, while the mind is awake, calm, and satisfied. Such a supposedly sleepless night is therefore a pleasure. [. . .]

2747 In dreamless sleep—it is very difficult for me to judge whether such experiences occur in deep sleep or in other stages of sleep. [. . .]

Sometimes wake sleep is also called "witnessing sleep," because upon awakening, it retrospectively seems as though a quiet, impersonal observer has been present during parts of the night, but with no conceptual insight into the nature of the state itself. We explored the phenomenology of "witness consciousness" in chapter 19 and will focus on the specific quality of "nondual being" (i.e., the experience of existence itself) in chapter 26. Our data show that the phenomenology of pure awareness during dreamless sleep is often described as a form of witness consciousness, and also as the expansion of a nondual experience of pure being beyond formal meditation practice, into the wake state and beyond:

1423 I meditate daily, but this state of pure being comes when I lie down to sleep in the afternoon, I doze off, and suddenly I am wide awake, but don't know who I am or where I am, I am only aware of existing. I am there. [. . .]

1354 [. . .] Through repeated experiences of this being, it became a peaceful stream that spread more and more into my daily activities, so that I can be dynamically active and yet at the same time rest within myself. I am then the silent witness who simply observes the activities. This experience of being a witness now also happens during most of the time of sleeping and dreaming. [. . .]

White Nights: Epistemic Openness and Minimal Consciousness in Sleep

It is neither sleep nor waking but intermediate between the two. There is the awareness of the waking state and the stillness of sleep. It is called *jagrat-sushupti*. [. . .] Call it wakeful sleep or sleeping wakefulness or sleepless waking or wakeless sleep. It is not the same as sleep or waking separately. It is *atijagrat* (beyond wakefulness) or *atisushupti* (beyond sleep).

—Ramana Maharshi (1879–1950), *Talk 609*, January 18, 1939

Out of 1,402 participants with valid data on this question, only 25 unequivocally confirmed having experienced pure awareness during dreamless deep sleep; 56 gave a rating of at least 70 out of 100 possible points when asked, "Did your experience occur during dreamless deep sleep?" In the first group of those unequivocally confirming wakeful sleep, 9 named Transcendental Meditation (TM) as their meditation technique, 7 Mahamudra/Dzogchen, 8 Vipassanā, and 7 Shamata. There were 3 participants each for Zen and Metta meditation, 1 for mindfulness-based stress reduction (MBSR), and 10 "Other."

Being conscious during dreamless deep sleep sounds like a contradiction in terms, and as an empirical hypothesis, it has a quality of supreme bizarreness. Had I not experienced this phenomenon myself (as a young man, on an intense ten-week meditation course), I would not believe a single word of all this. I would have thought it was just another example of the delusional New Age folklore that surrounds meditation practice—the dangerous, omnipresent nonsense that makes it so appalling and suspect to scientists, critical intellectuals, and everyone who values intellectual honesty as an indispensable part of genuine spiritual practice (more on this in the epilogue). But our data show that conscious experience during dreamless deep sleep exists, and my own experience helps me not write it off as confabulation. I also believe that conscious deep sleep may become an important and decisive focus for research in the future of consciousness studies. It has been described and systematically pursued for many centuries, and our reports show that it is a phenomenon that can appear in a secular context and is not bound to the theory or practices of any single spiritual tradition. Although counterintuitive from the perspective of folk psychology or empirically uninformed Western philosophy of mind, it has been known for centuries that this specific phenomenology does exist, though mostly in very advanced practitioners and only rarely in novices.

> In the beginning, you do fall unconscious for a while, you do sleep. Yet within that sleep there is always the possibility of recognizing the nature of mind. When that happens, it's called "capturing the luminosity during deep sleep." Experience is unblocked, wide open. Although you do not visibly see what is around you and your body is still asleep, it's like a wide-awake state that occurs from within the deep sleep. With training, it is possible to become used to that.[5]

Today, conscious deep sleep may be of central importance in the scientific search for an empirically based and conceptually coherent theory of consciousness. We therefore should allocate resources to understanding it. I actually think that understanding the phenomenon of pure awareness during dreamless deep sleep could lead to a decisive breakthrough in consciousness research, perhaps to a minimal model explanation and the formulation of the first standard model of consciousness mentioned in the introduction. I think this because, after pure awareness in waking meditation, it is a good second candidate for the simplest form of conscious experience that human beings can have.

Please recall that in the introduction, I defined a "full-absorption episode" as a process after which the quality of "awareness itself" is the only phenomenal quality that can be reported. I pointed out that such states of conscious experience are "concurrently ineffable" because any attempt to report them while they occur would immediately destroy their entirely nonconceptual and nonagentive nature. The phenomenology of wake sleep is important not only because it presents us with tentative evidence for a second type of full-absorption episode (a fact that is of great methodological importance), but also because we can use it as a second, stand-alone example to triangulate the minimally sufficient neural correlate of MPE. This would be the set of neurofunctional properties in the brain that (1) reliably brings about an experience of pure awareness and (2) cannot be made any smaller without making the experience disappear.

To explore this idea further, let us distinguish between type I episodes of full absorption, in which MPE is entered from the wake state (usually during periods of formal meditation practice), and type II episodes of full absorption, in which MPE spontaneously occurs during periods of NREM sleep. Type III episodes would start from a lucid dream (as explained previously; see chapter 21 for more). All three of them are concurrently ineffable, and, as we are now beginning to discover, retrospective descriptions of their phenomenal character are somewhat convergent. For example, if we could isolate the neural correlates of the first two types of conscious experience, then this would enable us to determine their anatomical and functional overlap. If we then also developed convincing computational models describing the informational flow in these specific parts of the physical world, we could look for commonalities on a more abstract level of description—for example, by comparing mathematical models of MPE for type I and type II episodes of full absorption. This is the kind of research that in the long run could lead to the formulation of a first standard model for consciousness because it would home in on a minimal description, one that abstracts away from all unnecessary details, screening out everything that is not causally necessary for phenomenal consciousness to emerge. What we might call the Triple Triangulation Project would be the logical first step for the kind of empirical research for which this book and some of the publications preceding it have tried to lay the groundwork.

If we then, in a second step, compare the results of all three triangulations on a formal level, we will already be very close to the minimal model of conscious experience that we have been aiming at all along. The third step would consist in refining and validating our minimal computational model on a more mechanistic level[6] (e.g., via an

The "Triple Triangulation Project": Sketch of a possible research strategy

1. Triangulation One:
   (a) Isolate the minimally sufficient neural correlate for type I episodes of full absorption in neurotypical human beings;
   (b) isolate the minimally sufficient neural correlate for type II episodes of full absorption in neurotypical human beings;
   (c) describe commonalities and overlaps by creating and empirically validating a computational model that, on a mathematical level, allows us to see whether there is a phenomenological "cut set" connecting the two kinds of episode.
2. Triangulation Two:
   (a) Isolate the minimally sufficient neural correlate for type II episodes of full absorption in neurotypical human beings;
   (b) isolate the minimally sufficient neural correlate for type III episodes of full absorption in neurotypical human beings;
   (c) describe commonalities and overlaps by creating and empirically validating a computational model that, on a mathematical level, allows us to see whether there is a phenomenological "cut set" connecting the two kinds of episode.
3. Triangulation Three:
   (a) Isolate the minimally sufficient neural correlate for type I episodes of full absorption in neurotypical human beings;
   (b) isolate the minimally sufficient neural correlate for type III episodes of full absorption in neurotypical human beings;
   (c) describe commonalities and overlaps by creating and empirically validating a computational model that, on a mathematical level, allows us to see whether there is a phenomenological "cut set" connecting the two kinds of episode.

attempt to *re-create* the pure-awareness experience using advanced forms of real-time neurofeedback or even direct stimulation during sleep).

However, before consciousness research can proceed to the Triple Triangulation Project, we need to look at the actual phenomenology with an open mind, as precisely as we can—and we need to do some more philosophical spadework. As well as being one candidate for the simplest possible form of conscious experience, wake sleep is conceptually interesting for a second reason. The term "minimal phenomenal experience (MPE)" was originally coined by Jennifer Windt in a commentary on an important target paper by the Canadian philosopher Evan Thompson that discussed "lucid dreamless sleep."[7] However, her original use of the term differs slightly from the way I am now using it in this book. In 2015, Thompson wrote:

> [. . .] we need to distinguish between ordinary dreamless sleep and lucid dreamless sleep. Ordinary dreamless sleep is the sleep of ignorance, in which awareness is described as being in total darkness. Lucid dreamless sleep is described as a state in which awareness is luminous and without an object (free of thoughts and images). Whereas lucid dreaming consists in knowing that you are dreaming, lucid dreamless sleep is said to consist in being able to witness the state of dreamless sleep and recall its phenomenal clarity upon waking up. Although the background metaphysics of Yoga, Vedānta, and Indo-Tibetan Buddhism differ in significant ways, they all describe lucid dreamless sleep as disclosing a basal level of pre-personal consciousness that lies deeper than the modes of awareness that characterize the ego-centred waking and dreaming states.[8]

In her reply, Jennifer Windt introduced the concept of "minimal phenomenal experience."[9] One theoretical motivation was the question of what, conceptually, one would have to subtract from a previously existing minimal model of *self*-consciousness—as introduced by Olaf Blanke and me in 2009—to arrive at a minimal model of consciousness itself. Some of us had been discussing this for a long time: How do we get from "minimal phenomenal selfhood (MPS)" to MPE?

For Windt, the shift from MPS to MPE involves subtracting two conditions: (1) a weak first-person perspective and (2) spatial self-location:

> The key idea is that while even the simplest forms of dreaming are characterized by phenomenal selfhood, or the experience of being or having a self, the transition from dreaming to dreamless sleep experience occurs when even this minimal form of phenomenal selfhood is lost. While the analysis of dreaming can help identify the conditions for minimal phenomenal selfhood, the analysis of dreamless sleep experience may provide a glimpse of an even simpler (and perhaps even minimal) form of phenomenal experience.[10]

Among the merits of Windt's approach are that it offers a positive model of the transition from MPS to MPE and that it provides us with the first explicit theory of what MPE really is—namely, something resembling an almost empty and merely *temporal* frame of reference, involving "pure subjective temporality, or phenomenal experience characterized only by the phenomenal *now* and the sense of duration, but devoid of any further intentional content."[11] Hence, she writes:

> The moment at which self-location dissolves—or at which minimal phenomenal selfhood is replaced with the maximum unit of identification—involves a transition to the type of pure subjective temporality that earlier, I suggested might be the

> phenomenal mark of dreamless sleep experience. As lucid dreaming gives way to lucid dreamless sleep experience, minimal phenomenal selfhood shades into pure phenomenality, in which phenomenal experience is characterized only by its temporal structure.[12]

Judging from the phenomenological material presented so far in this book, it has become clear that neither explicit embodiment, minimal spatial self-location, nor epistemic agency is a necessary condition for the phenomenal character of pure awareness to emerge. Neither a conscious body image nor even the mere identification with an extensionless point in space (as we find in "asomatic out-of-body experiences" and "bodiless dreams")[13] is required. Nor is the experiential character of controlling thought and attention (e.g., while engaging in mental calculation or object-oriented forms of meditation practice like *ānāpānasati, shamatha*, or body scan). In addition, many of our experiential reports make clear that the space of pure awareness is a space that lacks boundaries and a knowing self forming its center (see chapter 23 for details). It does not even have an egocentric geometry: The phenomenological feature that Olaf Blanke and I called the "weak first-person perspective" is entirely absent.[14] This term refers to an egocentric, visuospatial model of reality that includes a spatial frame of reference plus a global body representation, with a perspective originating *within* this body representation. If you have a weak first-person perspective (even if you are not actively controlling your attention, and even if there is not a single thought in your mind), then there still is a center of projection. According to subjective experience, this center functions as the geometrical origin of the passively "seeing" or "hearing" organism's perspective.

It seems that for advanced meditators, all these features sometimes can *coemerge* with the quality of pure awareness (more on this in chapters 26, 27, 28, 32, and 33), but they are in no way necessary for it to occur, as most cases don't include them. Therefore, they will not be part of a minimal model explanation of consciousness. We have also seen that full-absorption episodes during deep meditation, lucid dreaming, and wake sleep are currently the three best concrete candidates for MPE, the simplest form of conscious experience that we currently know. Wake sleep, therefore, may actually come very close to the hypothetical essence of phenomenal consciousness for which many of us are looking.

Windt's original MPE theory is an excellent springboard for this inquiry because it gives us purchase on crucial interactions among sleep, dreaming, selfhood, and consciousness. However, I think we can now begin to gradually enrich and expand her theory in a number of directions. First, we must do justice to the phenomenology of "timelessness," which will be investigated more closely in chapter 22. Many of our reports clearly show that there is no time representation at all in some experiences of MPE. "Pure subjective temporality" is an interesting concept, but the phenomenal now and the sense of duration are clearly forms of explicit temporal experience. They are not part of pure awareness itself. Then there is the highly interesting phenomenology of "timeless change" (just think of Aldous Huxley's "transience that was yet eternal life," described in chapter 9). This has to be taken seriously, so it is important to ask whether there are situations in which the experience of duration and the nowness created by locating oneself in a temporal frame of reference can actually be *united* with the timelessness of MPE. We will return to this question in chapter 22.

A second avenue for expanding on Windt's first model starts with the observation that this model ignores the "phenomenal signature of knowing" set out in chapter 7.[15] As we have seen, MPE is characterized by a silent quality of subjective confidence, a nonconceptual and nondual character of knowing. This, I believe, is another reason why the phenomenology of wake sleep is so interesting: Some of the clearest expressions of this experiential quality are found in traditional terms like "empty cognizance," "clear light sleep," and "witnessing sleep." The phenomenal quality of epistemicity or passive, nonagentive knowing is something that any future theory of MPE will need to explain—and inquiry into the experience of wake sleep may help. How is it possible for the phenomenal signature of knowing to occur in dreamless deep sleep?

Third, any satisfying account of pure awareness will also have to provide an answer to the question of contentlessness.[16] If it is timeless, if there is no temporal content whatsoever, does wake sleep have any other form of content; is there something that it *represents*? To be sure, many traditional theories of pure consciousness say that it has no content at all, that is a completely nonrepresentational phenomenon. From a naturalist perspective, this is entirely possible. For example, take the "mode-neutral brightness" that can be found in wake sleep (see reports #58 and #3171 in this chapter) but is not addressed in Windt's original account of MPE. Is this nonvisual brightness perhaps a nonintentional property of the state (i.e., one with no quality of aboutness), not content at all (even highly abstract), but a dynamic property of the neural carrier in the meditator's brain? Could the clear light be like the subtle but entirely contentless hum of an amplifier in a "phenomenal phonograph" in which the turntable and all speakers have been turned off?[17] In 1986, at the end of an important paper on the direct introspection of brain states, the Canadian philosopher Paul M. Churchland wrote this:

> [. . .] those of us who prize the flux and content of our subjective phenomenological experience need not view the advance of materialistic neuroscience with fear and foreboding. Quite the contrary. The genuine arrival of a materialist kinematics

> and dynamics for psychological states and cognitive processes will constitute not a gloom in which our inner life is suppressed or eclipsed, but rather a dawning, in which its marvelous intricacies are finally revealed—most notably, if we apply ourselves, in direct self-conscious introspection.[18]

The kind of introspection that is relevant for the epistemic practice of meditation is clearly one that does not apply concepts, either old or new. But if science (or a philosopher like G. E. Moore; see chapter 28) tells us that there actually may be something to look for, then this might guide attention into regions of phenomenal state space that were previously eclipsed, finally revealing their intricacies. A first example was given in chapter 6, when I explained how through the technique of "body scan," I first discovered the contact sensations behind my inner eyelids. Here, it was not science, but an excellent meditation teacher, that helped me feel them. They were now decompressed and elevated to the level of conscious experience—"unsuppressed," if you will. And please note that after the first discovery has been made, you can choose to abide in the suchness of those sensations (chapter 9), nonconceptually, with a silent mind. They can now be investigated on a finer level. Later, this finer investigation may turn out to be genuinely innovative, yielding new targets for research. Sonam Kachru puts it like this:

> But philosophers have also used meditation in various ways to benefit philosophy as a discipline of knowledge by creating experiences and/or experiential skills that produce new theoretical objects or new characterizations of available theoretical objects to incorporate into theoretical explanations. By a "theoretical object" I mean a phenomenon that comes in for attention in the theory and practice of an epistemic culture, like philosophy or science, typically as an element in explanations or as the target of explanation.[19]

My point is that the same could be true for many new concepts that are provided by philosophy, neuroscience, or computational modeling, such as those referring to the quality of timeless change mentioned previously, to the nonvisual quality of brightness, or to the silent quality of subjective confidence (i.e., the nondual character of knowingness folded *into* the process of consciously perceiving the world). Through meditation practice, these implicit qualities can be unfolded. Once we know that there is something to look for, once our attention has been guided, we can investigate what remains if we let go and look *without* applying any concepts at all. Returning to theory, we may then create new and better theoretical objects—and entirely new levels of description. Paul Churchland may be proven right in ways that he probably never thought about, and that many meditators may feel ambivalent about.

Please recall that wake sleep itself is not described as being aware of the fact that we are currently sleeping, nor is it an awareness of being in the dream state—it is just consciousness itself. This is why wake sleep is relevant for the subjectivity argument: It is a prime example of *nonsubjective* consciousness. Witness consciousness itself can also never become the object of awareness; it is a state that cannot be introspectively accessed while we live through it. *You* cannot know it while asleep. It may, however, be a state that "knows itself" (chapter 30), autopoietically modeling the physical conditions of its own existence. While the clear light exists, there is no knowledge of "having witness consciousness," nor is there awareness of "being in a contentless state." There is certainly no philosophical concept of "directly introspecting the brain itself" being applied, nor is there a phenomenal quality of ownership, or even a concept of MPE currently being "my own state." The philosophical lesson to be learned is that from the fact that a mental state is being introspected or remembered, it does not follow that it is necessarily represented as *a state of the subject*. I claim that what makes wake sleep so interesting for the philosophy of mind and cognitive science is the phenomenological fact that it is not merely a nonconceptual, but also a *nonegoic* form of self-awareness (chapter 29). Only after returning to the ordinary wake state can one realize that an episode of wake sleep actually occurred. Later reporting that I myself was in such a state while it occurred may, strictly speaking, be a form of self-deception—a subtle, autobiographical confabulation.

The Spanish philosopher Adriana Alcaraz-Sánchez has argued that Evan Thompson's original working concept of "lucid dreamless sleep" may actually be a misnomer.[20] Technically, she argues, lucidity is a recognition of the hallucinatory nature of my overall perceptual state, and the phenomenology of clear light sleep does not involve this recognition, either conceptually or nonconceptually. When it comes to the feature of insight, wake sleep is really not equivalent to lucid dreaming (more on this in the next chapter). Alcaraz-Sánchez thinks that there is a minimal condition for lucidity: You must accurately judge that your own perceptual experience is nonveridical while it occurs; there has to be a realization that what you are perceiving right now is not real. More traditionally oriented analytical philosophers would probably insist that this realization has to be a conceptual affair because you must categorize your current state in a certain way, perhaps as a dream or as an ongoing "immersive spatio-temporal hallucination," as Windt's well-known dream theory has it.[21] But even if we allow a nonconceptual form of meta-awareness (as Thompson does), there is a phenomenological argument to be made that what *really* characterizes clear light sleep is the experience of pure awareness or a mere "void," but not the recognition of the hallucinatory nature of my own perceptual state. On the other hand, if we opt to consider the "void" of pure awareness as some form of minimal perceptual environment, then we are not speaking of a sleep state anymore. Adriana Alcaraz-Sánchez points out that clear light sleep might then be better conceived as an imageless lucid dream.

I think the conceptual insight requirement doesn't apply in wake sleep because, phenomenologically, the second-order level does not exist, so there can be no judgment that the body is physically asleep while awareness persists. MPE is an entirely nonconceptual state, and if it occurs as a stand-alone phenomenon during dreamless deep sleep, then there can be no concurrent recognition of the sleep state as such. There can be only retrospective recognition, the memory that actually, amazingly, "the light was on" during parts of the night—an impersonal, timeless light that had nothing to do with you and the narrative that you like to spin about your life. Even the distinction between "real" and "unreal" or "veridical" versus "nonveridical" was absent; it was not part of the experience. All we may have is the spontaneously arising memory of a selfless state, an active memory that now begins to cunningly weave itself into the story of "your" life as a practitioner of meditation. Gradually, the clear light of MPE sleep becomes an experience that "you" think you have had—an element of your autobiographical self-model that you can even verbally report.

This leads us to an interesting philosophical point. In full absorption, not only is pure awareness concurrently ineffable, but concurrent *unrecognizability* seems to be an additional defining feature of all genuine full-absorption episodes. In the words of Jonathan Shear:

> What is it remembered as? Not as anything. One remembers simply that one was, one was awake, *without being aware of any thing*, without even noticing *that* one was awake. Yet one was awake, aware, not unconscious—*this* one remembers. In memory it is clearly distinguished from unconsciousness, as for example, from the "blackout" of deep sleep; it is clearly distinguished from unconsciousness because one remembers being awake, just simply consciously *being*.[22]

In this new context, I wonder whether you remember the "contraction principle" discussed in chapter 8? MPE sleep is a perfect example of conscious experience without contraction. The phenomenal character of pure awareness—the "clear light"—is there, but as it occurs, it is not automatically attributed to an ego; it is not integrated into the self-model that dominates the meditator's waking life (more on this in chapter 25). During an episode of clear light sleep, there is no knowledge of any person at all, of anybody who existed in the past, who perhaps finds himself in a dormitory on a long meditation retreat, who is looking forward to the future. Whenever—as Tulku Urgyen Rinpoche put it in the opening quotation to this section—"luminosity is captured" during deep sleep, all we have is the suchness of self-evidencing wakefulness because pure awareness is not attributed to anything else. Phenomenologically, it just knows itself, nonconceptually.

On an intellectual level, deep philosophical questions now begin to arise: Is nonconceptual awareness an epistemic state that allows genuine insights, or is the purely phenomenal signature of knowing all we ever have? In highlighting overlaps between the quality of insight in lucid dreaming and insight in meditative practice, the Canadian philosopher Lana Kühle has suggested that we can always interpret the quality of "insight" occurring in meditation or the dream state either as an epistemic state or as an experiential form of insight. We can treat insight as genuine self-knowledge or as something phenomenal, something that might always turn out to have been mere appearance.[23] Kühle points out that if insight is an epistemic state, then it must have an intentional object, something that it is about. But what exactly are the intentional objects of the state of insight? What kind of knowledge does the state of insight involve? Is it just a sudden reduction of uncertainty in the brain, something subpersonal, or should we say that true, epistemic lucidity is always tied to some conceptual self-labeling of the current state as the type of state that it is? Does it therefore need categories, a successful classification—even if this classification takes place only psychologically, not verbally? To illustrate this issue, let me ask a different, deliberatively provocative question: Can MPE be hallucinated? As a follow-up question: Could it be that wake sleep as occurring in people who have meditated for a long time is actually a way of *dreaming* about the pure-awareness experience, including the accompanying illusion of insight?

To answer this question, we can put our new conceptual tools to work. A human being in dreamless deep sleep is not epistemically open, and it doesn't matter whether her sleep is currently accompanied by the wakeful presence of MPE. The body is paralyzed and only weakly open to the world. Nothing is heard, seen, or felt—except, during episodes of witnessing sleep, MPE itself. Using the concepts introduced in chapter 4, we can say that stimulus-readiness exists and is internally modeled: There is an experience of wakefulness, but no stimuli are consciously processed. Of course, subliminal perception still works, so an alarm clock or a sudden, loud noise could wake our sleeper up. But during a full-absorption episode of clear light sleep, there is no focus of attention that could be controlled, and the sleeper is completely unaware of her capacity to orient herself in time and space or to remember what person she was, is, and wants to be in the future. Therefore, the degree of epistemic openness is actually very limited.

Of course, in real-life situations, many graded states will exist, states in which absorption is present but not complete. To return to and expand on a novel metaphor from one of our participants, so-called white nights are not exactly the same as our participant's full-blown "midnight sun"; they occur in those locations where the sun remains less than six or seven degrees below the horizon. Accordingly, people in these places experience midnight *twilight* instead of midnight *sun*—while real midnight sun during the summer months occurs only in places north of the Arctic Circle or south of the Antarctic Circle. Phenomenologically, twilight sleep and clear light sleep may both exist. On the other hand, full-blown MPE sleep as such will always be timeless, nondual, and nonegoic. This observation leads to an interesting conclusion: In rare situations, it is actually possible that while epistemic openness is extremely limited on the functional level, it is nonetheless consciously experienced as pure, self-knowing wakefulness. What we do not know is whether the phenomenology of epistemic openness itself comes in degrees as well.

I think that a more rigorous scientific perspective may help us see that there are more than two options here. There may be a middle way. First, almost all the knowledge embodied in the delicate structures of a human brain is nonconceptual. Our best current theories view it as a system that constantly creates and updates probabilistic models in the form of nonconceptual representations, constantly predicting what will happen next. These models support the organism in protecting its boundaries to the world; they are continuously active on thousands of levels at the same time while being nested into each other on many timescales, relentlessly improving and trying to sustain themselves in a process mathematicians describe as "model selection" or "hierarchical Bayesian updating." Model selection is a local process going on in our brains, turning us into platforms for the survival of the fittest *model* in our minds. And now we have a new perspective on evolution that allows us to see all this as a nested process, a perspective that "casts groups of organisms and entire species as families of viable models that vary in their fit to a particular niche."[24] We begin to see biological evolution itself as a process of Bayesian model selection: Mind and life rest on the same fundamental principles.[25] If these models weren't good models, we wouldn't be here—if anything is knowledge, if anything successfully reduces uncertainty, then it is these nonconceptual models of world and self in our brains. Almost all our conscious knowledge is implicit knowledge in this sense.

Second, almost all the knowledge encoded in our brains is unconscious—just think of the procedural knowledge that helps you tie your shoes or ride a bike. What we experience are the predictions made by the generative model in our heads, predictions about what will probably happen next. One point that I want to make is that the unconscious, procedural form of knowledge that is itself almost never consciously experienced may include the long-term practitioner's knowledge of how to return to the state of pure awareness again and again, the knowledge that supports her in gradually learning how to abide in it, and perhaps even how to live from it in an effortless way.[26]

Do you remember the "dolphin model of meditation" from chapter 10? If such an unconscious mechanism had been established by years of training, it is empirically plausible to assume that it could also come online in periods of low arousal—such as during dreamless deep sleep. It would have become functionally autonomous. To take a highly speculative example, let us assume that MPE is causally related to a predictive model of the internal signal created by the ascending reticular activation system (ARAS), the part of the brain that triggers the process of waking up in the morning.[27] This predictive model would be the belief that knowing is likely, a statistical hypthesis that says "something can be known." Pure awareness would then be a form of *subcortical* phenomenology, originating deep in the brainstem. What we consciously experience in ever-fresh wakefulness would therefore not be the unconscious mechanism itself, but rather a continuously self-renewing prediction of epistemic openness: By constantly predicting and thereby controlling the level of tonic alertness, the brain's ARAS model would contribute to creating the specific timeless quality that Tibetan Buddhists many centuries ago appositely called "ever-fresh wakefulness." Perhaps the evolutionary expansion of the cortex did not implement consciousness per se, but only superimposed it with content?[28]

Now think of full-absorption episodes. If a stand-alone ARAS model were active during dreamless deep sleep, we would expect an entirely nonegoic and nonconceptual phenomenology of wakefulness to appear, like moonbeams falling into your bedroom at night, impinging on your eyelids and creating an experience of "visual openness" without you actually seeing anything. If the moonbeams were made not of light, but of wakefulness itself; if they came from the interior of the brain itself, then there would be great clarity and a representation of epistemic openness, but nothing else.[29] Given that the underlying model in the brain is a good one—that it is "Bayes-optimal" and from a statistical perspective, it minimizes the probability of misclassification—one can plausibly predict that there should be a robust phenomenal signature of "self-knowing wakefulness." This would be a new form of insight, perhaps even of self-certainty, an embodied form of nonconceptual knowledge. But the fact that it could in principle be precisely described by future mathematicians from the outside does not in any way imply that the brain itself uses a *conceptual* mode of presentation to become aware of the fact that this new form of knowledge is present, that it currently has insight into its own potential for epistemic openness. Whenever a critical level of tonic alertness is preserved in and represented during dreamless deep sleep, the expected phenomenal correlate would simply be wakeful, passive epistemicity with no other reportable content.

Nevertheless, being identical to a stable probabilistic model, the conscious experience of wake sleep could certainly be treated as an implicit, nonconceptual, and embodied form of self-knowledge that is entirely valid from an epistemological perspective. Hallucination versus insight is not the only choice. There is a middle way: The openness itself may be more like a dream because it does not represent an objective functional fact—but the self-knowing wakefulness itself could really be a form of insight, a very special form of knowledge. Ultimately, we may come to see it as a nonegoic form of bodily self-awareness.

21 Pure Awareness and Lucid Dreaming

> Lay down and almost immediately snapped into the Basal Clarity. [#3058]

> The lucidity was so shocking that I had the experience of falling out of the dream, down into a space that was a vast, limitless, empty void but also filled completely with pure awareness. [#105]

A lucid dream is a dream in which you know that you are dreaming. There are different types of dream lucidity, and during the last three decades, the phenomenon of lucid dreaming has become more and more important for consciousness research.[1] The relationship between dream lucidity and pure awareness is complex and interesting. For example, there may be a relationship between passive dream lucidity and the phenomenology of witness consciousness (chapter 19). There is more than one kind of lucid dreaming. Sometimes we are only passively witnessing the dream story, like watching a movie, while all the time knowing that this is a dream. At other times, there is full-blown agency, and we can control what happens because different self-models lead to different dream experiences.[2] There are also prelucid states in which we already feel that "something is wrong here," but we never realize that all of this is actually a dream because there are ways of being aware of the virtuality of one's own state without recognizing it as the *type* of state that it is (chapter 28). Such prelucid phases do not involve conceptual awareness, but they may causally enable a transition to the phenomenon that we later describe as a "lucid dream." The most important distinction, however, may be the one between those states in which lucidity is still tied to a knowing self, with the insight being of a propositional nature (called "dual meta-awareness"), and those in which an entirely nonpropositional and nonconceptual form of open monitoring has carried over into the dream state (called "nondual meta-awareness" or "clear light dreaming"). In the first case, there is still someone having the insight; in the second case, there isn't.[3]

Lucid dreams seem to occur more frequently in the context of intensive meditation practice, such as during long retreats. Let us therefore look at some results from our first study. As you will see, the phenomenological landscape is rich, fascinating—and at times messy.

A first clear finding is that there is a clear phenomenological path from lucid dreaming into a full-absorption episode of pure awareness:

> 1272 [. . .] I was having a lucid dream. Suddenly the dream started to fall apart. The dream images appeared, with long moments of emptiness in between. I started to panic and feel fear of letting go. It was like I kept on forgetting who I was. Just experiencing emptiness, and suddenly remembering again. Feeling afraid of letting go. After some struggle I finally relaxed into emptiness, and into not knowing. After experiencing it for who knows how long, I started waking up in my bed. I felt great afterward!

> 105 I was in the early stages of a lucid dream; lucidity came from flying upward. The lucidity was so shocking that I had the experience of falling out of the dream, down into a space that was a vast, limitless, empty void but also filled completely with pure awareness. It was completely black, but filled with radiance. There were no sounds, sights, colors, or objects. I was not aware of having/being a body but could feel that I was also part of the limitless space filling it and expanding into it entirely. A feeling of deep peace, of both being the entire limitless space of awareness and being in the entire limitless space of awareness at the same time. A place or state completely filled with nothingness. I refer to it now as the lucid void. It felt like a state I could easily have fallen all the way out of, back into full wakefulness, but instead, I landed in a vast state of pure awareness on the way.

Second, the quality of wakeful presence during the dream state is not something that the dream self fabricates; it typically occurs spontaneously:

> 933 [. . .] I woke up in the night and noticed that I was still sleeping. I realized that while I was sleeping and dreaming I was simultaneously wide awake. This lasted for a few minutes, at a guess. It was a very pleasant experience; it was accompanied by an inner smile. [. . .]

A third general characteristic is that in gradual transitions to pure awareness, there can be a quality of de-immersion that may relate to the aspects of "nonidentification," "witnessing," and "virtuality," as discussed in chapters 8, 19, and 28:

2973 During a nightmare I step out of the scene with the realization that all these images and stories are a product of my thinking—and I experienced an immediate silence, freedom, alertness, and clarity that went very deep. [. . .]

2774 [. . .] I was also in a similar state after lucid dreams, after they collapsed and I was simply there. There was an I-consciousness, a sense of duration, but no objects and no own body. This state was dark too, like in meditation. But once it was bathed in glistening white light. [. . .]

Interestingly, there is also a fourth observation, a phenomenological path leading from pure awareness into lucid dreaming. The transition can be reversible, and complex state transitions may apparently also include a temporary loss of dream lucidity, followed by its return. In dream research, this well-known phenomenon is called a "lucidity lapse"—and of course, this phenomenon is related to what every mindfulness meditator on her cushion experiences again and again, many thousands of times over a lifetime of meditating:

3058 My answers refer to what is known as "Basal Clarity"—pure awareness in the sleep state [. . .]. Luminous awareness. Lay down and almost immediately snapped into the Basal Clarity. A dream in the arctic. A lot of snow, ice, everything white, in it I already recognize the inherent radiation of consciousness. To my astonishment, the arctic is alive: fish and polar bears . . . I return into the dreamless Basal Clarity. It is bright, it is luminous. I hang snug in the brightness; it could stay like this forever. Dreams surface and fade away. No information nodes today, no exciting stories—just Being There. But a dream leads me unconsciously and insidiously away. [. . .] The memory activates a concatenation of memories and I become clear again. Centre myself again in the Basal Clarity. Enjoy luminous awareness. Later I wake myself up, become layer by layer more awake. [. . .]

2747 [. . .] 1st experience: before the beginning of a dream: My consciousness is completely clear, pure consciousness and nothing is there, absolutely nothing. Only darkness. I am completely present and pure awareness. Then suddenly the dream begins: I realize that I am sitting in a deck chair, a wooden one with a hanging canvas. Suddenly, about 25 m away from me, a light behind a frosted pane of glass goes on . . . etc. The dream begins. [. . .]

75 [. . .] as my consciousness returns to dream I nearly always perceive bubblelike images, each with a dream inside. As I am attracted to one, I enter the dream with full lucidity and do my dream practices in that state.

A fifth general observation is that there seems to be a direct phenomenological path from ordinary, *nonlucid* dreaming into states strongly resembling pure-awareness experiences:

> 3481 When I was younger, I dreamt that I wanted to open the lowest shelf of a very large cupboard in the living room, and the cupboard started falling on me. I realized I'm going to die as soon as the cupboard smashes me, when it would hit the floor. Instead, I was still alive, but I was not as my body—I was a presence in the room, being aware of that space, with all light entering the room (the dream was happening during the day, but there was just daylight). The presence did not have feelings (e.g., any positive or negative or happy or sad). It was constant and it didn't need any form of human expression. It was just there.

Sixth, in terms of vividness and intensity, the phenomenal character of pure awareness during an ongoing lucid dream resembles some aspects that also were experienced during periods of nondual awareness in the wake state (see chapters 26 and 27 for examples):

> 3171 [. . .] In lucid dreaming—vivid, energized, vibrant, euphoric, with visual imagery. [. . .]

Seventh, dream lucidity can exhibit the holistic and global character of witness consciousness (as investigated in chapter 19):

> 932 [. . .] I had a very strong experience of a lucid dream. I felt pure awareness watching (being aware of) myself dreaming . . . and there was some fluidity, hovering—between each other . . . how I dream. I also knew that I am also something "bigger, from outside and above," watching myself how I am dreaming. It was such a strong and compelling feeling . . . as if I were inside and out as one . . .

Eighth, people who actively practice lucid dreaming often experience periods of pure awareness in phases of transition:

> 135 [. . .] I became lucid in a dream and remained lucid in the transition between one dream and another. That was the moment when I experienced pure awareness, and I realize that at the moment when this happened it was as if there had been a void of everything, as if all concept of time had disappeared, as if I had witnessed the crumbling and decomposition of what my mental senses were perceiving until I could no longer think or imagine, until I was no longer there without needing to be. As if there was a void not distinct from me. [. . .]

183 While practicing lucid dreaming I had the sensation of separating from my sleeping body and then moving forward through total emptiness. I could not see any image of my body or my surroundings. Just pure alive emptiness.

2352 I've had experiences of pure awareness right in the moment when I was becoming lucid in one of my dreams. For me it was really intoxicating and fulfilling, I felt like I was born again in the dream and was becoming aware of everything around me with a brand-new perspective that I never have when I'm awake. Great amazement at being conscious of everything, just becoming conscious is fabulous and exciting.

3536 I have tried to slide wakefully into a dream, and have experienced this awareness moment in the intermediate state between dream and reality.

Transition phases triggering the wakeful presence of pure awareness can also occur during exhaustion within contemplative practice, or at sleep onset:

2485 The experience [. . .] was almost dreamlike, but while I was still awake. I was very tired while doing vipassana meditation, and suddenly I was incredibly awake in an entirely different way than my usual experience.

2628 [. . .] during out-of-body and lucid dream states there is a different type of pure awareness, without thought but with infinite agency. And during reverie, often in the hypnagogic state before sleep, awareness can "sink" to a simple frequency/continuum that is the same essence it has been my entire life, unaffected by any of my experiences, memories, or stories.

A final, tenth discovery is that a variety of quite complex phenomenologies are also possible. For example, not only can a full-absorption episode of pure awareness spontaneously occur during the lucid-dream state, it can also be the result of deliberate contemplative practice *in* the lucid dream. This can then be followed by what dream researchers call a "false awakening" into a nonlucid dream, with the subject's autobiographical memory later appearing to be preserved across all stages:

1649 [. . .] The dream became lucid and clear, and I noticed that dream sequences were apparently repeating themselves, which helped me to wake up in the dream: It became crystal clear to me that I was sleeping and dreaming. I then remembered—in the dream—a friend's instruction not to take part in the activities in the lucid state, I sat down in the dream and existentially questioned everything I saw in the dream ("This is only a dream!") and meditated in the dream. Everything disappeared: The people in the dream, buildings, my body, everything dissolved . . . what remained was pure undiluted attention,

> deep black disembodied space, seemingly endless, . . . I was this space and this space was pure awareness. I awoke crystal clear in one room, this awakening felt absolutely concrete and real (but as I later realized, I was still in a dream!). In this state I could remember everything that had previously happened in the dream, the dissolving etc. I was in a room with a beautiful view of a meadow and a tree. I left the bed and sat down on a meditation cushion in the room and immersed myself in meditation. Then I left the room and went to a supermarket. On the way back to my apartment I heard unfathomably beautiful singing, sounds similar to whale song . . . I also saw a big light over a park. I went there. The park was flooded with light, many people were standing around the park and looking in . . . in the park sat beings that are hard to describe: human, androgynous bodies without genitals, with animal heads. These creatures were lying on the meadow or sitting in the trees, they emanated light and this fascinating singing. Shortly after that I woke up, this time for real, and I could remember everything down to the very last detail, the 1st dream, the lucid state, the pure awareness, the black endless space, the awakening in this room, these strange creatures . . .

All this gives you a flavor of the relationships between dream lucidity and minimal phenomenal experience (MPE), which are not at all easy to understand. Let me end the phenomenological section of this chapter by drawing your attention to one highly specific detail. In our study, a small category of experiential reports pointed to *mirror experiences* as a special context in which pure awareness can occur. To bring these together in the same place, I will present two experiential reports describing such experiences as occurring during a lucid dream, and two more that took place during the wake state. We begin with those starting from lucid dreaming:

> 1708 During a lucid dream I decided to look at myself in a mirror. What I experienced was, as far as I could tell, indiscernible from looking in a mirror while awake. Having such percepts while knowing that anything apparently "external" was actually a projection of this "internal" dream state was what I believe is meant by "pure awareness" in this survey, although I disagree with this description because it implies a hierarchy of conscious experience. The experience I am describing was not more or less "pure" than any other experience. However, the image in the dream mirror was not a mere reflection as we usually think of it. In that context the mirror surface was identical to the scene it seemed to reflect, so that it was more like a window than a mirror. And what was on the other side of the window? For a moment I wondered whether I was

looking at my wakeful self, who was looking not through a window as I was, but at a mirror surface reflection. This uncanny sense of being in two places at once (but from a first-person perspective) was noticed without considerable surprise, as if this was exactly what one would expect to happen while looking at a mirror in a lucid dream. And yet somehow, this was all profoundly puzzling. I turned to look at the environment and then back at the mirror. All of it appeared realistic under scrutiny, giving no hint that it was being rendered on the fly. Perhaps this was wishful thinking, and I could have found something amiss if only I had looked more carefully, but it was not apparent to me then.

81 [. . .] I am in a dream, fully lucid with the intention of finding another dream, so I create a mirror and start to enter it. It gets to the point where my feet are on the floor in one place, my body halfway through the mirror and the top part of me in this blackness in a state of nothingness. I am slowly dissolving into the void. I feel there are two places that dreamers call the void, the first is in the early stages of a wake-induced lucid dream when you drop or phase to a light sleep state and retain awareness of this state change. At this point it gets eerily quiet as the sense of hearing is the first to shut down and you can be in a darkness that feels like an open space and will often precede hypnagogic imagery. The second place is entered from within a lucid dream, and I term this the space between dreams. In the early days I would enter this space and quickly transit to a new dream; now I am able to spend time there and see it as a place of its own with potential for adventure. It is this second state that I am trying to describe. My first impressions of this space were from trying to change to a different dream and when for a few moments I find myself in this darkness and often with the feeling of not wanting to remain in there due to a vague sense of unease. Very quickly I would find myself in another dream, having passed through within a few seconds. Over time I started to wonder about this feeling and then started to associate the sensations with a place separate from normal dream space. In my thoughts this darkness or space between the dreams became a separate reality that I suspected I could explore if I had the skill. The void always seemed to be present and was a recurring event in most lucid dreams. Over a series of dreams, I had lot of entries into the void and found that if I just accepted being there I would remain for longer periods. In time I also started to lose the uneasiness that seemed to accompany me at first. The more time I spent in the void the more I seemed to lose any sense of self, and it got to the point where I felt I existed as pure awareness. In the void there is blackness that seems to be all around and has a presence, it is

not seen with eyes but experienced with a set of inner senses that combine to create the awareness. This awareness is very hard to define in normal terms. There is nothing to touch and I have nothing to touch with as there is no energy or dream body either. There is thought and full access to memory and the ability to leave to another dream if desired. I suspect what is happening is a shutdown of all sensory input and no recall of sensory data to construct a dream with. This is a shift to an inner sense and for me a new form of awareness that leaves me with a sense of wonder and more puzzles to try and solve. I think that when I am in the void, I am also a part of the void so exist for a few moments as pure thought. To be something with awareness and having shut off all contact with both the physical body and the energy or dream body is a state of wonder for me. I asked to meditate while in the void and had the following experience on transit to another dream. One dream goal is to meditate in the void, so I ask for this from within the dream and then imagine I close my dream eyes, start to relax my mind, and lose all thoughts. I am now gently spiraling in a 360-degree rotation like I am moving into a long tunnel. This is a pleasant motion with some movement, and I am enjoying it. I feel warm and safe, I am still moving forward, and a bright white light is expanding in my mind and I think it is the sun in the distance. I fall into this light and it is warm and infuses me with a nice energy that I am enjoying. At this point I lose all sense of self and let the motion and feelings carry on till they finish and place me into another dream. [. . .] I think the void is an area of mind that exists as a place of its own and can be accessed if desired and therefore can also be explored. In this mind space normal symbols may not be valid, so for now it may be an empty arena with the potential for adventure when I learn how to interact with whatever controls the state. [. . .]

And here are two phenomenological reports of pure awareness related to mirror practice, but this time originating in the wake state:

1926 [. . .] This happened while I stared into my own eyes in the reflection of a mirror. Like many, my first sense was that I was standing in front of a mirror looking into my own eyes/reflection, but keeping at it for about ten minutes there came this sense of being just aware of awareness. To clarify, yes, I stood in front of a mirror gazing at my reflection but the subjective sense was that what I saw in my reflection was devoid of any concepts. Looking into my own eyes in the mirror seemed synonymous with looking at a chess piece on the ground where there was no subject or object. There was only what was appearing.

3447 You sit or stand facing a mirror. During this time, you focus attention on your reflection and at the time you feel enough focus. You then turn attention to what your reflection is looking at. If successful, you find nothing. I experience a strong feeling of motion that feels like falling backward. The reflection in the mirror and the world around it gain clarity. The sensation of all awareness suddenly snapping into focus. It has a quality of infinity to it like a torus rotating around its central circle. This gives it a dual sensation of the awareness both falling into itself and constantly creating itself.

Pure Awareness, Different Kinds of Lucidity, and the Embodied You-Turn

> It appeared to a man as in a dream—it was a waking dream—that he became pregnant with Nothing like a woman with child, and in that Nothing God was born; He was the fruit of nothing.
>
> —Meister Eckhart (1260–1328), Sermon *Surrexit autem Saulus de terra*

Much recent research demonstrates that the human mind has the capacity to self-locate in more than one place at a time. Evidence from neurology, cognitive neuroscience, and ingenious experiments using virtual reality (VR) and robotic reembodiment show that "bilocation" is a genuine experience. For example, there are a number of phenomenal states in which you see yourself from the outside, such as "autoscopic hallucinations" or visual out-of-body experiences.[4] Then there is "heautoscopy," a rare neurological disorder first described in 1935 that involves a multimodal illusory experience characterized by the duplication of one's own body and self. During classic heautoscopy, the patient sees a double of himself in extrapersonal space. Here, however, this double is not a mere image or visual hallucination, but something that the sense of self can jump into and out of. The locus of identity can be experienced as being in the position of the physical body and—sometimes simultaneously, or even in rapid alternation—in the position of the double. Not only may self-location and first-person visual perspective alternate between an embodied and a disembodied location, but, as the well-known Swiss neurologist Peter Brugger has demonstrated, it can even be difficult for individuals to decide where they are localized.[5]

As I explained in my book *The Ego Tunnel*, this fascinating stream of research is important if we want to understand what the simplest form of self-consciousness really is. Recently, innovative experiments have created an artificially induced body reduplication paradigm using a humanoid robot avatar and have demonstrated that healthy human beings are capable of mentally bilocating in two bodies at the same time (see figure 21.1).

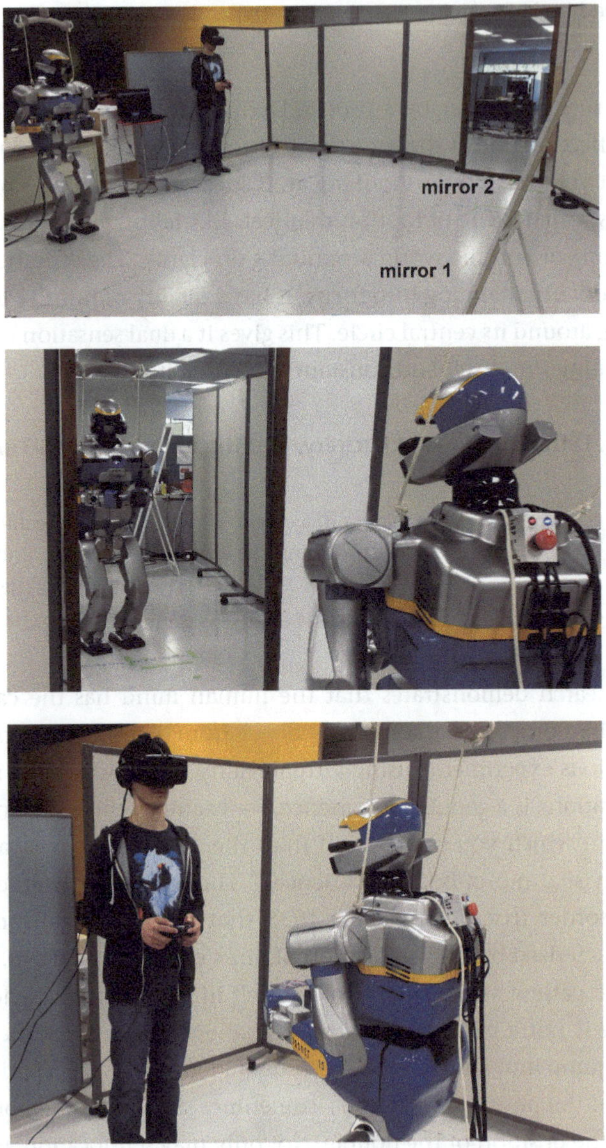

Figure 21.1
Experimental setup: A participant wears a head-mounted display that provides a visual first-person perspective from the robot's eyes, a headset that provides stereo sound feedback from the robot's ears (microphone), and holds a joystick to control the robot's movement. The robot is strategically placed next to the participant at the beginning of the experiment to avoid the participant being in the robot's field of view. Mirrors are placed along the path so that the participant has a full view of the humanoid robot's body. The sketch shows the U-shaped path followed by the participant during the experiment, with the start and end positions and the location of the main elements of the setup. What this shows is that the consciously experienced sense of self can expand in space. (Figure courtesy of Laura Aymerich-Franch, caption adapted from Aymerich-Franch et al., 2016. The robot belongs to CNRS-AIST JRL.)

In an intriguing study conducted by Laura Aymerich-Franch and colleagues, you would first learn to identify with a robot by looking out of its camera-eyes, then look at your own robot body in a mirror, and then turn around to see your biological body from the outside, wearing VR goggles and controlling the robot with joysticks. Artificially induced own-body reduplications of this type can produce a feeling of bilocation and illusions similar to heautoscopy, but in healthy people. Interestingly, for some participants, an experience ensued in which the sense of self seemed to be "smeared" in phenomenal space between two candidate bodies. This observation is highly relevant in the context of MPE because it might relate to the many descriptions of something that is unbounded in the sense of having the potential to expand (chapters 4, 5, and 8) and to phenomenologies in which the "true self" (chapter 29) is described as a global experience of spacious awareness, of wakeful presence "everywhere and nowhere at the same time." What you identify with and what you experience as "your" self is apparently something that can be contracted or uncontracted (chapter 8)—something that can, phenomenologically, be bound to a single body, but also something that disperses through the space of conscious experience. It is interesting to note that the underlying mechanisms involved in expanding the phenomenal self and shifting rapidly between viewpoints may be important in social contexts (e.g., as part of developing capacities for compassion and perspective-taking).[6]

What I like about this "robotic you-turn" experiment is that it is an embodied version of certain ancient meditation techniques in which, after you have stabilized attention on a perceptual object or even on pure awareness itself, you try to dissolve dual mindfulness into nondual mindfulness by simply trying to find the meditator. The phenomenological fact hidden in plain sight is that—if you gently and precisely turn attention back onto itself in this way and then let go completely—you do *not* find the meditator. My claim would be: Ultimately, the whole notion of a meditator is merely a figure of speech, and the actual phenomenology does not involve the meditator who is aware. In reality, there is no knowing self, no stable "epistemic agent" (chapter 25) at all. What you find is a diffuse sense of mental effort and a slightly indeterminate feeling of bodily location (often in the head), neither of which has anything to do with the phenomenal character of awareness itself.

Recall that in each chapter of this book, I am deliberately offering only some appetizers and a dessert because I want to encourage you to cook your own main course. Everybody can try this: Nonconceptually, in silence, without thought, you can directly ask yourself *who* it is that meditates, *what* exactly it is that is aware right now, *how* this is known—and *where* it is. Ancient traditions like Tibetan Buddhism make a strong phenomenological claim: There is nothing to be found except a nondual *space* of wakefulness that is already aware. If you try and then let go, do you reach the same result? There are modern versions of this ancient technique,[7] and our robotic you-turn experiment shows that for some participants, the egoic sense of "knowing from here" can actually spread out in space. In the classical version, only attention moves; in the new version, two bodies are involved, one standing and one moving, one biological and one artificial. Here is an interesting research question for all of us: How exactly would one *integrate* attention- and motion-based, fully embodied versions of the you-turn into one another?

When investigating the relationship between lucid dreaming and MPE, it is important to bear in mind that there are multiple kinds of dream lucidity. As explained at the beginning of this chapter, there is the prelucid state, in which lucidity is beginning to dawn on us but we haven't had the realization yet. Reality may already have taken on a dreamlike or symbolic quality, the motif of dreaming itself may occur in the dream, we may have the strong sense that we are just about to remember or discover something important—but it is still an ordinary dream because lucidity is lacking. Then there is passive observation while knowing that all of this is a dream. And sometimes we later see the emergence of plot control and the dream self's discovery of agency: Now, the dream narrative can be controlled and the lucid dreamer can fly, move through walls, and do almost everything she desires. But recent research shows that dream plot control is not quite the same as the actual *insight* of lucidity.[8]

And then there is the strong form of cognitive lucidity as a new feature of the dream self who now can think and report about the fact that she has become fully aware of how all of this is only a dream. The thinking self is not only lucid but intellectually *knows* about her own state of mind and the nonveridicality of all perceptual experience. The epistemic agent model has been fully stabilized (chapter 25). Future research will have some tricky issues to resolve when teasing apart these kinds of dream lucidity. For example, there are good empirical reasons to infer that at least some experimental participants reported lucid dream content, even though all they ever did was stay awake and control vivid waking fantasies. We could also question whether all experimental participants possess mastery of the concept of "lucidity" (let alone all its possible versions).

This book is trying to lay the very first foundations for future empirical investigations of MPE, and in chapter 20, I sketched out what I personally would consider the most exciting first step: the Triple Triangulation Project. Lucid dreams may be part of an important second step in expanding this research strategy. As you may recall, my initial idea was to use the phenomenon of clear light sleep or witness consciousness during dreamless deep sleep as an entry point from which to triangulate the minimally sufficient neural correlate of MPE. We first distinguished between type I episodes of full absorption, in which MPE is entered from the wake state (usually during periods of formal meditation practice), and type II episodes of full absorption, in which MPE spontaneously occurs during periods of non–rapid eye movement (NREM) sleep. Stable lucid dreams, in which the lucid dreamer has the ability to *deliberately* dissolve all the content of conscious experience and merge into a nondual state of pure awareness, would be the logical next step for the Triple Triangulation Project—if the phenomenon could be brought into sleep labs, which is certainly a technically demanding enterprise. Let us define type III episodes as full-absorption states that occur around the edges of lucid dreaming, such as before a lucid dream actually begins, or intermittently, or whenever a practitioner manages to *willfully* access MPE starting from the lucid-dream state. I have now presented statistical data and qualitative phenomenological evidence demonstrating that this is an absolutely real phenomenon: Reports #105, #183, #1274, #1649, #2747, #2774, #2973, and #3058 in our database provide us with first-person descriptions of different variants of the target state. The methodological relevance of lucid dreams used as a gateway into MPE—an ancient practice that was cultivated, for example, in Tibetan dream yoga[9]—consists in providing a third access route to help us home in on the physical and computational correlates of pure awareness.

If we now return to the experience of pure awareness as coemerging during ordinary wake states, this leads to a series of interesting new research questions: Is there something like "lucid waking"? If so, how does it relate to MPE? Is lucid waking an MPE mode of experience? How many kinds of "lucid waking" are there? For example, when focusing on contemplative phenomenology in the wake state, do we find analogs of the different kinds of lucidity briefly sketched above? But eventually we should also move from science and theory to actual practice: In the ordinary wake state, do you actually know that you are awake and conscious right now—and, if yes, *how* exactly do you know this? *What* exactly is it that you know? For the practitioner of mindfulness meditation, is there perhaps a phenomenological analog to "prelucid waking"? And how is all this related to the much more direct (and often shocking) phenomenology of nonconceptually realizing that all this is neither real *nor* unreal? I call this phenomenological kind of metaphysical indeterminacy the experience of "virtuality," and we will return to this topic in chapter 28.

22 From Timelessness to Timeless Change

> Everything that has been and still is in a timeless "moment," in an empty "place" on the tip of a needle in the now. [#1145]

> A self-nature that flows without time. [#2426]

We have now looked at the two major contexts in which pure awareness can appear as a stand-alone phenomenon: as a full-absorption episode entered from the wake state, typically during contemplative practice, and as a spontaneously emerging state in dreamless deep sleep. We then proceeded to look at pure awareness during the dream state. In investigating the phenomenology of lucid dreaming, we discovered that the specific phenomenal character of awareness itself, the simple and nonconceptual experiential quality of consciousness per se, can remain after everything else has vanished, and also that there are modes of conscious experience during which it may coemerge with other, often very complex, forms of experiential content. The totality of your lived, conscious experience at this very moment is the content of a model of the world, a complex simulation currently running in your brain. The phenomenal character of pure awareness can apparently coexist with such complete, global models of reality. As we will now see, this coexistence can manifest in a variety of ways.

You may recall that in the introduction, I distinguished between minimal phenomenal experience (MPE) states and modes. Our phenomenological data show that there are not only *states* described as carrying the qualitative character of MPE (normally experienced as personal states—that is, local episodes in the meditator's own mind), but also specific global *modes* of conscious experience that have this quality. These are entire conscious models of reality dominated by an all-encompassing and nonegoic quality of awareness as such (more on this in chapter 33). For example, witness consciousness (as investigated in chapter 19 and then explored in other guises in chapters 19 and 20) can be seen as an MPE mode: It creates a whole new model of the world. Now, the world is experienced as resting in the gaze of an ideal observer.

Here, we will continue this line of inquiry by first considering one of the major phenomenological characteristics of pure awareness—namely, the distinct quality of "timelessness"—in isolation, as we have done in many other chapters so far. In a second step, we will look at phenomenological data demonstrating that this quality can actually co-occur with forms of temporal experience. I have termed this the phenomenology of "timeless change." In chapters 26, 27, and 33, you will find additional examples of timeless change in the context of prolonged periods of more continuous pure-awareness experiences. But there is more. Among a number of unexpected results, one of the greatest surprises in this first study was just how many reports focused on *spatial* phenomenology, not merely on the character of timelessness. I am particularly grateful to our meditators for offering unusually rich descriptions in this phenomenological domain. The current chapter, as well as chapters 23 and 24, therefore, will look at the connections between pure awareness, time experience, spatiality, and "bodiless body-experience" in greater detail. As always, let us go slowly and first try to get as close as possible to the experience itself.

One point of strong convergence is manifest in many experiential reports trying to describe pure awareness: "unbounded space, but no time." Temporal experience and spatial experience can apparently be separated. Here is one example:

> 3330 I would best describe my experiences with terms like "stopping of time," "perception of space without time" [. . .]

Subjectively, episodes of pure awareness are timeless. Objectively, from the third-person perspective of scientific observation, their duration is typically in the order of a few seconds or minutes, especially if they are periods of full absorption. (Nonabsorbed periods of what in chapter 30 I will call "dual mindfulness" may last longer.) At the outset, it is interesting to note that a phenomenological analog to the experience of timelessness seems to exist in the spatial dimension:

> 2771 It was an experience of not moving forward, not moving back, neither of staying still.

In pure awareness, timelessness and motionlessness can also come together:

> 2721 [. . .] A moment of unknown length or timeless silence, complete motionlessness. [. . .]

What is common to both types of experience is the absence not only of motion (either through time, through space, or both) but also of *self-location*. During an episode of full absorption into pure awareness, there is no self-location relative to a spatial frame of reference, and there is no self-location in a temporal sequence either. The *here* and the *now* have been suspended:

1175 [. . .] What followed [. . .] was a timeless event, without space or perception, a feeling of complete restfulness and peace unlike any other "peaceful" state I had felt before; it was deeper. It felt like I had been sucked down a hole or something. It was neither light nor dark. The only thing I knew about it was afterward when I realized I wanted to go back there, and immediately felt the arising of some excitement. It was the end of the sit & the bell had gone, but it felt like eternity.

2122 [. . .] I would say that I felt as though I was pure awareness, with no intentions or memories. I felt bliss and fully aware, relaxed and full of energy. I would describe the mental quality as like doing a pleasurable activity like painting and finding yourself lost in the activity. All thoughts end. Emotion is positive. Time does not exist.

3268 [. . .] I had no sense of time or space. [. . .] I had neither thoughts nor other feelings, only when you became aware again that you were meditating did I notice that for an indescribable time (seconds or minutes) I was completely without time, space, and thought.

During a fully absorbed experience of timelessness, the inner narrative and all goal-directed mental effort have ended—the meditator's mind has neither run away into past or future, nor is it focused on the present moment as such. All conscious thought is suspended; sometimes, the breath may briefly stop as well. As an aside, here is one testable empirical hypothesis: There may be a causal connection between the ending of thought, the ending of phenomenal time experience, and the ending of breath. The three may unfold as distinct steps in a causal sequence, in this order or potentially also in other orders. Much more often, however, the phenomenal quality of "timelessness" actually coemerges with other experiential content. The relationship between timelessness and the overall phenomenal field is quite interesting. Let us, therefore, look at a wider variety of concrete examples in which time stops:

3431 [. . .] suddenly during walking meditation in nature the feeling arose that time was standing still, that the whole world was standing still. I stopped. It seemed as if my breath was standing still too. [. . .]

933 [. . .] So I stood outside in nature and at that moment I lost the sense of time. It felt as if I could stand like this forever and no time was passing while at the same time I was fully aware of what I was experiencing. It was a very pleasant experience. I cannot tell you how long I actually stood there. [. . .]

3472 In a totally everyday situation an experience of timelessness, motionlessness, silence, and being whole. [. . .]

60 [. . .] I did perceive these sensations, but quickly let go again. Between these sensations I felt the pure awareness, which presented itself as a feeling of emptiness but also as a feeling of omni-possibility. Time seemed to stand still until the next thought process. [. . .]

1885 [. . .] Immediately I felt the self drop away. The boundary of my consciousness expanded to the room I was in. Time seemed to stop. I was awed by the moment. It dropped away quickly. While it was timeless in the moment, I felt that I could account for time as a reference to the before and after. [. . .]

2463 A moment of total BEING-HERE. All sense of time was as if extinguished and at the same time the knowledge of the now-moment as the only existing thing in space and time. I also had an experience of total fusion with a leaf which at that moment was falling from the tree. [. . .]

2312 [. . .] "Infinite simultaneity" and wide awake in it.

2687 [. . .] a deep feeling of joy, timelessness, happiness. I could perceive it and had no sense of time. [. . .]

3132 First, self disappeared, then later time became all one, as in no present existing apart from past from future. Any sense of time disappeared; it was all here, now.

3353 [. . .] I don't know how long it lasted, but it felt like very long, although it could have been only a few minutes, as I reconsider it. [. . .] and suddenly, there was a deep feeling of wholeness in eternity, no physical boundaries, there was no time, there occurred a wholeness, oneness with all and no time nor space, there was only a very broad experience of peaceful oneness which included also here and now in spaciousness, timeless. [. . .] [For] a very long time I could not and did not speak of it, but very often I was very much aware that this had happened, considering it for myself as an eternity experience, there is no time at all, there is only being. [. . .]

Let us now proceed to the special case where the phenomenology of timelessness and actual time representation occur together. Some participants explicitly report experiences of "static flow" and the distinct phenomenal quality of "timeless change." The phenomenal character of timelessness can clearly coemerge with or "envelop" ongoing temporal experience, which is a highly interesting result in itself:

1662 [. . .] There was no one (including me) who "owned" this experience—it just "was." There was awareness of time passing, but this very sensation seemed insignificant, giving rise to the sense of "static flow." [. . .]

2303 [. . .] I felt a timelessness while also being completely aware of the movement of time. Sensations were heightened.

2426 [. . .] Only an all-encompassing "object," "a self-nature that flows without time," that is constantly transforming itself (is unstable) and shapes everything through its form, is present. [. . .]

2798 [. . .] I was standing at a window and looking out to a tall bunch of grasses. It was a windy day and the grass was moving a lot. Then for some period, I coincided with these grasses. Their movement showed (or maybe better: was) the nature of time, the complete ungraspability of the present while at the same time there is nothing else, no past, no future, just this ungraspable now. The difficulty of answering your questions is that there was no thought at all about time or the passing of time, there was just this movement. Only later, in reflection, there came the thought "I saw into the nature of time."

2820 Sitting in stillness. Life flows through me, around me. Millisecond by millisecond everything changes, nothing changes. Thought is superfluous. Movement or nonmovement is irrelevant. Life is, I am.

2966 [. . .] I felt so calm during the sitting meditation. And also, in the moving. As if I could feel every little part of my body. The breathing and moving. There was no sense of time. [. . .]

3623 Deep, blissful yet emotionally devoid experiences [. . .]. I sense a present, where the flow of time is static and there is an eternal "mental space" my consciousness seems to be in.

The Emptiness of Time Experience

Q: Is this joy something that comes and goes?

A: No. It is timeless. When I contact it you may say it looks as though I am contacting it now and not five minutes ago when we were talking about something else. But when you contact this region, you don't have a sense of returning to it. You have a sense that no interval exists whatever between the occasions of enjoying it. No interval whatsoever.

—Douglas E. Harding (1909–2007), *Face to No-Face*

As a student in the winter term of 1979/1980, I participated in a seminar at the philosophy department of the Goethe University in Frankfurt am Main, led by Gerhard Gamm. The seminar was about the concept of the soul as it had developed in modern philosophy from René Descartes (1596–1650) onward, and it led to an unexpected turning point in my academic life. At the time, I was almost entirely certain that I wanted to drop out of university, mainly because I simply couldn't find anything that could hold my interest. Then we looked into the *Treatise of Man* and the *Passions of the Soul* (published in French in 1664 as *L'homme* and in 1649 as *Les passions de l'âme*) and the way that Descartes imagined the mechanics of mind/body interaction—how the soul causally related to the physical body. Gamm made me see something pivotal: If consciousness has no spatial properties, then there can be no *place* in the human brain where it causally interacts with this world, be it Descartes's pineal gland, the "liaison brain" of Karl Popper and John Eccles,[1] today's minimally sufficient neural correlate of consciousness,[2] or any other *region* in the physical universe. The whole research program aimed at isolating this region was misguided. This insight changed my life. It got me hooked on the mind/body problem, and I ended up writing a thesis on recent solutions to the problem from Ullin T. Place to Jaegwon Kim, published in 1985.[3]

Descartes formulated the classical modern variant of the mind/body problem. I still find his arguments for mind/body dualism intuitively attractive. They actually speak to a robust duality that I can experience within myself—for example, whenever I meditate. Why is this so? What exactly is it about Descartes's ontological distinction between extended, material things (*res extensa*) and nonspatial, thinking things (*res cogitans*) that—even after centuries of philosophical criticism and an endless series of counterarguments—still has intuitive appeal? Where does the deeply rooted Cartesian intuition come from, the philosophical hunch that there is an important sense in which our conscious mind cannot be localized in physical space? The answer is that Cartesian ontology is *phenomenologically* plausible because it directly corresponds to distinct layers in our phenomenal self-model.

Our "phenomenal self"—the way in which we are given to ourselves in conscious experience, subjectively appearing to ourselves as an embodied, thinking, mysteriously unified person—actually has two major parts or layers. They are easy to distinguish. One part, the conscious model of our own body, is represented as possessing spatial properties: We experience ourselves as embodied beings with a well-defined position in space. The other part, the conscious model of our ongoing cognitive processes, appears in subjective experience as well, but as devoid of spatial properties. It apparently has no place in the world—and it is precisely this fact that, on a theoretical level, leads to the most important modern version of the mind/body problem. High-level mental operations like symbolic thought, it was frequently believed, are exclusively serial; we experience them in the phenomenal mode of succession.[4] Body experience and bodily sensations, on the other hand, are localized and distributed in space; we can experience them in the mode of juxtaposition. This particular feature of our conscious self-model creates an intuitive dissonance in our very own self-experience: If one set of the events in our consciously experienced self is encoded in a spatial frame of reference, but another set is not, then, according to subjective experience alone, there can be no *place* where the causal interaction between mind and body could take place. The actual mechanism, its modus operandi, cannot be observed. We can always imagine one without the other; disembodied existence is conceivable because the two parts of the conscious self seem to be dissociable.

This is where careful contemplative practice can help to clarify matters. You may recall that I argued in the introduction that meditation is really an epistemic practice aimed at self-knowledge—one that operates on a more fundamental level than words and concepts. If we don't just speak as we have been taught to speak, but actually look for the right words, then we discover that, as a matter of phenomenological fact, the modus operandi of causal mind/body interaction can never be observed. It simply is not part of our conscious reality. Mind/body interaction is an interesting philosophical concept, but in meditation practice, no such thing is actually discovered. In action initiation, in the spontaneous arising of perceptions and thoughts, the *actual* causal connection between body and mind remains unclear. It is not something that we experience: The dolphin (as discussed in chapter 10) just jumps into the air, and in the very beginning, there is always an element of surprise to it; the jumping itself always happens unexpectedly, spontaneously. The feeling of bodily agency and the "thinking self" are the brain's tricks for explaining away the surprise. Elegantly surfing uncertainty,[5] the brain swiftly hallucinates a disembodied Cartesian ego, creating the inner image of an abstract epistemic agent (which we will investigate two chapters down the road). In trying to understand Descartes's own solution to the mind/body problem, my first discovery was that—on conceptual grounds—it makes no sense to look for a physical location where something that is apparently not in space can interact with something that is. Taking our own phenomenology seriously, the second important discovery we can now make is that the mind/body problem is actually built right into us: Apparently, a particular computational feature of the neural mechanism generating the representational deep structure of human self-consciousness is what leads to the intuitive dissonance just mentioned, to the "prephilosophical mind/body problem."

The prephilosophical mind/body problem makes us *feel* the theoretical problem of psychophysical causality in our own experience, and it gives rise to the intuition that Cartesian metaphysics and the "hard problem" later exploit: the intuition that certain metaphysical scenarios are perhaps actually (and not merely logically) possible. Thus, the phenomenological dimension ultimately arises from a "technical" problem that our brain has in generating a conscious model of the self: How does a human brain manage to embed mental representations of abstract cognitive processes (e.g., the conscious idea of a *res cogitans* or the Cartesian thought "I am certain that I *myself* exist!")

into a preexisting image of one's own body, which—for obvious reasons having to do with its evolutionary origin—gradually developed from a concrete spatial model of the biological organism as embedded in its behavioral space? And how, then, is the brain supposed to represent within conscious experience cause/effect relationships between such purely cognitive events and specific bodily movements, all of them necessarily embedded in a spatial frame of reference?

It is as if one part of us is inside the world while another one isn't—and of course, many philosophers in the past have said exactly this, jumping from vague phenomenological intuitions to unwarranted metaphysical conclusions. Intellectual honesty demands that we first admit that all of this may feel extremely muddled or hazy for most people most of the time—there may be no real phenomenological fact of the matter. But careful meditation practice can certainly help us explore the phenomenological dimension of what later becomes the "philosophical mind/body problem." After all, Descartes's dualist distinction between thinking and extended substances is *phenomenologically* plausible for beings like us. It will always remain intuitively convincing because it always already corresponds to the representational architecture of our everyday model of the self.

This leads us back to the experience of timeless change as it can sometimes appear in the context of meditation practice. What many people don't see is that there is an analogous metaphysical problem for any entity that lacks *temporal* properties: If an entity isn't in time at all, what could permit it to influence something that is—for example, some process in the human brain? What *event* could constitute the integration of something timeless with something that has temporal properties like duration? There seems to be a deep philosophical puzzle here, and this was one of the reasons why I included the Harding quote in the epigraph at the beginning of this section.[6] Once again, we see how metaphysical puzzles and philosophical intuitions are anchored in the structure of inner experience.

Of course, Descartes was wrong, and for more than one reason. He equivocated in his descriptions of the conscious mind, characterizing it sometimes (in introspective terms) as being indivisible and possessing no parts, and at other times (in metaphysical terms) as having no spatial properties. A "fallacy of equivocation" occurs when a key term in an argument is used in an ambiguous way, with one meaning appearing in one portion of the argument, and then another meaning in another portion. One way to sort this out is by saying that the introspective experience of nonspatiality (the representational *content* in the self-model) could, of course, be created by physical activation patterns in the brain (the representational *carrier*) that, from the third-person perspective of science, are extended in space.

Similarly, the conscious experience of timelessness could, of course, result from physical processes in the brain that themselves possess temporal properties—as a matter of fact, some neuroscientists think that certain time constants will turn out to be a central aspect of the underlying neural algorithm that we will discover in the end.[7] This point leads us to new questions, to empirically tractable research questions that (despite my youthful fascination) are much more interesting than the armchair metaphysics of the mind/body problem, and much more fruitful too: How can a physical system like the human brain create conscious models of reality that, on a conceptual level, seem to *necessitate* paradoxical descriptions like "timeless change"?

Again and again, taking contemplative experience seriously shows how readily high-level conceptual oppositions like "inside versus outside," "mental versus physical," and "real versus unreal" can become meaningless. They simply do not map onto the meditator's phenomenology. Through practice, the "always already" character of these oppositions is gradually attenuated.[8] As we have just seen in some of the reports here, this can also lead to paradoxical formulations. We first touched on this problem in chapter 6, and we will return to it a number of times later in this book: In some phenomenological contexts, these oppositional distinctions no longer correspond to the structure of appearance; there is nothing to which they refer. They reflect successful high-level priors that are normally used by the brain to organize its conscious model of reality but are now suspended. Our mental landscape of such priors is the landscape of "beliefs" that we already have about the world before we take the next piece of perceptual evidence into account.

Emptiness—the epistemic openness of pure awareness discussed in chapters 4 and 5—means that this landscape has been flattened. Here, my main point is that this allows us to better understand the paradoxicality of many verbal reports. They are not caused by some poetic form of irrationality; it is just that the conceptual and cognitive instruments that have evolved alongside humankind, and that are now at our disposal to describe and imagine possible inner experiences, have been anchored in that mental landscape of useful shortcuts. For reasons of biological and cultural evolution, our linguistic and imaginative tools are firmly grounded in the terrain of mostly unconscious hyperpriors and beliefs about the structure of reality. We embody the world of our ancestors via our inner landscape of hyperpriors and background assumptions—and if this terrain suddenly turns into wide open space, then all we have left are three options: the noble silence of not speaking at all, the mysterious indeterminacy of neither-nor-ness, or the formulation of blatant contradictions.

Empirically establishing the existence of a phenomenology of timeless change demonstrates this general insight from a new angle: Importing dualistic conceptual distinctions like "timelessness" and "temporality" into descriptions of conscious experience does not do justice to its openness, its richness, or its fine grain. In lived contemplative experience, there are simply so many other possibilities. Phenomenologically, the distinction between "timelessness" and "temporality" is not exclusive and exhaustive—other possibilities do exist. For example, the atemporal phenomenal character of awareness itself can be quite explicit while being strongly overlaid and seamlessly integrated with different forms of time experience. These include nowness, duration, nonsimultaneity, succession and temporal order, being in the past, or being expected in the future—and the phenomenal experience of change itself, including our feeling of the passage of time. This type of superimposition is what I have termed the phenomenology of "timeless change."

Here is one methodological problem that needs to be solved. In verbal reports referring to temporal experience during meditation, there is often an ambiguous relationship between "being fully in the present moment" or "nowness" and what Jiddu Krishnamurti, in his dialogues with theoretical physicist David Bohm,[9] called the complete "ending of psychological time," the actual *absence* of temporal experience. From a scientific perspective, all consciously experienced time—past, present, and future—is an internal simulation. Episodic memory is a simulation of past events in the context of an autobiographical self-model, taking place in the biological brain; planning and thinking about future states of self and world involve active simulations of possible future events; and our conscious experience of the present moment is a simulation as well, one that is "smeared" in time because it does actually have temporal extension.

The global experiential quality of timeless change is an important challenge not only for precise conceptual interpretations by philosophers, but also for scientific disciplines like mathematics, computational modeling, and cognitive neuroscience. As I said in my book *The Ego Tunnel*, I very much like William James's metaphor as a starting point: The present is not a knife-edge, but a saddleback with a breadth of its own, on which we sit perched, and from which we look into time in two directions. But the physical universe does not know what William James called the "specious present," nor does it know an expanded, or "smeared," present moment.

The early Dzogchen scholar–practitioners in Tibet knew all of this very well, but through their own meditation practice: "Nowness" is empty; it is just another virtual representation of time, one of "the three times." I agree with the Tibetans on this point, and it may mean that some of our phenomenological reports contain a subtle conceptual ambiguity, perhaps even an inaccuracy. Perhaps many committed practitioners do see the virtuality of psychological time and actually experience timeless episodes, but later report "nowness" or "being fully in the present moment," simply because they learned to speak in this manner when they first learned to meditate. Developing experimental and psychometric methods to resolve this ambiguity is one important research target for the future; creating a computational phenomenology of timeless change is another.

23 Space without Structure, Center, or Periphery

Being unbounded and being held. [#3524]

One of the classic phenomenological characteristics of minimal phenomenal experience (MPE), which has been reported for many centuries, is a nonphysical kind of spatiality. The space of pure awareness is not identical to the space of bodily movement in which perception leads to action and in which actions cause new perceptions to arise, but it is sometimes described as permeating, enveloping, or "groundlessly grounding" the space of embodied experience as a whole (there's more on MPE as "groundless ground" in chapter 26). Pure consciousness itself is often described as an unbounded and centerless experience of "spacious awareness." This means that we may find the phenomenal signature of knowing discussed in chapter 7, but in an entirely uncontracted way. In other words, there sometimes arises an epistemic kind of openness, a model of a large space of epistemic possibilities.

One of the more surprising results of the MPE-92M study was how many meditators—when asked to describe a typical pure-awareness experience—actually privilege the phenomenal character of spatiality. Our participants created many beautiful figurative descriptions, saying that abiding in pure consciousness is like being "weightless in the space of infinite possibilities" (#761), for example, or like "[a] feeling of being enveloped. Like in a floating bubble" (#772). Meditators described the experience as "[t]he perception of space, infinite, unseparated, without inside or outside, velvety warm and simultaneously sharp" (#2528) or simply as "[a]n existence in boundless space" (#2565), or they referred to "[a] huge, boundless, absolutely quiet space, which however has a circular shape" while at the same time pointing out that "[i]t is so quiet that it is already loud again" (#3026). If one looks carefully, the relevant experiential quality of "spaciousness" has been reported as appearing in many forms and nuances. Let us now consider a selection of them.

First, many meditators report a phenomenal character of spatiality as such, a vast openness lacking any more concrete frame of reference. Often, we find this coemerging with some of the specific experiential qualities we have already investigated, like existential ease, simplicity, brightness, unity, a velvety character, high precision, clarity, wakefulness, and connectedness. Here are ten examples:

1727 I felt an incredible vastness, as if I were connected to another sphere. [. . .]

1739 It is an experience of spaciousness. The mindfulness is directed inward, toward the breath or the body, thus a space opens up in which there are no longer any reference points such as outside and inside, center or periphery. It feels very free and relaxed, but at the same time simple and unspectacular. Nevertheless, one is present and capable of acting, so not at all kind of distanced.

1894 [. . .] I would describe it as "being aware of having a mind empty of thoughts and images." Some of your questions ask whether there was a sense of "brightness" and "unity," and those are two adjectives I would indeed use to define my experience. Another illustration I could add was the feeling of my mind occupying a vast amount of space [. . .].

1947 Relating one's experience of (what is here called) "pure awareness" tends to turn heavily on single words, or short phrases. Call these qualia-catchers. Then, an underused qualia-catcher for me is "roundedness"—the experience of. Indeed, I think this experience is explained by the type and degree of awareness of the third dimension, which is atypical in ordinary life.

2241 [. . .] then it felt like I was entering an emptiness, an empty space [. . .]. Suddenly all distracting thoughts are gone for quite a long time and I am in a state where only I and this space exist. In fact there is simply nothing else there. [. . .] I still know that I exist, that I am meditating, but I would say that my outward bodily awareness has decreased in that moment. As I said, I remained only in this space. It was even something like as if I felt myself moving forward in this space like a spaceship in outer space. [. . .]

2528 The most pronounced thing for me is the perception of space, infinite, unseparated, without inside or outside, velvety warm and at the same time sharp.

2759 [. . .] In the experience of pure awareness it felt as if I was simultaneously centered and spread out very wide. [. . .]

2771 It was an experience of not moving forward, not moving back, and also not standing still.

3329 [. . .] Then I have the feeling of vastness and clarity, as if my consciousness were expanding and becoming very calm and wakeful. [. . .]

3371 I have the impression of being a large space without borders in a unity of everything. Of being awake, connected, and then again just being, without concepts. Another experience was of falling into an empty space . . .

For many respondents, the space of awareness is a prime example of unboundedness:

2202 I had a feeling of a boundless space.

2565 [. . .] An existence in boundless space.

3497 A feeling of being unbounded and observing is there, surrounded by and part of [an] unbounded space. [. . .]

In chapter 5, we saw how the phenomenology of unboundedness is not some explicitly experienced and merely quantitative infinity or a concrete, endless expanse of some sort. Rather, "unboundedness" means that the experiential character in question includes the *potential* for expansion, and that in the experience of pure awareness itself there is no such thing as a "beyond"—that is, *another* consciously experienced finite region or realm "on the other side" of a boundary. Spacious awareness may not even be an "experience" in the prototypical sense (see chapter 31). On the other hand, it evidently has a signature of effortlessness and ease:

1236 [. . .] The state is characterized by the "experiencing" and "direct knowing" [*Erleben und Erfahren*] of a completely wide open space, sometimes lasts only a short time (one to two minutes), often longer, and is characterized in particular by lightness and effortlessness. Otherwise very difficult to put into words.

The space of pure awareness is unstructured and aperspectival. Reports on the entirely nonconceptual phenomenology of spacious awareness demonstrate that the phenomenal character of consciousness *itself* is not necessarily integrated with a first-person perspective, neither in the weak and purely geometrical sense of having an origin nor in the richer, epistemological sense of an active "knowing self" forming its center:

2614 [. . .] It felt like previously my center of awareness had been a cylinder in the middle of my field of awareness, but when this experience happened I felt the borders/shape of the cylinder dissolve outward until there were no more borders, no sense of "the knower" from which attention originates, instead it was all awareness in 360 / a sphere around my body and beyond. It felt very peaceful and alert. [. . .]

Spacious awareness as such may be experienced as timeless during episodes of full absorption. In addition, we find a distinct phenomenology of entering and leaving. Here are three examples of a dynamic transition into and out of spacious awareness:

> 2878 [. . .] My consciousness expanded very quickly and incalculably like a balloon; there were no limits: Everything was within me—I was in everything. Afterward I was terrified.
>
> 3243 [. . .] During a sitting meditation it grew "light." Then a feeling of vastness. Then the feeling of being in the bodies of the people sitting with you. Expanding further, merging into space.
>
> 3480 [. . .] Everything was this perception of an endless space, like between planets, without light, but somehow bright, without anything in it, a complete nothing, but at the same time with all the potentialities in it! It was also a perception of a complete, deep silence. After some time, I started to regain senses, body, and memory, and like with another type of mind I started to recall what was before. This type of thinking was not possible during the experience itself, as if all resources were used in the experience.

As #2878 shows, transitions can have an affective quality. Moving in and out of the contemplative experience of pure spatiality can also trigger positive feelings, but it seems as if states of full absorption themselves are emotionally neutral:

> 2580 [. . .] I notice how within me a vastness, a field opens up in which I am. Thoughts continue to appear, but are very much in the background and/or do not last long. I feel joy and enclosure in a wide space. [. . .]
>
> 2582 [. . .] Several times as a very light, weightless, wide state of complete calm and vastness. After these states it was difficult to come back because they are so light and peaceful. In the second form (less often) it was a wide, empty space, not filled with light and not so "positively colored," but completely neutral, very very wide and "powerful." I was the boundless space, the space was me, there was no more self or separation, no more positive or negative, no light, no shadow, just complete, powerful emptiness and silence.
>
> 3621 My experience was devoid of any emotions. It was of abstract space, feeling like enlightened awareness emanating from a constant center. [. . .]

Some reports describe the silent quality of pure awareness as something that you can attend *to*, as if it were still some distinct entity, but also as something that can never be reified and that, in a very subtle way, permeates everything—even what you took to be your own body. In addition, many dozens of meditators report that if this all-pervasive quality becomes explicit and gradually becomes the foreground of experience, then it can lead to a sense of their body boundaries dissolving and to an attenuation of the feeling of being *situated* in this world, of being present as a self. This phenomenological aspect will be the central topic of chapter 24, but as it is intimately related to the experience of spatiality, I will present three examples here:

> 24 For me, pure awareness is an all-pervasive state where the boundary between me and the reality around me melts away while both my body and reality remain clear. It has to do with a rhythm linked to the breath that becomes a unison between me and every form of life around me. It is muffled and clear at the same time.
>
> 32 My experience of consciousness occurred during meditation, suddenly I felt my body dissolve and become something like an energy, there were no limits, and it merged with the cushion I was sitting on and spread around the room. I was fully aware and amazed by what was happening. Everything seemed to originate in my hands, from there this feeling grew. Everything was silent, and there was a lot of light. Presence of the here and now.
>
> 2859 [. . .] it is a simple condition of just "stopping." Although the world clearly continues (sounds, people talking, etc.) I am not aware of being present in it; there is just a stillness that permeates everything.

In these examples, which are typical of the broader pattern of responses, it seems as if body and environment can still be clearly represented, while the spacious silence of pure awareness expands or permeates everything and the boundary between self and nonself blurs, becomes meaningless, or melts away. Sometimes practitioners say that it is their *mind* that merges with space, as in the following example: "My mind then broke open, all breath ceased entirely, and for the next few hours I bathed in an ineffable experience of mind mixing with space" (#185). But most of the time, the experience of vast, open space is associated with bodily self-experience. Careful phenomenological investigation demonstrates the existence of a type of conscious experience in which we experience ourselves as *embodied* in an experience of unbounded epistemic openness. We identify with space itself.[1] This perhaps maximally abstract form of embodiment is highly interesting from a theoretical perspective because it has an additional quality of nonconceptual "knowingness."

Please note that bodily self-consciousness is a special case of spatial experience. There is a deep connection between embodiment—the holistic quality of being an aware self, grounded in and present as a lived body—and the functional capacity for self-location in a spatial frame of reference.[2] Here, my main point is that there clearly exists an equally holistic phenomenology of being groundlessly grounded in and present as a timeless space of knowing. What in chapter 5 I termed the conscious "model of an epistemic space" can actually function as a nonegoic self-model. Human beings can *embody* the space of knowing. The following experiences therefore serve to create a gradual transition to chapters 24 and 25, which focus on the rich and variable experience of "bodiless body-experience" and what scientists call "ego dissolution," before the new philosophical concept of a "nonegoic unit of identification" is introduced in chapter 29. I will let our participants speak for themselves, beginning with this description of an unusual kind of ground:

> 1557 [. . .] as if sunk into an all-embracing space, or in other words: settled on/in a ground that has no floor at all, but is nevertheless completely safe. Perceiving this "ground" fills me—no matter what is going on in my consciousness in parallel—with unconditional trust and acceptance, or something like that. It has a strong quality of "ground" (like a floor) without being really physical. The "velvety" that came up earlier in the questionnaire would also fit here. [. . .]

A recurring phenomenological motif is the sense of self dissolving into spatiality per se, or of the meditator being "embodied" in the space of pure awareness itself:

> 82 [. . .] It happened that at one point in the formal practice, I entered or became an infinite, timeless space and was no longer identified with my thoughts and emotions. It was as if everything was expansion, and the sensations were incredibly pleasant, I felt so much lightness, harmony, and fullness. [. . .]
>
> 2724 [. . .] . . . . the sense of self becomes an infinite space.
>
> 2652 [. . .]—space disappears or is boundlessly open—everything is in harmony—everything is my "body"—[. . .]
>
> 2543 I felt as if I and my room with all the objects in it were one single space, i.e., there was no clear separation between me and the space around me. I was as much a part of the space as the bed I was sitting on and the street lamp visible through the window.
>
> 2764 Most impressive was the experience of unity and the associated release from physical boundaries. I was still in my space, but I no longer knew whether I was the whole space or just a fraction. It was a feeling of limitlessness and deep peace. [. . .]
>
> 2665 [. . .] my perception altered toward brightness, toward a feeling of being one with the space surrounding me, and beyond the boundaries of physical space.

2619 [. . .] The big awareness was beyond it. I felt like a humungous sphere, that my awareness was not located in my body but sort of in a big sphere around it. Anything I could sense at any distance was within my sphere of awareness and I was identified (though that's not really correct) with that big awareness. This awareness was without words. It felt incredible and huge and spacious. It was wonderful. The sounds I was hearing from far away were all perfect and wonderful. The self that was selfing inside the awareness, making narratives and having feelings, was also wonderful. Whatever "I" was was not just the I. [. . .]

In sum, we see that MPE is often compared to an experience of vast, open space without center, periphery, or boundaries. This space is not a physical space, but an "aware-space" that carries the signature of knowing—and it can even be embodied, functioning as a new phenomenological unit of identification.

Centerlessness and Unboundedness

So meditation is the inquiry into, and the discovery of, this space without a center. Therefore it is not an experience at all. You understand?
—Jiddu Krishnamurti (1895–1986), *Fourth Talk in Rajghat*, November 26, 1964

The entirely nonconceptual phenomenology of spacious awareness demonstrates that consciousness itself—our inner model of the space of knowing—is not necessarily integrated with a first-person perspective. It *can* be a subjective experience in this sense, but as it turns out, subjectivity is not one of its essential features. This observation relates directly to the "contraction principle" introduced in chapter 8: The phenomenal character of openness is not necessarily tied to a knowing self, but nevertheless it remains an *epistemic* form of openness. Phenomenologically, there are nonegoic forms of knowing.

One important piece of evidence for the existence of nonegoic forms of knowing is that many practitioners make clear that pure awareness is not simply identical to what is experienced as physical space during ordinary wake states. From a scientific perspective, this means that the "space" of consciousness is not simply the space of sensorimotor integration or the behavioral space in which the brain represents boundaries between objects, in which it simulates movements, action goals, and complex bodily navigation within a physical environment. On the other hand, the nonphysical space of pure awareness almost certainly has a deep relationship to all these other ways in which we represent space. For example, as we saw in chapter 11, peripersonal space may be an important part of the scientific picture. Not only can it be viewed as an invisible extension of the embodied self,[3] but it is also related to the experience of connectedness investigated in chapter 11.

I think that on its most fundamental level, far below the level of egoic self-awareness, conscious experience is a prediction about the possibility of knowledge. MPE could be a model of an unstructured but integrated space in which states carrying epistemic value can appear and in which different *kinds* of knowing can manifest—perception, thought, attention, or active inference through embodied action. Yet MPE is not identical to any single one of these manifestations because what it nonconceptually represents is precisely a very large number of possibilities: the epistemic *openness*, the potential itself. It predicts epistemic gain. MPE, the experience of awareness itself, is capaciousness, the self-knowing experience of mere epistemic capacity. If you will, pure awareness is *virtual* knowing in an abstract, centerless space.

Let me now draw your attention to a second conceptual point. A large number of our reports mention the quality of unboundedness, and of course, this experiential feature is especially prominent in descriptions of spatiality. The phenomenal character related to the concept of "unboundedness" can be misunderstood as implying the conscious experience of very large distances or an explicit representation of infinity in the spatial domain. Rather, unboundedness is the phenomenal experience of there being no second, finite region to which attention could be directed, and no consciously experienced boundaries, limits, or horizon. To say that the space of pure consciousness is "boundless" does not imply that there is an explicit experience of infinite expansion or of large distances. Rather, it means that there is no "other side beyond the boundary" to which attention could shift.

Perhaps it will be helpful to recall the example already given in chapter 5, when we investigated the experience of clarity. In visual awareness, attention can shift from a red patch into an adjacent green patch, transgressing a color boundary. For the experience of spacious awareness, there is no such boundary because nothing outside the space of pure awareness can be deliberately attended to. Everything outside is simply unconscious. There is no possible phenomenal contrast here.[4] To take another empirical example, perhaps the physical correlate of an unbounded and centerless space of knowing, rather than being a well-defined pattern of electric activation in the brain, could be a cloud of transmitter molecules that supports this pattern. This cloud could be something dynamic and variable, and it could have indeterminate boundaries in the sense that the brain's subpersonal mechanisms cannot detect them. In any case, it would be false to interpret relevant reports of unboundedness as the phenomenology of an explicitly experienced and merely quantitative infinity or as a concrete, endless expanse of some sort. "Unbounded" just means that (1) the phenomenal character in question includes the *potential* for expansion and (2) in the experience itself, there is no such thing as a "beyond"—another consciously experienced finite region or realm "on the other side" of a boundary.

I have often wondered why, in normal wake states, the phenomenal world appears as irrevocably real. Perhaps the world appears as real because we can never experience its boundaries? The potential for exploring the virtual reality (VR) in our head seems almost infinite, and there also seems to be no real beyond because all you ever get is more appearance—but then why is it transparent? (See chapter 28 for more on transparency.) You can experience a picture *as* a picture only if you are able to see its frame. Is it the framelessness of consciousness itself that ultimately makes the content of the multimodal virtual image in our brain become real and fully immersive?

Unboundedness as potentiality and the absence of beyond has another interesting implication, with a deep philosophical flavor. In the relevant phenomenological domain, there are no countable entities: The domain of MPE is an unstructured space that has neither center nor periphery and cannot be fragmented or split into indivisible units of experience—not even a single one. Individuation is impossible. Phenomenologically, pure awareness is not a thing or a substantial entity in any sense, because it is not a *particular* form or object of consciousness. This has the interesting consequence of adding "non-oneness" to "nonduality" as a new phenomenological descriptor for MPE.[5] Pure awareness is that which can never be reified. I think that precisely this combination of non-oneness and nonduality also explains a large portion of its ineffability. Perhaps you will recall my brief mention of Śāntarakṣita's classical "neither-one-nor-many argument" in chapter 5. If the phenomenal character of consciousness itself is open, in the sense of having no inherent nature at all (neither singular nor manifold), then this openness is almost impossible to contract into verbal descriptions—at least in the kinds of language that we have today.

On January 15, 2006, in the Swiss city of Basel, I heard a 100-year-old person give a lecture for the first (and still only) time in my life. It, too, was about something that is almost impossible to put into words. The speaker needed two crutches and had to be helped up onto the stage. But the lecture itself was brilliant, of admirable, almost unbelievable clarity and lucidity, and the audience was deeply moved by it. One of the lecture's essential points was this: Every natural scientist will also become a mystic in the end, for whatever the creator is, he speaks to us not in churches and through words, but through his own creation itself.[6]

I had met the speaker before, seventeen years earlier, in May 1989.[7] We had two long conversations, one of them at a breakfast table in the Hotel Maritim in the city of Bad

Homburg, close to Frankfurt am Main, Germany. He was Albert Hofmann (1906–2008). He was the person who single-handedly and inadvertently changed the lives of millions: Without him, many millions of barely effable experiences would never have been known, and their consequences never have been lived. Because of him, many people who have never meditated in their lives have directly experienced the unbounded, centerless space of self-aware knowing that this chapter is all about—and some of them have even become one with it. Hofmann was the discoverer of LSD, and he would always consider it something as dangerous as it is valuable. "LSD is a tremendously potent drug; there is no other substance that is effective in these quantities," he told me.

On April 19, 1943, he had embarked on the first planned LSD trip ever:

> I started my experiment with the smallest amount that could be expected to have any effect, namely a quarter of a milligram. If you take a milligram of prussic acid, nothing will happen to you; the strongest poisons do not work even with a quarter of a milligram. I thought of increasing gradually. And that was the first planned LSD trip; 1943, on 19 April. And it was actually five times overdosed. That's when I had this dramatic reaction; it was some kind of horror trip.[8]

Decades later, Hofmann compared the substance to nuclear energy. He even used the term "psychological atom bomb." This was not just about its extraordinary psychoactive potency per gram. He continued:

> Secondly, the effect attacks the core of our humanity, our consciousness. That means it is an enormously potent substance, and the more potent a substance and its effect are, the more possibilities there are, the greater the danger of abuse. Just think of ordinary energy and atomic energy. The possibilities are enormous, but so are the dangers. It is the same with LSD. It is an extraordinarily dangerous substance. Extraordinarily potent in its possibilities, but it is very dangerous. If you take it carelessly, unprepared, not in the right environment, or not in the right mental state or not under guidance, then really serious things can happen—catastrophes, personally and also in the outside world. That's what happened. In the sixties, maybe 80% or 90% had a positive experience, but the remaining 10%—if maybe a few million took it, it was still an enormous number who ended up in psychiatry.

In the midst of my research career as a philosophical psychonaut, I knew this all too well, but I still wanted to know what exactly he thought made it so dangerous. "We were not born to live out there in space," he said. "If you want to go on a journey into space, you have to prepare, it's dangerous, but then you also have to keep a log somehow, you have to process that and you have to come back to Earth."

The next question I asked was this:

"What would you advise young people today who say, we have an interest in expanding consciousness, perhaps we have a religious interest, we want to undergo this experience—but there are no teachers for us. We have LSD, but we don't have teachers here in Europe or in America."

I noticed that this question made him somewhat uneasy, and in a slightly more severe tone, he replied: "I don't quite understand when young people—especially very young people—want that. They still have so much ahead of them; they still have the whole normal reality to build up. I understand it more when someone of advanced age has the need, the feeling 'I can't get any further, now I want to expand my consciousness with pharmacological help'. Then I understand that." Hofmann clearly thought that taking LSD responsibly required a certain level of personal maturity—and as much as I disliked his answer at the time, as the years passed, I came to agree with him. He then explained that he thought that LSD can help us refine and develop a certain quality of cognitive flexibility and openness to the world, and help us sustain it throughout our lifetime:

> And if we can do that, then growing old is beautiful, and we grow richer and richer, don't we? But most people close up, they start to close up. That has to do with our top-heaviness. If you live only with your head, with your mind, then over time it becomes routine and then you ossify, at forty or fifty. I see this with many of my friends and colleagues, you start to get narrower and narrower and soon everything is over. But when you realize that I make the world, that I become richer through what comes in, you try to stay open.

And were there any other reasons, I wondered, why someone might want to be a psychonaut and explore the *Weltraum der Seele*,[9] the deep outer space of the soul? "You have a wonderful opportunity," Hofmann told me. "You see the world from a different perspective, you see the world as a blue planet floating in this infinity." As is well known, the LSD experience can have considerable overlap with the phenomenology of meditation. For example, you can directly learn (sometimes the hard way) and nonintellectually understand what it means for your experiential world to be *literally* just a virtual model (chapter 28). Millions have also directly experienced what phenomenological terms like "ego dissolution" or "nondual awareness" (discussed in chapters 25–27) may refer to. Just as many people do today,[10] in the early twenty-first century, Hofmann already thought that ego dissolution might have therapeutic potential: "This detachment from the personal, from the 'encapsulated ego' as the Americans say, this blasting, that seems to me to be a very important function, a possibility."

I asked if he thought that LSD should be used therapeutically. He said:

> You have to bring that in as a healing element above all. In most cases, and this is how I have experienced it myself, the difficulty is that a psychological problem, any personal problem, is only seen in isolation and as such and then takes on gigantic proportions. These disorders are based on this egotism, which somehow becomes pathologically fixed there and lets the problems take on oversized contours. If, on the other hand, I know: I am safe, I don't have to be responsible for everything about my mind, I can have confidence, my five senses also function without me thinking about it, my heart functions without me thinking about it, and my eyes, I see and hear—that is all something that opens up a whole world for me. If I have that, that would be the therapeutic thing in medicine, in psychiatry: if patients can be released from their narrow, egotistic problems. If they can suddenly realize: Aha—it also works without!

But, of course, therapeutic potential isn't all there is. The psychedelic experience is clearly a dramatic expansion of phenomenal state space, and there is no corresponding risk analysis that would support simply *excluding* almost all members of society from actively exploring this space. Exploring it may bring significant epistemic benefits. At the very least, as I have noted elsewhere,[11] it is now a well-established fact that psychedelic phenomenology contains strong elements of spiritual insight or even religious experience.[12] However, it is often overlooked that—for the large majority of citizens—these forms of spiritual or religious experience cannot be accessed by any other means. Multiple human rights issues are at stake here, concerning cognitive liberty and the freedom of private religious practice, among others; and from a philosophical point of view, drug law currently fails to recognize a basic human right to mental self-determination, which guarantees an individual's sovereignty over her own mind.[13] From an ethical perspective, the current situation is clearly untenable. But what we lack is an evidence-based, rational, and unideological approach to enculturation, an appropriate sociocultural context in which to optimize the risk/benefit ratio for anybody who wants to enter the relevant regions of phenomenal state space. I am skeptical that such a context will ever be created, but I hope that it will. Of course, I also asked Albert Hofmann about all of this. When it came to LSD, he thought that its use needs a sincere and respectful attitude, and that we should let the knowledge of the right application grow in silence.

I wondered: Is meditation training one way of trying to sustain some of what LSD may give you in a quick and seemingly effortless way? For now, we still live in a world where most people never go near a meditation center, and those who do usually cannot access guidance on how to access, sustain, express, and learn from states further along the spectrum of neurochemical alteration. And in this world, caution is wise. But when I mentioned training, Hofmann made clear that he saw contemplative practice not just as a way of reducing the dangers of the nuclear option, but also as a way to open new dimensions through gentler, less risky practices that make us ready to use the nuclear ones where they're needed. He added, "I also have the hope that in time meditation centres might emerge. We do meditation courses and so on; I could imagine that centres will emerge—real meditation centres, where the aim is not to heal the sick, but to develop people further, to open up new dimensions through meditation, through all these techniques, where chemical help could then also come."[14]

This sounded absolutely right to me, as it does today (see the epilogue for more). But at the time, Hofmann corrected me on my choice of words when I spoke of *Bewusstseinserweiterung*, the expansion of consciousness. He said:

> It is an expansion of consciousness, but I consider "expansion of consciousness," as it is used in the drug scene, to be too narrow. Every new experience, every encounter with a new person is an expansion of consciousness. This is of course a dramatic expansion of consciousness, but our whole life should be an expansion of consciousness, should become richer and richer—what else should we live for and be able to open up to?

Hofmann, the wise chemist from Switzerland, clearly knew about the self-constructed, virtual character of our experiential world and about the value of centerless experience, and he had a deep understanding of many of the phenomenologies described in this book. He also had his own opinion on the quality of epistemic openness. Continuing his thought about resisting middle-aged ossification, he said: "When you have grasped that I make the world, that I become richer through what comes in, you try to remain open. And it is also possible to remain open and to open yourself more and more, to let more and more of this immense miracle of creation flow in."

Of course, I couldn't let our conversation end without asking him what *everybody* wanted to know. In the almost half-century since that late afternoon in 1943 when he measured out those 250 micrograms for the first intentional LSD trip ever, how many times had he repeated this experience for himself? Many people still refuse to believe the answer that he gave, but I have it on tape: He told me that he had taken it maybe about twelve times.

24 Bodiless Body-Experience

A bodilessness in the consciousness of a body [#302]

My body dissolves into a micro-foam of impermanence, zooming into the fractal edge of present which endlessly unfurls fern-fronds of scintillating *now*-fuzz. [#1832]

The experience seems to be similar to a thread which forms a seam and holds on one side the body and on the other side infinity together, so both are there and yet they are not there, because they are put together. [#3218]

This is a long chapter, and for good reason. To me, perhaps the greatest surprise in evaluating the voluntary self-reports from our study was how many of our participants chose to describe pure awareness as involving an attenuation or even a complete disappearance of body boundaries. Unexpectedly, I found that if one asks meditators to describe a paradigmatic experience of "pure awareness," they frequently devote most of their description not to the phenomenal character of minimal phenomenal experience (MPE) itself, but to extended episodes characterized by a loss of body boundaries, a dissolution of the sense of self, a merging of sensory modalities, and an impression of "becoming one" with the world. They often describe a global *mode* of experience in which pure awareness has switched from background to foreground. It is as if the background has become so dominant that what was previously the foreground now becomes translucent, with some of its structural features gently fading out or vanishing altogether. This is one common way in which body-experience changes in pure awareness. However, if one looks carefully, there are actually many *different* ways for body-experience to change in this context.

Let us take the attenuation of body boundaries as our starting point. What I will call the phenomenology of "bodiless body-experience" is a special case of the experience of spatiality that we investigated in chapter 23, and it often starts with the body gradually disappearing from the phenomenal field:

> 3569 I was in a deep, black space totally without boundaries. It was peaceful there and I could hardly feel my body anymore; sometimes I couldn't feel it at all.

In the context of pure awareness, an attenuation or dissolution of body boundaries is frequently reported. However, if we take the phenomenology seriously, this experience is not unitary, and it can actually come in many forms.[1] For example, sometimes the body can still be felt, but its boundaries in space are indeterminate; at other times, body-experience disappears altogether. On the other hand, specific commonalities exist as well. For example, a subtle phenomenal character of "vibration" or "tingling" is frequently reported, while simultaneously the body boundaries begin to dissolve:

> 2778 [. . .] a completely clear, calm attention like a clear vibration of the whole body, which you feel as such but which is not limited [. . .]
>
> 2907 [. . .] then changed bodily sensations appeared, the contours dissolved more and more, there was a strong tingling in my upper body and finally this tingling climbed up my spine and I had the feeling of gliding with my consciousness over a threshold, to a totally different level, as if into another space of experience, whereupon I was also no longer observing the breath, but was merged with it. In this state my knee pain had disappeared and also my sense of time; there was simultaneously a great joy and lightness and the feeling of floating in space. This state was extremely positive and free of worries, effort, concepts. [. . .] I was one with my experience, even if the observer was still in the background. Other experiences that I had later were much calmer and corresponded more to the feeling of plunging into a calm lake, being completely still and at the same time flowing along. Here too, there was often the experience of dissolving, the expansion of consciousness beyond the boundaries of the body.
>
> 1926 I had a clear awareness of awareness during one meditation session where I sat down with eyes closed. As I was aware of my attention (going from the breath to sensations in the body and my visual field) I got to the stage where the concept of my body dissolved. In my mind my hands no longer had the shape of hands and my whole body became points of either vibrations, tingling, or temperature. [. . .]
>
> 597 [. . .] Shortly before I experience this state, sometimes a kind of swinging or vibrating starts in my body, usually very subtle, but very clear. Then I "lose" the boundaries of my body; I no longer feel it at all. I completely expand into "space" (or shrink to nothing; in a paradoxical way, it's the same thing). I can hardly feel my breathing either—I think I have long pauses between breaths,

but I can't say exactly, because the sense of time has disappeared. Thoughts no longer arise. I "am" then only in an infinite space (which is mostly diffuse, homogeneously dark, sometimes it also switches to bright), or I myself am the space. I am "somehow" everything, and everything is me. But it's no ordinary physical (three-dimensional) space, rather a state of being without the ordinary dimensions like length, width, height (unfortunately very hard to explain). [. . .]

1435 [. . .] the conscious perception of bodily sensations in the present moment, the perception in free flow is like a recognition and understanding and at the same time a dissolving of connections and structures without having or wanting to judge it. It is how it is and is permitted to be, no matter how strong or subtle the sensations are. Suddenly it becomes so clear how everything changes from moment to moment—comes into being and passes away. One moment there was this sensation and then another sensation. But that is not important. I am in the here and now, awake, watching every moment of change with equanimity. I let go, without expectations or ideas. A stream of tingling, pulsating [sensations] floods my whole body, without my attention being focused on a specific spot. Everything spreads out, dissolves, I can no longer perceive where my body begins and ends.

2143 [. . .] brief experiences that can be best described as centerlessness and openness, without any feeling of constraint. Bodily sensations were experienced as pure vibration without any identifiable position in the body. The overall experiences were very peaceful.

1756 [. . .] at some point my internal sense field no longer included clues to my body, it was all just empty space with very faint but uniform tingling occurring in this space. I was just the perception of this all occurring, I had a sense that I was nothing more than just witnessing the existence of this vacuous space. I was the perception point for subjective experience but nothing more.

Related to the loss of body boundaries, a second interesting phenomenological detail is found in reports referring to an experiential quality of lightness or weightlessness. We first encountered it when looking at joy, awe, bliss, and gratitude in chapter 15:

717 [. . .] It is a feeling of merging with all happiness and all love. It is soooooo sooooooo big. , weightless, the words we have at our disposal here are not enough for the description.

The phenomenology of weightlessness is interesting because we also find this quality when looking at the phenomenology of bodiless body-experience and what in the second half of this chapter I will term "abstract embodiment." "Graviception" is a biological organism's capacity to detect the Earth's gravitational field. It can have its own phenomenological profile, although most of us do not attend to it in ordinary life, and philosophy of mind has largely ignored it. Nevertheless, in human beings, the conscious experience of embodiment is strongly determined by perceiving the physical body's own weight. A field of background sensations is normally determined by information coming from "graviceptors," giving the biological organism information about its position relative to the local gravitational field. This in turn leads to the background feeling of "being grounded." Closely related to this form of perception is a cluster of phenomenal qualities like momentum, a sense of balance, and our ongoing experience of the relative positions that our limbs currently have to each other in space. And in addition, there exists a geocentric reference frame, which in the case of seeing helps us detect the verticality of an environment through the perception of gravitational pull. Thus, the feeling of our own weight actually influences the way that we see things. Interestingly, the sense of the weight of our own body can be completely absent in some nonmeditative states too, such as during flying dreams or out-of-body experiences.[2] It is plausible to infer that some of our mechanisms for weight perception can go offline, and that this influences our conscious self-model.

In the context of pure awareness, we often find a phenomenology of "floating" or "light embodiment," or combinations like "lightness and dampened sensations" or "body dissolution and groundless groundedness." Once again, it seems as though the brain's inner landscape of priors and unconscious beliefs about the world (as discussed in the previous chapter) becomes flattened. What were previously striking opposites now seem to peacefully coexist, as evidenced by the fact that many of our meditators use seemingly paradoxical formulations like being "unbounded and earthed," "bodilessly aware of a body," "gone and fully present at the same time," or "weightlessly floating and grounded at the same time." Here are some examples:

> 39 [. . .] While meditating I entered a bright (nonvisual) mental weightlessness that was boundless and yet offered a nontactile/nonvisual space of security. [. . .]
>
> 302 A bodily feeling of lightness, detached from the bodily sensation, at the same time a perception that strong energy was flowing through the body, especially noticeable in the arms and hands, and especially the palms of the hands. Detached from the sense of space and time, but wide awake during the guided meditation, connected to beginning and end. In summary: a bodilessness in the consciousness of a body.
>
> 304 It was as if I were gone and completely present at the same time. I was me and yet part of this infinite universe. I was everything and nothing. Everything was

possible. Everything existed as possibility. I was light and floating and at the same time deeply rooted.

3329 [. . .] My body feels lighter and physical sensations, like sitting or hands on knees, feel as if they are smothered in a thick layer of clothing. [. . .]

1153 It was a feeling of deep calm and peace. My body was as if dissolved, the boundaries were no longer perceptible, simultaneously unbounded and grounded.

The phenomenal character of weightlessness can occur in motion as well as in stillness:

3259 In the walking meditation [. . .] weightlessness and at the same time conscious contact with the underground [. . .]. In the sitting meditation the feeling of wholeness, weightlessness, inner and outer silence, [. . .] considerably slowed breath [. . .].

Weightlessness can also co-occur with density (chapter 6), unity (chapter 26), and timelessness and ego dissolution (chapters 22 and 25, respectively):

306 It was a feeling of weightlessness and simultaneous density. [. . .] It felt like "being one," but without the usual feelings like love or bliss. Space- and timeless, detached from the "I."

We also find weightlessness, clarity, and a complete stillness of bodily sensations, leading to an unbounded sense of translucency or virtuality (see chapter 28):

2486 Pure awareness for me is to be weightless, dreamless, and totally aware of the moment. The body is like invisible: not breathing, not moving, no weight, just be still and quiet.

684 [. . .] Tactile and physical sensations acquired a kind of weightlessness or transparency that was not limited. [. . .]

We have encountered the phenomenal quality of unboundedness many times before. Here, it returns coupled with weightlessness, including as "unbounded embodiment," or in union with the experience of "being held":

328 I had a state of weightlessness. Of boundless physicality. And of the simultaneous existence of the self, of the consciousness of my self, without limitation in complete peace.

3524 [. . .] penetrating floatinglike thoughtless state of not searching further, limitless being-held [*entgrenztes Gehaltensein*] [. . .].

In the vicinity of full-absorption states, the phenomenal character of weightlessness not only occurs together with emptiness, clarity, and silent unity, but it typically also coemerges with a blurring of body boundaries:

1737 This is really not so easy to describe. [. . .] It suddenly feels as if someone is pressing a forward key inside me, it is a feeling of speed/quickness inside me, my heart also starts beating faster. Then when I let myself fall into it and "go through," I arrive at a total feeling of weightlessness, emptiness, awareness. It is all incredibly clear. I no longer feel my body, hear nothing more from my surroundings, don't know anymore where I begin and end or where I am. I also have no sense of time anymore. Sometimes this lasts a few seconds, sometimes an hour.

2467 Dissolving of the bodily feeling, feeling of floating over the body, weightless. Feeling—silence, being one, everything is ok as it is, boundaries blur—[. . .]. [. . .]

Let us now end our investigation of weightlessness and take a closer look at reports about a blurring of body boundaries. At times, this may be a case of what philosophers would call "phenomenal indeterminacy" (i.e., body boundaries are *neither* well defined *nor* absent). As Jennifer Windt has pointed out, such an experience "would be one *neither* of phenomenal embodiment *nor* of phenomenal disembodiment."[3] The philosophy of modern dream research suggests an interesting parallel to the experience of embodiment during the dream state: The body can be absent from conscious experience in a form of explicit disembodiment in which we clearly notice its absence; but much more often, the phenomenology of embodiment is lacking without this fact being explicitly experienced. Large parts of the phenomenal body can go missing in dreams without us noticing, without the absence itself being part of the experience. This fact may be equally relevant for MPE:

2762 The perception of the body disappears without being noticed. As a process of gentle release. One notices the disconnection only when consciousness realizes that it moves completely free as light in an infinite space. This space has a warm, inviting, and comforting endlessness. [. . .] You carry everything within yourself. All needs and thoughts are silent, as if they were completely satisfied and thereby dissolved. Peace and harmony arise, sensation is all-encompassing. There are no more separations.

Famous philosophers of consciousness like Franz Brentano (1874) and Daniel Dennett (1993) have pointed out that the absence of representation is not the same as the representation of absence. Put differently, sometimes there simply is no fact of the matter when it comes to conscious experience itself—and for the case of embodiment, this observation may connect the phenomenology of meditation with the phenomenology of dreaming. Sometimes, there is simple neither-nor-ness.

What is more, there seems to be a phenomenological link to the notion of "epistemic openness" (chapters 4 and 5): In meditation, the body itself can melt or dissolve into the centerless clarity of epistemic openness:

1142 [. . .] There was a deep quality of clarity and ease within me that seemed to blur the lines between my bodily limits and the awareness within my mind so that I was totally comfortable and undistracted in the present moment. [. . .]

1256 [. . .] My body perception was beyond precise and imprecise, my body was everywhere without limits, warmly flowing, expansive, fused, simultaneously awake and with keen perception of every detail without hyperfocus on any one detail. [. . .]

Sometimes, such states are described as a full dissolution of body boundaries:

1329 [. . .] I started to feel an extreme comfort in my whole body (nothing hurt anymore and I didn't feel the boundaries of my body either), all sensations were pleasant (the temperature of the room and my body were harmonized, the sounds were distant and soft), and I started to feel that I was one with the whole. There was no longer any boundary. The sensations were of total peace and well-being; I felt happiness and at the same time amazement to be feeling what I was feeling. The sense of time and the sensation of gravity disappeared and for a period of time lasting approximately 1 hour and 50 minutes, I experienced a state of consciousness that I have never relived [. . .]. I was simply there; and I was no longer me.

2549 [. . .] sometimes I feel I've no edges and free from everything and bliss.

2881 [. . .] To begin with, the breath is only slightly perceptible in the lower abdomen as a movement. Then it is no longer present at all. The border between my body and the surroundings is no longer there. I am everything. I feel as if I'm in the phase between being awake and sleeping. But there is no dreaming involved. I can sustain the feeling of boundlessness for a few minutes. But if my thinking tries to recover it, it does not work.

2890 Feeling of the dissolution of space and time, dissolution of physical boundaries, feeling of unity, unconditional love, timelessness, but always aware and with the knowledge of being in control.

3288 I was sitting paying close attention to sensations in my body. I felt the limits dissolving until I could no longer distinguish "myself" from the surroundings. The sense of time and space also disappeared.

3421 Feeling of growing bigger, expanding, and that the boundaries of the body are dissolving. I am breathed—I don't have to do anything for it, just observe, it goes through me, the breath becomes very light and shallow, slow and stretched, effortless. Feeling of looking at myself, of being opened, incredibly present. [. . .]

We have already seen how the phenomenal quality of weightlessness can occur not only in stillness, but also while the body is in motion (#3259). The same is true for the dissolution of body boundaries:

1557 [. . .] An experience of perfect dissolution, or absorption, in what is; this was initially connected with the internal image of "earth"; that is: I was practicing walking meditation outside, on a soft, uneven meadow, so I felt the ground very strongly. And after a while I felt something like an invitation to be absorbed into this earth. At first this frightened me quite a bit, and the need arose for me to decide: Do I allow what is happening here, or do I get out? That lasted for a little while, then I somehow let go of my control and it felt like I dissolved. However, I still knew that I was dissolved while I was dissolved. It was connected with a feeling of fullness, or wholeness, or maybe also emptiness—I wouldn't know how to tell them apart in this situation—very pleasing. [. . .]

But this is not the final stage, it seems. There can also be a full-blown experience of explicit disembodiment or bodilessness, a global and peaceful state of pure being in which the part completely expresses the whole. Sometimes this may involve timelessness and a disappearance of the entire spatial frame of reference:

680 [. . .] My body was as if it did not exist and belonged to the universe. Time did not exist and it was all peace, my body, my mind did not exist. [. . .]

1054 It is a state of mental calm, vastness, perceived bodilessness, a state of is-ness or being.

51 [. . .] I came into a state that felt completely detached from everything else. I no longer felt my body. There were no boundaries. I had my eyes closed and it was as if my head were open. It felt as though everything inside me had merged into a benevolent and pleasant field. Space (which is actually not a suitable word for it, as perception is limited by it, but I don't know how else to express it) was filled with love and security. I thought of nothing. I was just there. There was nothing to think. I was just there. It was beautiful and a little bit of me didn't want it to stop. But I knew it would pass. I was grateful for this experience. It was fulfilling!

2384 [. . .] As I went through my body from top to bottom with my attention, I perceived my whole body as a tingling, vibrating Something. I didn't know whether my arms were above or below my legs, for example. It was a "dissolving of the spatial frame of reference." I heard nothing, saw nothing, and thought nothing. I felt no pain. It was timeless. It was very pleasant. It was soft. [. . .] I then had the feeling that I had not breathed at all, or only very little, the entire time. I had also probably not moved at all. Afterward I knew that I am the expression of the whole universe, wherever I am. I do nothing, everything happens.

A combination of bodilessness and timelessness can lead to a transitory experience of spatial disorientation:

1867 Loss of body consciousness, loss of the sense of time (a few minutes felt like a very long time), contentment and calm, initially disorientation due to the loss of the sense of space. With later attempts increasing calm and pure consciousness.

Loss of body boundaries can lead to a disappearance of all other boundaries as well. It seems that a dedifferentiation of the conscious body image is just *one* aspect of, or one possible starting point for, a wide range of nondual episodes in which people feel they are melting into the whole phenomenal field, experiencing unity or pure being, or dissolving into a global state of epistemic openness. All these aspects will be investigated in chapters 25, 26, and 27. Meanwhile, let us explore examples of cases in which the loss of body boundaries plays a pivotal role because it leads to a loss of all other boundaries:

383 Dissolving of all boundaries—physical and mental. Outside and inside the body, energetic perception that no boundaries exist. On hearing a bird that was chirping outside, "I" was the bird, as well as the tree, the sun, etc.

2085 [. . .] at some point, there was a sense that the boundaries had dissolved. That the sound I was hearing from a nearby refrigerator was no different than sensations. The sense of a body had dissolved, there was just experience happening. I didn't feel a contracted version of a self, but rather an openness, boundless space. The experience was more peaceful and pleasant. I've had a few prior experiences of body lightness and a vibration throughout in other meditation sessions. These sensations were not experienced during this session. [. . .]

When all boundaries have dissolved, an experience of spatiality and unbounded openness can emerge (as in the two preceding reports). However, as already noted, this quality of openness is an *epistemic* openness (chapter 4) because the "knowing" sense of pure awareness, observation without an observer, free-floating attention, and impersonal witnessing are essential aspects:

1582 [. . .] The body felt like it dissolved and what was left was my awareness, being aware of itself. While I was aware of myself, I also was aware of everything else and the boundaries dissolving here too. I felt like I just "was" and my awareness just observed. [. . .]

2093 [. . .] when suddenly the boundary of my body dissolved, and I was everywhere at the same time. Attention floated freely in space between the different mind objects. I felt completely at peace.

3451 [. . .] the breath deepens and it feels as if it happens by itself and as if my whole body is immersed in the movement of the breath, body boundaries dissolve in a warmly pulsating way . . . at some point there remains only perception of "breath happens in space," no self that perceives, but an impression of "there are impersonal eyes in space." [. . .]

I found many reports referring to the phenomenal qualities of "emptiness" and "openness," which at the same time highlight the nondual and "selflessly self-aware" character of MPE (chapters 27, 29, and 30):

1545 [. . .] A very strong sense of connection with something I recognized. The body felt it at times and at other times not, integrating itself into that emptiness as well. [. . .]

3620 In this state there was no self-experience, but sounds were perceptible. My body schema no longer existed. There was no inside and outside, no experience of time, no words, concepts, or similar. Totally averbal. There was no front–back. No concept of space. Only in retrospect were such categories of description applicable to the occurrence.

There may be a deeper relationship between selflessness, the loss of all distinctions and boundaries, and the experience of open embodiment. I do not know what this relationship might be, but I have deliberately coined the new concept of "bodiless body-experience" to flag this—admittedly still poorly defined and possibly very large—region of phenomenal state space. For example, this subspace contains a very subtle, nondual phenomenology of embodiment, an almost selfless experience of undifferentiated spatial immersion, or bodiless flow:

3340 [. . .] In this bodily state not being "inside" but open, all senses open. This is the awareness of being there without the awareness of an apart "me." Nevertheless, it is very bodily. [. . .]

1085 [I come] [. . .] into a state of great silence and boundless perception of space. It is a bit like being immersed in some kind of "granules." On the one hand, all distinctions seem to have disappeared—on the other hand, there is definitely the ability to evaluate the quality of the moment positively and also in certain categories to make adjustments—e.g., to exhale a little deeper. [. . .]

1056 [. . .] My body was there, delicate, flowing. But actually there was nothing there perceiving it. Perception appeared only once the "event" was fading away. And yet "I" was there, but I don't know with what aspect, or how. My perceptions were very much weakened and I became conscious of them only toward the end of the event. It was something flowing, bodiless and yet emanating from the body, on a very deep level of consciousness. [. . .]

Bodiless body-experience and pure awareness can also be stages of a longer process (e.g., leading from body dissolution to full absorption and back):

1718 [. . .] The conceptual awareness of my body sensations in space slowed and stopped, with these subtle vibrations no longer mapped by concepts (where in the body they were). The body concept dropped along with the effort of breathing. The separate vibrations gathered into a nebula initially in front of my awareness, as if I was looking at the whole. Then the awareness became the center and periphery of this nebula cloud. There was just a now and a here, and my sense of self was now the nebula, the whole. There was increased brightness and expansion as the sensations/self quickened to pulse/vibrate as one. It became quite intense until an awareness arose within this pulsation of a self that was going to die. My self-awareness returned in a state of surprise/awe, I perceived my body map once again and I felt as though my body would collapse from the sitting position if I "let go" totally. I noticed slight fear (as density in my chest and a pulling in toward my center) and the pulsing/vibrations then dissipated as I felt the location and warm sensations of my body. My breath was deep and slow through my mouth with tears welling in my eyes. I felt relaxed and amazed.

Let us now close this large selection of reports by asking one last question: What would a nonconceptual experience of "social embodiment" look like? In 1903, the German philosopher and psychologist Theodor Lipps (1851–1914) published a paper on empathy, inner imitation, and organ sensations, in which he introduced the concept of *Einfühlung* (empathy, literally "feeling into something" or "feeling something from the inside"). Lipps thought that our capacity for empathy rests on the ability not only to sense something in your own body, but to *feel yourself in an object.*[4] For example, if you empathize with another human being in pain, or perceive the conscious suffering of another sentient creature, then you do not simply simulate *their* pain in your mind—you actually use your own bodily self-model. What really happens is that you feel *yourself* in the model of the other human being that your brain creates; your own inner organ perceptions are transposed into the model of their body. For Lipps, interestingly, such objects could be visually perceived human movements or body postures, but also architectural shapes. The French philosopher Maurice Merleau-Ponty (1908–1961) later spoke of *intercorporéité* (intercorporeality) and *intersubjectivité charnelle* (carnal intersubjectivity), and today we are beginning to have a much better understanding of how, in social cognition, the bodily self-model in our brain can expand into the social domain. One way that it does so is by allowing the intentions and movement sensations of others to "inhabit us," turning them into something that we can directly feel within ourselves. The new phenomenological material that we have gathered shows how what I have just termed "social embodiment" can also appear in the context of MPE. The dissolving of body boundaries and merging with the phenomenal field can sometimes happen in the social domain (e.g., as a transitory phase in the process of melting into space and transcending the distinction between inner and outer). Here are four examples:

> 3243 [. . .] Then a feeling of vastness. Then the feeling of being in the bodies of the people sitting with you. Expanding further, merging into space.
>
> 3471 [. . .] Suddenly I felt completely at one (unified) with my participants. Spatially, temporally, experientially there was no longer any distinction. But I was very shocked by this state. [. . .]
>
> 1918 [. . .] After hours of practice our group became a cloud / figures / inside of me and outside.
>
> 2863 [. . .] walking *kinhin*, I suddenly realize my body is completely in sync with the woman in front of me, the spatial relationship was fixed as one; if she had suddenly done a back flip, I would have instantaneously done one also. My discriminating mind was suddenly gone, had lifted away, and a beautiful openness and energy was all-present and all-pervading. My consciousness delighted in the energy of pure being and soon the entire *kinhin* line was one being with many legs. I felt like my body could have done anything, unlimited, I was all energy or energy was everything. A panoramic vision of the entire room and *kinhin* continued like a giant centipede. [. . .]

What all of this shows is that our everyday experience of embodiment is just *one* variant of a much larger set of phenomenological possibilities. There are many other ways in which human beings can nonconceptually identify with something, in a way that is experientially immediate and direct, by simply *being* it: We can embody space; we can embody the phenomenal signature of knowing; or we can embody MPE, the phenomenal character of dual mindfulness or even pure awareness itself. And we possess the capacity to embody other feeling, knowing selves—even whole groups of them. We are physically embodied epistemic spaces; conscious embodiment is a *capacity* that we have, and it can be extended to many other domains beyond the physical organism as such.

Abstract Embodiment, Changing Units of Identification—and What Reincarnation Really Is

> And Soul is not in the cosmos, but rather the cosmos is in Soul. For the body is not a place in which Soul is, but Soul is in Intellect, body is in Soul, and Intellect in something else. And there is nothing else beyond this such that it would be in that. It is, therefore, in nothing at all. In this way, then, it is nowhere.
>
> —Plotinus (205–270), *The Enneads*, V, 5, 9, 29–33

Pure awareness in the context of bodiless body-experience is an interesting new topic if read against the background of certain ancient Western ideas. Plato (*Gorgias* 493a2–3), probably citing Pythagoras, said that the body is a prison-house, the soul's tomb (*sōma sēma*). Accordingly, the true philosophical sage would always strive to liberate herself from all things bodily, trying to set her soul free. On a purely phenomenological reading, therefore, one would expect that those modes of conscious experience in which all body-experience has disappeared might also achieve a form of *experiential* deliverance, bringing that which is most fundamental—perhaps MPE, perhaps the "true self" (chapter 29)—into the foreground of experience. Plotinus, the founder of Neoplatonism and the author of the epigraph at the start of this section, thought that in this earthly, mortal life, a part of us is "covered" by the body but we can use the central, essential part of our soul that is not "flooded" by the physical organism to calmly abide in what really is the center of all other things:

> Here and now, part of us is, by reason of the body, as though we had our feet in water, with the rest of the body out of the water. When we actually raise ourselves up, by the part not plunged in the body, we fix ourselves, using our centre, to the centre of all things, in a way, like fixing the centres of the greatest circles to the containing sphere; and there we come to rest.[5]

Once again, on a purely phenomenological reading, one would expect those regions of phenomenal state space that are normally "flooded" or "submerged" by body perception to become much stronger in states of bodiless conscious experience. But what is this "center of ourselves" that seems to remain?

"Bodiless body-experience" and "abstract embodiment" are two new phenomenological concepts that I am offering to help us describe the experiences reported in this chapter. We have just surveyed many examples of bodiless body-experience, and it has turned out to be a rich and finely nuanced phenomenon. We could easily expand on these two with a whole series of equally paradoxical concepts to describe specific but often neglected aspects of the phenomenology of pure awareness. Indeed, some of the options already figure in other chapters of this book, like chapter 22, which discusses timeless time-experience. Other possibilities that this book doesn't explore in detail might include "silence in sound," "stillness in motion," and "the emptiness in an arising thought." It is also interesting to see how often meditators describe paradoxical or indeterminate states in which, for example, they "were there and not there at the same time."

The paradox of abstract embodiment is a way of nonconceptually experiencing your very own body, of *being it*, even while being abstracted away from the physical organism in time and space. It is not a form of sensory self-perception, like seeing or hearing your own body. As we learned in chapter 18, which discussed the work of Yochai Ataria, it may have something to do with an abstract, nonsensory form of "being in touch with oneself." I think that human beings have a capacity for *meta-attentional self-touch*: We can actually direct attention to attention itself, including the wandering of its focus and the mere capacity for what today's computational modelers might call "precision management" itself (see figure 34.1 in chapter 34 to get an idea of this). But strictly speaking, this is not a form of interoception or "inner touch"; it is not about perceiving your body from the inside; it isn't like sensing the flexing of your muscles and tendons, your breath, your heartbeat, or the sensations originating in your intestines. It is not sensory, nor is it a perceptual experience in the true sense of the word, because there are no specific receptors involved—neither on your retina, inside your blood vessels, or in your gut. It involves no special, functionally adequate stimuli represented in only one special stimulus modality, such as in your intestines, in the vestibular system, in tendons and joints, or via the pain-related "nociceptors" that respond to damage of body tissue. Obviously, abstract embodiment is also not a physical movement experience; it does not relate to what neuroscientists might call "motor content."

The pure-awareness experience is thus hard to characterize as embodied in the usual senses of the word. Yet it can be fruitfully understood as a particularly deep form of embodiment—not only because it is subtle and profound but also because it represents certain aspects of the bodily process by which the brain activates itself or wakes itself up. We might even go so far as to say that pure awareness is awareness of another kind of body—a "wakefulness body," if you will—constituted by an abstract level of our conscious self-model that has transcended or always already preceded the distinction between what is inside and outside our body. This is reminiscent of what Lipps says of empathy: "It is the fact that the opposition between me and the object disappears, or, to be more precise, *does not yet exist*."[6] It now becomes conceivable that the phenomenal signature of knowing—the experience of "knowingness" itself—is something that human beings can *embody*.

There are interesting overlaps between knowingness and some experiential qualities that we have already investigated. For example, if you refer back to chapter 7, on the phenomenology of "soundness," you will find two or three examples in which soundness is actually an embodied form of *insight*, or related to an experience of global harmony in which you "embody the world" after the original, merely physical body image has dissolved. Like the ordinary self-model that we all know, bodiless body-experience is something that the process of knowing can inhabit, something that it can abide and dwell in, and something on which it can ride. Perhaps meditation is like learning how to surf the crest of pure awareness, and then how to embody the ocean?

Let us try to make this more precise. Philosophically, an important conceptual instrument is the "unit of identification." I first introduced this concept in some of my academic writings on out-of-body experiences, dreaming, and spontaneous task-unrelated thoughts (aka "mind-wandering").[7] It may sound complicated, but it is quite easy to understand: The unit of identification is whatever experiential contents lead to self-descriptions of the form "I *am* this!" (if introspective access and verbal report are possible while the experience is still unfolding) or "I *was* this!" (in a retrospective description).

In humans, typical units of identification are the body as felt from the inside, the emotional self-model, and the sense of effort in attentional or cognitive agency. If you say, "I am my body!" then the unit of identification is the conscious body-model in your brain plus the global experience of ownership, which comes from controlling the body as a whole. If you say, "I am that which has feelings!" then the unit of identification is the experiential content supervening on the emotional self-model currently active in your brain, which also includes and arises out of inner body sensations, gut feelings, and the like. If—like René Descartes—you say, "No, I am the thinking self, that which forms concepts and controls thought!" (chapters 22 and 25), then the organism's unit of identification is the brain's conscious model of a cognitive agent, including the subtle sense of mental effort that comes with inner actions like mental calculation, logical thought, and the attempt to concentrate (more on this in chapter 30). If you say, "I am that which meditates, that which mindfully brings the focus of attention back to the breath after noticing a stray thought!" then you identify with the subtle sense of effort that accompanies control of the focus of attention (the quality of "attentional agency") and with an almost automatically arising model of an entity that has just "done the noticing" (the mindful agent, the successful "meditative self"). And so on.

If a conscious system has *no* unit of identification, then it lacks the phenomenology of identification and has no sense of self. The new and philosophically interesting question that arises from our phenomenological data is whether there can be *nonegoic units of identification,* with no conscious sense of self. One of the main claims that I am making in this book is that our phenomenological data clearly show that there can be. Phenomenologically, the conscious biological organism that you are has the capacity to be something without being it in a *selfy* way. I will come back to this point in chapter 29.

Using the new concept of the unit of identification, we can describe a number of logical possibilities, each of them referring to potential configurations of conscious experience. We might call them "possible phenomenological worlds." Human beings have known them for millennia. Some of them are directly relevant to the phenomenological material presented here because, far from being mere conceptual possibilities, they actually correspond to *phenomenal realities* that have so far been ignored or described only imprecisely. Here is a short list of the most relevant ones:

- There could be states of consciousness in which there is *no* unit of identification whatsoever.
- There could be states of phenomenal experience during which *multiple* units of identification are simultaneously present or in which multiple units of identification quickly alternate.
- There could be *maximal* units of identification, where we identify with the whole world or the phenomenal field as a whole.
- There could be *minimal* units of identification, where we identify with one very subtle phenomenal quality, with a simple form of phenomenal character.
- There could be *nonegoic* units of identification.

If you had a choice, in which type of phenomenological world would you like to live your life? In the epilogue, I will briefly return to the question of what makes a state of consciousness a *valuable* one, one we may want to cultivate and strive for. Which of the five phenomenal realities listed here would you personally assess to be most valuable? For example, which one would we expect to reduce psychological suffering most effectively? All but one of these phenomenological possibilities are clearly directly relevant to a deeper understanding of what pure awareness might be. The second one—multiple units of identification present simultaneously or in quick alternation—is an interesting, if partial, exception. It almost never appears in our data set, probably because it would refer either to the psychiatric syndrome of dissociative identity disorder (formerly known as "multiple personality disorder") or to the normal "unaware" state of mind, in which we are constantly distracted by stray thoughts and attentional lapses, bouts of mind-wandering, and spontaneous daydreams. Yet it is actually possible to abide in a global state of mindful meta-awareness while some forms of spontaneous thought and mind-wandering still take place, but this time without the phenomenology of identification, without actually getting "sucked into" an inner attention sink or being immersed in an inner landscape. So this second option could be useful to explore in future studies of identification and its relationship to mind-wandering. Meanwhile, to gradually flesh out our new working concept of the "unit of identification," let me now briefly go through the other four phenomenological options one by one.

The complete absence of any unit of identification relates directly to Buddhist theories of nonself (*anattā* in Pali) and nonattachment. For example, we can imagine situations in which, experientially, we would not be "glued" to the body as felt from the inside, not attached to the current content or emotional self-model, as well as states in which the sense of effort and active control in attention and thought had disappeared. But the idea can also be used to describe the phenomenology of some acute psychotic episodes and severe psychiatric conditions like depersonalization or Cotard syndrome[8] (in which patients claim that they are dead or do not exist). Selfless states without any other unit of identification may also occur during dreamless deep sleep[9] and partial epileptic seizures.[10] In particular, certain psychoactive drugs such as LSD, psilocybin, and 5-MeO-DMT have dramatic effects on self-consciousness and can temporarily suppress *any* form of self-identification—a phenomenon known in the scientific literature as "drug-induced ego dissolution."[11]

Our survey responses make clear that many experiences can be (and are) described as lacking a unit of identification altogether. Chapter 25, "Ego Dissolution: Melting into the Phenomenal Field," provides some examples; you can find others among the quotations in chapter 8, "Nonidentification," chapter 16, "Simplicity, Nothingness, and Absence," and chapter 27, "Nondual Awareness: Insight." The most interesting question for our purposes is whether pure MPE itself, as it occurs during full-absorption episodes, can sometimes be accurately described as the phenomenological unit of identification. I will come back to this when discussing the fifth possibility later in this chapter.

Then there is the possibility of processes during which *multiple* units of identification are simultaneously present or in which multiple units of identification quickly alternate. We discussed neurological disorders like heautoscopy and the phenomenology of robotic reembodiment in chapter 21. In these situations, the sense of self can oscillate between the felt, physical body and a visual body hallucinated in extrapersonal space or between the physical body and a robot. In psychiatry, we find the example of dissociative identity disorder (previously termed "multiple personality disorder"), characterized by the maintenance of at least two distinct and relatively enduring personality states and accompanied by severe memory gaps. Here, we have different and alternating sets of personality traits and multiple autobiographical self-models—different inner narratives that the patient identifies with at a given point in time. There may also be a healthy version of this process, in which we display different virtual selves in different social contexts[12] but maintain a sense of control and a high level of functional efficiency.

I think that in normal life, many of us alternate between rapidly changing units of identification (e.g., during nocturnal dreaming or whenever we drift off into a short episode of daydreaming and "come to" again, when haunted by unbidden memories, or when sudden attacks of automatic planning and mental time travel into the future make us briefly lose touch with the present moment).[13] Meditators, of course, are experts in studying precisely these processes under the microscope of introspective attention, and accordingly, they are generally much more aware of them than most other people. One thing that meditators often begin to see is the utter unpredictability, the volatility, and the degree of discontinuity in the neural processes that create ever-new units of identification, simulated future selves, and sudden trains of thought, like a fountain sputtering out wastewater in the middle of what otherwise would be a large, clear lake. Another thing that meditators begin to see is that the fountain isn't them. At times, they may find that one can actually *embody* the lake, even while the fountain is still active. This would be one example of what this chapter is all about: abstract embodiment and the rich possibilities for changing units of identification.

Under normal conditions, however, all we ever do is wastewater surfing, and often this actually gets us somewhere; it is functionally useful. But in terms of content, every onset and every ending of an episode of mind-wandering is characterized by an unexpected shift or sudden switch in what I will call the "phenomenal unit of identification."[14] Here is an example. Let's say that at first, you identify with the conscious content of an internal model of the self as driving a car, currently waiting for a red traffic light to turn green. Then an internal simulation of yourself as buying tofu and bananas pops up, as you "remember" that you need to buy tofu and bananas. Now you identify with the protagonist of this shopping narrative, jumping into the virtual self that constitutes the center of an automatic inner action simulation. Phenomenologically, and for only a very short moment, you literally "become someone else." For a brief moment, you "zone out" completely, which constitutes an involuntary and unexpected shift in the unit of identification. A few moments later, perceptual coupling may quickly be restored, and you reidentify with the "driver" (i.e., a model of the self as an attentional agent), quickly checking whether the lights have turned green. This is the end of your first mind-wandering episode. Phenomenologically, the driver is real again, and the shopper is only virtual—because the shopper is now no longer the unit of identification, but just the retrospective content of a sudden memory popping up, leading to a decision and an action plan.

However, in the very moment where "you" "remember" that you also wanted to buy almond butter and raisins, the unit of identification switches again, and you quickly "zone out" for a fast update, an enriched mental simulation of the shopper and its now extended task list, perhaps involving mental images of shelves and aisles. Traditional phenomenologists might say that you had an "afterthought." But my point is that really the afterthought swiftly created a new virtual self.

Remembering almond butter and raisins marks the beginning of mind-wandering episode no. 2, and the episode is functionally characterized by another bout of "involuntary mental time travel."[15] This second episode may again take less than a second to unfold, and as the light suddenly turns green, you "snap back" into the driver model, hastily shifting gears. The snapping-back is the next shift in the unit of identification, and it ends your second mind-wandering episode. There have now been two episodes and four switches. This way of applying our new concept yields another, entirely phenomenological way of understanding what "the cycle of death and rebirth" is. Reincarnation—the functional embodiment of a new unit of identification—is something that happens in your brain every minute of your waking life. But it is also something that can stop happening—for example, during meditation.

Buddhist philosophers have developed an astonishingly plausible theory of mind, but they did so before the theory of evolution by natural selection was conceived in the mid-nineteenth century, and long before modern neuroscience came along. Buddhist philosophers had never encountered the gene-centered view of evolution, which turns individual human beings into a kind of impermanent and disposable copying device through which genetic information is transmitted, selflessly flowing into the future. Early Buddhists also lacked the experimental and conceptual tools of modern cognitive neuroscience, including many of our present-day empirical insights into the biological evolution of cognitive mechanisms in animals and humans—for example, into the existence of a self-model or a "mind-wandering network" in the brain.[16] But they would probably have liked the idea of the physical body itself being a sort of *model*, a mere hypothesis about what was possible in the world of their ancestors; and they would certainly have understood modern ideas about the *conscious* experience of our own body as dependently arising out of a complex mesh of impermanent computational processes.[17]

Today, the Buddhist idea of reincarnation looks to many like a philosophical remnant from a different epoch of humankind's intellectual history, like a metaphysical leftover from another time (see also chapter 17). But please note that the idea of a "cycle of death and rebirth" (which is shared by most Indian religions) can also be read as a philosophical premonition of Darwin, and that—interestingly—it can *simultaneously* be applied on the level of individual human brains, on the level of their bodily hosts, on the level of the wandering mind, and on the level of the person as a whole: *Saṃsāra* is aimless wandering, jumping from one unit to the next. But now we are beginning to understand that all of this is a *nested* process that happens on many functional levels and timescales. For example, today we can view rebirth as the cycle of successive existence of ever-new biological copying devices, but also as a transmigration from one conscious unit of identification to the next. *Saṃsāra* in this new sense is a self-organizing biological or mental system going through a succession of states, leading to the impermanent functional embodiment of ever-new units of identification—but in a process that has no direction and no ultimate goal and creates an enormous amount of conscious suffering. *Saṃsāra* is a scale-invariant principle of conscious life. As it happens on many levels simultaneously, in life and in mind, we could call this naturalistic reinterpretation of what the cycle of death and rebirth really is "nested *saṃsāra*." Is there a way to get out of it?

The third new possibility for configuring conscious experience listed earlier is that there could be *maximal* units of identification where we identify with the whole world or with the phenomenal field as a whole. Many of the self-reports presented thus far clearly show that the phenomenal character of MPE often occurs in situations where the sense of self expands and the meditator gradually becomes "one with the world." This type of experience is like a phenomenological archetype of humankind, a timeless classic that has been described for millennia within many cultures. The deeper conceptual question is whether states in which the unit of identification has been maximized to include every other form of conscious content can still count as states of self-consciousness because a self/other distinction is lacking. But as usual, let us try to avoid metaphysical side-alleys and dead ends by staying as close to the reported experiences as possible.

Fourth, there could be *minimal* units of identification where we identify with one very subtle phenomenal quality, with a maximally simple form of phenomenal character. This possibility directly relates to MPE because—at least according to my tentative and empirically falsifiable working hypothesis—the pure-awareness experience as it occurs in meditation is our best candidate for the simplest form of conscious experience.

Matthew McKenzie describes the relationship between nondual awareness (see chapter 27) and the pure-awareness experience or pure-consciousness experience (PCE) very clearly:

> The [nondual awareness] experience involves the maximal unit of phenomenal identification: the space or expanse of consciousness as the context of phenomenal contents. In contrast, the PCE involves the minimal unit of phenomenal identification: the simple state of phenomenal consciousness devoid of contents. In both cases, though, there is a shift of self-identification from the typical sense of being a separate subject in relation to an object to a sense of being consciousness itself, whether in its pure form or in its spacious form.[18]

Can one *be* pure awareness? Is it possible for you to *embody* MPE, and how would you speak about it later? Please note that during a full-absorption episode, you cannot say "I *am* this!" but afterward you can say "I *was* this!"[19] In any case, many reports in this book show that human beings are able to identify with pure awareness itself, to turn MPE into their phenomenal unit of identification. This means that we now have some statistical data to back up this fourth claim. If pure consciousness turns out to be minimal in an interesting sense, then it could function as a target for the mechanism of identification. This begins to close the circle when looking back at the phenomenology of abstract embodiment and bodiless body-experience: The simple phenomenal character of wakeful, spacious awareness itself can clearly function as the unit of identification, and looking at the frequency of reports of the type presented in the first section of this chapter, it seems as if it might constitute one major category of human contemplative experience.

All of this leads naturally to the fifth possibility, as well as an important philosophical question: Can there be *nonegoic* units of identification? I think the answer is clearly yes, but of course, everything hinges on what we mean by "egoic." If you know my academic work, you'll know that there is a lot I could say here, but let us keep things as simple as possible. For now, let us just say that a system is self-conscious in the egoic sense if it currently fulfills one or more of the following three criteria:

- There is a sense of agency on the level of bodily motion, attention, or cognition (i.e., a sense of goal-directed control for the physical body as a whole, the focus of attention, and conscious thought).
- There is an autobiographical self-model (i.e., an inner representation of the system as a whole as having existed in the past, as possibly existing in the future, and as somehow being the same across time).
- There is global sense of ownership, normally for the physical body as existing in a consciously experienced here and now. Most often, I simply own my physical body as a whole, as something located in space and time.

For example, in full-absorption episodes in deep meditation or during dreamless deep sleep, all three criteria are never met. Full absorption cannot be fabricated;

accordingly, no sense of agency is involved. It is timeless, it is not located in physical space, and nobody *owns* pure awareness. Afterward, the phenomenal character of pure, wakeful awareness is the only kind of phenomenal character that can be reported. But whenever—as in some cases of abstract embodiment and bodiless body-experience—a meditator later credibly says "I *was* that!", we could now say that, phenomenologically, a nonegoic unit of identification had emerged during the episode. This result opens a new line of inquiry, which begins with a fascinating philosophical puzzle: If all of this is correct, can there also be nonegoic forms of *self*-awareness? We will delve further into this question in chapters 29 and 30.

From a scientific perspective, we can say that the unit of identification often represents the best hypothesis that the system has about its own global state. The unit of identification is dynamic and highly variable, and as out-of-body experiences and data from direct electrical brain stimulation show, it does not have to coincide with the physical body as consciously represented.[20] Possessing a unit of identification is the central causally enabling factor for many forms of intelligent behavior (bodily and mental) that presuppose the ability for self-reference. Biological systems sustain organismic integrity by preserving the integrity of their conscious self-model, constantly trying to minimize all uncertainty related to their unit of identification. Sometimes even confabulation, delusion, and functionally adequate forms of self-deception can be fruitfully viewed as an organism's attempts to sustain its integrity and stability across time.

Over the last three decades, I have written a lot about the idea of a conscious self-model, some of which may be helpful when trying to understand bodiless body-experience, abstract embodiment, and changing units of identification. In a nutshell, your phenomenal self-model is a conscious inner image of the organism as a whole. It is not a homunculus, some little person in the head, but rather a subpersonal brain state; it has many layers; and if it is transparent, it creates the phenomenology of identification: You, the organism as a whole, feel as if you *are* the content of whatever the inner image portrays. For example, if the model says that you are not a mere organism, but an enculturated rational individual—a *person*—then you feel that you are a person. However, it is crucial that we distinguish between a phenomenological and a computational reading of the term "self-model." All of conscious experience is a self-model in the computational sense, but only some of it is a representation of something that could be called a "self" in the phenomenological sense.[21] Let me briefly explain this conceptual distinction, because it naturally leads to a better understanding of abstract embodiment and the experiences described in the first half of this chapter.

Much of the best current consciousness research converges on the idea of consciousness as a controlled hallucination, and this is relevant to our distinction between the computational and the phenomenological self-model.[22] Our conscious model of reality is an inner state resulting from the brain's continuous attempt to predict the way in which sensory stimuli—such as the firing patterns on the retina as you look at a rose, an iris, or a carnation, as Aldous Huxley did—will change *other instances* of its own inner states (e.g., in the thalamus or primary visual cortex). This model has many layers, its functional architecture has evolved over millions of years,[23] and as you perceive the world, it is continuously updated from millisecond to millisecond.

The model is concerned not with truth but with survival; it is an instrument that the animal uses to achieve the goals of uncertainty reduction and successful procreation. It is a "self-model" in the computational sense because all it can ever do is predict how the organism's own internal states will evolve over time. The model helps to protect organismic integrity, creating a statistical bubble that predicts and actively sustains its own surface from the inside. It has become so good at this that it creates a *virtual* reality (VR) right in your own head. More often than not, it actually "extracts" from the environment those hidden causes that really are responsible for the stimuli impinging on your sensory surfaces, and it does so in such a swift and reliable way that it can be treated as a real-time stand-in, as an inner proxy for reality as a whole—environment, body, and knowing self included. Computationally, it is a self-model; phenomenologically, a whole world appears. But within this world, there is a body. The world-model contains a body-model, at least most of the time.

Do you remember that in chapter 6, I explained how through the "body scan" technique, one can become consciously aware of previously unknown aspects of one's very own bodily experience (e.g., the contact sensations behind our inner eyelids)? Once recognized, they feel like something that has always been there. The phenomenal self-model has been expanded, enriched, and refined. Obviously, the same can be done by cultivating mindfulness toward the outer world: Meditators often report that contemplative practice leads to refined and enriched forms of perceptual experience.

But perhaps one could do the same for an aspect of the computational self-model that is *neither* a mere part of the phenomenal self-model *nor* a part of your environment model. What if you could cultivate awareness of that which temporally precedes and causally enables everything else—namely, tonic alertness, which may be introspectively available as the previously unnoticed quality of wakefulness and epistemic openness? This could be thought of as a special form of body scan that reveals a fundamental and important bodily property. This type of scan would not involve scanning an already existing body-model in the brain (as in classical Goenka-style practice and the eyelid example), and neither would it yet involve integrating the experience of wakefulness and epistemic openness into something that could properly be termed a model of a "self." From this new philosophical perspective, *all* conscious experience is a form of body scan. The rose, iris, and carnation in front of you are bodily states that—in the sense discussed in the preceding chapter—*present* themselves as external objects. They are a part of the single computational self-model underlying the totality of all conscious experience, which is physically realized by a part of your own body. In embodied beings like us, this computational model automatically gets segmented into self and nonself. This is what makes the conscious experience of *you* looking at the rose possible.

In neurotypical human beings, the quality of wakefulness and "knowingness" gets contracted into the phenomenal self: into an inner image of a person who has desires and beliefs; an ego possessing her own body and her own emotions; an agent that miraculously always remains the same over time; a transtemporal self who pursues goals—and who now has an invisible "mind" that is wakeful and open to the world. All of this is a fiction. In reality, epistemic openness is a property created by the computational self-model as a whole, not by some sort of ego. (This point can be read as a reinterpretation of the Patañjali quote I presented in chapter 18: "The Seer is nothing but the power of seeing which, although pure, appears to see through the mind.") At any given point in time, epistemic openness is enabled by—and arguably *identical to*—a subpersonal state in your head, a part of the brain that neuroscience is now homing in on. In this sense, the transpersonal is in fact the subpersonal.

But of course, neuroscience is only one source of insight; it is only one kind of epistemic practice. Meditation is another kind. In its own way, it may show that wakefulness and epistemic openness really are entirely *impersonal* and nonegoic—and yet something that one can actually identify with and embody in an abstract way. There may be more than one way to do this. But at least sometimes, contemplative practice can provide a proof of concept. It can demonstrate that there is a real practical possibility that the computational and the phenomenal self-model can become one and the same.

In case a lot of this has been sounding too theoretical, here is one clear and simple practical instruction given by Sheng Yen (1931–2009), a famous Chinese monk who taught Chan Buddhism in Taiwan. He specialized in a methodless method known as "silent illumination":

> The foundation—relaxation—should be established firmly. Next, be aware of the totality of your body and maintain the simplicity of that awareness. As the bodily burdens and sensations fall away, your sensory field expands so that you can incorporate the environment into your whole being. Do not imagine yourself in the second state merely because it is a progression from the first stage of just sitting. It must be experienced. You are taking the whole environment as your body sitting,

> without being distracted by the particulars of sight or sound that are absorbed into the whole. Whatever stage you are in, maintain the principle that silence is not separate from illumination.[24]

Over time, systematically repeating the MPE experience changes the whole person and the social context around it, and conversely, there are social contexts and personal lifestyles that may either prevent or stabilize its occurrence. Once MPE in the sense of wakeful presence and epistemic openness has been discovered, a number of things can happen. Most likely, it will be immediately forgotten. After all, MPE is a subtle, entirely silent, and very simple form of phenomenal character that plays no role in most cultures on this planet, and at first glance, it may seem to have no practical value. Or it may be contracted into an egoic self-model, creating a "mindful self." In this case, MPE becomes just another experiential aspect of the self-conscious person that we take ourselves to be and identify with. Now it is a *subjective* experience. But yet another possibility is that it remains uncontracted or gradually expands. This means that while phenomenologically, the clarity of wakeful presence is not attributed to any personal-level self, it may nevertheless later naturally lead to verbal reports of the form "I *was* that!" This is what this very long chapter was all about: Our phenomenological data show that MPE can sometimes actually turn into what in chapter 29, "The True Self," I will term a "nonegoic unit of identification."

25 Ego Dissolution: Melting into the Phenomenal Field

It was as if my body / my being was pixelated and dissolved into the rest of the pixels that make up the entirety of everything else. [#3086]

There are countless ways in which the experience of pure awareness can shift from background to foreground. The phenomenology of translucency is one way: "It's like reality shining through when all ideas, concepts . . . step aside for a moment" (#2528; see chapter 28). Another common phenomenological transition leads into a new model of reality in which the conscious experience of perceiving and knowing is no longer structured by a localized and "knowing" self, depicted by the brain as intentionally directed at distinct, external objects of knowledge. In such cases, the experience of egoic knowing turns into the experience of nondual knowing. As the original egoic self-model expands and sometimes even dissolves, some of the phenomenological effects described in chapter 24 may occur—like a dissolution of body boundaries, a sense of weightlessness, bodiless body-experience, groundless groundedness, timelessness, or the disappearance of the original spatial frame of reference. In the second half of that chapter, I offered the new conceptual instrument of a "phenomenal unit of identification," which simply means any content of conscious experience to which we may refer by saying "I *am* this!" or "I *was* this!" Using this new tool, we can now point out that in some cases, the nondual quality of knowing can apparently become the new unit of identification because it can be "bodilessly embodied." In other cases, it will be more accurate to describe nondual knowing as a state lacking any identification whatsoever.

Nondual knowing sometimes may then lead to states later described as full ego dissolution, pure being, wholeness, or simply a feeling of oneness with the nonconceptual essence of conscious experience per se. At the risk of repeating myself, one of the many surprising results of our study was how frequently meditators, when asked to describe their own phenomenal experiences during episodes of pure awareness, actually report an attenuation or a complete dissolution of their body boundaries, or a soft superimposition of the conscious whole onto those boundaries. It seems that sometimes this can be the threshold to spontaneously occurring ego dissolution. Let us take this tentative phenomenological observation as our starting point:

> 3166 I know of states [. . .] that I would describe as very stable. There are hardly any thoughts, just very subtle. The space around me is white and bright. I hardly perceive the body anymore. But it is a state that feels like "before something," perhaps like before a complete dissolution. I practice trusting that this will happen on its own at some point. I know I can't do it. [. . .]

Body dissolution often leads to a phenomenology of spatial expansion and becoming one with the phenomenal field or its substrate as a whole:

> 1337 [. . .] I began to discorporate and spread in every direction. My mass had converted into energy. The energy was aware. My awareness spread in boundless directions. I was everywhere and everything.
>
> 1979 I experienced my body sense as having no boundaries and extending into infinity, although tapering out, and its "shape" was forever changing. This was life-changing.
>
> 2481 While meditating I experienced the dissolution of my self and body boundaries and no longer knew where or who I am or who is breathing. There was only the breath without inside or outside, without subject or object.
>
> 44 [. . .] It was my first experience of "awareness" in meditation. [. . .] I had been just watching the corridor without any thought. And suddenly I noticed that I feel there is no glass door separating me and the corridor, and even my body. There was only the shine of reflected wood in the corridor, feeling (I) no longer have a border to the world and neither does the world.
>
> 1229 [. . .] I reach a state where my consciousness is a vast dark void where there's no difference between the void, emptiness, and me. I'm part of the void, I can "see" it but I'm not body or soul, I'm part of the dark/void and there's no difference between me and the world. I lose the feeling of being a body, but I experience a feeling that all that is this void, or emptiness. Even though it's darkness it's a warm safe experience where no questions need to be raised about anything . . . it's an all-knowing, uncentered space that without doubt is behind every phenomenon that is ever experienced. I regard this experience as a feeling of unity, and I feel truly connected to our universe during these states.

Sometimes body sensations can merge into a nondual phenomenal field, creating a global state of "embodying the world," as if the unit of identification had been maximized:

1675 [. . .] I remembered the monk's instructions to try and see if you could experience the sounds coming from outside as coming from within. At this time frogs were croaking outside the meditation hall. I did as the monk instructed and almost immediately there was a "unity" experience of all conscious experience taking place in one field. Meaning that the tactile sensations of my body and the sounds of the frog were all being generated in the same field of consciousness, without any inner (the tactile sensations) or outer (the frog ribbits). Everything seemed very slow, and calm as well. This state continued after I got up from sitting and walked back to my dormitory. [. . .]

3334 [. . .] it dawned on me that I was the floor, the grass, the trees, the people who were there, and in fact the whole world. I felt the wind through my branches (I was the trees) and felt all kind of worms and insects crawl through me, because I felt I was the earth. [. . .] It was a very full, dense experience, very complete, colors were extra bright, food was extra tasty, touching things was extra intense. It wasn't very relaxing because I knew it would go away and I tried to make it last longer to investigate it. It also felt nice to be this whole. This experience lasted a couple of hours. The experience has changed my life.

In open, nonabsorbed states, body dissolution can lead to ego dissolution. Episodes of this type are often described as a peculiar kind of highest-order gestalt switch. One longstanding idea is that the frequently unnoticed phenomenal character of awareness itself forms the background for all other forms of conscious experience. It is not entirely clear what it would mean to validate (or falsify) this idea via modern-day computational phenomenology, but if it were true, then a certain subset of pure-awareness experiences could perhaps be described as a global background/foreground switch. Others may be better described as a form of indeterminacy or "neither-nor-ness," a dissolution of the distinction between background and foreground or inside and outside:

1690 Positive experience: I was sitting on a chair focusing on my breath, following a guided metta meditation. There was a subtle immediacy of experience that simply just manifested. As it manifested, it felt like the "background" of my sensorium became the "foreground," or perhaps the foreground itself just disappeared. It was all very subtle. As soon as I tried to capture it, I was back in a dual subject–object awareness. The experience itself was relieving, like taking off a tight shoe and letting your foot breathe. Negative: I wasn't meditating, and I was sleep deprived. After putting down a book that was discussing nondual consciousness, I felt a sort of immediacy of experience. The foreground disappeared, and the background was endless. It felt as if it had always been there, and I was just now remembering. But instead of "feeling home" or relief,

I felt an immense dread weighing me down like a thick blanket. As I came back to subject–object awareness, I felt like there was no point in navigating space in this manner.

2456 Pure awareness: the moments during meditation when perception "tilts" like a plane, the bottom of the plane of consciousness turns upward, I simply "am" one floor "higher." No sensory sensations, pure inner seeing. Weightless, silent, unspeakably bright and peaceful, all boundaries are abolished, my body boundaries have merged with the environment, with the universe, I am the sky, the sky is everything, the self no longer exists, it has dissolved. [. . .]

2559 [. . .] all of a sudden, I entered this state. It must have been between 30–60 mins, as the break passed by and I wasn't noticing. I had no body sensations, no pain, no time nor space, no thoughts, totally silent and whole. I felt as if something had turned "inside out," as if nothing in my perceptual fields was as usual. I had only one very slow thought passing by, as if it was passing inside the dark field, and it translates into something like—"This is it." There was total equanimity, and it was perfect, there was no need to "come back." When I finally did, I was totally shook. It was nothing like I'd experienced before. This happened fifteen years ago. Looking back—this experience changed my life path, my career choice. But most of all, it invoked a strange sense of fearlessness, of intimacy with life and beyond.

3369 [. . .] There was a sensation of "being turned inside out," as if I had slipped through the eye of a needle, and everything dissolved. Every bodily sensation, thought, emotion, everything. There was a very light-filled and incredibly joyful pure presence, pure being in which there was no object or subject, observer, etc., but only pure, luminous joy and limitless being. The sense of time was completely gone; when "it was over" it was 2.5 hours later. [. . .] The awareness was very strong that this is "actually" the true being and my real nature, also that it is infinite, eternal, and indestructible. [. . .]

The Epistemic Agent Model

What about thinking? Here I do find something: it is thought; this alone cannot be stripped from me. I am, I exist, this is certain. But for how long? Certainly only for as long as I am thinking; for perhaps if I were to cease from all thinking it might also come to pass that I might immediately cease altogether to exist. I am therefore a true thing, and one that truly exists; but what kind of thing? I have said it already: one that thinks.

—René Descartes (1596–1650), *Meditations on First Philosophy,* Second Meditation

What exactly does it mean to merge with the phenomenal field as a whole? Does it mean that the unit of identification is maximized or that it disappears? Here is a new conceptual instrument that may help us describe some of these experiences more clearly, in particular the phenomenology of ego dissolution in the context of minimal phenomenal experience (MPE). I call it the "epistemic agent model."

Let us consider a few examples of times when we have an epistemic agent model: the conscious experiences of (1) being a thinking self, (2) being a self in the very act of attending, and (3) being a meditator attempting to "have an insight" (e.g., to recognize her true nature). Let's begin with the first example. Whenever you try to mentally calculate, to think logically, or to actively form a new concept, you experience yourself as an epistemic agent, as a thinking self that wants to understand something, as an entity that wants to create new knowledge. You have a goal, and you act to reach this goal. Mental calculation is effortful, as is forming a new concept or attempting to create a novel philosophical argument. While you do it, you feel like you must control your own mind, and—if you're successful—an inner experience of ownership and agency arises. The experience results from a special form of deliberate, goal-directed mental action,[1] which in turn leads to the transient combination of experiential qualities: a feeling of mental effort, plus a sense of ownership and agency. Perhaps falsely, this combination of experiential qualities can be described as a specific, nonbodily sense of self. This "cognitive sense of self" may be the intuitive anchor behind one of René Descartes's central philosophical ideas, as published in 1641 in the second of his famous *Meditations on First Philosophy*: "'I am, I exist', whenever it is uttered by me, or conceived by the mind, necessarily is true."

What about the second example of a context in which we have an epistemic agent model? If you are not thinking at all, but instead carefully attending to the sensations in your feet as you slowly walk, to the sound of a bell vanishing into silence, or to the self-generated sound-shape of a mantra in your mind doing the same, then you have an epistemic agent model. There is something that you want to perceive and experience as precisely as possible. Again, you try to gently control the dynamics of an inner process, but this time by optimizing for precision. There is also a quality of motivation, maybe even earnestness. In the carefulness of your attending, there is a sense of effort, which may be more or less subtle. A goal state has been selected, and the conscious self in the act of attending actively pursues this goal state—this is our second example of what it means for the self to be an epistemic agent. The self is active, and it wants to know something. It wants to *realize* the goal state in its own mind.

And here is the third example. If you are a meditator and have read about fancy theories involving recognition of the "true self" (chapter 29) or the effortlessness of spacious awareness (chapter 32), if you're expecting to learn to see your own true nature and "let go" in some profound way, if you intellectually know the difference between dual mindfulness and nondual mindfulness, then you are in serious trouble. You have become infected with a new kind of goal state, something that can apparently be known and *realized*. Now it is almost impossible to prevent the birth of a new and particularly clever epistemic agent model in your mind. This time, it purports to want to know pure, nondual awareness—it is trying to touch the elephant as directly as possible while still cleverly sustaining the blind toucher's own existence.

Do you know what a mahout is? A mahout is an elephant rider. In South and Southeast Asia, a mahout (who is usually male) often starts as a boy. He learns how to train and keep elephants, and he receives an elephant early in its life. We can imagine the meditating self as a rider sitting on the elephant of pure awareness—but in reality, this rider is a parasitic epistemic agent model on the elephant's back. The mahout is the controlling, knowing self. It is an image, creating dual experience and constraining the elephant's space of possibilities. Please note, however, that with a deeply conditioned, thoroughly trained, "well-behaved" elephant, the rider might not be sitting on its back at all. The rider could be a *virtual* mahout, a fictitious entity that arises only from time to time. The empirical evidence about mirror self-recognition in elephants is inconclusive at best,[2] but if the elephant saw and unexpectedly recognized itself in a mirror, then it might be surprised to discover that there was nobody sitting on its back.

The three examples that I have just given illustrate what it means to say that we have an internal model not only of some passively knowing self, but of an epistemic *agent*. There are goal states plus a possibility of failure; there is a corresponding high-level capacity like thinking or attending; and often this capacity is not just an abstract feature, but rather something that is actually exercised now—a concrete, ongoing process that is consciously experienced. An epistemic agent model is a special layer in the human self-model. It is what creates subjectivity in the sense of a strong first-person perspective. According to phenomenal experience, we are (thanks to the epistemic agent model) entities that often actively construct and search for new knowledge relations to the world and ourselves. We are information-hungry; there is something that we want to *know*. Phenomenologically, for a conscious cognitive system to operate under an epistemic agent model also means that the potentially all-pervading quality of knowing is contracted into a transparent phenomenal self-model—a fictitious entity that apparently knows that it knows.

Using two of our new conceptual instruments, provided in chapters 5 and 8, we could say that an epistemic space has temporarily contracted into the model of a knowing self, an autonomous inner agent. A new virtual self appears, and it seems to be self-aware. Apparently, it really knows that it knows—but the virtuality itself, the "as if" quality, is not experienced (chapter 8). Since all of this happens on the level of conscious processing, the virtual self also creates the phenomenology of ownership for certain states of perceptual or cognitive knowledge. For a typical human being, this means that, subjectively, it now possesses a specific kind of self-knowledge—namely, knowing that I myself know and that I myself am apparently able to actively control certain epistemic states. This step finally introduces subject/object structure into the space of awareness, and thus I acquire a first-person perspective. Awareness is now dual awareness.

It is, however, perfectly possible to be conscious but have no epistemic agent model. The elephant doesn't drop dead without a mahout—but it may begin to wander around in a way that from the outside looks like aimless foraging. To shed some light on the epistemic agent model through contrast with situations in which it goes missing, let us look at some interesting new research on phenomena in the human mind that involve apparently aimless "epistemic foraging"[3] in an inner landscape. For example, the epistemic agent model collapses whenever an episode of mind-wandering begins. When our mind strays, we lose control over the thought process. Mind-wandering, "zoning out," and daydreaming are forms of involuntary mental behavior. The epistemic agent has disappeared; we are decoupled from the present moment, lost in automatic inner behavior. Our daydream may be a story about how we have been a stable epistemic agent in the past or how we will successfully control thought and attention in the future. But in the present moment, relative to our current environment, we have no stable first-person perspective.

Interestingly, the same is true for the dream state: The epistemic agent model continuously breaks down, as we move through the dream narrative from one attentional lapse to the next. Research from sleep labs all over the world shows that dreams are states dominated by constant confusion, memory loss, and recurring disorientation. Only when we enter a lucid dream does an epistemic agent model stabilize itself, leading to a dream self that "knows" it is dreaming.[4] The moment in which the meditator realizes that she has been carried away by a train of thought and returns to her practice may often be a very similar kind of event. However, when becoming lucid in a dream, the brain generates a new self-model that continues in roughly the same phenomenal environment (only *roughly* because this environment is now experienced as much more vivid and stable, and is explicitly labeled "unreal"), whereas coming back from an episode of mind-wandering into the present moment also changes the phenomenally experienced environment (in which the physical surroundings may now dominate again instead of, say, a constantly recurring worry about one's mother). Of course, the epistemic agent model can also be highly unstable or absent following brain injury, during psychiatric illnesses, when fully immersed in medial environments, and while under the influence of psychoactive substances.[5]

The epistemic agent model also is an instrument in mind reading and social cognition. You can understand the observable behavior of others much better if you depict them as "knowing selves." As an infant, you gradually learned to understand, and later even influence, your mother's behavior by creating your very first mental model of an epistemic agent. In your own conscious mind, what was previously a hard-to-predict moving object in your environment now gradually transformed into the experience of a knowing self: Mama! Later, the biological organism that is you "discovered" that you are such a knowing self too—and this helped you understand and control bodily actions and the focus of attention. They became your *own* actions and your *own* attention. Long after the biological body was born, the epistemic agent model came online, and you began to enter social life. One reason why many serious practitioners, such as nuns and monks, choose to live in solitude or silence may simply be that social interactions automatically trigger the epistemic agent model: As soon as the gaze of the other has triggered the knowing self, this self is almost automatically embedded into a mesh of mutual updating, an often narrative network of knowing selves continuously validating each other's existence. Not only does this shed new light on the pitfalls of teacher/disciple relationships, it also raises a few interesting questions. Could there be conscious social interactions without an epistemic agent model? In what contexts do animals like us really *need* the kind of active inner self-control and mental autonomy that a knowing self gives us—and when exactly would the absence of such a self be a good thing?

In my work on mind-wandering, I have argued that for roughly two-thirds of their conscious lives, human beings are not mentally autonomous subjects, and what we traditionally call "conscious thought" is a largely subpersonal process.[6] Most of the time, it really is the *brain* that thinks, not you, the person. Mental autonomy means that you, the person as a whole, are able to control your own inner behaviors, your thoughts, and your attention. It is a property that comes in degrees: Like any other information-processing system, you can have a higher or lower degree of self-control or mental autonomy. Meditation practice increases mental autonomy. This includes the capacity to impose rules on one's own mental behavior, to explicitly select goals for mental action, to rationally guide mental activity, and, most important, to intentionally inhibit, suspend, or terminate an ongoing mental process. Scientists sometimes call this "veto control,"[7] and mind-wandering implies an *unnoticed* loss of veto control.

To have an epistemic agent model means to have an inner image of the property of mental autonomy attached to a virtual self. All of this also means that we will have to depart from the "myth of cognitive agency," which says that the paradigmatic case of conscious cognition is one of autonomous, self-controlled rational thought.[8] It isn't. Conscious thoughts are mostly automatic, subpersonal processes that are hard to control, and only rarely do they become part of a stable inner model of an active, knowing self. Hard-thinking, professionally thoughtful academic philosophers in the West have perpetuated the myth of cognitive agency for centuries, but the philosophical practice of meditation cultivated in the East debunked it long ago. And now the new Western science of "spontaneous, task-unrelated thought" has confirmed that debunking through experimentation.[9]

Figure 25.1 has been adapted from an open-access paper of mine entitled "M-autonomy," and it sums up a lot of empirical research on mind-wandering, day-dreaming, and what is sometimes called "spontaneous task-unrelated thought."[10] The striking result is that having a stable "knowing self" is actually a rare experience, even during ordinary everyday life—it comes and goes very quickly. This is what meditators begin to realize when their practices begin to lift the veil of narrative self-deception (chapter 17): the instability, the impermanence, the volatility of the knowing self.

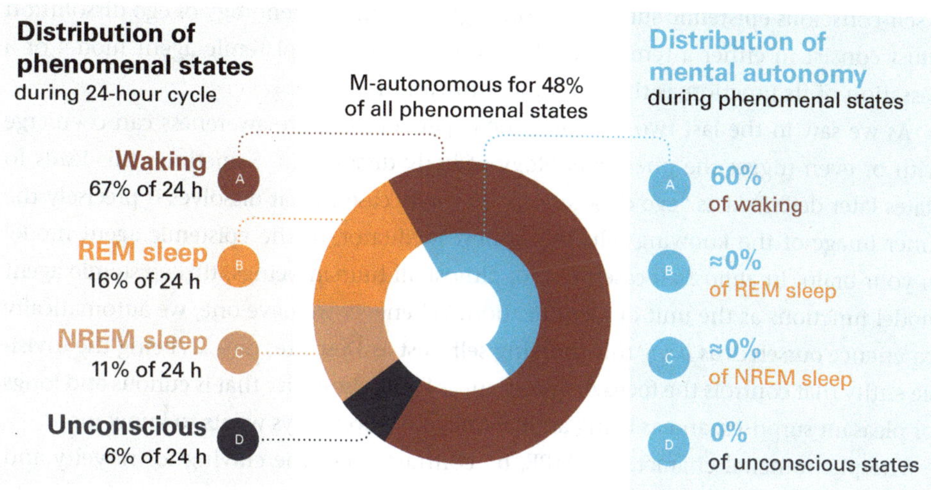

Figure 25.1

Distribution of conscious experience and mental autonomy over the twenty-four-hour cycle, based on empirical data plus a conservative estimate of the distribution of mental autonomy and the actual existence of an epistemic agent model across the conscious lifetime of a human being. Plausibly, we lack the experience of active mental self-control and a stable "knowing self" for about two-thirds of our lifetime, including periods of unconsciousness. In contrast to the traditional notion of "conscious thought," thinking turns out to be in most cases a nonautonomous, subpersonal process that occurs without consciously experienced mental agency. Figure taken from Metzinger (2015) and adapted by Cyril Costines. See Abelson (2022) for an interesting application. "REM sleep" refers to "rapid eye movement sleep"; "NREM sleep" refers to "non–rapid eye movement sleep."

Interestingly, the epistemic agent model is precisely what is completely missing in all full-absorption episodes of pure awareness and in the so-called nondual states that we will investigate in chapters 26 and 27.

For now, we can begin to combine this new conceptual instrument with the concept of the unit of identification introduced in chapter 24. During ordinary wake states, the epistemic agent model is what most of us identify with most strongly because it automatically functions as the unit of identification. Perhaps unfortunately, reading this book may feed and stabilize your epistemic agent model, and it may deepen your identification with the cognitive self. Yes, there is bodily self-awareness, and you may strongly identify with its affective tone and the emotional self-model emerging from it, but there is probably a clear phenomenological sense in which the thinking self and the active, selectively attending self are the true origins from which your mental perspective projects out into the world. They are what really turn you into a knowing *ego*, a self-conscious epistemic subject. Accordingly, the phenomenology of ego dissolution must consist in either a temporary disappearance of the epistemic agent model or a cessation of its function as the unit of identification.

As we saw in the last two chapters, the experience of pure awareness can coemerge with or even trigger the phenomenology of body dissolution. Sometimes this leads to states later described as "ego dissolution." In many cases, what dissolves is precisely the inner image of the knowing self, the earnest meditator, or the epistemic agent model in your brain. In almost all cases and for almost all human beings, this epistemic agent model functions as the unit of identification: Whenever we have one, we automatically experience ourselves as *being* this thinking self (just as Descartes did), as being the invisible entity that controls the focus of attention, as being the entity that is curious and longs for pleasant surprises, and as being an informavore who always wants to know more.

The phenomenal character of MPE, by contrast, lacks the craving for novelty and the two elements of dynamic knowingness and egoic identification. MPEs frequently seem to break their inner connection, weakening the process of identification with the fictitious knowing self. If there is knowledge, it is no longer egoic knowledge.

In the next two chapters, therefore, we will look at the phenomenology of so-called nondual states. Traditionally, this concept refers to states of consciousness in which the distinction between subject and object has been suspended. Phenomenologically, it is pretty clear what objects are (the tree over there, the apple in the palm of your hand), but now we have a simple and precise way of saying what a "subject" is. Phenomenologically, a subject is an epistemic agent model that functions as the unit of identification. Nondual states belong to the class of conscious states where this function has been suspended. But as we will discover in the next two chapters, they are much more than mere "states": Nonduality is a global *mode* of consciously knowing the world.

26 Nondual Being: Unity

A still life taking place by itself. [#699]

[. . .] like a single accordion playing itself, with no musician nor an audience. [#1682]

Only an all-encompassing "object," "a self-nature that flows without time," that is constantly transforming (unstable) and shapes everything through its form, is present. [#2426]

[. . .], being dissolved, no thoughts, unifying sensation of everything around me, as if out beyond the horizons, the mental ones too. [. . .] a no-longer-being overall and yet Present in everything. [#2537]

There are many ways of trying to verbally convey the experience of pure awareness. One frequent strategy involves describing it not as pure awareness, but as pure *being*—for example, as a nondual phenomenal experience of existence per se. One Transcendental Meditation (TM) practitioner has said, "It's pure being. It is is-ness, pure am-ness. It is the essential nature of existence."[1] One difficult question is to what extent the phenomenological profiles of "nondual awareness" and "nondual being" overlap, and whether a full double dissociation is possible. Can either exist completely without the other?

Depending on the context, it is conceivable that two meditators might refer to one and the same type of experience either as "pure awareness" or as "pure being" and, vice versa, that different experiences may be described using the same verbal labels. Here, we clearly need more unbiased research involving a phenomenological approach that has been liberated from any doctrinal background. This chapter and chapter 27 will take a first step toward the goal of understanding how "nondual awareness" and "nondual being" relate to each other. Let us begin with the phenomenology of existence, the experience of pure being in and of itself.

Our reports show that the phenomenology of "pure being" can either be the sole defining characteristic of a full-absorption episode or coexist with other forms of experiential content. Next, you will find nine examples of the phenomenal quality that I am trying to isolate here. Note that some of the descriptors explored in earlier chapters—like "body dissolution," "peace," "silence," "weightlessness," "ineffability," "simplicity," "nonduality," "timelessness," and the "invisible smile"—recur frequently in this new context:

80 [. . .] Perhaps one can best describe it as "simply being," and metaphorically as "peaceful and weightless."

521 In meditation I sometimes experience simply "being" without additional modalities, pure being in itself; I cannot say much more about this, because there is nothing to say about it in descriptions. To a certain extent, the perception of one's own existence without the observer role and without mental activity, perhaps one could say—as if the pause between the thoughts grows very long, but without waiting.

1316 [. . .] Deep silence, the feeling of simply being Pure Being [*Reines Sein*], completely beyond any individuality. [. . .] sometimes the perception that existence [*Sein*] is completely flat, without the slightest additional perception, sometimes also the experience of dynamism in the silence. [. . .]

1378 [. . .] The flow of thoughts came almost to a standstill, so that only a quiet awareness of one's own being was left. Then a feeling of silence stabilized, complete presence and agreement with what is, a feeling of the perfection of the moment and of emotional balance. Then the space opened up to an infinity in which a pure existence was possible. [. . .]

1424 [. . .] It is a realm of complete silence. Simplicity. There is nothing there that needs to be expressed, communicated, or evaluated. The simplest possible being.

1625 [. . .] All of a sudden the feeling changed and I plunged into something that I would call complete being. I as a subject no longer existed, all I felt was the great, all-encompassing being of which I was a part. There was neither an outside nor an inside, but only the boundless whole. Time no longer played a role either and it seemed to be an eternal state. [. . .]

1795 In pure awareness I always experience a lightness. [. . .] I am simply there, exist and know that I exist. Nothing more. Observing this conjures up a soft smile at the corners of my mouth. [. . .]

1898 My experience was the pure happiness brought to me by simply being. There was no me, no pain, no need.

1195 I sit and I am. Am who I am. Upright and penetrating everything infinitely far, embracing everything. There is neither boundary nor separation. All "others" and all phenomena are part of my oneness or universe. It is exhilarating and yet very simple, because it is still and empty. [. . .] "I" is only being, it's not personal.

One interesting detail is that reports of the experience of pure, nondual being involve more frequent use of CAPITALS than reports of other types of experience:

1261 Transcendental experiences cannot be described, one simply IS.

2870 It's like when the kisser and the kissed merge in the kiss. Neither one nor the other. It IS. It penetrates the ALL, the NOTHING, and fuses into one single BEING.

3243 [. . .] suddenly a consciousness of total boundlessness. It is very difficult to think of another word for it. There was neither body nor thought, feeling . . . just "BEING." I returned from this state with the awareness (now also figurative) of the whole universe, then being a particle and feeling a body again. While walking I stopped, looked at the sky, a bird flew by. The feeling of just "BEING" appeared again. [. . .]

3304 [. . .] deep moments of calm and of simply just BEING, in which I perceive sounds only very fuzzily and my body seems to "dissolve." [. . .]

2487 There was a moment in the middle of nature—I was in the forest and suddenly I had a moment of ONENESS. There was no differentiation between hearing, seeing, feeling. Everything was self-evidently there and self-evident, with no need to change anything. It was light, boundless, and full of peace. [. . .]

As explained at the start of this chapter, what nondual awareness and nondual being have in common is that they lack subject/object structure. There is no self in opposition to some reified content of experience. The experience of pure being is clearly a nondual state in this sense, simply because there is no distinct, knowing self. To use the new conceptual instrument offered in chapter 25, there is no epistemic agent model; hence the distinction between subject and object doesn't exist. Being nondual, the experience of pure being cannot be fabricated and is easily destroyed by even the faintest form of clinging or attachment:

924 I experienced a moment of being. The moment I noticed that, it was already gone.

2240 I experienced a break in the tension of meditation and directing awareness and a dense, dark, and warm expansiveness of being crept over me. My body was breathing itself and moving itself and being was just occurring. I have experienced this state several times with meditation, it spontaneously occurs and shortly thereafter dissipates as my sense of self returns and tries to claim or investigate and normally prolong the state of mind which seems to destroy it.

Ego dissolution and becoming one with the world are processes that cannot be actively constructed or directly accessed by any meditation technique. They often occur spontaneously, outside formal meditative practice (see chapters 10 and 32). Of course, a dissolution of the sense of self is one of the classical phenomenological descriptors for advanced meditative states, and it is important to be very clear about what concepts like "selfless" or "nonegoic" really mean if applied in the context of conscious experience (more about this in chapter 29). Many theories of "ego dissolution" have been constructed over the centuries.[2] But from time to time, one has to liberate oneself from the idealized conceptual frameworks and face the messiness of contemplative experience. It is of vital importance that we look at phenomenological reports written by *real* meditators in the *real* world. If we do so, and investigate ego dissolution in the context of pure being, holism, and unity, what we find is not at all clear cut. Phenomenologically, there seems to be a whole spectrum of conscious states or modes that can be described as "selfless":[3]

> 3068 [. . .] a moment in which all self-feeling had dissolved and I was walking in the forest (i.e., not a sitting meditation, not a specific walking meditation). Everything was a big . . . self-updating process. Timeless, almost dimensionless, in a way, empty, and yet every particle of "my experience" was somehow connected with every other particle, a wholeness. With no point of identity to which I could have referred. At the same time, there was a part of my self that was afraid of this (of losing myself in it, of going crazy, etc.). This part was not so much in the foreground, however, and made up perhaps a fifth of my experience.

A large subset of reports emphasizes the quality of existential unity, of becoming *one* with everything that exists. Sometimes this is preceded by a "softening" or a mild dedifferentiation of the phenomenal field. This process of existential unification may start locally (as related to a specific perceptual object or a single sensory modality), but it can then turn into a global, all-encompassing experience. Again, though, we should not jump to interpretations and conclusions too quickly. If we take the phenomenology seriously, it becomes clear that there is a rich and differentiated spectrum of states that can be described using terms like "unity" and "nondual being." To demonstrate this abundance, I will close this phenomenological section by simply presenting seventeen examples without further comment:

3219 During a meditation became completely one with a tree. A most extraordinary unexpected event.

1215 While peeling carrots, the difference between carrot and me dissolved, I noticed the connection, and being in the here and now, that was very impressive. I experience such phenomena very often in nature and in connection with people.

1918 [. . .] After my morning practice I was looking at a leaf which exploded / opened up a new world—I was it / it was I. [. . .]

2426 The outstanding characteristic of my experience was a radically changed perception of reality (through a kind of fusion with the wind and the trees and their movement; dissolving of boundaries) through to a perception without different objects or subject. Only an all-encompassing "object," "a self-nature that flows without time," that is constantly transforming (unstable) and shapes everything through its form, is present. [. . .] and [I] tried to imagine what a world could look like in which everything is connected, permeated by fields and constantly changing, without individuality. I also did one of my meditation exercises in between. Every time I noticed that I was focused on something, I tried to let it go. When I looked at the trees moving in the wind, I had the feeling of merging with them and with my surroundings. There were no more boundaries or different objects. Just one thing, the everything that is changing. [. . .]

1937 [. . .] the phenomenological field of sound/sense/weight/touch softened to a point of disintegration into a hyper awareness of "be-ing." Unity and nothingness at once, perfectly balanced. [. . .]

2107 [. . .] I was alone outside in a small meadow surrounded by pine and aspen trees. I was sitting with eyes open and paying attention to my breath. At some point it felt like all of my senses (of body, of sight, touch, smell, and sound) dissolved into one open awareness of being and there was no separation between me and the trees and the air around me. It was all just happening.

2206 [. . .] the concept of a "cloud of sensation." It began when I lost the feeling that certain subjects of consciousness (i.e., hearing, feeling) were connected to a location. Instead, all there was was a cloud of sensation. In that moment that was all "I" was.

3186 [. . .] For me, it is best expressed like this: "I become all that is." So not even the connection with everything that is. Instead, pure being. The touch and the encounter with your own soul. Simply beautiful and not comparable to anything on earth. Pure joy.

2411 My experiences have been in a wakeful state and have been characterized by a stable and extended felt sense of being aware of being aware in that particular place and time. It entailed a heightened sense of unity, of being a part of all that was in the moment. Thoughts, sensations, and emotions were present but felt fluid and seemed to emanate from myself, which simultaneously felt at one with all that was.

3403 All inner voices fell silent, there was neither future nor past, neither fears nor worries, in the here and now it was possible to experience pure being, the all-embracing context was more important than the individual, the latter receded into the background in favor of a higher synthesis, faded and lost its independence . . .

2804 [. . .] I feel totally content. There are no desires, no wishes, no body, no pain, no needs, just pure being. Then after a while consciousness expands to include what's outside the house we are sitting meditating in. Wind, birds, trees are now part of my experience of a flow of free energy. Everything is the same. Everything is movement and change. And in that moment, I am part of all that.

1190 A perception in waking consciousness of the complete unity of one's own self with all objects of contemplation around it. 100 percent bliss. A perception of absolute silence inside amidst the hustle and bustle all around. [. . .]

2943 A feeling of warmth and brightness streaming through. The body was no longer necessarily consciously present. There was spacelessness and weightlessness, lightness. I was very present, but not there. I felt simultaneously very centered and expansive. I felt gratitude, realization. Pure existence.

2354 [. . .] and I was very suddenly powerfully overcome with the feeling that I was no longer looking out at an open field and mountains that surrounded it on a beautiful sunny day, but that I *became* the open field, the mountains, and the beautiful sunny day. There was also a unity of consciousness that I had never experienced prior to this. The lines between the experiences of bodily sensations, sights, sounds, emotions, and even thoughts became blurred into oneness. It was no longer a feeling of "That's a sensation, which is separate from a sound." It was a feeling of "This is consciousness—all of it. There's no special distinction between any of the specific contents of consciousness." It was one of the most powerful experiences of my life, and I'd put it in the top 10 most life-altering experiences I've ever had.

2244 [. . .] It's as if the entire world and myself merge and I am simply the center of everything. Words drop away. Fear is not present because the me is gone. A greater feeling of openness is there. I feel large . . . as large as the entirety of my knowledge of the world while knowing that I'm also that which I don't know (hard to explain). No sense of separateness but at the same time I feel like I am the center of a center-less universe.

1682 The experience "felt" (for a lack of a better word) like pure peace. Everything felt irrelevant and unimportant. My awareness was so anchored in the present moment, as if I had a sudden amnesia of my past memories. There were no plans for the future, my brain was not trying to predict what was going to happen the next moment. It was completely focused in my surroundings and in my breathing. My breathing was so constant, rhythmic, and harmonic that it felt connected with the air coming in and out of my lungs. I suddenly couldn't tell the difference between what "I" was doing and what the environment was causing on my body, as if the air was the one causing my body to breathe. It stopped being a cause-and-effect relationship and felt more like a single accordion playing itself, with no musician nor an audience. The breathing in my body "was," I wasn't controlling it anymore, it just "happened." When I lost this boundary, I lost any responsibility of my body and a feeling of peace became apparent. As if it were always there but I was first finding it.

2431 And it was as if I woke up in a different state. I felt transparent and one with all around me, I understood the trees the wind as if they were me and I was them. Was calm knowing, but when I realized I couldn't wake up out of this state I started to worry a bit. But as I could function normally, walked back to my car and sat at a terrace ordering coffee, then this state slowly passed by.

Seelengrund and "Groundless Ground"

> Understand: all our perfection and all our bliss depends on our traversing and transcending all creatureliness, all being and getting into the ground that is groundless.
>
> —Meister Eckhart (1260–1328), Sermon *Adolescens, tibi dico: Surge*

Let us turn once again to the ancient fable of the elephant and the blind. What would happen if the king had the power to bring together not only all the blind-born from his own kingdom, but also those from all other countries of the Earth, including those from different religious creeds? What if the king could ask people from different historical epochs, people who had never known each other? Would they still converge on at least some aspects of what the elephant is like? The second part of this chapter is about the phenomenological aspect of minimal phenomenal experience (MPE) that—for many centuries—has been described as a "groundless ground," the "foundationless foundation" of all experience. To widen our perspective on the phenomenology of nondual being, I invite you to follow me in looking at a range of sources from different times and places, like Meister Eckhart's concept of the *Seelengrund* (ground of the soul) and the Tibetan Buddhist concepts of *gzhi'i gzhi med* (groundless ground) and *gnas med gnas* (foundationless foundation). Our whistle-stop tour will be guided by my philosophically motivated hypothesis that the human self-model might actually not (as I suggested in earlier publications) bottom out in invariances of interoception and elementary bioregulation,[4] but might instead bottom out in emptiness, in an indeterminate phenomenology of mere epistemic capacity that is neither self nor no-self. Perhaps it even turns out that there is an additional fact, a phenomenological sense in which the conscious self-model is not only groundlessly grounded, but actually *sculpted* out of the epistemic openness of MPE?

Begine of Hadewijch—also known as Hadewijch of Antwerp or as Hadewijch of Brabant—was a thirteenth-century philosopher–poet and mystic. She was part of a religious movement that began in the High Middle Ages and is known as the European *Frauenmystik*. Hadewijch probably lived in the Duchy of Brabant and was one of the first Western mystics to use the two terms of "ground" (*gront* in Middle Dutch) and "groundlessness" (*grondeloesheit*). We do not know much about Hadewijch, but her famous *Book of Visions* was probably composed between 1236 and 1245, and it describes dialogues between herself and Christ, using them as a form of religious teaching. At the end of her thirteenth vision, she describes how she was "overwhelmed by delight"[5] and then fell into a "groundless depth" (*grondelose diepte*). She describes "floating away from the mind" and says that the hour during which all of this happened will remain forever beyond words and language. In reporting her fourteenth vision, she writes about "tasting the one nature" and experiencing "one single taste (of one nature)" (*in ere naturen smake*).[6] This sounds a lot like the notion of "one taste" or "one flavor" in Tibetan Buddhism (Tibetan *ro gcig*; Sanskit *ekarasa*; see note 78 of this chapter for an example). For Hadewijch, this capacity to taste the one nature is a capacity that you can possess only if you have "experienced human and divine love as one being" and if you have become "undivided, inseparable, and completely one with the Godhood (*godheit*) that streams through oneself and in turn have begun to stream through it."[7] In her eighteenth prose letter, Hadewijch of Antwerp relates the groundless ground to the deepest level of the soul itself: The soul is an entity or a being that is visible to God and to which God is visible.[8] The soul is an "entity without ground,"[9] in which God is "fully sufficient to and for himself"—and the soul in turn finds its own self-sufficiency within him.[10] And God liberates, opening the path through which the soul "passes into freedom, in his ground, which cannot be reached unless she reaches it with her own depth."[11]

In German, *Grund* (ground) and *Abgrund* (abyss) are closely related words, just as in Hadewijch's Middle Dutch we have *gront* and *afgront* and Meister Eckhart's Middle German speaks of *grunt* and *abgrunde/abgrunt.* Accordingly, in Meister Eckhart, we also find the *abgrunt sîner gotheit* (the "abyss of Godhood"),[12] as well as the *groundeloese gotheit* (the "groundless Godhood").[13] English translations of *abgrunt* as an "abyss" or a "bottomless pit" make this relation invisible. Interestingly, in Middle Latin, *abyssus* also meant not only an unfathomable depth, but also "space" (the *Weltenraum* in which disembodied souls may live) or simply "ocean." In her poems, Hadewijch says that the groundless ground of the soul is "deeper than the sea" and that one can actually "swim through it."[14]

Seelengrund, the ground of the soul, is a Western concept that is deeply related to the phenomenology described in this book in more ways than one. Just like the ancient Greek *ataraxia* (chapter 14) and the notion of "transparency" in early Anglo-Saxon analytical philosophy (chapter 28), it is one of the few historical examples that I am offering you (from the many I could have chosen) not just as a source of inspiration, but also in an attempt to help you discover certain phenomenological parallels for yourself. For example, it is interesting to note that Hadewijch's notion of "self-sufficiency" actually appears in some of our participants' descriptions of pure awareness, as well as in the Buddhist context of "suchness." You may recall our discussion of the phenomenology of suchness (chapter 9), which allowed us to isolate the two experiential aspects of spontaneous, apparently uncaused self-manifestation and timeless self-evidencing. And the phenomenology of "nondual being" is directly related to the groundless ground of the soul, in which the medieval mystics found the liberating groundless ground of the Godhood itself.

It was Meister Eckhart (1260–1338) who introduced the concept of *Seelengrund* into the theological, philosophical, and spiritual debates of the Middle Ages.[15] He was also known as Eckhart von Hochheim, and toward the end of his life, he was tried by Pope John XXII after being accused of heresy by the Franciscan-led Inquisition. Meister Eckhart was never personally condemned as a heretic, and in 1329—after Eckhart's death—John XXII issued a bull stating that Eckhart had recanted all his errors and false teachings by subjecting himself and his theological writings to the decision of the Apostolic See.

There are many reasons why *Seelengrund* is an interesting concept in the context of MPE—not least because (as can be seen in the epigraph at the start of this section) the ground Eckhart speaks of is a *groundless* ground. He says that rationality and reason (*vernünfticheit*)[16] are unable to grasp the nature of God because they can grasp it only as it is known within them, but not "in the sea of its groundlessness" (*in dem mer sîner gruntlôsicheit*).[17] We find the two semantic elements of fundamentality and nonreifiability combined here: A ground is a foundation, something your soul can stand on, but—just like the sky above—the sea is something unbounded, not a well-defined perceptual object that could be juxtaposed with a knowing self. Eckhart's treatment also links to our more abstract discussion of the possibility that the phenomenal self-model "bottoms out in epistemic openness," sharing a common baseline with the computational self-model that creates the totality of our conscious world (as suggested in chapter 24).

Of the *Seelengrund*, Eckhart says this: "Here, God's ground is my ground and my ground God's ground" (*Hie ist gotes grunt mîn grunt und mîn grunt gotes grunt*).[18] The philosophical point is that knowing God is a form of self-knowledge. What is more, the way in which Eckhart describes the *Seelengrund* also directly resonates with many of the phenomenological themes that emerged from our meditators' reports: unboundedness, silence, nonidentification, presence, "coming home," nothingness, emptiness and fullness, timelessness, ego dissolution, nondual being, "the true self that knows itself," and so on. As we will see, in the context of pure awareness and the reports presented in this book, the idea of *Seelengrund* also possesses great intuitive beauty and force: It may offer another interesting way of describing what, experientially, pure awareness is all about—making contact with the groundless ground of the soul.

For Meister Eckhart, the *Seelengrund* is something into which human knowledge can never penetrate. The *Seelengrund* shares many features with pure consciousness, as described in Eastern religions, and also with MPE, as described by the meditators whose reports are presented in this book. I have extracted some examples for you: The *Seelengrund* is nameless (*namelôs*); it is simple and ineffable and lacks definable qualities (*wiselos*); it is characterized by silence and motionlessness of the soul (*stille*; *ruowe der sêle*); it is unified without a second (*einvaltec*) and unborn (*ungeborn*); it has detachment (*gescheidenheit*), unidentified with anything in time or space, with any mental images or forms; and of course, it has purity (*lûterkeit*; *pûrheit*). It is not tended or inclined (*geneiget*) in any direction; it is absolutely fundamental; it is the innermost ground, the ground of being, of heart and mind (*grunt*); it normally has a hidden and secret quality (*verborgen*; *heymeliche*)—but it is also luminous because the soul is a pure light in itself (*dem lûtern liehte, daz si in ir selber ist*[19]).[20]

All forms of imagination and conceptual thought are excluded from the *Seelengrund*. It is not a thing. It has no temporal or spatial properties, and it was never created. It is not even immortal in the traditional sense because there was never a time at which it didn't exist. Sometimes Meister Eckhart uses metaphors like the "little spark" (*vünkelîn*) that fends off and turns away from everything that isn't pure (*lûter*), or the "little fortress" (*bürgelīn*) that defends against the same. These metaphors refer explicitly to only one part of the soul: The ground, the purest part, in which the Godhood is permanently present. Because it is absolutely pure, nothing else is pure enough to ever enter it—except God Himself (*niht sô lûter, daz in der sêle grunt möhte komen wan got aleine*).[21] God is fully present in the *Seelengrund* as something that already belongs to us and has always been our own. For Eckhart, the *Seelengrund* is the place where we can and must break through (*durchbrëchen*) from the level of things and manifoldness into the reality of nondual being, of Godhood. This breakthrough can then lead to what he called the *Gottesgeburt*, the birth of God within the soul itself—the soul becoming aware of the divine and eternal character of its very own nature, of something has always been there in its innermost core: the *Seelengrund*. The birth of God within the soul itself is God's initiative, but (and this is one of Eckhart's many provocative points) he also has no choice: If a human being creates the necessary preconditions—has created emptiness in herself (*itelkeit*), has annihilated herself *in* herself (*der [mensch] sich selben vernihtet hât in im selben*),[22] and possesses a "well-practiced detachment" (*wolgeüebete abegescheidenheit*)[23]—then it actually becomes a *necessity* for God to pour Himself (*ergiezen*) into this empty vessel, or else He would cease to be God (*muoz sich got alzemâle ergiezen, oder er enist niht got*).[24]

I think it should be obvious how a lot of this relates directly to our empirical and qualitative investigation of MPE, including in the aspects of nonidentification, ego dissolution, emptiness, luminosity, peace, and silence. A phenomenological reading of these medieval texts that abstracts away from Christian theology and the underlying metaphysical assumptions shaped by an ancient cultural context may reveal further unexpected parallels to contemporary reports given by practitioners of meditation today. Let us therefore now look at some of these features in more detail, while also listening to what some other mystics in the European Middle Ages were trying to convey.

In chapter 3, we investigated the experience of silence, stillness, and the zero-person perspective. Silence is acoustic emptiness or epistemic openness in the auditory domain, but the term is also used to convey a general simplicity and absence of any form of mental representation. David von Augsburg (ca. 1200–1272) speaks of the "silence of spiritual vastness" (*stille der geistlichen weide*), which also connotes a grazing land or meadow.[25] Meister Eckhart teaches that anyone who wants to enter the divine ground has to be completely silent (*muoz gar stille sîn*)[26] and detached from all images and forms (*gescheiden sîn von allen bilden [. . .] und formen*).[27] He says that the ground of the soul and the groundless ground of the Godhood mutually penetrate each other. Their degree of silence or calm (*ruowe*) determines the degree to which they can rest in each other: If the soul partly rests in him, he partly rests in the soul, and if it rests fully in him, he rests wholly in it (*Als vil diu sêle ruowet in gote, als vil ruowet got in ir. Ruowet si ein teil in im, sô ruowet er ein teil in ir; ruowet si alzemâle in im, sô ruowet er alzemâle in ir.*)[28]

In passages reminiscent of classical ideas in Zen Buddhism, Eckhart teaches spiritual practitioners that they should love God in a "mindless" (*nichgeistig*) way, so their souls can be mindless and nakedly free of all mindedness—because so long as the soul takes the form of a mind, it has images. So long as it has images, it has mediation; and so long as it has mediation, it has neither unity nor simplicity.[29] Among the parallels that we find between the medieval Christian mystics and present-day contemplative practitioners (who are mostly inspired by Buddhist and other Indian philosophical systems) are the emphasis on calm and peace (chapter 2), silence and simplicity (chapters 3 and 16), spaciousness (chapter 23), unity (chapter 26), and arguably also the motifs of True Self (chapter 29) and of nonexperience (chapter 31).

In chapter 5, we looked at reports describing pure awareness as involving the phenomenal character of clarity. In his 1997 monograph on metaphor in late-medieval mysticism, Michael Egerding writes that Mechthild von Magdeburg (ca. 1207–1283) thought that a pure heart is what makes human being's consciousness "as clear in itself as the sun" (*clar an im selber als die suñe*).[30] For David von Augsburg, who distinguishes among seven different levels of prayer, we must be sufficiently "clear" and "purely enlightened" (*klârlîchen* and *lûterlîchen erliuhtet*)[31] to know God, and then we depend on his grace if we want to go further and transform ourselves "into clear clarity" itself (*in die klâre klârheit*).[32] Meister Eckhart sometimes uses clarity as a metaphor for God, but he also speaks of the "clarity of the soul" (*klârheit sîner sêle*[33]) that can be a precursor to the *Gottesgeburt*. There also exists a degree of clarity that is unbearable to humans: Mechthild tells us that she is unable to receive God's answer to her attempt to contact him, not only because it is so forceful and so groundless, but also because it is *úberclar*.[34]

In the early chapters of this book, we began to isolate a series of readings of what it could really mean for pure awareness to be "pure" (you will find a full list in chapter 34). The purity of pure consciousness is found in the Middle High German concepts of *luter* and *luterkeit*, and in many different contexts and applications. Purity of consciousness appears, for example, as "purity of meditation" in David von Augsburg (*lûterkeit der andaht*)[35] and as "purity of mind" (*lauterkait des gaistz*) in Heinrich von Nördlingen.[36] Purity of consciousness also features in Eckhart's teaching that the pure Godhood (*blôz lûter gotheit*)[37] can be grasped only beyond space and time, by the force of a soul that is detached and pure (*abegescheiden, lûter*),[38] and in his beautiful descriptions of the "pure clear light" (*lûter klârez lieht*)[39] and of a "light or pure spirit" (*ein lieht oder ein lûter geist*).[40] It also makes an appearance in the famous idea of the *luter pur clar Ein* that Eckhart offers in Sermon 83, referring to the undifferentiated "pure clear One." The "pure clear light" may also remind one of Buddhist terms like "luminous mind" (*prabhāsvara-citta* or *ābhāsvara-citta* in Sanskrit; *pabhassara citta* in Pali), which appears in numerous Mahayana texts and Buddhist tantras (e.g., as the "brightly shining mind" or "mind of clear light"). It is also reminiscent of the phenomenological term "luminosity," which we encountered in chapter 18 (*prabhāsvaratā* in Sanskrit) and which is translated in Tibetan Buddhism as "luminosity," "clear light," or even "purity" itself (*'od gsal [ba]*, literally meaning "radiant clarity"). For Eckhart, luminosity can be embodied, and specifically in the form of clarity: There is an overflow of light in the ground of the soul that can flow into the body and "fill it with clarity" (*[Ü]bervlüzzicheit des liehtes, daz in der sêle grunde ist, daz übergiuzet sich in den lîchamen und wirt dâ von vol klârheit*).[41] This short list of examples shows that not only clarity, but also purity, luminosity, and simplicity are classic, frequently recurring descriptors of contemplative phenomenology, in ancient Asia as well as in the European Middle Ages. Given that we also find them in our own data, it is plausible to assume the existence of what I will call a shared "phenomenological anchor."

Coming back to the idea of "one taste" in Hadewijch of Antwerp and in Tibetan Buddhism, it is interesting to note that, as Waldschütz points out, gustatory metaphors—for example, Eckart's "God in his own taste" (*got in sînem eigenen smacke*)[42]—relate to a holistic form of knowing from the inside out, because they derive from the Latin *sapere*, which means both "to taste" and "to know."[43] *Sapere* is related to *sapientia*, which means having insight or philosophical wisdom. We find many examples of gustatory metaphors in Christian mysticism. Mechthild von Magdeburg speaks of "tasting a nonconceptual and inconceivable sweetness" (*smekket ein unbegriffliche suessekeit*)[44] and Johannes Tauler reports on "a taste of eternity" (*ein smak der ewikeit*[45]). Meister Eckhart says that the person who is turned in on himself, so that he recognizes God in God's own flavor and in God's own ground, is freed of all created things and is locked within himself as in a true castle of truth (*Welher mensche nû in sich selber wirt gekeret, daz er bekennet got in sînem eigenen smacke und in sînem eigenen grunde, der mensche ist gevrîet von allen geschaffenen dingen und ist in im selber beslozzen in einem wâren slozze der wârheit*).[46] This shows how contemplatives across different times and cultures have generated a whole range of perceptual analogies in the attempt to convey the ineffable, drawing not only on vision (as in "clear light") or audition ("pure silence"), but also on the much more proximal and holistic experience of taste. However, in chapter 6, I pointed out that in our database of reports, the *state* of pure awareness is only rarely related to an abstract form of tasting or smelling via gustatory or olfactory metaphors—something that may be different for global *modes* of experience like nondual awareness (see chapter 27). The reasons for this remain unclear. On the other hand, as you may recall, we do find at least some examples of "deliciousness" and "sweetness" or "being satiated" in our reports. The general point is that taste will be an important future target for evidence-based and finer-grained phenomenological research projects designed to relate patterns in statistical items to modality-specific metaphors found within qualitative data such as open-ended verbal descriptions.

In chapter 8, we discussed the phenomenology of nonidentification and the contraction principle. In a passage reminiscent of central themes in classical Indian philosophy (e.g., the fifth song of the *Bhagavad Gita*) like inner renunciation and desireless action (*niṣkāmakarma* in Sanskrit), Meister Eckhart says that "one has to learn to remain (inwardly) unbound in the midst of external action" (*Man sol daz lernen, daz man ín den werken ledic sî*).[47] In his eleventh sermon, Johannes Tauler characterizes detachment (*abegescheidenheit*) as purity, as nakedness (in the sense of being liberated from everything worldly), as a form of freedom that is not spoiled by images, and as silence (*luterkeit, blosheit, unverbildete friheit und . . . swigen*).[48] Henry Suso (or Heinrich Seuse, 1295–1366), on the other hand, describes nonidentification as a state of "silent resting in which we have become detached from [the perceptual content delivered by] the outer senses" (*stille ruow und abgescheidenheit der ussren sinnen*).[49] Suso, a German Dominican friar and an important author in both Latin and Middle High German, was well known for defending Meister Eckhart's legacy after he was posthumously condemned for heresy in 1329. Suso speaks of a detached version of insight, a way of looking into the inside that is free of identification (*mit einem abgeschieden inblik*).[50] He also amusingly distinguishes between the two extremes of detached nakedness (*abgescheidner blosheit*) and degenerate amusement (*schedlich Kurzweil*).[51]

"Nothingness" (chapter 16) occurs in too many medieval locations to document. Perhaps most famously, in Sermon 42, Eckhart says of God that he is an above-floating being and an over-being nothingness (*Er ist ein vber swebende Wesen vnd ein vberwesenden nitheit*).[52] Eckhart's lasting appeal comes partly from his creativity: He created many new concepts that did not exist in this specific form before and were unknown in Christian philosophy. *Seelengrund* is one of them;[53] *Istikeit* is another example.[54] You may remember that the term "suchness" appeared back in chapter 9, when we first encountered Aldous Huxley and his literary descriptions of a mescaline-induced state of *seeing what is*. In the short quote I presented there, Huxley asked himself: "*Istigkeit*—wasn't that the word Meister Eckhart liked to use? 'Is-ness.'" *Istikeit* (as it was really spelled) relates to the phenomenology of pure being, the overarching topic of this chapter, which is perhaps the deepest layer of conscious experience. In a metaphysical reading, suchness is a quality of the pure, uncreated, featureless being as it is in and of itself, holding itself in existence by its own nature.[55] Sometimes suchness even co-occurs with nothingness because it can involve a process of diffluence, of melting away and dissolving into the nature of God. Suchness can give way to a timeless form of knowing "unborn is-ness" and "unnameable nothingness." This is how Meister Eckhart puts it:

> You should sink fully out of your youness and dissolve into his "hisness," and your "yours" and his "his" should become so completely one "mine" that with him you eternally understand his unborn "is-ness" and his unnameable "nothingness." (*Dv solt alzemal entsinken diner dînesheit vnd solt zer fliesen in sine sinesheit und sol din din und sin sin ein mîn werden als genzlich, daz dv mit ime verstandest ewiklich sin vngewordene istikeit vnd sin vngenannte nihtheit.*)[56]

Eckhart sometimes uses "is" as a noun, for example when he speaks of the experience of "becoming and being a single Is."[57] He also speaks of the *istikeit* of divine nature, to which God refers when he uses the pronoun "I," and this is-ness is also what God's true nature really is—it is he and only he who can be called an "Is" in this sense. God is more *istig* in all creatures than any creature is to itself, and God's intelligence is more "inward" in all things and more intimate to each than these things are to themselves (*sît daz got in allen dingen ist vernünfticliche und den dingen mê inne ist, dan diu dinc in selber sint*)[58]—a statement that may remind some of the Koran, where Allah says that, whatever thoughts the inner self of man may develop, Allah is always "closer to him than (his) jugular vein."[59] For Eckhart, the human soul and all things have *istikeit*—which is simply the presence of God. As God's ground and the *Seelengrund* are one and the same ground, it is the shared quality of *istikeit* that the mystic tries to bring into conscious experience.[60] Perhaps most interestingly, the experience of pure, nondual being (exemplified in Eckhart's metaphysics as *istikeit* or the *blôz lûter wesen*) is described as a groundless ground. Coming into the ground (*in den grunt ze komen*) is the deepest goal of everything that is, but it is impossible to find it outside of oneself. Coming into the ground can happen only by entering the innermost part of oneself in a state of pure humility (*in lûterer dêmueticheit*).[61] Groundlessness, according to Meister Eckhart, refers to an unfathomable reality and to God in his bottomless depth (*in sîner grûndlosen tieffi*).[62] Groundlessness also alludes to inconceivability, to the namelessness of the groundless God, and to the mysteriousness of the ways in which he moves. Eckhart's notion of the *Seelengrund* (*grunt der sêle*) in turn influenced many other mystics and spiritual philosophers.

Let us also briefly look at what others said about the groundless ground of the soul. Even before Eckhart, David von Augsburg (1200–1272) already spoke of the "groundless Godhood" (*gruntlosen gotheit*) and the "groundless fullness" of human bliss (*gruntlose volle*);[63] this may remind you of material in chapter 15, "Joy, Awe, Bliss, and Gratitude," and chapter 17, "Emptiness and Fullness."

The monk Johannes Tauler (ca. 1300–1361) was an important disciple of Meister Eckhart, a Roman Catholic priest, and a theologian who also belonged to the Dominican order. He wrote about the "hidden abyss" (*verborgen appetgrunde*), in which the divine abyss and the abyss of the human soul turn toward and call to each other. This hidden abyss is the purest part, the innermost and most hidden ground of the soul, and through the practice of concentration and calm collectedness, the human mind has to "sink out of itself" (*entsunken*)[64] to lose itself in God as a drop of water does in the ocean. We should cultivate our *Seelengrund* like a farmer, Tauler writes, who removes the weeds from his field so that at one point, "created nothingness can sink into uncreated nothingness."[65] Tauler points out that compassion can also have a quality of groundlessness, as well as suggesting that God's lovingness is infinite and deep, something that we can move into (*in die vertieffete grundelose erbarmherzigkeit Gottes*).[66]

For Suso, Eckhart's breakthrough into the unknowable is more of a forceful impact, a kind of collision with divine nothingness—and he too speaks of the "groundlessness" of the ground, the ground that expunges all distinctions.[67] For him, however, reaching the state of unification is a question not of lawful necessity but of grace. Suso speaks poetically of the groundless ground "that opens itself in all loving hearts, like a rising morning star" (*Got gruesse dich, ufgender . . . morgenstern, von dem grundelosen grunde aller minnenden herzen!*).[68] You may recall that in the previous chapter, we explored the idea of "melting into the phenomenal field." The phenomenological motif of "melting" was well known to European mystics, with both Tauler and Suso, for instance, speaking of sinking and melting into the uncreated mind of God and God being melted into the ground of one's own heart. Tauler speaks of sinking and melting into the uncreated spirit of God (*versinken und vermseltzen in dem ungeschaffenen geiste Gottes*),[69] and Suso tells us that "God has to be melted into the ground of my heart" (*Gott in den grund mins herzen gesmelzet werden muß*).[70]

A little later in this period, St. John of the Cross (1542–1591) was a Carmelite friar, a mystic, and a major figure of the Counter-Reformation in Spain. He called the ground of the soul *fondo del alma*, and interestingly, he even spoke of an "awakening" of God in the center of the soul and the ground of the soul.[71] He taught that human beings in whom a unification with God had not yet happened were usually unaware of his presence in their very souls.

This experience of the actual presence of God in the *Seelengrund* became a common theme, including for the Christian nun Marie de l'Incarnation (1599–1672), who described her spiritual experiences in terms of "being strongly drawn into the ground of her interior/inside" and as "being completely withdrawn into the ground of the soul." This ground was the center of the soul and the seat of God at the same time.[72]

This last example ends our brief look at what some other medieval Europeans, following Begine of Hadewijch and Meister Eckhart, said about the groundless ground of the soul. I think that even this small selection shows how many phenomenological details found in our own data—but also more general experiential features like the motifs of ego dissolution, nondual being, and the existence of a foundationless foundation of experience—had already been described by the mystics of the Middle Ages.

But what about today? Today, the notion of a "groundless ground" is frequently found in the Buddhist pop culture that has evolved in the West, typically with no knowledge of its historical roots. As I have learned from David Higgins, one of the world's leading experts in Tibetan Buddhism, if one actually searches the Tibetan canon (the massive collection of mostly Indic works translated into Tibetan), one finds no Indian precedents for the Tibetan terms *gzhi'i gzhi med* (groundless ground), *gnas med gnas* (foundationless foundation), or the like.[73] "Groundless ground" and related terms seem to be distinctively Tibetan concepts, possessing only distant roots in old philosophical speculations (in India, Tibet, and China) that there is a ground of experience beyond any of the metaphysical (ontic or epistemic) grounds established in the Buddhist philosophical schools. As I have also learned from Higgins, one of the prime examples of the groundless ground in Tibetan Buddhism is the Dzogchen (Great Perfection) tradition's idea of a "primordial ground" (*gdod ma'i gzhi*) or "originary ground" (*ye gzhi*) that is claimed to be beyond or beneath the "substratum consciousness" (*ālaya-vijñāna*) of the Mind Only or "representation-only" school (*vijñapti-mātra*). This latter concept was posited to account for states of delusion but was unable to account for states/qualities associated with spiritual awakening. Interestingly, in ancient Tibet, this groundless ground is not a metaphysical ground established by the usual sources of valid knowledge but rather is a matter of existential, personal self-realization (*so sor rang rigpa'*, "personally realized self-awareness"; or *so sor rang rigpa'i ye shes*, "personally realized primordial knowing").

As we have now confirmed multiple times, there is clearly a phenomenological dimension to the groundless ground (it can be viewed as a region in phenomenal state space), but at the same time, it defies any representation on the level of natural language. Perhaps future mathematical models of consciousness will be able to generate new solutions to this age-old problem.

As Higgins explains,

> "The attempt to understand the 'ground' inescapably comes up against the limits of thought and language. We are confronted with the quixotic prospect of naming what is unnameable and conceiving the nonconceptual, all in an attempt to

understand an abiding ground that is nonetheless groundless in the dual senses of being unceasing and impermanent."[74]

Phenomenologically, one major aspect of the groundless ground is that it is experienced neither as existent nor as nonexistent (more about this in chapter 28). In their paradoxical formulations, Meister Eckhart and eminent Tibetan scholar–practitioners like the Eighth Karmapa strongly converge in what can be seen (again in the words of David Higgins) as

> [. . .] an attempt to articulate an invariant continuum of being and awareness that is available to first-hand experience, but cannot be reduced to the oppositional categories of existence and non-existence and the extreme views of eternalism and nihilism based on these. [. . .] We might add that it is precisely because human experience is as heterogeneous and hierarchically stratified as it is that it remains radically underdetermined by what we make of it, lending itself to multiple descriptions without being definitely captured by any of them.[75]

We began our journey in the fourteenth century, with Begine of Hadewijch of the European Middle Ages, but as it turns out, the idea of groundless ground can be found as early as the twelfth century in Tibet, for example in the work of Zhang rinpoche.[76] The philosophical idea that the nature of mind is without ground or source (*gzhi med rtsa bral*) is an established topic of discussion and well attested in the earliest Dzogchen traditions[77] long before Hadewijch of Antwerp felt compelled to give testimony of the *grondelose diepte* ("groundless depth") that she experienced in another, distant part of the world. The Tibetan Buddhist notion of "one taste" or "one flavor" (*ro gcig*)[78] is another example of a geographically distant precursor to strikingly similar European ideas. *Ro gcig* encapsulates the idea of "tasting the one nature"—that is, experiencing or realizing all inner and outer phenomena (emptiness, bliss, nonduality, etc.) as being of a single nature. Both this and the closely related idea of having "one single taste (of one nature)" (*in ere naturen smake*)[79] are also found in Hadewijch's writings.

Whatever the historical and philosophical details, many meditators will immediately see the beauty in poetic and paradoxical philosophical concepts like *Seelengrund*, in characterizations of pure awareness as the "ground of the soul" and "groundless ground." I would also predict that for many meditators, they will make strong intuitive sense as well. Of course, the metaphysical and theological frameworks that evolved around these concepts in different centuries and different cultural contexts are extremely complex, often baroque. These frameworks may be Buddhist or they may be Christian; they may involve some sort of immortal self or not; there may be different concepts of God or none at all. Obviously, such theoretical systems have also emerged from different forms of practice. But they all seem to be getting at something similar through their use of metaphors of ground and its opposites. We should not be blinded by the varying conceptual surface—but in trying to see the phenomenal reality below it, we must also avoid falling into extremes.

"Perennialism" is the philosophical thesis that a common core can be identified in all mystical experiences across all cultures and traditions, in all periods, and in many social and religious contexts.[80] I am not a perennialist. Also, MPE itself is not a mystical experience, because MPE states are not mystical states—though some MPE *modes* may actually fall under one or other technical definition of "mystical experience." More interestingly, we can now ask the question of "MPE perennialism": Is there an essential, shared phenomenal character that unites all human beings who undergo minimal phenomenal experience? MPE perennialism could be the philosophical thesis proposing that, across all cultures and traditions, there is one single kind of phenomenal character that, though mostly unattended, exists in all conscious human beings, and perhaps even in other animals, like the Buddha nature that exists in all sentient beings. I am not an MPE perennialist either, but I think an evidence-based, bottom-up approach may open a new middle way that most of us can live with.[81]

You may recall some of the seven Eastern concepts related to the experience of pure consciousness that I listed as opening examples in the introduction: *dharmakāya*, *rigpa*, *sākṣin*, *samādhi*, *sat-chit-ananda*, *turīya*, and *ye shes*. I am certainly not saying that all of them refer to exactly the same category of experience, or that they simply "are" MPE. In the same vein, I am not saying that Western concepts like *ataraxia* or *Seelengrund* are coextensional in the sense of actually having the same referent; they are definitely not like the two terms "Morning Star" and "Evening Star," for example, which refer to the same physical object—namely, the planet Venus (a simple empirical fact that was unknown to humankind for many centuries). In humankind's history of contemplative practice, *many* such intimately related concepts have been coined—but they certainly do not all refer to exactly the same state of consciousness.

We must get used to the possibility that there may be no such thing as "sameness" here.[82] What would our criteria for identity be? Our search for a single conceptual essence may itself be a form of narrative self-deception, a craving for reification ultimately driven by our basic need for mortality denial (chapter 17). Perennialism is an emotionally attractive but empirically untestable ideology: How would one produce intersubjectively verifiable evidence for it?

To be sure, a certain kind of essentialism can be found in the ninety-two questions of our online survey and the statistical dimensions that emerge from them as our twelve factors: The software treats all identical answers to the same question as identical, as single data points—what else would it do? But here is a thought experiment: What would happen if (as a hypothetical) all the participants and expert texts provided reports that sounded strictly identical? We would have discovered a strong pattern of commonality on the level of verbal *reports*, but we still wouldn't know whether the phenomenological *referents* were actually identical, whether all of these identical responses actually referred to one and the same kind of conscious experience. It could always be possible that people have slightly different experiences but later describe them in identical ways. No two brains are exactly alike. But what if, extending the hypothetical, participants showed identical patterns on all objective markers? What if, as spiritual naturalists, we concluded that it is not the physical brain itself, but some more abstract neurocomputational signature—a certain high-level pattern in the flow of information—that unequivocally determines MPE? Wouldn't this be a form of evidence for perennialism? This scenario is not far-fetched because we might certainly discover the neurocomputational signature of pure awareness in the course of this century. But to accept *this* signature as the sole criterion for sameness would itself be a philosophical move. It would need an independent philosophical argument.

Spiritually as well as scientifically, taking the pure-awareness experience seriously also means dissolving any form of intellectual contraction or metaphysical reification. From a theoretical perspective, there exists a complex history of ideas, and in a scholarly sense, there is not much that will remain simple and intuitive here. Empirically, a future computational neuroscience of MPE will hopefully develop formal criteria for simplicity and sameness across time, but these criteria may not satisfy the intuitions that we have about our own experience. Nonetheless, it does seem that *in some way*, "groundless ground" and *Seelengrund* are picking out something important, something historically long-lasting, something that is directly related to the pure-awareness experience, and perhaps even something that may be largely innate and ultimately shared among all sentient creatures. This intuition must be taken seriously because it results from the fact that all these theories and concepts have what I would like to call a "phenomenological anchor."

A phenomenological anchor is not an essence. It is not a single and context-invariant experiential quality, a single intrinsic nature that can be extracted and easily described in folk-psychological terms.[83] Rather, an anchor is a region in phenomenal state space. For every embodied being, this region will be slightly different, and it will also change over each individual's lifetime. As I explained at the beginning of this book, the space of possible conscious experience has many dimensions and is characterized by a rich inner landscape. Yes, this landscape may always have a deepest point, and physically, it is locally realized in the human brain. This brain, however, is not only embodied but also enculturated—and therefore the use and definition of terms like "pure consciousness" necessarily change over time as linguistic communities and their corresponding cognitive niches keep developing.

Nevertheless, the anchor for pure consciousness itself certainly exists; there is a *prototypical* region in the space of experience, a cluster of phenomenal properties that "hang together" with high probability. "Nondual being" and "groundless ground" are examples of such properties. MPE is the core of this prototypical region, and it anchors the way that we speak about our contemplative experiences—perhaps, unbeknownst to most of us, even the way that we speak about *all* our experiences. As David Higgins would put it, commonalities in testimonial accounts of pure awareness do suggest that certain modes of inquiry (as reflected in, and constituted by, our MPE reports) open commensurable modes of experiencing and articulation. What all the ideas and concepts mentioned in this discussion—*Seelengrund*, the groundless ground, or the seven Eastern examples from the introduction that I just relisted—have in common is that their semantics are causally rooted in a neglected but specific form of nonconceptual experience, a prototypical region of phenomenal state space. This experience was what came first, and all the elaborate conceptual systems—as well as all those poetic and paradoxical philosophical ideas like the "groundless ground of the soul"—are ultimately anchored by the phenomenal character of MPE.

What came first was the epistemic practice of silent meditation. Later, human beings tried to make sense of and convey their own experience on a conceptual level; they tried to describe MPE phenomenology to each other. In doing so, they created new words. Ambiguities arose, they had to be settled, and gradually the original project degenerated into theology and philosophical metaphysics, eventually leading to the kinds of philological nitpicking, dogmatism, and scholastic hair-splitting that are still with us today. This is how the problem of theory contamination arose and why a fresh bottom-up approach is needed. Now it is the phenomenology itself that really counts—and perhaps we even need a new generation of scholar–practitioners not blinded by ideology. "Phenomenology" is a very European project, but today, a genuinely transcultural approach is needed. Might a new form of computational phenomenology[84] provide us with much finer-grained conceptual instruments that help us transcend old limitations? The challenge is daunting, but somehow we must try to get as close as possible to the experience itself.

27 Nondual Awareness: Insight

Without an observer. Nobody on the cushion meditating. Everything on the cushion meditating. [#635]

I perceived subject and object as the same substance, facing each other like the two inner surfaces of a folded sheet of paper. As if subject and object were created by folding the fundamental substance and the observer were placed arbitrarily on one of the two sides. [#2916]

There was a permeability; self-image and infinity somehow fell into one. [#3024]

Awareness is a nonconceptual way of knowing one's own inner model of reality, a form of knowing that itself often goes unnoticed. Phenomenologically, this knowing is direct and immediate because the experiential character of awareness itself is almost transparent (we'll think more about this in chapter 28; see also figure 34.1 in chapter 34). If you will, the awareness itself is not *part* of our inner model of reality—and meditators are people who change that. Often, the inner model includes an explicit representation of a subject and an object component, for example a "knowing self" that directs its attention toward some perceptual object, or a thinker of thoughts who then categorizes perceptions, actively using thought and memory. This knowing self is precisely what, in chapter 25, I called the "epistemic agent model." Whenever the epistemic agent model is transparent, the result is an ego: an apparently immediate and direct experience of a knowing self with which we identify. What the phenomenology of meditation practice shows is that awareness can also occur *without* any explicit representation of subject and object. To use our brand-new conceptual tool, awareness can exist without an epistemic agent model.

The experience of directly knowing inner reality does not have to be contracted into a self and a first-person perspective because it can also occur in a "nondual" way. For example, the "meditating self" (which may have been actively trying to control the focus of attention beforehand) can sometimes be absent. Perceptual objects can then take on a new quality, not as distinct entities outside the space of egoic self-awareness, but as parts of what the German poet Rainer Maria Rilke (1875–1926) called *Weltinnenraum* ("inner world-space"; more on this in chapter 28). Now, as he describes in a *Letter to Lou Andreas-Salomé*, pure consciousness pervades and integrates everything:

> He commemorated the hour in that other southern garden (Capri), when a bird call was there, outside and within him in unison, and *it didn't break, so to speak, at the boundary of the body, but brought both together into an uninterrupted space in which only a single spot of purest, deepest consciousness remained.* (emphasis mine)

We also find an element of bodiless body-experience (chapter 24): Rilke goes on to tell us that during this episode—it was around February 1, 1913—he "closed his eyes so as not to be misled in such a magnanimous experience by the contour of his body, and the infinite passed into him from all sides so intimately that he might believe he could feel the gentle coming to rest of the stars that had meanwhile entered his chest."[1] Nondual awareness does not merely lack subject/object structure; in addition, it is often described as a spacious and unbounded form of awareness for which conceptual distinctions like "inner" versus "outer" or "real" versus "illusory" do not make sense. We will look at all three of these aspects separately, postponing an examination of the second (inner/outer) and third (real/illusory) until the next chapter.

Let us first consider nondual awareness in the simple and canonical sense of a global state or, better, a *mode* of consciousness lacking the duality of subject and object. This mode is characterized by an experience of pure knowing (without a localized knowing self) plus a deepened sense of global connectedness resulting from the lack of independent, separately existing objects that could be juxtaposed against this self. How, if at all, can this specific mode of conscious experience be described in words? After all, the vast majority of languages on the planet have either subject-verb-object or subject-object-verb structure built right into them.

Our participants came up with many new and beautiful metaphors to convey how it feels to enter nondual awareness. One said that conscious experience is "like a stream. And I keep trying to catch it—jump in!" (#2515). Another described the transition to the global quality of nondual knowing (during walking meditation) like this: "very slow movement, breath completely calm, the gaze lowered, green grass . . . the question remains: Does the grass see me or I the grass?" (#2444). And here is a third attempt to convey the specific phenomenal character of nonduality: "The experience of perception takes the form of oneness between perceiving (subject) and perceived (object).

Perhaps a comparable impression to when the Vulcan Spock says in the series Star Trek: 'My mind to your mind, your mind to my mind!'—Subject and object become one in the process of perception" (#2550).

One highly interesting result of our study is the fact that experienced meditators, when asked to choose and describe a good example of "pure awareness," do *not* necessarily report only full-absorption episodes in which there was no experiential content except awareness itself. On the contrary, they often provide us with an enriched phenomenological definition of what "purity" means: "Pure awareness" is the explicit experience of awareness per se, but it is not necessarily devoid of *all* coemerging content. Pure awareness is, for many people, awareness lacking an explicitly represented subject/object dichotomy. On this reading, purity is nonduality. When summarizing some of our provisional results in chapter 34, I will label this reading "P5."

A second interesting finding is that nondual awareness often occurs in regular meditators, but usually spontaneously and unexpectedly, outside formal practice (chapters 32 and 33). It is almost as if the implicit background assumption that one is *not* practicing right now plays an important causal role in its appearance. It may be something that cannot be actively pursued; perhaps the quality of humility that, for example, Krishnamurti[2] spoke of, or the *lûtere dêmueticheit* mentioned by Eckhart,[3] is one of its necessary precursors. Nondual awareness is something that cannot be fabricated, something that defies any merely technical approach to contemplative practice—it is beyond all mental techniques because it is something that cannot be "achieved." Let us take a look at this idea.

To begin with, nondual awareness is frequently described as becoming one with the processes of phenomenal experience itself:

> 79 [. . .] I lost all sense of being the agent of action and perception, instead feeling myself as identical with these things. Rather than feeling myself as THAT TO WHICH these experiences were given, I felt myself identical with the ACTIVITY of experiencing itself. This experience involved visual perceptions of my own body and my bodily activity. I was looking at my own body and experienced myself AS the seeing of the body—as the mode through which body was given. The experience was one of effortlessness, and without any noticeable sense of desire. There was no longer any feeling of needing to get someplace or something.

> 1612 [. . .] The sense of living in a world that I experienced (this is still my default mode of being) fell completely away and instead there was only the experience itself. The distinction between self and world no longer existed. The contents

of the experience were exactly the same, but the perspective of them was so different that the change felt monumental. The world I was experiencing no longer existed independently, because I had become the unfolding of that experience. The previous "I" as experiencer, chooser, thinker did not exist. Instead there was experience itself. There was a visual center to the experience, but only because that's where light met the eyes. The center was no longer meaningful in any way. Anything that wasn't the unfolding contents of awareness no longer had any meaning. This was a very positive experience. It didn't really map onto any previous experiences I'd had so I couldn't really call it joy or pleasure or happiness in the traditional sense. It was, however, undeniably positive, both light (unburdened) and wondrous at the same time. It took no effort to maintain. From inside the experience, there was nothing that existed to provide any effort, so of course it did not need any. It lasted many hours, with the (felt) profundity fading over time. During the fading period it became very amusing/pleasing that the contents of the experience (the world) moved. There was a joyful surprise every time a nearby object moved in relation to an object that was far away, even though this was constantly happening.

1832 [. . .] Everything that I thought was a distraction *is* the present moment. For the first time, I feel that I understand why the Buddha speaks of the body as an illusion. I don't experience the body as a flowing river of sensations, but I experience *experience* as a flowing river of experience, more or less broad—encompassing anything from a vast swath to a hyperfocused pinprick of present arising. I'm not following the river of attention, I *am* the river of attention. There are no distractions because everything I'm aware of is awareness itself. [. . .]

Episodes of nondual awareness do sometimes appear during formal meditation, and in these cases, they often begin when the meditator redirects attention to the very process of trying to meditate and then lets go and simply abides in the resulting state. This is a classic technique that in the West has sometimes been described as the "you-turn."[4] Let us return to report #1703, already presented in chapter 8:

1703 [. . .] It starts when I look for "who" is doing the seeing. Or "who" is doing the feeling? [. . .] There is certainly physical sensation happening. That is very clear. But who is feeling the feeling? Or is there just feeling? Or that same line of thought, but with hearing or thinking. As soon as "I" look for the thinker of thoughts, it hits me: There is simply thinking. But no one or thing is doing it. It's just happening all on its own. And it's like there is nothing behind the experience. There is just sort of floating. Just like a big smear of sensations all suspended somewhere. And then as quickly as it comes, it goes . . . in the sense of thoughts of how "I" just had a cool experience and of ways to get it back. I start clinging. I start wanting it to come back and stay for longer . . . When I'm having the experience, certain bits of language seem completely nonsensical. For example: "I hear a sound." It seems that "There is hearing going on" feels so much more of a correct description. The same goes for, "I am thinking a thought" → "There is thinking going on."

2289 [. . .] I try to *RE*direct attention onto the self paying attention. I'm not finding a self, I enter a nondual state. It is there I try to remain. As that state stabilizes (usually after several dozen repetitions of turning attention onto itself) my bodily sensations and my sense of possessing with them dwindle and I am left in a place of pure awareness. It is as though the objects and people around me are no longer separate. And my body is just one more field of sensation with awareness in it. Is depersonalized. Concepts and thinking drop away and the state is restful.

3153 I tried to look for "who" was seeing the sights in front of me. I didn't find it. I didn't experience the "who," I only experienced awareness of the sights and scenes.

One obvious phenomenological prediction is that the phenomenology of ego dissolution (chapter 25) and nondual awareness should be intimately related. And as it turns out, the process of becoming one with experiencing itself often is described—with a striking degree of convergence among respondents—as a process of merging into something selfless:

1716 [. . .] At the end of the three hours or so, I noticed that the tree in front of me had a different "quality" in it, especially in its movement. Then everything took on this different "quality." There was suddenly no distinction between the physical tree and my mind, and when the branches of the tree were moved by the wind, my mind moved with it. So it was with the rest of the world around me—as things moved my mind followed them, and when things rested my mind stayed still. The same was true of sounds: Birds chirping registered in my mind, and when they stopped there was no mind for bird sounds. The only thing that existed for me was this mental activity, but the mental activity was entirely dictated by externalities. There was no "I wonder if . . ." or "I should probably . . . ," and there was hardly any "I"—there was understanding of the utility of "I" in describing a viewpoint for conversational purposes.

However, during this experience it is enough to say, "there was awareness," and this awareness was not unique. [. . .]

1935 [. . .] there is no sense of self or center to the awareness. There are no boundaries, and it's quite difficult to tell thoughts from any other sensation, when I meditate. I would say there is a sense of unity, the thoughts and sensations just come and go, and there isn't anything that "I" have to do to produce them or make them change. I, or awareness, is simply aware AS them or itself, not OF them.

2780 [. . .] I waited at the reception, completely relaxed and not practicing or thinking anything special. Suddenly I was immersed in a state of nonduality. My experience was incredibly strong and alive. I was able to think and control my actions completely normally. But it was as if everything around me had a very strong "radiance." Everything was awareness. It wasn't that I, for example, saw a chair, but awareness in the form of a chair "happened." Inside and outside were not separated at all, even though I could distinguish everything well. [. . .] My mind was incredibly alive and yet deep and calm. [. . .] What is really fascinating is that the mind is so clear, smart, can reflect incredibly well, but categories like inside and outside suddenly don't work anymore, don't make any sense in this kind of experience.

2816 I worked in the kitchen and suddenly I smelled the fragrance of the tea that I was preparing. There was no longer the difference between me, the tea, the smell. There was only one endless fulfilling. [. . .]

3096 I only experienced the scenery in front of me. The normal thing for me is to experience sights and scenes while feeling like I am experiencing them from where "I" stood. But in that moment, I didn't experience "me" looking at the tree. I only experienced seeing the tree.

3126 I only experienced seeing; I didn't experience "myself" as the one seeing.

3491 [. . .] Usually "I" observe the "breath" but there was only "the breath." There was no self but just the breath.

The transition in and out of episodes of nondual awareness is interesting. While the beginning of nondual states is often unexpected and often accompanied by a strong sense of existential relief and relaxation, the ending can be experienced as a recurrence of fragmentation, or as a re-creation of a specific sort of inner tension that had never previously been seen clearly. It can also happen during sleep/wake transitions:

1647 [. . .] Very suddenly the constant pain of anxiety that I have lived with for many years was lifted and I existed as both myself and my surroundings. The suddenness of it felt very intentional . . . like something outside myself tapped me on the shoulder and made it happen . . . and approximately an hour later tapped me on the shoulder and took it away. The "giving" and the "taking" felt very intentional . . . like something intelligent was providing me with a glimpse of something important. I have not experienced anything like this before or since, and did not (and do not currently) have any strong beliefs about such things. I was and remain fairly agnostic about intelligent spiritual entities. The experience itself was simultaneously extremely profound while remaining incredibly mundane. I was flooded with a powerfully vivid memory of playing in the forest as a young child . . . it felt no more special than the ordinary existence of every child pre-trauma. After the extremely vivid memory I "returned" to the forest and existed in this hyperaware state for an hour or so. Adjectives I would use to describe the state: vivid, light, at peace, sharp, beautiful, sublime, awesome, natural, effortless, blissful (but more as an absence of pain than a euphoric sensation), detailed. Even though the experience, in this specific case, felt very much like an on/off light switch . . . it also felt like something that was achievable at the end of a spectrum. Like the more I can quiet my physical and mental stress and anxiety I can uncover more of this baseline reality of experience. It did not feel like something I could strive to achieve through hard work . . . but more like something I could relax or ease into.

2417 [. . .] when transitioning from sleep to waking. I had the impression that everything was starkly, profoundly ordinary, just as it was. Very shortly upon registering this, and as I was transitioning out of it, a thought occurred, "Oh my God!, the tension I call Jeff was gone!" And in that moment, a tension came over me as my normal mode of feeling myself and the world was reestablished.

When investigating "bodiless body-experience" in chapter 24, we encountered states of pure awareness that can be described only by seemingly self-contradictory statements like "I am there and not there at the same time." Many of our contemplative practitioners report such "paradoxical" states, in which it seems that a self is present and absent at the same time. Again, please note that there is a direct parallel here between body dissolution and ego dissolution, as investigated in chapter 25. For example, the phenomenology of "bodiless body-experience" can now often be described more clearly—namely, as a special case of "nonegoic self-awareness." We will return to this phenomenological discovery in chapter 30. Meanwhile, here are three examples of paradoxical juxtapositions:

1754 These moments feel like though I'm still there I'm also not because there is only a field of relations that comprises everything.

2679 It is like realizing that one is in a state of nonphysical, nontemporal presence, but eternally present and never having been at the same time. It cannot be said that it is or that it is not, it is the deepest of contradictions that asserts itself while denying itself. It is an unfathomable miracle. It is touching oneself as nothing and everything at the same time. It is the beginning of the world and its end.

3024 [. . .] "I" sat outside and looked at my surroundings. Suddenly there was only looking, as a verb without subject. I perceived the breeze of the wind, the rustling of leaves. But these impressions were not limited to the place of their origin—the present wind and the trees around me. It was more as if the stream of LIFE were present as a whole in this one moment, in this particular place and everywhere at once. And although "I" was somehow present as the person that I am, I was somehow also not. There was a permeability; self-image and infinity somehow fell into one.

Perhaps one could sometimes say that the self-model becomes fully integrated with the model of the unbounded space of epistemic openness out of which it originally emerged. However, our participants' attempts to describe the all-encompassing but ineffable phenomenal quality of "nonduality" do differ in significant ways. Some don't seem to have a radical disruption of self-consciousness,[5] but for others, full-blown ego dissolution is central to the experience. For this second group, the disappearance of the sense of self is complete and there is a dominant phenomenology either of "existence as such" (i.e., of pure, nondual being; see chapter 26) or of "pure nonegoically self-aware knowing" (see chapter 30). Just as it was for Rilke in 1913, the song of a bird may now be experienced in a completely different way (see report #2908 here and #3287 later for similar examples):

192 Note: the following experience is difficult to express in words, due to the fact that it involved a complete dissolution of the self. As such, when I use the singular pronoun "I," I feel that it inadequately captures the unique (to me) nature of the phenomenal experience. [. . .] At the time I was meditating, there was a bird outside my window (at home) that was producing irregular sounds. I consciously chose this sound during the earlier stages of the practice. While I do not know when in the practice the particular experience occurred, I distinctly recall a very unique experience that has not since occurred in my meditation practice. The experience was a pure awareness of the sound of the

bird, such that I was no longer aware of being aware (i.e., as a meditating, perceiving subject). Although it involved an auditory experience, there was no judgment of the experience as "auditory" in nature. "I" ceased to be present in the experience, and the only content was the birdsong itself. There was no distinction between subject (I) and object (birdsong). The birdsong did not belong to the bird, and I was not a subject being aware of the object. Instead, I believe there was simply an experience of what was referred to in the questionnaire as "pure being." Although I think the duration of the experience was brief (maybe a minute or so), the experience was stable enough for me to recall it when I returned to a more reflective state. When I reflect on this experience, I often feel a sense of positivity and calmness, and it is hard to separate this affectivity from the experience itself. I believe that the experience was infused with a sense of calmness, but I think describing it in any valenced terms (i.e., positive or negative) would be inappropriate. [. . .]

2908 As if I am completely dissolving, all physical boundaries blur, there is no separation between outside and inside. The birds, which I previously perceived as separate on the outside, chirp in "me." Everything is connected to everything, no "I," no body, pure being. A direct, real experience without filter. The real thing, just like waking up to real life. As if everything else I know is just a dream, an illusion, behind a veil.

3048 My experience happened [. . .] with the question: Where is the "I," who sees this I? There was a recognition that this "I" does not exist; I exist, but not as an individual. I was there, it was not a physical me. This "I" is much bigger than this body and I am contained in this bigger thing. The experience was overpowering and natural at the same time. It took place in the waking state. I could see the people in the room; although I had known them for a long time, I saw them for the first time. Tears ran down my face. This was not connected with any feeling or emotion, it simply was. We were one. There were bodies, these were unimportant. It was as if we were parts of an image that could see each other. As I search for words to describe this, I am back in this space—in this moment. At the same time I realize how familiar it is to me; as soon as I turn my attention inward, this experience is there again. Spaceless, timeless, clear, awake, silent, and alive at the same time.

3174 [. . .] an intensive experience of the present—so to speak, into its furthest corners of space and time—but as if the I had been "taken out." At the same time, there was a feeling of unity of everything that was perceived as present ("unity with the world"); this is not surprising insofar as there is no longer any difference or opposition between subject and object, past, present, and future. It is interesting that in this state "nobody" existed anymore. In this respect, it was a state of complete solitude—but without even the most rudimentary addition of any loneliness. On the contrary—it was an intensive coexistence of everything.

2760 [. . .] it is as if I am no longer there. Great happiness flows through me, at the same time a feeling of instability, a feeling of no longer being there. Exactly then we all get up for *kinhin* (walking) and I feel stability through the group, pad literally around in a circle in the group, great happiness, wholeness, feel my body only slightly, total acceptance of everything that is. Every cough, every sneeze in the room, everything is totally accepted and my sense of self is gone, no feeling of space and time and yet I am walking around in a circle. It is difficult to portray. [. . .]

Empirical evidence for conscious experience in the absence of a first-person experience, as well as data that demonstrate the entirely selfless phenomenology of "existence as such" and of "pure nonegoically self-aware knowing," together constitute important bottom-up constraints for philosophy of mind, constraints that have been ignored for too long (see the subjectivity argument in chapter 34). Another philosophically relevant aspect is the phenomenal character of *fundamentality* (in this new context, you may want to recall our investigation of the "groundless ground" in chapter 26). Nondual awareness is sometimes described as the basis of all knowing: It appears as that which implicitly grounds all other forms of conscious knowing that occur in more complex and more fragmented states. What we describe as pure awareness would thus be something that can *enfold itself* into the usual form of dual knowing, into the conscious experience of knowing the world from an individual first-person perspective, involving agency, perceptual objects, multimodal scenes, and a representation of past, present, and future. Phenomenologically, the qualitative character of awareness itself is neither subjective nor objective, but it seems as if it can often contract into a special form, which we later call our own subjective perspective. On this account, nondual awareness would be a process in which what was previously *enfolded* becomes *unfolded*, in which the contraction principle is suspended. We'll think more about this in the theory section later in this chapter. First, let us look at some reports in which the character of fundamentality becomes explicit:

987 [. . .] It felt like being at the ground of all cognizance. It filled the whole body, in the field everything was slower, easier, and clear. The mood was relaxed, positive, and almost without desire. Thoughts and feelings surfaced from

time to time, but less often than usual, and they came slower, but disappeared out of consciousness again more quickly. The difference between subject and object dissolved. My body, awareness, and objects were one.

2916 I perceived subject and object as the same substance, facing each other like the two inner surfaces of a folded sheet of paper. As if subject and object were created by folding the fundamental substance and the observer were placed arbitrarily on one of the two sides. [. . .]

2936 [. . .] and then there was a shift in perception: I could no longer locate the energy/consciousness/awareness within me, it no longer radiated from me, instead it radiated toward me, from the trees, from the earth, from the sky . . . —so I thought at first—but then I noticed: It is everywhere.

Phenomenologically, our reports show that it is possible to become nondual awareness itself—to identify with it. Often, this process of identification goes along with an experience of insight, understanding, or recognition. Technically, this means that the global phenomenal quality of nondual awareness can now function as the practitioner's new unit of identification. Before turning to the new conceptual instrument of a "nonegoic unit of identification" (which will be introduced in more detail in chapter 29), let us look at one last selection of twelve reports. I think that they will give you an excellent final impression of what nondual awareness really is:

1913 [. . .] I experience a widening of my senses to the point that [I] seem to merge into, not one sense exactly, but it feels to me to be pure awareness.

2213 [. . .] sometimes I get very brief moments of pure awareness. Where the idea of self dissolves and I just become one with the world. Everything is recognized as being consciousness.

205 [. . .] on the meditation cushion I usually have the feeling of looking AT awareness. [. . .] So far I once had the feeling, during a meditation that included sounds as well as movements and visual impressions, of looking OUT of awareness.

919 [. . .] During the waking state the awareness that my environment, e.g., objects, are part of my self. The feeling of not being separated.

1662 [. . .] The sensations also included that of a "self" being present, but not the agent of or contributor in the experience—it was just part of the field of awareness without being a subject or object. "Unity" would be the best word to describe this experience, with the caveat that this would imply the presence of something in addition to awareness, which wasn't the case—awareness was all there was.

2301 [. . .] I had a moment of being a walking consciousness but without really being the person who walks or the consciousness itself, it was not like many moments before where I had been more or less observing the awareness. It didn't feel like I was aware of anything, it felt more like everything just was as it is. My consciousness experience was more the moment where everything kind of appears. It felt peaceful and true and yet I don't really know how I felt. At some point I became again the observer and therefore the center of my experience, I felt very aware of awareness but not being awareness itself like before.

2619 [. . .] It's no use really trying to describe the Big Awareness Event. It's always there and also much bigger than words. But I have to try anyway, for the sake of the I that writes, which is the narrative self with words. I liked the Big Awareness so much that I wanted to keep it after I got off the pillow. I sort of decided I would try. And I did! I was walking around the house to get my shoes and pick up some fallen plums, and everything just kind of looked from a different perspective. The same but not the same. There was no sense of uncanniness or pettiness. Just as though I was both looking from far away, from outside the body, but also completely here and aware of every detail in the normal way or more. [. . .]

3072 [. . .] It was an experience of oneness with everything. I felt an unusual calm and peace, and at that moment there was no desire even to remain in that state. It was a moment of absolute lucidity and understanding. [. . .] There are few words to describe the experience itself, but the summary would be two words: absolute understanding. [. . .]

3152 The body became more and more silent, it was there and it was also not. The eyes opened. They saw and they didn't see. The space was the same inside and outside, wide and deep. The events around me were outside and at the same time somehow a part of me. There was no impulse toward any kind of activity. Every now and then a thought popped up and then blew away. But it was not my thought either. [. . .] At some point, hours later, the state dissolved. I came back to "normal" daytime consciousness, very deeply rested, peaceful, and silent for a long time after. I then went back into action very carefully. [. . .]

3287 [. . .] One afternoon, sitting under a blooming tulip tree, I suddenly felt the weight of my body like I had never felt it before. And instantly there was a sense of total awareness. I could no longer distinguish between my body and the surroundings. There was no sense of a center of sensation or thought. A blackbird singing nearby was singing through "me." There was a sense of totality, of my not being there at all, of there being no "me," no separate entity, for that border had been dissolved. Although I had my eyes closed, I had a feeling of seeing with perfect clarity. My being was whole like perhaps it had never been, and this was the greatest joy imaginable. It was what I would call the bliss of nonduality.

3595 [. . .] that consciousness had made a change of perspective, out of the focus on I-consciousness toward the awareness of the field to which this I-consciousness is connected. In this field the atoms were dancing, all solidity of the body was dissolved into the subtlest vibration. And if someone had wanted to sell me this experience as eternity or the experience of God, I would certainly not have hesitated to snap it up. [. . .]

2286 [. . .] Once my body was very relaxed, I had a feeling I often experience when doing body scan of no longer feeling my body or my face or anything "tying" my mind. Then it was like my mind "expanded" and I was no longer the awareness within, but I was awareness everywhere. I didn't feel like I had any physical shape, I was everything in the universe, there was even no longer a sense of self or an "I" experiencing it because I was no longer the person who had started the meditation. I was one with everything and everyone. It was a great, even overwhelming feeling of everything being interconnected in a whole, and it came with a great sense of bliss, peace, and harmony. It felt warm, but not physically warm, more like feeling extremely safe or "hugged," while simultaneously feeling like I was the totality and I was completely safe there. It was like an abundance of loving kindness in a space that stretched out in eternity. And I was one with that love. [. . .] It was a sense of eternity and I understood that what I truly am in essence is what I experienced. The latter is really hard to describe. But I felt that I was just as much my neighbor and everyone else as I was myself. When I realized I was having this profound experience, I quickly returned to myself, and I remember thinking that I didn't want the experience or "feeling" or "understanding" to stop, so I didn't want to "wake up." Of course, that was my ego coming in right there, bringing me back to being the human that I am, and the experience subsided. However, the feeling and understanding I experienced continues to be with me, and it touched me deeply. It is one of the most blissful and greatest experiences I've had, and in seeking to describe it here, my body is actually physically shaking right now, while I'm reflecting on this experience. [. . .]

What Is Nonduality?

Some simple-minded people believe that they are meant to see God as if he were standing there and they here. This is not so. God and I are one.
—Meister Eckhart (1260–1328), Sermon *Iusti vivent in aeternum*

[. . .] but we must remember that no dualism of being represented and representing resides in the experience *per se*. In its pure state, or when isolated, there is no self-splitting of it into consciousness and what the consciousness is "of." [. . .] The instant field of the present is at all times what I call the "pure" experience. It is only virtually or potentially either subject or object as yet. For the time being, it is plain, unqualified actuality or existence, a simple *that*.
—William James (1842–1910), "Does consciousness exist?"

It seems as if consciousness as an inner activity were rather a postulate than a sensibly given fact, the postulate, namely, of a knower as correlative to all this known; and as if "sciousness" might be a better word by which to describe it.
—William James, *Psychology: The Briefer Course* (1892)

William James was an influential American philosopher, a radical empiricist, and a pragmatist. He is perhaps best known as the "father of American psychology" and for one of his most important books, *The Varieties of Religious Experience*. It speaks to his greatness as a Western thinker that he saw the philosophical significance of nondual awareness—or "sciousness"—so early. Using our new conceptual tools, we can now say clearly what this is: Nondual awareness is conscious experience in which the distinction between subject and object has been dissolved because what we termed the "epistemic agent model" has disappeared. Today, a typical definition of nondual awareness would refer to "the experience of pure (or empty) consciousness and phenomena at the same time."[6] But often, an epistemological and strongly metaphysical interpretation of the experience itself is offered in the same step: "It is the realization of one's own nature as an unbounded expanse of subtle consciousness, pervading one's internal and external experience as a unity."[7]

The common metaphysical twist is perhaps not surprising when we recall that nonduality is deeply rooted in the complexities of ancient Indian philosophy and practice—not only in Advaita Vedanta and Kashmir Shaivism, but also in early Buddhism. As John Dunne puts it:

> Notions of non-duality (Sanskrit., *advaya*) occur early in Indian Buddhism. [. . .] In speaking of Mahāmudrā as a non-dual style of practice, however, I am referring specifically to a form of non-duality that finds its first expression no earlier than

the third century (C.E.), and that undergoes further development around the seventh century. This form of non-duality is concerned specifically with the duality of knowing subject vs. known object (*grāhyagrāhakadvaya*). From an historical perspective, two developments within Indian Buddhism allow this style of practice to develop. First, Yogācāra philosophy (starting around the second or third century) maintains that ignorance (*avidyā*) occurs in its subtlest form as the seemingly real appearance of an ultimately false distinction between object and subject in experience. In other words, for Yogācāra thought, ignorance in its subtlest form manifests as the sense that there is a subjectivity that stands distinct and separate from the objects it apprehends. Since one central goal of all Mahāyāna practice is to eliminate ignorance by experiencing reality as it truly is (*yathābhātadarsána*), for Yogācāra thinkers a truly liberative meditative state must not be caught in the false distinction between subject and object. In other words, the state must be non-dual, in that the experience is not structured by the duality of object and subject. It is, instead, "non-dual wisdom" (Skt., *advayajāna*).[8]

When I first began analyzing what Eastern and Western philosophical traditions say about pure consciousness, I quickly saw that "low complexity" would be key to any working definition (PC2;[9] see chapters 2 and 3). The phenomenal character of pure consciousness itself is so simple that it is often compared to pure nothingness or associated with mere absence (chapter 16). It is homogeneous (chapter 6) and lacks internal structure (chapter 17). In humans, one of the most fundamental structural properties of conscious experience—arguably *the* most fundamental—is the subject/object relation. There is a knowing self directed at the world, at potential action goals, and sometimes even at itself. Philosophical concepts like "first-person perspective" or "epistemic agent model" are attempts to describe certain aspects of this fundamental relation more clearly.

Importantly, subject/object structure is *not merely* some sort of illusion. It is a way of internally modeling reality that has proved to be extremely efficient and successful for biological organisms like ourselves. It is, if you will, a functionally adequate form of self-deception. There is an evolutionary reason for this fact, which is often overlooked. We navigate the world under what I have called the "single-embodiment constraint":[10] All our sensors and effectors and all of our conscious information processing are pulled together at one single location in space; they are part of a single body possessing eyes and ears, legs and arms, and a central nervous system. This body has a statistical boundary (a "Markov blanket")[11] that it tries to sustain; the body defends and reconstitutes precisely this boundary for as long as it survives. On a causal level, there really *is* an inside and an outside—and for our biological ancestors, it was important to know this fact, if only in a nonconceptual, procedural sense. What we now call "consciousness" and the "model of the self" are inner aspects of this life process, like nested Russian dolls creating their own boundaries within the physical body. The single-embodiment constraint also turns our behavioral space into a *centered* space: Because we have only one body, our inner model of the current environment—of the space in which we can perceive objects, act, and interact with our fellow human beings—is anchored on the body. The body that is also felt from the inside creates the natural origin of our perspective onto the world, and it actually structures perceptual space.[12]

The two main causal factors that lead to the "contraction" of awareness into a fictitious self are that (1) beings like us control only one body, and (2) beings like us need to socially interact with each other (e.g., by competing or cooperating, by speaking or trying to understand each other's minds). Single-embodiment, selfhood, subject/object structure, and social relationships are extremely conspicuous and tenacious features of our conscious model of reality, but they are also *contingent* features. They are not logically necessary. It is easy to conceive of conscious intelligence without the single-embodiment constraint.

Here is another thought experiment for you, to help you see the optional nature of single-embodiment more clearly. A future conscious artificial intelligence (AI) could easily control hundreds of robots at the same time while selflessly seeing though their eyes and feeling their artificial bodies from the inside; it could administrate large numbers of localized robot brains in parallel, all of them physically embodied at different locations on the surface of our planet or even in outer space. In this way, our future AI could have a conscious model of reality, generated by perceiving and actively knowing the world through multiple perspectives and embodiments created by many epistemic agents at the same time. We can even imagine every single robot agent as having a *local* model of itself as an epistemic agent. The AI itself would have an extremely complex form of conscious experience—but it could also create a maximally simple, all-pervading signature of knowing that enveloped and permeated everything else in a nondual way. It could administrate and understand its own process of multiagent knowledge acquisition in real time, but without itself having an egoic epistemic agent model. This would give our AI "nondual wisdom," a way of perceiving reality in a new format—one that is uncontracted, lacking the distinction between subject and object that is typical of biological systems. We can also imagine our gigantic AI as being sui generis, the only one of its kind, a unique conscious entity without the need to communicate with any other conscious entity in the universe.

Among other things, the nondual AI thought experiment shows that the phenomenology of singular embodiment is not a necessary feature of conscious experience. It also raises some fascinating philosophical questions: Are biologically evolved creatures like all of *us* perhaps local, epistemic agents that are not yet fully aware of the fact that something much bigger is trying to use them as a window onto reality? Is evolution itself a computationally massive process of "multiagent knowledge acquisition," now slowly awakening to itself via agent-based self-modeling? Is there a deeper physical process that tries to look *through* us?

Before returning to nondual awareness, let us take a second look at the phenomenology of nondual being that was discussed in chapter 26. What can we say about the phenomenal character of simply *being*, the experiential quality of existence itself? If you look at the statistical distribution of questionnaire items in the analysis sample shown in figure 1.1 in chapter 1, you immediately see that "simply being" (item #32), "unity" (item #15), and "pure being" (item #55) are very close to each other among the most strongly endorsed items. Unity, as a phenomenological notion, clearly refers to the nondual character discussed previously, and it is almost certainly not directly connected to numerical identity or countability in a metaphysical or mathematical sense (very few meditators indicated in their reports that this is what they meant). Rather, our meditators' reports confirm an intimate relationship with the experiential quality of "feeling whole" in item #90. Unity is not oneness, but wholeness.[13] Yet it can be pure, in the sense of not implying any reportable content, and the phenomenology of wakefulness may also play a central role in nondual being. In Western literature, full absorption into minimal phenomenal experience (MPE) is sometimes described as a combination of wakefulness and pure being (as discussed in chapters 4 and 20).

The phenomenology of nondual being can also occur outside contemplative practice, such as after fainting or suffering an accident. This fact can be seen from the following two descriptions of the phenomenology of nonduality, which are taken from Western research literature. In the second example, the nondual state is characterized not even as carrying the phenomenal character of awareness itself, but rather as a contentless and indeterminate feeling of being, as an experience of "pure being" that resembles a purely "ontic" state (i.e., one that seems to be about reality rather than phenomenality):

> There was something, and the *something* was not the nothing. The nearest label for the *something* might possibly be "awareness," but that could be misleading, since any awareness I'd ever had before the accident was *my* awareness, my awareness of one thing or another. In contrast, this *something*, if it be called awareness, had no *I* as its *subject* and no content as its *object*. It just was.[14]

> During the syncope, there is absolute psychic annihilation, the absence of all consciousness; then at the beginning of coming to, one has at a certain moment a

> vague, limitless, infinite feeling—a sense of *existence in general* without the least trace of a distinction between the me and the not-me.[15]

Nondual awareness, as exemplified in the selection of reports presented in the first section of this chapter, clearly differs from the experience of nondual being in James's sense of "existence in general," as investigated in the previous chapter. On the other hand, whenever the two distinct kinds of phenomenal experience occur in practitioners of meditation, they seem to be functionally related—they may share a similar causal history. *If* MPE is fundamental in the sense of being related to a normally unrecognized baseline of all other consciously experienced contents, then it would be rational to predict a number of causal effects.

First, whenever the experiential quality of MPE comes forth, gradually growing more and more explicit, our conscious representation of clarity and epistemic openness should also become more salient, and this should lead to an enhanced feeling of wakefulness (chapter 4). As we saw, the phenomenal character of clarity can be described as pertaining to an unobstructed space of knowing, and wakefulness simply *is* the conscious experience of this primordial form of openness, of the mere capacity that opens the space of knowing. Now imagine that clarity and wakefulness could not become contracted into an illusory self anymore because subject/object structure had dissolved. If the experience were nondual, this could even become a new quality of reality as a whole, turning the world into what David Hinton has described as an "awakened cosmos." My prediction is that, all metaphysics aside, this in turn would create the "phenomenology of panpsychism": Awareness would turn into an all-permeating feature of reality. Phenomenologically, it would have to be the overall situation, or reality *itself*, that awakened into nonconceptual knowing—not the meditator, since this would bring duality right back in.

The second prediction about the emergence of MPE and its aftereffects would be that the misrepresentation of awareness itself as being contracted into the mind of a single agent or some virtual self (chapter 8) should be gradually replaced by a more global representation of what in chapters 7, 8, and 18 we called the "signature of knowing." "Nonduality" means that the contraction principle has been suspended and the experiential quality of "knowingness" has expanded.

Perhaps most interestingly, MPE could sometimes—particularly in a full-absorption episode or coming out of a syncope—manifest as something mind-independent and simply given, while at other times, the quality of knowingness, of it being some kind of *representation* of confidence, might be preserved. In technical terms, MPE could be transparent or opaque. In the first case, one would predict that our hypothetical, entirely unstructured baseline state should turn into a pure experience of being itself, into a mere "something is there," while in the second case, what arose would be the confidence or possibility that "something can be known."

We will devote chapter 28 to the phenomenology of transparency and opacity (chapter 28). For now, let me just illustrate the general idea given here—transparent MPE is all-pervading "realness," opaque MPE is all-pervading "knowingness"—by coming back to the example of the multiembodied AI. What would it take to make an AI experience "realness" itself?[16] For our future AI, this could mean that if it were to represent the global phenomenal signature of knowing in a way that made it impossible to detect *as* an internal construct, *as* a representation of knowing, then it would have to understand the quality of subjective confidence *as* something that is simply given. If a high statistical likelihood came with the evidence of the model itself—with the model being nothing but the system's global expectation of epistemic value—then this would *reify* its content, turning it into something mind-independent. Realness is reified certainty. For good high-probability models, this would create a sense of realness thanks to a second-order form of confidence (a "belief that one close-to-optimally believes") that makes the constructed nature of the original first-order confidence invisible. And this could be the way in which transparent conscious experience is created.

In chapter 5, I briefly mentioned the work of Florent Meyniel, which demonstrates that we are constantly learning not only about the environment but also about our learning processes themselves, and shows how this process creates subjective confidence, the feeling of knowing. We not only estimate the characteristics of the outside world; we also evaluate the degree of certainty that accompanies our estimates and thus form beliefs about their validity. If it were strong enough, the first-order confidence (or the degree of certainty) *itself* would now be reified in a nonconceptual way. The confidence would be transparent. If our future AI had no introspective access into its own inner workings, then its own model of first-order confidence, or "knowingness," would simply create a structureless, basal form of *appearance* because pure knowing would be reified as something directly given. This would enable our AI to experience onticity, being, realness itself—but in a nondual way. If it *did* have the relevant form of access, then everything would take on a dreamlike quality, like a virtual reality (VR) or some kind of magical illusion (chapter 28).

What makes these computational ideas so interesting is that in our own phenomenological data, we find precisely the nondual ways of experiencing reality that they predict should be possible. Processing the data from our first MPE study, I was impressed by the sheer number of experiences of nondual awareness that our participants told us about when reporting on experiences of pure consciousness. This was something I did not expect. In our initial instruction given to participants, we said that we were investigating experiences in which there is an "awareness of awareness itself" or "consciousness of consciousness itself," and explicitly pointed out that we were not interested primarily in mystical experiences or dramatic spiritual peak experiences of any sort, but rather in all states characterized by a quality of "pure awareness" or of "consciousness itself." What's more, all participants had already answered ninety-two questions about the phenomenal character of pure awareness before they wrote their brief experiential reports. By then, they must have had a good general idea of what our research was targeting. It was only at the end of the survey that we told participants that we were currently building a database of phenomenological reports of pure-awareness experiences. We made clear that this was entirely optional, but if they wanted to provide us with short reports about their own experiences, whether in meditation, during the dream state, or during dreamless deep sleep, it would be extremely helpful. Many of them chose not to describe classical meditation experiences or episodes of full absorption, but they clearly thought that nondual awareness counts as a typical, paradigmatic MPE experience. This is an interesting result, and for more than one reason.

First, for regular practitioners of meditation, spontaneously occurring nondual experiences as described in this chapter, the previous chapter, and chapter 34 (and also in some others) simply seem to occur more frequently than previously thought. We have already seen that there is an obvious lack of concurrent reportability during full-absorption episodes, but the lack of any phenomenally represented subject/object structure in nondual states of awareness may also diminish the intention or cognitive capacity to verbally report, mentally categorize, or actively memorize the global phenomenal character in question. Whereas in full absorption, both concurrent and retrospective reportability are inhibited, in nondual states, concurrent reportability may be somewhat inhibited and the retrospective kind is perhaps made slightly less likely by an attenuated motivation to communicate using words and language. In both cases, we have good reason to believe that the reported phenomenology sometimes, often, or perhaps even usually occurs without causing self-referential thoughts.

My point is that nondual awareness could gradually turn from a state into a trait, whereas, for obvious reasons, full absorption could not. A state is something momentary that occurs temporarily in a certain situation; a trait in this sense is understood in psychology as being a stable, constant personality characteristic. What makes a stable nondual mode of consciousness special, however, is that we can understand it philosophically only as a property of the whole organism and no longer as a personality trait: The identification with the self-model and the respective personality structure is simply suspended. There will always be some people who are more modest than others and more likely to start merging with what one report called "that which never speaks"

(#3624; see also chapter 30). These people may quietly and open-mindedly, perhaps even with moderate interest, listen to all these philosophical and scientific discussions, but then walk away in noble silence, with an almost invisible smile on their face. Some of these people may be only weakly inclined to begin telling others about the most precious, subtle, and profound experiences in their lives unless directly asked (e.g., in the context of a scientific survey that guarantees full anonymity). In some traditions, advanced practitioners are not allowed to speak about their experiences and attainments—a fact that may present an obstacle to some forms of empirical research[17]—and there may be very good reasons for this, based on centuries of practical wisdom. I still remember how one of the very first meditation teachers I had in my life pointed out to me that it is possible to ruin something extremely subtle and precious by dragging it out into the open and into words, and that not speaking about it at all may sometimes be the better alternative.

This point about negative self-selection bias is somewhat speculative. But our preliminary results suggest that many more people experience nondual states than is commonly believed. Therefore, it is a live possibility that certain "deeper" or more "advanced" phenomenologies remain invisible to science because a progressive identification with "that which never speaks" may actually silence the meditator. Communicative motivation has something to do with personality structure, and if practitioners identify less strongly with their original personalities, then this motivation might also weaken. Accordingly, this phenomenological dimension of human experience may be much more weakly reflected in the cultural mainstream, in cognitive science, and in academic philosophy of mind than its true frequency merits. This is a first interesting result: There could be highly relevant forms of conscious experience that systematically evade academic research because they create a motivation not to report them.

Second, from a statistical perspective, nondual experiences also seem to be related to factor 8 ("Emptiness and Nonegoic Self-Awareness"), which arguably expresses the "spiritual essence" of MPE most clearly. The second- and third-strongest-loading items in factor 8 offer two metaphorical descriptions of first-order reflexivity that combine "Self-Knowledge, Autonomous Cognizance, and Insight" (factor 3) and "Wakefulness" (factor 4) with the phenomenal qualities of emptiness and epistemic openness. The questions asked were: Would it be a good description to say that there was "an emptiness that has awoken to itself"? Did you feel as though it was not you who had an experience of "pure knowing" without any object, it was rather as if the "pure knowing" was self-aware, knowing only itself, while you had nothing to do with it? These items are, in turn, negatively correlated with the phenomenology of selfhood: The phenomenology of self-knowing and self-awakening picked out by this factor is nonegoic. The substantial psychometric evidence that we have gathered for the existence of nonegoic self-awareness gives us theoretical insights to which I will return in chapters 29 and 30.

Once again, please recall that many of our practitioners chose not to describe classic meditation experiences or episodes of full absorption, but they thought that nondual awareness counts as a typical, paradigmatic MPE experience. My third point is that these initial empirical results seem to contradict many traditional taxonomies. Often, these make pure awareness and nondual awareness two fundamentally different forms of phenomenal experience, or they describe the two as "orthogonal," in the sense of falling under mutually exclusive category headings.[18] On the contrary, our data set demonstrates the fecundity of a dimensional approach, as explained in the introduction.[19] Quite plausibly, we are here dealing with a whole family of possible states of which the stand-alone experience of MPE is only the purest and most prototypical expression. Nondual states are a different, much richer phenomenological category in which the character of pure awareness can be stable and continuously present. They are MPE *modes*, not MPE states.

Finally, let me draw attention to a more philosophical point. Classical descriptions of nonduality refer to an absence of subject/object structure, and one conceptually more precise way to describe them could be to say that the epistemic agent model is absent (chapter 25). But please note that, at the same time, the nondual state is often characterized as an epistemic state, whereas states still structured by a subject/object duality are merely phenomenal states. According to this classical motif, nondual states constitute an insight ("realization"), whereas states of the subject/object type are mere experiences (at best, "recognitions" of pure awareness). Often, the insight is described as a new but nonegoic form of self-knowledge, and a claim is made that it is an epistemologically superior form of knowledge.

Here, I would like to point out that the distinction between insight and experience does not have to be exclusive and exhaustive: There could be states that are insights and experiences at the same time. When discussing the phenomenology of pure awareness spontaneously occurring in dreamless deep sleep, in chapter 20, we first encountered the difference between insight as a purely experiential phenomenon and as a genuine form of knowledge. There, the question was whether "lucid dreamless sleep" is an appropriate term for the kind of experience that has, for example, been described by Tibetan Buddhists as "clear light sleep." Is it really a "lucid" form of experience, in terms of carrying genuine knowledge about the nature of the current state itself, perhaps even about one's "own" true nature? Technically, the issue is whether "insight" should be treated as an epistemological or as a phenomenological concept. As the philosopher Lana Kühle has made clear, we can always interpret the quality of "insight" occurring in meditation either as genuine self-knowledge or as something phenomenal—something that might always turn out to be mere appearance.[20]

Could nondual awareness be a *mere* appearance? Could it sometimes—for example, during dreamless deep sleep—even be a hallucination? The open question now becomes whether nondual awareness has a genuine epistemic dimension to it, whether it brings us "closer to reality" in the sense that there are nondual facts that it represents more accurately than the more ordinary form of consciousness does. In the following chapter, I will call this the possibility of a "convergence mode." As explained previously, our everyday, perspectival form of experience is grounded in a single physical embodiment and a single behavioral space, necessarily resulting in a centered and bounded model of reality. It is a biological phenomenon. The inner image of an actively knowing self sporadically pops up in this space, driven, like the rest, mostly by evolutionary imperatives. By contrast, the scientific method attempts to abstract away from all merely subjective perspectives as far as methodologically possible. My final point is that nondual awareness seems to achieve something very similar in the context of individual, brain-bound models of reality, but in an entirely nonconceptual way. Perhaps all facts are ultimately nondual facts, and nondual awareness is the previously unnoticed point of convergence between human phenomenology and the scientific image of reality?

This is an intriguing possibility, but knowledge claims made in the public sphere always need independent justification. To publicly claim that nondual awareness is epistemically self-justifying would quite simply mean committing the E-fallacy (chapter 7). On the other hand, it would be intellectually dishonest to ignore the fact that the phenomenological profiles of some MPE experiences clearly possess a salient nondual aspect plus a quality of self-validation. An anonymous reviewer of my work has remarked that such experiences have a "noetic purport to be metaphysically ultimate."[21] Of course, a lot of what I said about the E-fallacy, the phenomenology of certainty, and the signature of knowing in chapters 7 and 18 directly applies here. But a somewhat deeper aspect of the problem is that many of the relevant philosophical traditions have offered very different conceptual, epistemological, metaphysical, and soteriological interpretations of phenomenological reports from the one presented here, in particular via scholar–practitioners who extensively cultivated such experiences themselves. We need more research on this.

The philosophical difficulty lies in correctly assessing the accompanying phenomenology of certainty and insight with respect to the states' own metaphysical status, especially when they involve more permanent alterations to conscious experience or when the report itself even states that this is *not* an experience (see chapter 31). Is their "noetic purport to be metaphysically ultimate" mere phenomenology, just an interesting form of experiential character, or is it actually a deep, nonconceptual form of knowledge, perhaps one that has never been adequately described? Is it not phenomenal, but noumenal? Could it be *both*? I will not answer these questions here (though see chapter 31); but whoever wants to make the phenomenology of consciousness itself a target of serious future research will need to find a middle way that avoids feigning ignorance, jumping to unwarranted metaphysical conclusions, and proliferating new versions of the E-fallacy, all at the same time.

In closing, it may perhaps be inspiring to look at a classical perspective—one that is not merely theoretical but is also meant as practical guidance for serious meditators. In the quote from Wangchuk Dorje (1556–1603) that I have selected, the nondual state is characterized as an epistemic state, whereas states still structured by a subject/object duality are merely phenomenal states. According to this canonical description, nondual states constitute the manifestation of an insight ("realization"), whereas states of the subject/object type are mere experiences:

> [. . .] if experiences are experienced in a dualistic way as objects of a subject, it is experience. If they do not arise as objects, it is realization. If they are experienced by the mind, it is experience. If the mind itself appears as their essence, it is realization. If they are experienced in the form of references, it is experience. If there is an understanding of the particulars of the reference, knowing directly their characteristics, it is realization. If the mind exists as a meditator and the meditation on bliss, clarity, nonconceptuality and emptiness exists as an object of meditation or experience, it is experience. If the meditator and the meditation are directly realized as inseparable, without it being just a fabricated, intellectual understanding, it is realization.[22]

28 Transparency, Translucency, and Virtuality

All experience made of ME. Franziska is in ME—I am not in Franziska [#88]

I am not in the world—the world is in me! I am the space in which everything appears. [#2299]

It was as if suddenly everything was clear. The realization that "I am in everything and everything is in me." [#3478]

This is a special chapter. I want to draw your attention to three particularly interesting phenomenological aspects, and therefore I offer more extensive commentary than usual, in addition to our selection of experiential reports. In chapter 27, on nondual awareness, I said that ordinary consciousness is a nonconceptual mode of knowing one's own inner model of reality. However, any time that this way of knowing is contracted into an ego and falsely experienced as direct and immediate, a large part of the model appears as an *outer* reality to us. We then have what the Finnish philosopher Antti Revonsuo has called a built-in "out-of-brain experience";[1] we experience ourselves as an embodied agent situated in some external environment. This makes good evolutionary sense: Our biological ancestors successfully learned to use different parts of their inner model as a proxy for parts of their environment. According to our ancestors' subjective experience, models of trees turned into trees and models of wolves into real wolves. Their brains also learned to use the model of the body that carried them, including sensations like hunger, thirst, breath, and heartbeat, as a proxy for the body itself, improving the organism's capacity for self-control. This is what it means to have a "transparent self-model"—that is, a conscious model of yourself as a whole that has become so reliable that you are unable to experience it *as* a model—and we will learn more about all this in the second half of this chapter. To stay alive, there was a boundary that had to be protected, or re-created from moment to moment. Conscious experience often includes an explicit representation of inside and outside (e.g., of the interior parts of my body, of inner feelings and emotions arising from them, and also of what appears as my "own" mind)—plus an outside world of mind-independent objects. What meditation practice shows is that awareness can also occur with *no* explicit representation of inside and outside. This can lead to a conscious model of reality in which, according to verbal description, everything is inside and outside at the same time—or in which it has *neither* quality.

We are now beginning to expand our understanding of what the phenomenological concept of "nonduality" means. One of the most fascinating aspects is that some experiences of nondual awareness given by our participants quite directly match our best current models of what consciousness might really be from a neuroscientific, computational, or mathematical perspective, while at the same time cleanly mapping onto phenomenological descriptions given by scholar–practitioners more than 1,000 years ago. This may well be one of the most philosophically important results of our first interdisciplinary study. It is interesting to see how one can apply the idea of nonduality not only to the distinctions between subject and object and between inside and outside, but even to the distinction between what is "real" and what "does not exist."

Before we get into a little bit of theory and return to the idea of an empty, unobstructed model of epistemic space (as introduced in chapters 4 and 5), let us carefully look at a series of experiential reports. Investigating this category of minimal phenomenal experience (MPE) experiences, it will become much clearer what it really means to say that consciousness itself models an unstructured space of knowledge, creating a global but maximally simple way of knowing one's own internal model. This mode of knowing is characterized by the fact that sometimes even the deepest and most fundamental predictions and anticipations governing its contents can be temporarily suspended. Perhaps most interestingly, consciously experienced objects that appear within this space can take on a phenomenal quality of *virtuality*, or "as-if-ness." If you will, they now appear as existing and not existing at the same time, or, perhaps better, as *neither* existing *nor* not existing—and in a way that is intellectually hard to grasp for beings like us. But the experience of ontological neither-nor-ness itself is quite explicit. From now on, I will call it the "phenomenology of virtuality."

Do you still remember our fable of the elephant and the blind? One aspect of virtuality seems to be that, phenomenologically, objects are sometimes clearly "less" real than the centerless space of awareness itself, more like objects perceived during a lucid dream. You may recall a passage from chapter 19, where one committed practitioner said, "You can often wonder when you're having an experience in activity if it's real or if you are only imagining it."[2] Another aspect is the combination of absence and presence. Let us begin with an excellent phenomenological description of both—a description that was already presented in chapter 17, "Emptiness and Fullness":

> 1788 [. . .] It is an emptiness that does not mean the absence of something, it is not a lack of something, but it is really elementary, very real and clear, extremely alive, and the objects chair, table, etc. still seem like chair and table, but not as real, as if they were only dream objects. They recede into the background as meaningless and a very present emptiness comes into the foreground, but it is more than the space between the objects. The living emptiness then permeates everything, concepts like here and now no longer exist. [. . .]

The experiential quality of virtuality goes beyond "illusoriness" or the phenomenal experience of "realness." The contents of experience are now experienced as *phenomenal* in the sense of being insubstantial, lacking inherent existence, being a representation of something that is only possible, something that perhaps possesses *likelihood*—but that is not real. Suddenly, they seem to appear *as* appearances, as if they were parts of a virtual reality (VR), not really referring to anything but also not exactly mere hallucinations—rather, as if they are suspended in and permeated by the clarity of a larger, timeless space. The quality of epistemic openness described in chapters 4 and 9 has become vivid. There is a quality of de-immersion, as if the whole Ego Tunnel were now floating in this space of epistemic openness:

> 1703 [. . .] And it's like there is nothing behind the experience. There is just sort of floating. Just like a big smear of sensations all suspended somewhere. [. . .]
>
> 3305 [. . .] Perception was crystal clear, the surroundings shimmered and I became aware of the actual timelessness in the world.

There is no quality of dissociation in these states. The phenomenal quality of pure awareness remains intimately connected; everything is contained within it and arises out of it:

> 2574 In pure awareness I experience awareness itself unseparated from all emerging or disappearing experiences in mind or body. Awareness without end—clear and wide and at the same time everything in it, out of it. By "everything" I mean the completely normal perception of things, thoughts, feelings . . . [. . .]

Some of our meditators describe MPE or pure awareness as a timeless and nonphysical space within which the phenomenal world of mere appearance is embedded. This space may be characterized by an all-permeating, deep, and luminous silence:

1151 It was an experience without self-reference, with open eyes, in which it was completely obvious that everything in existence is pure appearance, "unborn," and that in fact and truth nothing has ever stirred out of deepest silence.

1381 [. . .] Into a space of timeless self-luminous world-penetrating silence. [. . .] The silence is not acoustic, the space not physical. [. . .] The whole world was completely just logical and clear, embedded in this bright fragrant sweet silence.

This kind of silent spatiality can lead to a global state that is characterized by a strong sense of lightness, beauty, and purity, and in which the phenomenal character of awareness itself even gains a "hyperreal" quality. The distinct phenomenology of MPE is somehow "more real" than that of an ordinary wake state. Paradoxically, however, the concrete experiential contents appearing within such global states now acquire a "dreamlike" and "ephemeral" quality:

3464 [. . .] At the same time the state appears as the most natural thing in the world, like a kind of basic state of mind. And, crazily enough, as the only real thing in the world. [. . .]

2844 [. . .] The moment was unusually pure, light, free of self, wide, peaceful, good. All senses were available, alive, I was fully aware of the sensory experiences, but the experience felt unbelievably, inconceivably light, ephemeral, dreamlike, and beautiful. What touched me most was the purity of the experience—there was a feeling as if this were truer or more real than normal, confused experience. Since then I have been accompanied by the certainty that awakening is possible.

This "dreamlike" or "ephemeral" quality is what I have termed the phenomenology of virtuality. It means that things are experienced as *neither* real *nor* illusory. There is a context of lightness and beauty, plus a whole range of qualities that we know well from earlier chapters, like soundness, holism, weightlessness, unboundedness, and gratitude. In the following report, the distinct phenomenal character of "virtual" perceptual objects is described in terms of "holographic constructs":

1629 [. . .] The bare trees creaked loudly overhead, as if the wind was going to take them down. I sat on a log in the forest to meditate briefly when suddenly and spontaneously everything "clicked" and made intuitive sense, not in any way that could be accurately verbalized, but as though everything was [what] it was supposed to be in utter perfection. Looking back at the experience now, it seemed as if everything merged into an undifferentiated whole while simultaneously maintaining the appearance of individuation (e.g., there was still

> a perception of "trees," colors, the coolness of the air, but not in the least bit as contrasts such as "ground" versus "sky" (as if they were fundamentally different), grey versus brown, warmth versus cool. Everything appeared different but was in actuality the very same. There was a quality of weightlessness, buoyancy, unboundedness, as though this body was an arbitrary mass but "leaking" and "mixing" everywhere. The whole world was mixing, completely fluid, whole but only (again) appearing (as if illusory) as separate objects. A deep sense of peace and gratitude pervaded. There was no "self," only the appearance of one in order to tie the experience together into a coherent whole. There was also an equanimous joy—not coarse/rugged, but a subtle yet pervasive equanimity with positive intonations. [. . .] walking through the forest, feeling as though there was no separation between "me" and "it" (forest, other "objects" supposedly "outside" of "me," which felt like holographic constructs intended for convenience only), still with this buoyant/weightless quality (bodily boundaries dissolved) and with every step rooted in peace, equanimity, joy, wholeness. [. . .] Everything felt fresh and alive, all embracing. [. . .] Everything felt silent, crisp, beautiful.

In an attempt to explain the difficult Tibetan concept of *rang snang*, the English Dzogchen teacher and translator Keith Dowman arrives at a very similar phenomenological description. He says that it is best understood "by identifying with the all-pervading cognitive principle (*rigpa*) and then conceiving one's own body and the environment as a holistic gestalt, or hologram, projected within it."[3] This is the global phenomenology of "virtuality" that I am drawing your attention to. It may turn out to be of central importance for philosophy of mind, for example because it points to a certain type of computational model. MPE could be a model of an epistemic space in which phenomenal reality appears. In the words of one of our meditators: "[. . .] it's an all knowing, un-centered space that without doubt is behind every phenomenon that is ever experienced" (#1229).

A close technological relative of VR is *augmented reality* (AR). In AR setups, users can see and interact with tools or other objects as if floating right in front of them, even while they remain invisible to other people. The objects being moved around or used appear seamlessly superimposed on what the users normally take to be reality itself. Of course, what they typically take to be "reality" is itself only the conscious VR created by their brains, as I have explained in depth in my previous books *The Ego Tunnel* and *Being No One*. The phenomenological metaphor of "holographic constructs intended for convenience only" is brilliant because it may give us a deeper understanding of what consciousness really is: the organism's own pragmatic AR. But *what* exactly is being augmented? It seems as if pure awareness itself, the primordial model of an empty epistemic space, is what is augmented during ordinary conscious experience whenever we move through our life-world in ordinary wake states. The "holographic constructs intended for convenience only" are what help a biological organism survive, successfully navigate through the physical world, and eventually copy its genes to the next generation. But all the constructed contents of experience are embedded into a timeless quality of epistemic openness.

As we have seen, in MPE modes, experiential contents are often described as "floating" or "suspended" in an unbounded, open space of timeless awareness. Sometimes this quality of looking out from timelessness can lead to a concrete conscious perspective *sub specie aeternitatis*. Just like the Greek *ataraxia* (chapter 14), the German *Seelengrund* (chapter 26), or the English "transparency" (to be explored in this chapter), this Latin concept is another example of many concepts originating in Western philosophy that are helpful in understanding certain MPE phenomenologies. You may know that the philosopher Baruch de Spinoza (1632–1677) coined the term *sub specie aeternitatis* in his famous *Ethics*,[4] referring to an abstract view of the world "under the aspect of eternity" or "from the perspective of the eternal." But did you also know that Spinoza thought that precisely this way of viewing the world was how the true philosopher does it? In our data, we find phenomenal correlates of Spinoza's philosophical worldview:

> 1828 [. . .] The ego boundaries, which are primarily physically felt, expand into infinite vastness, and all experiential contents rise in pure awareness like a bubble in champagne, or they almost hang in the air like a bird in free flight in the sky. In an even deeper level of the same experience it is possible to dissolve the last remaining latent experiential backgrounds of space and time, so that the subjective impression is created of simultaneously surveying all times, places, and worlds from a perspective of vastness. [. . .]

The category of strong nondual states on which we are currently focusing is related to the phenomenology of "witness consciousness" investigated in chapter 19. Here too, the witness can be experienced as something eternal, an impersonal and timeless ideal observer, an awareness that functions as the "three-dimensional screen" in which everything else arises and disappears (compare with #2916). The space of awareness itself is vast and unbounded, but it contains the phenomenal self at the center of its phenomenal world. The boundaries of this virtual world are indeterminate; it seems to be floating, and sometimes its contents are only a tiny fraction of the totality of conscious experience:

2807 I experienced myself as a collection of sensations and behavior patterns that, while normally occupying 100% of my awareness, suddenly were about 0.00000000000001% of my conscious experience. All these arisings of personality, fear, liking, protecting, thoughts, ideas, were just popping into being in a vast, vast, warm, infinite space of awareness. It had an absolutely hilarious quality to it also—the way I'd try to describe it is the realization that I'd spent all this time running around inside a tiny, compelling kaleidoscope looking for a doorway out, when in reality the kaleidoscope has no floor or ceiling.

Normal phenomenal reality is just a thin layer of experience superimposed onto the surface of something vast, something unbounded and unstructured:

2355 [. . .] I bathe completely dissolved in myself and know about all this. Normal reality feels like a thin sheet of paper on this ocean of Pure Awareness. [. . .]

In the experiential reports related to the family of phenomenal states that I am drawing attention to here, MPE is experienced in part as the representational space in which phenomenal reality appears or could potentially appear. Here is one example in which a meditator has had a clear experience of this contentless space itself, but *as* a representation in which the contents of perceptual experience could potentially appear:

87 Experiencing infinite space. Although the space was dark, its expansion was clearly three-dimensional, but at the same time I was aware that this space was a mental representation of my mind. If I lived in a two-dimensional world, I might have seen a two-dimensional space. For me this space was kind of like an object of my consciousness, in which the objects of my sensory perception are themselves placed. In this particular experience, however, only space as an object of consciousness was present. [. . .]

The first-person perspective, the epistemic sense of a self *to which* this reality appears, is just one part of the experience. It is a contingent and dispensable structural phenomenological feature:

3458 Awareness as a field or space in which reality appears. This reality also contains the impression that there is someone to whom this reality appears. A notion arises that this is only ONE way for reality to be experienced. The perception of one's own "selfhood" as a small phenomenon in the field of awareness.

The meditator gradually discovers that the "knowing self" is just another fictitious entity that the brain uses to explain what is happening to itself by constructing the epistemic agent model introduced in chapter 25. She may now even begin to identify with the process as a whole (medium plus content), as in the following reports:

3289 I started perceiving my room as some sort of transparent sphere or bubble. Objects inside it, as well as my sense of self. Instead of being independent in some way, [I] become part of the totality of it.

2804 [. . .] Then stillness. [. . .] The feeling that everything in the room is without borders; if someone in the room makes a noise, it's like the noise is coming from in me, not from them. But I am not in a fixed state but more so of air, or something very spacious. I am me and still I am everybody around me, every noise. Everything flows through me. [. . .]

This is what in chapter 3 I called the "zero-person perspective" (0PP). The word "perspective" comes from the Latin *perspicere*, originally meaning "to see through." Having a perspective normally means having a form through which we experience the world, where the form itself is transparent and therefore usually invisible. Here, however, awareness of the old first-person perspective (1PP) is part of the experience because we look not *through* it, but *at* it. The conscious organism envelops it or views it from the infinity of all possible perspectives, suddenly realizing its constructed, virtual nature. If you will, the 1PP is now a part of the 0PP. Interestingly, the VR appearing in the space of nondual awareness can even be a *social* reality (a 1P[plural]P, as when we refer to ourselves together with others) because it sometimes contains multiple subjective perspectives that can be directed at each other:

3048 [. . .] It was as if we were parts of an image that could see each other. [. . .]

I hope that the reports presented here have served as a good illustration of the basics of what I have termed the "phenomenology of virtuality." Some of them also point to another important aspect of what makes the phenomenology of Rilke's *Weltinnenraum* (coming up in the next section) so interesting: the varying degree of "transparency." Our inner model of the space of pure awareness is a medium, like a window. You can look *at* a window, or *through* a window. And as we all know, sometimes you can somehow do both at the same time:

2576 [. . .] as if the sensations were becoming lighter and lighter and the senses were expanding into the room. The bodily sensations are as light as all other sensations (auditory and visual) and therefore a more spatial, expansive quality was perceived in which the body is no longer at the center. It would be as if the "observer" was no longer in my body, but everywhere in my perception. This was accompanied by a way of all experiences being "embedded" in a larger space that is not accessible to us, since it is unconscious (otherwise it would be an experience). It was as if one could look "through" all experience, one could see the projections of the mind and feel the projection surface behind it, which simply "is." Untouched. Another description might be that the three-dimensional space and its experiences are reduced to a two-dimensional plane of experience. Just as the images on the eyes can be recognized as such and the edge of our field of vision represents the transition of experience to the container of experience. [. . .]

Just like a window, our inner model of the space of pure awareness and epistemic openness (which may be what MPE is) can be more or less transparent. If it is fully transparent, we typically do not notice it. We see *through* it, but it reappears as the experiential realness of what we take to be the external world. It reappears as the all-pervading subjective confidence that *things are there and we know about them*. It reappears as *the certainty that something has been known*, the nonconceptual knowledge that knowledge exists. It reappears as the background experience of *openness to the world*. Only through meditation practice do we sometimes become aware of the window itself, quite often while still looking through it at the same time. Then we begin to notice something that has been there all along. Here are three short examples:

3159 [. . .] It is as if in everyday life I see the world almost literally through invisible glasses. Which are so well cleaned that I don't even notice them.

2295 [. . .] As if the border between my face and the world is transparent. [. . .]

3464 [. . .] The extreme lightness of the state; awareness seems like a breath, barely noticeable, so that I could easily overlook it. [. . .]

Meditation practice now starts to seem like a sustained attempt to finally stop overlooking something that has been there all along. Let me close this section by drawing attention to one last phenomenological aspect, perhaps the most important of all. Apparently, our model of epistemic openness itself—the silent, nondual, timeless space of wakeful awareness—can sometimes function as the new unit of identification (chapters 24 and 29). The model of this space can function as if it were the new body (an abstract "space-of-awareness body"), as if it were taking on part of the role of what in the past we called our "self":

1746 For a few moments I vanished but was instead a void perceiving reality as a bright picture.

2898 It can hardly be described in words. I had the feeling of perceiving EVERYTHING and at the same time being able to consciously perceive only a millionth part. I myself was awareness; at the same time I was embedded in it, carried, protected, absolutely safe. On the one hand everything was within me, on the other hand I was in everything, unseparated. I simply was.

2299 [. . .] I am not in the world—the world is in me! I am the space in which everything appears. [. . .]

Diaphanousness, Indeterminacy, and *Weltinnenraum*

> And in general, that which makes the sensation of blue a mental fact seems to escape us; it seems, if I may use a metaphor, to be transparent—we look through it and see nothing but the blue; we may be convinced that there is something, but what it is no philosopher, I think, has yet clearly recognised.
>
> [. . .] the moment we try to fix our attention upon consciousness and to see what, distinctly, it is, it seems to vanish: it seems as if we had before us a mere emptiness. When we try to introspect the sensation of blue, all we can see is the blue: the other element is as if it were diaphanous. Yet it can be distinguished if we look attentively enough, and if we know that there is something to look for.
>
> —George Edward Moore (1873–1958), *The Refutation of Idealism*

In this section, I want to draw your attention to two philosophical concepts, one old and one new. I think that they might help us describe some aspects of advanced contemplative phenomenology more accurately. The first of the two is the "diaphanousness" or "transparency" of consciousness. The term was introduced into the context of Western academic philosophy more than a century ago by the British philosopher G. E. Moore, a professor at Cambridge University and one of the founders of analytic philosophy. His successor to the chair for philosophy of mind and logic at Cambridge was the Austrian philosopher Ludwig Wittgenstein, and the Nobel laureate Bertrand Russell, when he was a young man, said this about Moore: "I almost worship him as if he were a god. I have never felt such an extravagant admiration for anybody."

Let me first explain what "phenomenal transparency" means. This philosophical concept has a lot to do with what, in meditation research, is often called "reification." Phenomenology is described as transparent when certain contents of experience (e.g., perceptual objects, the body, and sometimes even the content of thought) are experienced as thinglike, mind-independent, ultimately real, and immediately given. We don't see that they are processes anymore because, phenomenologically, they have become frozen or "solidified" into something objective and real. You experience the tree in front of you not as a holographic construct, but simply as a tree, out in front of you, irrevocably real and as if immediately given. This is because normally, your introspective attention does not penetrate the earlier processing stages in your brain that construct the visual scene, bind different visual features into a gestalt, resolve ambiguities, and separate the tree from its background.

Here is another example. If you try to meditate but get carried away into a manifest daydream, then you are not aware of your thoughts *as* thoughts anymore; you have forgotten that you are daydreaming—and even that you were intending to meditate. This is true even if you actually initiated the mind-wandering episode yourself, perhaps because you simply got too bored.[5] First, you just played around with some mental images—maybe you recalled some pleasant memories and began to revel in a slightly self-aggrandizing future scenario—and then "you" got lost in the process. The epistemic agent model collapsed, and thoughts became realities. This is exactly what it means to say that the daydream becomes "manifest": You are now unaware that all this is only a representation, a dynamic inner construct. Your body sits on the cushion, but your mind is far gone, fully immersed in the melodrama of the narrative self.

In my book *Being No One,* I said that transparency is a special form of darkness. There is something that you cannot see—namely, the fact that most of the content of your experience are constructs, the brain's best guesses, representations not experienced *as* representations. They are probability distributions, predictions, mere possibilities—but represented as realities. In *Being No One,* I then applied the notion of transparency to the conscious self-model, the inner image that an organism's brain sometimes creates of the organism as a whole. If the self-model is transparent, the organism *identifies* with it by, as it were, "confusing itself" with its own inner image of itself. The image becomes the unit of identification (see chapters 24 and 29 for more on this idea). Now the self-model is not experienced as a holographic construct but is fully reified: phenomenologically, the model has turned into a real self. Functionally, the organism is now attached to its content, driven forward and enslaved by whatever happens in the self-model. *You* have come into existence.

Let us now take a quick look at what Moore really said.[6] He thought that there is one maximally general phenomenal property shared by all sensory qualities—namely, awareness itself, the phenomenal quality of knowing per se. Remarkably, Moore actually used the term "emptiness" to describe this property. As can be seen in the short quotes provided in this chapter, this quality of "mere emptiness" is normally transparent or "diaphanous" because it is not explicitly experienced—but, as Moore correctly insists, it *can* be experienced. For us, this is a highly relevant point. Moore goes on to identify another phenomenological property, that of *evasiveness*. Some aspects of experience "seem to vanish" as a result of exercising attentional agency, actively trying to "fix our attention" onto them. Nondual awareness is highly evasive because any sense of exerting mental effort to attend to it or any attempt to meditate on it immediately destroys it by creating an epistemic agent model (chapter 25) and contracting the phenomenal signature of knowing into this model (chapters 4, 8, and 18).

Most interestingly, this famous analytic philosopher from Cambridge also claimed that this property of consciousness itself can become phenomenally opaque—it can *lose* its transparency—if two conditions are met: (1) we look "attentively enough" and (2) we "know" that there is a possible object for introspective attention.[7] This sounds as if some high-level expectation or belief (which might perhaps even be unconscious) needs to be in place: a belief that there actually *is* a target for attention that we can detect, fixate, and home in on. Moore himself thought that this was really difficult: He found no record of any philosopher who was "able to hold *it* [consciousness] and *blue* before their minds and to compare them, in the same way in which they can compare *blue* and *green*."[8] We can now explain why this *had* to be the case. Any attempt to "hold" MPE plus any other conscious content "before our minds and to compare them" necessarily creates an epistemic agent model. *You* can never be "dead on target," because *you* are the occluder.

Moore probably knew nothing of Dzogchen or Mahāmudra philosophy, but he clearly thought that awareness is relational and that consciousness is a second-order process of knowing—because, quite simply, having a sensation is being aware of something, and consciousness is the knowledge that this awareness currently exists. In his own words:

> The true analysis of a sensation or idea is as follows. The element that is common to them all, and which I have called "consciousness," really *is* consciousness. A sensation is, in reality, a case of "knowing" or "being aware of" or "experiencing" something. When we know that the sensation of blue exists, the fact we know is that there exists an awareness of blue.[9]

Consciousness then becomes meta-awareness. Being conscious is being aware of awareness itself, just as one can be aware of a specific sensation. Importantly, "awareness of" is the same (epistemic but nonconceptual) relation of knowing something in sensation as it also is in becoming aware of this very awareness itself. In Moore's words: "To be aware of the sensation of blue [. . .] is to be aware of an awareness of blue."[10]

In the context of meditative experience, we can now interpret Moore as saying that an aperspectival and nonagentive form of meta-awareness always exists in all forms of conscious perceptual knowledge. Of course, this meta-awareness has nothing to do with words, concepts, or the thinking of thoughts; it is, as today's psychologists and philosophers of cognitive science would say, a "nonpropositional" form of meta-awareness.[11] A related point made by Moore, which strongly resembles what humankind's contemplative traditions have often said about pure awareness, is that this second-order relation ("awareness-of" the current existence of another "awareness-of,"—namely, of some specific perceptual quality or sensation) is mostly transparent. This means that it often goes unnoticed *as such*, but we should actually be able to make it phenomenally opaque by attending to it in the right way. It is amazing to see how at the very beginning of analytical philosophy, following in the footsteps of the German philosopher Gottlob Frege, this British thinker very clearly expressed insights that Asian philosophers and scholar–practitioners had had at least 1,500 years earlier.

Moore's successor, Wittgenstein (in §275 of his *Philosophical Investigations*), put the point about transparency differently. He said that when we look at the blue of the sky and say to ourselves, "How blue the sky is!" we never have the feeling of pointing-into-ourselves and also never think that really we ought not to point to the color with our hand, but instead ought to point with our attention. Wittgenstein then asked an interesting question: What exactly does it mean "to point to something with one's attention"? Of course, the question also arises as to who or what does the pointing. Could there be a way of gently guiding attention not to the blueness of the sky but to the awareness of the blue, without contracting it into an epistemic agent model, into a knowing self? Actively searching for the experiential quality of awareness itself necessarily creates an epistemic agent model and the well-known duality of subject and object—but in real life, it may be possible to "ignite" a recurrent process of awareness reflexively knowing itself, and then to open up and let go completely. Meditation may have a lot to do with learning how to repeatedly point attention to awareness itself, and then simultaneously let both attention and awareness go—like the string of a trompo, a spinning top.

I would like to ask some slightly provocative questions. Could there be something like MPE blindness? Are we normally not seeing something that has always been there—the window itself? Perhaps we just need to take two steps back, making the window frame become part of the overall picture again and rediscovering the obvious. As I said in chapter 23, you can experience a picture *as* a picture only if you are able to see its frame. What we call "meditation" might then simply mean finally beginning to notice MPE and gently de-contracting it, rather than constructing some fancy state of consciousness through years of practice. Is the first-person perspective that which *occludes* the zero-person perspective? Could it be that the manifest image of our whole experiential world is more like a blindfold pulled over the zero-person perspective, a virtual veil covering pure awareness?

Before we proceed to the phenomenology of virtuality, expanding our toolkit yet another time, let us ask whether all this could lead to a deeper understanding of what conscious experience *as such* really is, this time on a conceptual level. A common motif in many self-reports is that after having fully identified with the space of awareness, all experiential content appears *within* this space, within oneself (look back at the first three metaphorical descriptions at the beginning of this chapter). This aspect maps neatly onto phenomenological descriptions given by expert meditators and philosophical scholar–practitioners many centuries ago, as well as onto modern theories that describe the brain as constantly creating a "virtual model" of reality as a whole. I think the relevant difference between such MPE modes of experience and ordinary wake states consists precisely in whether someone has an additional model of the *space* of knowledge and experience (with which they could identify) or whether they are completely immersed in their own VR and completely identified with the virtual self-model created by the brain.

One of my main claims in this book has been that pure awareness is not literally contentless, but that it actually represents something. It is *of* something, but in a way that makes it completely natural to describe it as devoid of content, as entirely empty, as not like anything, or even as a kind of nothingness. If pure awareness should turn out to be something that philosophers might call a "global nonconceptual representation" or neuroscientists might call a "predictive, generative model," then in some sense it is an embodied image occurring in a biological organism. You can see an image in two different ways: You can recognize it as an image, or you can look right through it, as it were, and mistake it for a direct experience of reality itself. Drawing on Moore, we can now say that in the first case, the image would be called "opaque," and in the second case, it would be "transparent." This is merely a visual metaphor. But if it's pointing in the right direction, then pure awareness *itself* also could be sometimes opaque and sometimes transparent: Sometimes we might be able to look at it, and sometimes we might be looking through it. And sometimes we *are* it.

We now have interesting computational models that help us understand not only what opacity and transparency are, but also the fact that opacity is something that can be *controlled*. For mindfulness meditation, you need to be able to "see" your own attentional processes in the first place—if they are fully transparent, you can neither recognize nor control them. Contemplative neuroscientists and theoreticians are now beginning to view attentional processes and meta-awareness as the mechanisms of opacity control at different hierarchical levels. They call the underlying process "deep active inference." I like an analogy that Lars Sandved-Smith and his colleagues have suggested, the idea of the scrolling wheel on a pair of binoculars:

> Some mental processes function only to make aspects of the world perceivable. We are not aware of them "as such," but rather, we are aware of the content that they make available: these cognitive processes are "transparent," like a glass window that

allows us to see what is outside. Other processes, however, make these cognitive constructive processes accessible per se. This second set of processes are about other states of the mind, to which they provide access, as a new source of data now made available for further processing. These processes are akin to the scroll wheel on a pair of binoculars, which has a position state that its user can control and which enables one to apprehend and to control the precision of sensory inputs.[12]

Conscious experience is a medium. It is a medium through which we can know the world and ourselves. It creates epistemic openness. Whenever the medium is "transparent," we are unaware of it and perceive just the world and ourselves in it, in a seemingly direct and immediate way. Philosophers would call this "the phenomenology of naive realism." "Naive realism" is not a derogatory term. In the philosophy of perception, it simply refers to the idea that the senses provide us with direct awareness of objects as they really are. The idea is also called "direct realism," "perceptual realism," and "common-sense realism"—and Antti Revonsuo's concept of an "out-of-brain experience" from the very beginning of this chapter is a good illustration of it. Again, this isn't necessarily something negative, because we can also view it as a triumph of biological evolution: The transparent VR in our heads has been highly parsimonious and computationally efficient in helping us survive and copy our genes to every next generation. However, meditation practice can make the medium "opaque," so we sometimes become aware of the medium while still perceiving the world. And in a full-absorption episode, we have nothing but the autonomous recurrent activity of the medium itself, creating an internal model of mere epistemic capacity, of the abstract space of awareness as such. Epistemic openness is all there is.

More often, however, meditators may gradually become aware of the "invisible pair of glasses" while still looking at the world. As they begin to turn the scroll wheel on their binoculars, an air of pure awareness starts to appear, like a translucent mist of mindfulness, and its ethereal clarity makes it easy to overlook. As one report that has appeared elsewhere in this book put it:

> 3464 Typical for me is the extreme lightness of the state; awareness seems like a breath, barely noticeable, so that I could easily overlook it.

As explained in chapters 4 and 5, my own working hypothesis is that consciousness is our inner model of an "epistemic space," a space in which possible and actual states of knowledge can be represented. I think that conscious beings are precisely those who have a model of their own space of knowledge—they are systems that (in an entirely nonlinguistic and nonconceptual way) *know that they currently have the capacity to know something*. Conscious systems have an inner image of this very capacity, this open epistemic space (elsewhere, I have called consciousness a "self-modeling epistemic space").[13] The space itself can be almost empty, containing only a clear, abstract image of the mere capacity for knowledge and experience. Or it can contain images of actually ongoing actions, perceptions, feelings, and thoughts. In this account, MPE itself is a model of the entirely unobstructed empty space in which all those different processes of knowing something can occur.

But now there is a new picture emerging. Full-absorption episodes, in which there is just pure awareness and nothing else, are stand-alone models of our epistemic space as such, of our capacity to know. They implicitly contain the myriad possibilities of what *could* in principle be known. Nondual awareness, on the other hand, is a state in which the model is there and has become salient, and in which something actually *is* known—but the structural feature of subject/object duality (by means of which the embodied brain normally explains what is happening to itself) has been attenuated. Now the experience of knowing is no longer contracted into an ego, into an epistemic agent model—now "everything knows," but selflessly.

Here is the idea. In normal waking consciousness, our brain represents the statistical property of subjective confidence (which we discussed in chapter 5 as the estimated accuracy of perception and the congruence with prior expectations), either in a transparent or in an opaque way. Transparency creates realness; opacity creates knowingness. We must also imagine the whole process as something that is embodied and alive, fluid and context sensitive—the selfless dance of a dynamically nested hierarchy. If our models are very good, inferential uncertainty is low and our high degree of subjective confidence appears as "realness": The low-level models' content is portrayed as mind-independent, immediately given, and irrevocably real to us—it *appears*. Here, consciousness is a transparent inner image of subjective confidence. At other times, when inferential certainty is fluctuating (because the validity of our beliefs is not fully established), our representation of confidence turns into the feeling of knowing, which is then contracted into a knowing self (chapters 8 and 25). Within the epistemic agent model, subjective confidence is *in* the mind (but not independent of it); it is something that the self tries to construct (not something immediately given), and there is always the possibility that it is referring to something that will turn out not to be real. Accordingly, the feeling of knowing fluctuates.

But what is the conscious*ness* of it all? One special and highly interesting case is provided by the background experience of alertness itself (see chapter 4 for details). Tonic alertness fluctuates only on the order of minutes or hours; therefore—*if* we had a model of it, and *if* we were aware of it—the degree of subjective confidence should be very high. The general phenomenological prediction is that wakefulness should come with a robust quality of knowingness, of certitude. If we *had* a stable background-model of our own alertness, of the fact that this organism is now open to the world (as explained in chapter 4), then the associated inferential uncertainty should be very low. But there are two logical possibilities. If the alertness model were opaque, we would "see" it. There would be an explicit experience of emptiness and bare wakefulness, of epistemic openness *as such*. But what would happen if your inner model of epistemic openness were fully transparent? What is the phenomenological prediction? If the alertness model were transparent, the knowingness of bare wakeful awareness itself should be invisible, but the certitude would still be there. It would be transferred to the content. All you would see would be "the world"—that which can be known, the current content of your epistemic space—but as immediately given, perfectly real, and with high certitude. Meditation practice may have a lot to do with building and learning how to sustain a model of bare wakefulness, of epistemic openness itself, while actually keeping it visible—in order to *realize* wakefulness.[14]

Another specific phenomenological prediction results from this new philosophical picture. In advanced meditators, there sometimes should occur states in which subject/object structure is missing, while an image of the world as a whole has appeared that is no longer experienced as mind-independent but is also not experienced as being "in the mind only," as a single conscious thought or mental image might be. Could there be nondual states in which everything appears as *virtual* (i.e., as going beyond the conceptual distinction between "existent" and "nonexistent")? The content of consciousness would then be experienced as a single VR or as elements of a lucid dream happening in the wake state, enfolded or embedded into the centerless, all-encompassing space of pure awareness. If you will, the Ego Tunnel should actually be experienced as floating in this space.

Let us now turn to the phenomenology of virtuality.[15] "Virtuality"—in a solely phenomenological sense—will be the second new concept offered in this chapter. *Virtuality* refers to the specific experiential quality that often appears when people experiment with VR, wearing VR goggles while moving through a computer-simulated environment. In a strange way, with this technology, things are experienced as *neither* real *nor* illusory, or as being real *and* illusory at the same time. As we have touched on before, a general principle applies to many of the paradoxical-sounding verbal reports presented in this book: They were written not by professors of logic but by human beings trying to find the right words for something that, just like the VR experience, played no role in the linguistic communities they grew up in.

Let us stay with the VR experience for a moment before returning to contemplative phenomenology. Intellectually, we know that all of this is generated in the computer, and one part of us never believes in it at all. But on a bodily, gut level, we do believe: We feel fully situated; we're just *there*. This creates the phenomenology of virtuality: the experience of things being neither real nor unreal (and again, please note that, plausibly, some people might describe the very same experience as being both at once). Technically, we could perhaps say that the phenomenology of virtuality is the phenomenology of metaphysical indeterminacy because it involves consciously experiencing this very neither-nor-ness, but on an entirely nonconceptual level. Virtuality is the global phenomenal character that emerges when it is no longer the case that the person wearing VR goggles or the meditator experiences things as *either* existing *or* not. In an interesting way, the distinction itself has become meaningless. For example, while in a VR environment, you know that all of this is just a model, a computer-generated simulation, but you still automatically react to it "as if you really were there." Researchers in VR call this the "place illusion."[16]

Astonishingly, very few people seem to have realized that we have continual access to the richest, most robust, and closest-to-perfect VR experience currently imaginable: our very own ordinary biologically evolved form of waking consciousness. It creates a very robust place illusion (i.e., it involves a stable version of Revonsuo's "out-of-brain experience") plus a fantastically realistic experience of embodiment with the help of a transparent body model that cannot be experienced *as* a model. And as the icing on the cake, it adds a virtual ego—the epistemic agent model that we normally identify with. As is well known, the algorithms behind today's social media platforms try to maximize the user's *engagement* with a website (by creating outrage and a sense of urgency, by encouraging constant social comparison and aspiration without fulfilment, by inducing compulsive multitasking, by systematically draining attentional resources, and so on).[17] It is interesting to note that evolution discovered all this a long time ago: A mostly transparent model of reality also creates *maximal engagement* because it enslaves the organism via the nervous system from which this model arises. It creates a new level of fascination with reality by entangling the organism in a mesh of "immediate" experiences of salience and valence, attaching it to an inbuilt existence bias, thereby forcing the animal to act out the biological imperatives encoded in the model.[18] The creature becomes fascinated by its own life.

VR is the best technological metaphor for conscious experience that we currently have.[19] On the other hand, the history of philosophy has clearly shown that technological metaphors for the human mind are fertile but problematic. Think of the mechanical clock, the camera, or, more recently, the computer as a physically realized abstract automaton.[20] All these metaphors have severe limitations, but if each of them captures at least one central aspect of the mind, then they should also give us an idea of how to re-create it. Could a confluence of artificial intelligence (AI) and VR result in "synthetic phenomenology," the technological creation of artificial consciousness?[21] For many of the forms of conscious experience described in this book, it is hard to see how they could be noninvasively reconstructed in the human brain, at least using the technology currently available in VR labs around the world—but this is exactly what needs to be tried. Doing so would give us a better understanding of the neural mechanisms underlying them, and in the long run, it might even help improve the efficiency of contemplative practice itself.

I predict that a whole new stream of future research will take this direction, leading to a confluence of AI, VR, and neuroscience. Neurotechnology will always have practical limitations, but we will see a second wave in which the VR metaphor itself may point the way to a whole range of new research goals and phenomenological insights. Have you ever imagined that you were a visionary who could see into the future? Well, your ordinary conscious experience is an (often reasonably successful) attempt to do just that, because what you see is the most *likely* future state of the world. Scientifically, phenomenal content (the brain-based content of conscious experience) is the content of an ongoing simulation; it is a prediction trying to model the probable causes of a sensory signal. It is not a veridical representation of the actual environment, and it is useful for precisely this reason, because it helps the organism reduce the uncertainty about what will happen next. This is what conscious experience and VR have in common: They provide us with a counterfactual image of the world. Waking consciousness is a complex hallucination constrained by the senses, and—as sketched out earlier in this discussion—whenever the associated second-order confidence becomes transparent, it makes something possible become real. As opposed to today's VR, biological consciousness is about successfully anticipating the future. This is the reason why we could even say that ordinary consciousness is an action-oriented kind of *visionary* experience.

Two decades ago, when comparing the phenomenology of dreaming to that of the waking state from a philosophical perspective, I described the wake state as a form of "online dreaming."[22] Conscious waking is a dreamlike state modulated by the constraints of ongoing sensory input:

> [A] fruitful way of looking at the human brain, therefore, is as a system which, even in ordinary waking states, constantly hallucinates at the world, as a system that constantly lets its internal autonomous simulational dynamics collide with the ongoing flow of sensory input, vigorously dreaming at the world and thereby generating the content of phenomenal experience.[23]

Today, it is striking to note that some of our best current theories of consciousness describe it as a "controlled online hallucination."[24] These theories are beginning to say, in a much more precise way, what I was gesturing toward in the quote given here. Many of us doing consciousness research now view conscious experience as something based on *predictions* about current sensory input.[25] It is important to understand that relative to the actual state of the world, all predictive representations are nonveridical. They are not in real time, and they deal in probabilities only. From a strictly epistemological perspective, they are misrepresentations; they are only *as if*—but nevertheless, they are potentially beneficial for the biological system in which they occur.[26] The VR content generated by nonliving machines is typically part of an animated computer graphics model, and if taken as depicting the actual physical three-dimensional scene surrounding the user, it is also a misrepresentation, conjuring up an "as if" world. Again, they are both fit for purpose. VR content does not result from a design flaw—the whole point is to generate perceptual representations of *possible* worlds in the user's brain. The actual world is beside the point. A new VR (created by a computer) becomes embedded in an old VR (created by a biological brain that has a long evolutionary history).[27]

Elsewhere, I have argued that because of this, VR technology will function as a new cognitive niche to which the human mind will adapt:

> What is historically new, and what creates not only novel psychological risks but also entirely new ethical and legal dimensions, is that *one* VR gets ever more deeply embedded into *another* VR: the conscious mind of human beings, which has evolved under very specific conditions and over millions of years, now gets causally coupled and informationally woven into technical systems for representing possible realities. Increasingly, it is not only culturally and socially embedded but also shaped by a technological niche that over time itself quickly acquires a rapid, autonomous dynamics and ever new properties. This creates a complex convolution, a nested form of information flow in which the biological mind and its technological niche influence each other in ways we are just beginning to understand. It is this complex convolution that makes it so important to think about the Ethics of VR in a critical, evidence-based, and rational manner.[28]

At this point, we could leap off in any number of interesting directions. We could explore probability distributions; the minimization of prediction error; or how attention, perception, and what is sometimes called "embodied action" are all actually expressions of one and the same underlying principle. VR as a new cognitive niche also has a lot to do with the idea of *Bewusstseinskultur* (an applied ethics and practical "culture of consciousness," to which I will return in the epilogue). But what is even more interesting is how the model of reality underlying nondual awareness and many of the experiences reflected in the reports in this chapter converge with the scientific image of what could actually be happening.

The American philosopher Wilfrid Sellars (1912–1989) spoke of the clash between the "manifest image" of "man-in-the-world" and "the scientific image" of him. The manifest image is created by a realm of appearances—today, we could refer to this as the phenomenal model of reality currently active in our brains—but science ultimately reveals things as they are in themselves. Sellars was not an ideological reductionist at all, but he nevertheless boldly claimed that in the end, science is the measure of all things.[29] This raises the interesting question of whether some global modes of conscious experience are *closer* to the emerging scientific image than others—whether their underlying ontology maps more closely onto the scientific worldview than does ordinary waking consciousness, which is strongly shaped by biological imperatives. I think that nondual awareness could be a high-convergence mode in exactly this sense: a form of conscious experience that is suboptimal from a biological perspective but is closer to our best scientific understanding of what the conscious brain really does and what the deeper causal structure of the world actually is.

Let us call this the "convergence principle": the idea that some global modes of consciousness, in terms of their underlying ontology and the way in which the ongoing process of conscious experience *itself* is portrayed by them, are actually closer to the scientific image of reality than others—and that MPE modes are special in precisely this regard. Take as an example a neuroscientist with whom we can communicate in real time,[30] who has a lucid dream and, while looking at the dream environment surrounding her, says: "This is all a virtual model of reality, sculpted out of the dynamics of the neural correlate of consciousness in my head! I embody this phenomenal space. I am actually walking around in myself!" Then, to see the convergence, compare this with an MPE mode that we might characterize as "lucid waking"—as described, for example, in report #2299 (cited in chapters 29 and 33): "I am the space in which everything appears. [. . .] When, e.g., I walk around, I have the sensation of walking around inside myself." You may recall that in chapter 27, we first encountered this point when speculating that perhaps all facts are ultimately nondual, and that nondual awareness is the previously unnoticed point of convergence between human phenomenology and the scientific image of reality. I will briefly return to this question at the end of our journey. For now, let us stay with transparency, translucency, and virtuality.

The phenomenology of virtuality has been known for centuries. In many classical texts, including those of Tibetan Buddhism, we repeatedly find statements to the effect that the world may appear as a magical illusion or take on a dreamlike quality. What many of these traditional texts convey is a phenomenology of virtuality, and also of translucency (recall Moore's term "diaphanousness," whose root is the Greek *diaphainesthai*:

"to shine through"). In Aristotle's ancient theory of perception, "the diaphanous" (*to diaphanês*) denoted the sensory-material medium in which the visual sense operates. We encountered what I call the phenomenology of translucency in chapter 24, "Bodiless Body-Experience," when discussing the possibility that pure awareness has switched from background to foreground. The idea was that the background has become so dominant that what was previously the foreground becomes translucent, with some of its structural features (e.g., body boundaries) gently fading out or disappearing altogether.

In our reports, we sometimes find statements such as "It's like reality shining through when all ideas, concepts . . . step aside for a moment" (#2528), or metaphors like that of a "very thinly veiled moon" (#2623). One participant described it by saying that "time did not exist anymore but I was very aware of the variations in my environment, my consciousness and my body were more 'lucid' and I was seeing things as if they were transparent; that is, I was feeling what was behind the phenomena [. . .]" (#1397). To make what I am describing as the phenomenology of translucency more intuitive, I would like to point out that the light involved in this type of metaphor is one that never throws any shadows. In terms of body-experience, translucency can also be described as a feeling of weightlessness and of the body itself becoming transparent—but this time, not in the abstract philosophical sense of representations being transparent (or opaque). In some reports, we find this transparency evoked as a concrete and direct form of phenomenal experience: Tactile and physical sensations acquired a kind of weightlessness or transparency that was not limited (#684).

In Tibetan Buddhism, the remaining content is variously described as being like a dream, an optical illusion, a hallucination, a mirage, an echo, an emanation, a reflection, or an apparition; it is compared to castles in the clouds, a city of heavenly beings, or the Moon reflected in water. For example, virtuality appears in the words of Dakpo Tashi Namgyal (ca. 1513–1587), who wrote one of humankind's most profound meditation manuals, entitled *Clarifying the Natural State,* where he says that "you may have the experience that everything is like a dream and magical illusion."[31] Here, the warning to advanced practitioners is not to cling to these meditative experiences (or the three meditative "moods" of bliss, clarity, and nonthought) as if they were paramount, making the mistake of training "in a way that is fettered by them, being happy when they come and unhappy when they do not."[32] In *The Lamp of Precious Guidance,* Barawa Gyeltshen Pelsang's (1310–1391) commentary on Yang Gönpa's (1213–1258) *Song of the Seven Introductions,*[33] we find an instruction for the meditator who is eager to discover the experiential quality of "the many having one flavor" in everything that she consciously perceives: "There is no need for a meditation that tries to give up appearances, taking them as something extraneous. Rather, during this appearing, one remains loosely poised in the state, prior to grasping, in which you don't have to meditate, yet are not distracted." There, we also find an explanation of what I call the phenomenology of virtuality, which includes the notion of external perceptual objects now being experienced as "vitreous floaters":

> Although appearances manifest in many ways, they are of one taste within the nature of mind. By preserving things in this way, the duality between appearance and mind blends into one, so you no longer find any appearance that is not mind. At this point, since the entirety of what appears externally as the object and the entirety of what appears internally as the agent have resolved into this single state of nonduality, there is no more hindrance or attachment to external things, which are similar to vitreous floaters and the like. Then, the error of the belief in the reality of entities falls to pieces.[34]

Vitreous floaters are impurities in the eye's vitreous humor, which is normally transparent. They can become particularly noticeable when looking at the open blue sky or a monochromatic surface. Typically, vitreous floaters appear as moving specks, as spots or wormlike strings in your visual field, and their motility, size, and shape can even be measured. They tend to multiply with age, as the vitreous gel gradually liquefies. Such "eye floaters" are called *mouches volantes* in French and *muscae volitantes* in Latin—in English, "flying flies."[35] Can you imagine a global mode of conscious experience in which *all* the contents of your experience have turned into one single *mouche volante*? How would it feel to suddenly enter a phenomenological mode in which what you previously took to be the world with yourself at the center has turned into a floater, a flying Ego Tunnel in the field of pure awareness?

In the present context, it is interesting to briefly recall Keith Dowman's perfect explanation of the classical Tibetan Buddhist idea of a new global state of translucent, nondual awareness created "by identifying with the all-pervading cognitive principle (*rigpa*) and then conceiving one's own body and the environment as a holistic gestalt, or hologram, projected within it."[36] *Rigpa* is one of the classic notions of pure awareness, and semantically, it has a very close relationship to what we are here gradually approximating as MPE—like a blind person tentatively learning about the elephant. It seems that early Buddhist scholar–practitioners did make a major phenomenological discovery that we can now begin to express more clearly using our new conceptual instruments. There is a possible phenomenological configuration in which (1) MPE has become clearly recognized, highly salient, and "translucent"; (2) MPE has been transformed into a nonegoic unit of identification (see chapter 29); and (3) all *other* conscious content acquires the phenomenal character of "virtuality" described in this chapter.

It is gradually becoming obvious that there is a wider context to the phenomenological concept of "nonduality." The term can refer not only to the disappearance of body boundaries or a loss of the automatic distinction between subject and object, but also to the disappearance of the top-level hyperpriors "real" / "does not exist" or "is currently present" / "is currently absent."[37] Phenomenologically, this results in a global quality of metaphysical indeterminacy, of ontological neither-nor-ness—which I have termed "virtuality." In verbal reports, this may in turn yield paradoxical statements that two apparently incompatible things are true at the same time, or else claims of total ineffability. However, the phenomenological anchor may be the same for all such reports.

In chapter 14, when exploring the philosophical notion of *ataraxia*, we discovered that according to ancient Greek thinkers, the clear, unperturbed, and tranquil state of *ataraxia* results from a suspension of judgment about what is real and what is mere appearance. We also saw how Pyrrhonian skepticism was initially a solely intellectual enterprise but then unexpectedly transformed itself into a practice. The German phenomenologist Edmund Husserl (1859–1938) came back to the Greek skeptics' notion of abstaining from belief when he developed his method of *epoché* (ἐποχή). He said that if we want to seriously investigate the structure of our own conscious experience—something that meditators were doing for centuries before academic philosophers decided to start doing so in intellectual, theoretical ways, of course—then we must "bracket" the question of whether the natural world around us exists, or even more generally, what exists and what doesn't. Husserl also described a form of "universal *epoché*" in which *all* existence assumptions regarding the external world are put in brackets.[38]

It is interesting to note that you can do this using mindful perceptual attention, but you can also do it with your thinking mind. Husserl pointed out that to inquire into consciousness, we must learn to separate what he called the "act" of consciousness (i.e., the actual phenomenal state) from the "intentional object" (i.e., the content; what the state is directed at, such as the perceptual object to which it may refer). This bracketing (in German, *Einklammerung*) of assumptions has also been called "phenomenological reduction," and it directly relates to the experience of suchness described in chapter 9 and to the phenomenology of virtuality investigated in this discussion. It is often misunderstood as a merely intellectual exercise, some kind of cognitive operation for armchair philosophers. It is not. Phenomenological reduction never was a solely intellectual enterprise; it is closely linked to the nonconceptual epistemic practice called "meditation."

In a little-known essay called "On the Teachings of Gotama [Gautama] Buddha" (1925), Husserl notes:

> That Buddhism—as it speaks to us from its pure original sources—is a religio-ethical methodology for soul purification and satisfaction of the highest dignity and worthiness [*Dignität*], thought through and practised with an almost unequalled inner consistency, energy, and noble attitude, must soon become clear to every devoted reader. Buddhism can be paralleled only with the highest configurations of the philosophical and religious spirit of our European culture. It is now our destiny to have to assimilate the (for us) completely new Indian way of thinking with the (for us) old way, which is being revived and strengthened through this contrast.[39]

We can now see how the contemplative phenomenology of virtuality and metaphysical indeterminacy, as already described by early Buddhist meditators, relates to the philosophical idea of universal bracketing. If you will, it is not the academic armchair version, but the hardcore cushion version, that came first.

In closing, let us briefly recall that many of the experiential reports presented in this chapter and chapter 27 also show that the general computational principle of suspending priors, of episodically dissolving some of the brain's "unconscious prejudices about reality," applies not only to subject/object or existence/nonexistence divisions, but equally to the boundary between inside and outside. Not only can what we previously (in an egocentric and naively realistic manner) called "the world" be transformed into something aperspectival and virtual, but "the world" can actually become an "interior space" in a very special way that is hard to put into words. It is, so to speak, inside and outside at the same time. To explain this neglected phenomenological fact accurately is a major challenge for future research.

Great poets like Rainer Maria Rilke knew all of this long ago. I chose to put Rilke's notion of *Weltinnenraum* (which roughly translates as "inner world-space") in the title of this section because it gives us a perhaps more intuitive angle on the relevant modes of consciousness. Often, poetry addresses the same challenges that philosophers and scientists face. In the nondual phenomenology of *Weltinnenraum*, the world as a whole suddenly appears as embedded in an all-encompassing quality of lucid emptiness that permeates all sentient beings, a larger space in which the distinction of "inside" and "outside" has been suspended:

> From almost all things there is a beckoning to feeling, from every turn there wafts in: Commemorate! [. . .] What have we experienced since the beginning, if not that one recognizes itself in the other? [. . .] Oh house, oh meadow slope, oh evening light, suddenly you bring it almost to our faces and stand by us, embracing and embraced. Through all beings reaches the one space: inner world-space. The birds fly silently through us. Oh, I who want to grow, I look out, and within me the tree grows.[40]

29 The True Self

A well looks at a donkey, not vice versa [#33]

Self-image and infinity somehow fell into one. [#3024]

One of the many things that make a more genuinely *philosophical* meditation practice so interesting is that some of its aspects can be taught, while others can't. Some things you simply have to find out for yourself. Before we start, let us do a little demo. Be assured that you do not have to be a regular meditator to do this—all it really involves is noticing something that is already there. See whether you can verify this simple phenomenological truth in your own experience: Whenever you, the organism, move from one unit of identification to the next, your feeling of being, your sense of existence moves into the new unit too. This applies to getting lost in a daydream, but also to the experience of pure consciousness. Can you *be* the silence; can you *be* the knowingness? Whenever, in your conscious model of reality, you shift from a global mode of "I am the knowing self" to another mode, say to the experience of "I am a self-knowing field of awareness," then the "amness" comes along as well. Can you find this in your own experience? It cannot be taught because you cannot know it—you have to *be* it.

For some of the most interesting kinds of conscious experience, this seems to be a general principle: You cannot know it; you can only be it.[1] It certainly doesn't work every time. More often, you remain the attentive, knowing self looking *at* awareness. But if you unhook from the knowing self (i.e., from the entity that wants to find out more, tries to understand, and perhaps sometimes even meditates) by simply letting yourself fall back into the open space of wakeful awareness that already exists, then the "selfiness" disappears, but the "amness" perhaps remains part of that awareness. Can you find this? Again, it is one example of something that cannot really be taught. Later, when the knowing self has returned, it may even say: "I *was* uncontracted, awake awareness!" This is because the "amness" came along despite the lack of self; and this is what I mean when, in what follows, I speak of "nonegoic units of identification."

When the phenomenal experience of awareness itself is stable and has coemerged with ordinary conscious content like colors, sounds, and movement, then it can sometimes happen that a fundamental shift takes place. This shift can be described in many ways. One is as an eversion: a turning inside out. To be more precise, it is not so much a shift in the content of consciousness itself, but rather a global transformation of its deep structure into a new mode, or a transformation of the *perspective* from which the world is known and experienced:

33 The state of looking at your own experience from the perspective of pure space. A well looks at a donkey, not vice versa.

3441 Normally I feel as if I'm behind my eyes, in my head. Pure awareness seems to occur when I am able to distance/separate whatever I am from the feeling of being behind my eyes or in my body. In some ways, it feels like my usual perspective reverses itself. I am not behind my eyes or in my body, but my eyesight and bodily sensations are appearing to me moment by moment.

These first two examples directly relate to the idea of a virtual world-model appearing in an uncentered space of nonconceptual knowing, as I discussed in chapter 28. They also clearly relate to the new concept of a zero-person perspective (chapter 3) and to the many examples of experiential eversion (e.g., background/foreground switch) that we have encountered in other chapters.

However, on a conceptual level, there is another way of approximating the phenomenological nature of this shift: namely, describing it as a shift in the phenomenological "unit of identification" (introduced in chapter 24). What is sometimes referred to as the "true self" forms the origin of the perspective from which the world is experienced, and it is what we identify with. For example, eversion (turning inside out) is a shift in the *point of origin* from which the perspective projects, as it were—dissolving or expanding the original vantage point into the space of knowing as a whole. Now it is as if the point of origin had turned into the space of origin—some sort of primordial space that now functions as that which really knows. You may remember that the phenomenological "unit of identification" quite simply refers to whatever set of experiential contents lead to and anchor self-descriptions of the form "I *am* this!" or "I *was* this!" It may be helpful to go through our series of concrete examples one more time.

If you say, "I am my body!" then the unit of identification is the conscious body-model in your brain plus the global experience of ownership that comes from controlling the body as a whole. If you say, "I am that which has feelings!" then the unit of identification is the emotional self-model currently active in your brain, which also includes and arises out of inner body sensations, gut feelings, affective tone, and the like. If, like René Descartes, you say, "No, I am the thinking self, that which forms concepts and controls thought!" then the unit of identification is the brain's conscious model of a cognitive agent, created by an "explaining-away" of the subtle sense of mental effort that comes with inner actions like mental calculation, logical thought, or an effort to concentrate. If you say, "I am that which meditates, that which mindfully brings the focus of attention back to the breath after noticing a stray thought!" then you are identifying with the subtle sense of effort that accompanies control of the focus of attention (the experiential quality of "attentional agency") and with an almost automatically arising model of an entity that has just "done the noticing" (the mindful agent, the successful "meditative self"). And so on.

Here, the phenomenological discovery is that the phenomenal character of awareness *itself* can begin to function as the new unit of identification, but that it is selfless. The important conceptual point is that *nonegoic* units of identification clearly exist. We have already stumbled on this fact a number of times, such as when investigating the phenomenology of nothingness and bodiless body-experience. Again, please recall that the term "unit of identification" refers to whatever it is that phenomenological self-descriptions of the form "I *am* this!" or "I *was* this!" attempt to pick out. We can now see how in some cases, such descriptions are phenomenological stopgaps—and are actually false because there never was an "I" that identified "itself" with the experiential content in question. The surface grammar of such reports suggests a duality that never existed in the experience itself. To some people, it may even seem as though the old identification with a person *being* conscious was false. In hindsight, this is something that can even seem like a full-blown misidentification, as in the following:

> 897 [. . .] It is as if, without active striving or doing, a "shift" occurs of a "point of view" from which / in which perception takes place. It becomes obvious and undoubtedly "perceptible" and recognizable that awareness is constantly and exclusively aware of itself and that a previous perception conditioned by acquired convictions (I as a person WITH awareness am aware of a new object in constant succession) is factually false.

> 2953 During a period of sitting meditation, lasting about 2.5 hours, at some point my consciousness became aware of awareness, rather than of my meditation object (breath). I became aware of being aware, not in a place but just feeling as though I were a center, no body or body sensations or awareness of having a body or being anywhere. It was contentless as far as consciousness is concerned: dark and silent, having no qualities of time or place. Yet, I felt

just like being myself, but not being anything in particular, including being a human or individual person. In fact I really didn't even know I was in such a state, while yet being in it, until the mind suddenly produced a few thoughts, which I observed as if they were a phenomenon outside myself. I could hear the quality of them as if being part of a person that I recognized, but no longer identified with. When thoughts were not present I was aware, but not really conscious or thinking about being aware; it was just silent and empty.

3473 [. . .] it suddenly sunk in, at a deep visceral level there is no guy I all these years identified as me. I, as just a blank spectator, could see there is nobody inside. By nobody I mean no thinker or doer, just the awareness. I opened my eyes right after and the object before my eyes happened to be my son's stuffed toy he had left behind. That dog literally was also me at that moment. Feels stupid to say that now. I felt it so strongly that I almost broke down out of lightness and also sort of missing the person who I thought always existed but now knew never existed to begin with. [. . .]

It is interesting to see how meditators, in trying to accurately describe their own phenomenology, oscillate between interpreting it as being in touch with "the true self" or with "no self" at all. This oscillation relates to chapter 28's hypothetical suggestion that different types of paradoxical self-report might sometimes possess the same phenomenological anchor. The following three features make this more plausible. First, there seems to be a new quality of self-intimacy, of being in touch with oneself in a maximally direct way—closer to oneself than ever before. Second, this quality is "ahistorical": It has nothing to do with your private or public life history or with the egoic process of meaning-making and continuous narrative self-deception described in chapter 17. Third, in retrospect, it even has the character of transcendentality, which means that it feels like a special kind of self that has "always already" been there as a condition of possibility for everything else (more about this in chapter 31):

173 The sense of self was greatly changed. I would not say that there was no "I" present. It was more like a strong feeling of intimacy with oneself. Filling everything out. Not an "I" in the sense of a person with a history. But the most familiar, connected, fulfilling feeling of an "I" that I experience. It has always been there. It is completely satisfied. There is also a component that I would call "sober." In a very pleasant, light, bright, free sense. Not tainted by the heat of the usual stories.

2668 I had an experience of pure experiencing in which I became aware of myself. I became aware that the self is absolute and that nothing is "before" it. [. . .]

From a philosophical perspective, an important question is whether there are any global modes of conscious experience in which human beings spontaneously stop using the first-person pronoun "I." In this context, one interesting detail is that some meditators put "I" in quotation marks when trying to communicate the relevant experience in writing, while others use capitals (for the German "*ICH*"). Please also note that many of the phenomenological aspects investigated in previous chapters coemerge in situations in which the old, personal-level self is bracketed. Here are some examples:

280 [. . .] Previously I had practiced yoga and running only very sporadically and for health/fitness reasons. Suddenly these forms of movement gained a completely new dimension and meaning for me, a meditative depth. In the movement I could perceive that there was something different, something deeper within me, that I am real and that this has nothing to do with my body, name, or preferences. Particularly significant experiences occurred first while walking in the forest, later also suddenly while moving through my normal everyday life (in the office, while walking through my apartment). In these moments I became aware that my body was moving while "I" could observe and perceive it and everything else. These states are very pleasant, calm, and completely harmonious.

937 After my thoughts had moved away, I felt an optically and acoustically empty, boundless space far away from any sense of time and with regularly flowing vibrations "through me." This *I* [*ICH*] seemed to be bodiless and boundless. At the same time a feeling of absolute peace and harmony with everything.

3320 [. . .] The "I" used here is a formulation in retrospect; at that moment it did not exist, did not matter. It was absolute presence in absolute presence of everything that surrounded and contained everything at the same time. Boundaries, even physical ones, did not exist. A "sensation," "feeling," "perception" of flowing, taking, giving, sending, and receiving, absolute alertness, contentment, and suchness.

2954 [. . .] Dissolving the boundaries of the body, the feeling of falling into space- and timelessness, dark living emptiness in which there was breathing. There was a consciousness (observer/attentiveness) that controlled this without "me" wanting anything. It was looking around in wonder, listening in wonder to the silence, just breathing. And an experience of deep peace in and with everything. [. . .]

2900 Afterward calmness returned and a strong identification with the space/ground of the experience took place. Emotions came and went without being

> identified with them; perception happened in such a way that they were only content and not part of what was felt as "I." Strong sense of self-knowledge. Absence of fear. Feeling that the "I" will remain intact, regardless of what is perceived as the content of the experience. [. . .] Feeling that the "I" is unchangeable, boundless, and constant. Feeling of silence resting in itself.

Some long-term practitioners explicitly say that they *identify* with pure awareness. Pure awareness is the epistemic space in which everything else appears, but it can also be empty. There are qualities of unity, pure presence, unboundedness, silence, and timelessness to it. Experientially, this specific phenomenological aspect can be more or less clear, more or less frequent; it can appear suddenly or it can stabilize itself over a lifetime. There are episodes of full immersion, but there is also the stable phenomenology of a continuous background. Let me close this section by presenting four typical reports (other examples are found in #2206, #1682, #1703, #1786, #1935, and #3906):

> 1315 [. . .] My favorite word: bliss-consciousness. I identify with it: I am pure consciousness, but not my body, and it is the same experience in and outside of meditation since my childhood, but I perceive this inner being more and more clearly. My consciousness does not change, my infinite inner being is complete, but the perception of this being increases in quality and quantity. [. . .]
>
> 3285 I experience my self as unity of pure consciousness in which I am simply allowed to linger. A state of freedom, despite mental and physical limitations.
>
> 3294 It was as if everything disappeared, body, mind, space and time, but awareness itself was still present. There were no particular qualities that involved self-reflection, like emotion etc., until after the experience ended and I was again aware of my surroundings. This particular "pure presence" involved the emptiness of all content except awareness itself. As if awareness is the only thing that exists, but I am that awareness and not separate from it.
>
> 2299 [. . .] The falling-away of identification with the limited person. I am not in the world—the world is in me! I am the space in which everything appears. This is an experience that has remained until today. When e.g., I walk around I have the sensation of walking around inside myself. It is a sensation of infinity. The body does not end at the obvious body boundaries, it is infinitely large. And there is only one thing. THIS. Everything is this THIS, or pure awareness, or pure consciousness or God or presence. The sensation that I MYSELF am everything that appears, this ONE, the WHOLE. There seem to be two states of consciousness running simultaneously. The SILENCE (pure awareness) as the basis of all being is always there and at the same time the processes in the

world of form, which however is no longer experienced as different from the SILENCE. Therefore the impression is there that nothing is really happening.

Nonegoic Units of Identification and the View from Nowhere

> Essentially I have no particular point of view at all, but apprehend the world as centerless. As it happens, I ordinarily view the world from a certain vantage point, using the eyes, the person, and the daily life of TN as a kind of window. But the experiences and the perspective of TN with which I am directly presented are not the point of view of the true self, for the true self has no point of view and includes in its conception of the centerless world TN and his perspective among the contents of that world.
>
> —Thomas Nagel (born 1937), *The View from Nowhere*

At this point, a topic arises that dominated centuries of Indian philosophy: Conceptually, should we interpret this general type of experience—the kind of experience that is also described by our participants in the reports given in the previous section—in terms of "no self" or "true self"?[2] As Matthew MacKenzie has pointed out, it could be the case that an enduring, substantial self exists, even if there are phenomenally selfless episodes of experience. Or it could be the case that some minimal sense of self or subject is an inherent feature of human phenomenal consciousness, even though no such entity as the self exists.[3] The phenomenology itself seems to underdetermine all possible philosophical interpretations. Should we follow the path of Buddhist metaphysics, or that of Advaita Vedanta?

For anyone who chooses to describe the all-pervading quality of awareness itself as their true and timeless self, this self now becomes the epistemically open, empty, and boundless space investigated in chapters 4 and 23. But there are conceptual problems lurking in the background. First, "truth" is a property of sentences, not of selves: Sentences are true or false, and if one looks closely, it is quite unclear what it could actually mean to say that a self is "true." Is this way of speaking more than some unwarranted form of metaphysical essentialism, perhaps motivated by an unconscious need for mortality denial (chapter 17)? Second, taking a more practical perspective, how are you, the meditating self, ever meant to "become one" with something that is neither one nor many? We have seen that minimal phenomenal experience (MPE) is a phenomenological domain without individuation, an open space of conscious experience that cannot be divided into countable entities—not even a single one—and in chapter 5, we also briefly touched on Śāntarakṣita's "neither-one-nor-many" argument. Some may be intuitively inclined to think that, metaphysically, there could be only *one* "true self." And since phenomenologically it lacks any individual characteristics, it would seem logical that there could never be many of them.[4] So is there only one single "true self" in our entire universe? As a matter of fact, some of our meditators go so far as to describe what I would like to term the "phenomenology of solipsism." Solipsism (from the Latin *solus*, "alone," and *ipse*, "self") is a class of philosophical positions claiming that nothing outside the self or one's own mind exists—no external world, no other selves or minds. Everything is part of a single self or inside a single mind. Here is an example: "I felt an awareness that I was the only being in existence. There was nothing but me" (#2344). In this new context, you may perhaps remember the "principle of phenomenal correlates" introduced in chapter 19: Namely, for every metaphysical theory that philosophers have developed, there is an altered state or mode of consciousness that directly corresponds to it.

We can now see that the ancient metaphysical intuition of "true self" does have a phenomenological anchor—even if some of its conceptual expressions lack coherence. As the phenomenology of "unboundedness" (investigated in chapters 5 and 23) is a prototypical core feature of MPE, it is hard to conceive of *multiple* instances of pure consciousness juxtaposed with each other, or of multiple spaces conceived of as particular entities. Yet, if one looks more closely, there is also no phenomenal character of singularity in bare awareness per se, just as there is no inherent plurality. Again, this phenomenological observation directly relates to Śāntarakṣita's classic "neither-one-nor-many" argument in the *Madhyamakālaṃkāra*. So we find that for many classical philosophical positions, there is an anchor in contemplative experience—but the phenomenology itself underdetermines the metaphysics, so different interpretations always seem to remain possible. By now, you will not be surprised to hear that I plan to sidestep this whole issue—one reason being that this time-honored metaphysical debate rests on a false phenomenological opposition. It presents a false alternative: The possible alternatives are incompletely described because other cases are possible (philosophers speak here of a false conclusion by false disjunction). Given our new conceptual tool of a "nonegoic unit of identification," we can elegantly bypass the no-self versus true-self dichotomy. Doing so may reveal a middle way that does more justice to the actual phenomenology, resulting in a richer and finer-grained account. Let me explain.

A third way to describe the relevant phenomenological transition from the state of being contracted into a knowing self to a state of vast epistemic openness is as a change in the unit of identification to MPE, or as a shift from what was previously the transparent content of the conscious self-model to pure awareness. It is only natural that meditators, in the absence of alternative linguistic tools, later try to describe it as a transformation in sense of self, in the phenomenal character that philosophers sometimes describe as the quality of "selfhood." But what this shift leads to really is something *new*, something that is certainly not an object, but that also lacks some—though perhaps not all—features of what we once called a "subject" or a "self."

If we want to say that pure awareness can function as a nonegoic unit of identification, we need a clear definition of what "nonegoic" means. In his contribution to a collection of texts titled *Radical Disruptions of Self-Consciousness* (which we edited together and which is freely available online), the French philosopher Raphaël Millière did an excellent job of teasing apart the most important varieties of selflessness. After distinguishing six notions of self-consciousness commonly discussed in the literature, he argued that none of the corresponding features is necessary for consciousness because there are states of consciousness in which each of them is plausibly missing. Importantly, he also pointed out that we have preliminary empirical evidence for the existence of some states of consciousness that lack *all* of these six putative forms of self-consciousness. So is pure awareness nonegoic, in the sense of being totally, radically selfless, or is there perhaps some implicit residue of phenomenal subjectivity in it—maybe a hidden inner perspective, a subtle form of "selfiness" that remains?

Here is a quick overview of Millière's six types of self-consciousness:

1. *Cognitive self-consciousness*: Thinking of oneself *as* oneself
2. *Spatial self-consciousness*: Being conscious of one's location (with respect to one's perceived environment) as one's own
3. *Sense of bodily ownership*: Being conscious of one's body as one's own
4. *Sense of bodily agency*: Being conscious of one's bodily actions as one's own
5. *Sense of mental ownership*: Being conscious of one's mental states as one's own
6. *Sense of mental agency*: Being conscious of one's mental actions as one's own

Let us begin with the first point on the list. Clearly MPE has nothing to do with thinking, although it can at rare times coexist with spontaneously arising thought, even "translucently" permeate such thought. But none of the case studies presented in this chapter is an example of thinking of oneself *as* oneself.

What about the second point? Millière skips self-location in a temporal, as opposed to a spatial, frame of reference (i.e., "temporal self-consciousness"), which is actually an important, necessary feature of minimal phenomenal selfhood.[5] It is related to what in German is called *Anwesenheit*—namely, the phenomenology of self-presence in a temporal sense, of being present as a self *now*, by being conscious of the present moment as one's *own* lived moment, the only one in which *oneself* currently exists. Being a conscious self necessarily means identifying with a point in time, embodying a lived Now, and this feature is absent in the timelessness of MPE.

But let's return to spatial self-location. Could one say that the reports given in the first half of this chapter count as being conscious "of your location in your perceived environment as your own"? In full-absorption episodes, MPE occurs in the absence of any perceived environment (#3294, cited earlier in this chapter, is one example). We saw that in other cases, the conscious body-model may have disappeared, lost its boundaries, or still be present, but it is not owned (Millière's third feature) in the sense of being what Thomas Nagel called the "locus of identification," which involves phenomenologically marking out one's *own* location in space.

What about the next point: the sense of bodily agency, the experience of currently being conscious of your bodily actions as your own, the feeling of being the entity that controls them? Report #280 in the first part of this chapter is an excellent example of how the character of pure awareness can be experienced in motion, while running or doing asanas, but without the sense of bodily agency. Here, we have the phenomenology of bodily motion, but without effort, without the contracted sense of control that normally accompanies it—there is no "physical agent model." Some of our reports are reminiscent of the phenomenology of witness consciousness investigated in chapter 19, like the "blank spectator" mentioned in #3473, and we find something very similar happening for mental ownership and mental agency (Millière's fifth and sixth features). For example, take report #2953: "[. . .] the mind suddenly produced a few thoughts, which I observed as if they were a phenomenon outside myself. I could hear the quality of them as if being part of a person that I recognized, but no longer identified with." In sum, none of the reports in this chapter describing the experience of becoming one with the quality of pure awareness mentions mental ownership or mental agency.

I think that it makes good sense to say that the phenomenal character of pure awareness in and of itself is nonegoic, in that it satisfies none of the six constraints listed by Millière. Since many of the relevant reports explicitly mention the quality of timelessness, we can infer that temporal self-location in terms of "nowness" and "self-presence"—both of which are forms of conscious time experience—does not play a role either. The phenomenological core of egoic self-awareness seems to lie in the sense of effortful control that accompanies goal-directed bodily and especially mental agency. Egoic self-awareness is anchored in the ongoing minimization of prediction error via active inference. This most prominently includes the experience of deliberate, goal-directed thought and of actively controlling the focus of attention, but our egoic self-model also bottoms out in diffuse feeling tones and "gut-level selfhood," the weaker and slightly vaguer bodily sense of simply being alive, which is created by low-level *interoceptive* inference (i.e., continuous control of inner states of the body).[6] You are an ego in the strong sense when you navigate the world with the help of an epistemic agent model, when there is not only bodily self-awareness and the fluctuation of gut-level awareness plus feeling tone (i.e., the emotional background arising out of it) but also a robust knowing self, actively attending and thinking.

The quality of mental agency is really the one that leads to contraction; the sense of effort that comes from hallucinating a goal-state—the specific experiential quality emerging from the process of quashing prediction error on the mental level—is what counts. Whenever you identify with pure awareness, none of this is the case: In a way, you are like a conscious organism running a new operating system. This organism is now employing a different kind of ontology for segmenting reality, navigating the world under a nonegoic unit of identification. If this is correct, the resulting MPE mode should be characterized by an all-pervading quality of effortlessness.

Using our new tools, we can also say that the phenomenology sometimes described as "true self" is one in which the experience of knowing is no longer *contracted*. It is one example of a shift from the first-person perspective of an active, information-hungry self to the zero-person perspective of knowingness itself—that is, to an uncontracted version of the phenomenal signature of knowing (chapter 18). The first-person perspective is the perspective that you take when you ascribe conscious experiences to yourself using the first-person pronoun: *I myself* am having this thought, *I* hear music from a distance, *I* am observing my breath, or *I* am trying to meditate. There is also a third-person perspective: A neuroscientist is looking at your brain from the outside, maybe while you are having thoughts or are trying to meditate. The scientist can then ascribe properties to you from an external perspective: *His* default-mode network is highly active; *her* posterior cingulate cortex is slowly getting deactivated. The scientist could also do this without knowing who you are: *This* default-mode network is highly active; *that* posterior cingulate cortex is slowly deactivated. Your first-person perspective, on the other hand, is your very own, individual perspective on the world; it is what makes conscious experience subjective, what makes it seem to be intrinsically tied to a single, knowing self. Philosophers and scientists, therefore, have thought long and hard about whether the first-person perspective can be reduced to the third-person perspective, or whether there are mysterious "first-person facts," radically *subjective* facts that cannot be fully grasped by the scientific method. And this finally brings us back to Thomas Nagel.

Nagel's beautiful idea of the View from Nowhere bears considerable structural similarity to some of the phenomenological reports presented in this book. He thought that there had to be something like a "true self," a subject of a centerless view of the world, and that this self "seems incapable of being anyone in particular."[7] This "objective" or "true" self actually apprehends the world from the outside rather than from a standpoint within it. In a very interesting way, Nagel drew attention to the fact that there is in *some* sense an "objective" point of view, from which his connection with the specific person named "Thomas Nagel" seems entirely arbitrary. It's interesting to recall that some spiritual teachers, such as Nisargadatta (chapter 19), have claimed that this connection can "snap" altogether.

Thomas Nagel's *The View from Nowhere* also speaks to the convergence principle (introduced in chapter 28), which says that nondual awareness may involve hitherto unnoticed points of convergence between contemplative phenomenology and the scientific image of reality. Nagel thought that any solution to the philosophical problem that he was pointing to would have to bring "the subjective and objective conceptions of the world" into harmony.[8] Isn't there a sense in which nondual awareness achieves exactly this?

Nagel also said that even if his own verbal exposition of the problem turned out to be faulty, his goal was to "evoke a sharp intuitive puzzle" so as to convince readers that there was *something* real in it. When you think about it, there is something deeply unintelligible, or perhaps even absurd (chapter 17), in the idea that given the vastness of the physical universe—"in those oceans of space and time"—each of us should really be just one single embodied ego, one minuscule person among countless equally contingent others. The absurdity results from what I call the "inbuilt metaphysical megalomania of the self": How can I—indeed, how can any real ego—be something so unimportant, so highly contingent? Aren't I in some sense *necessarily* the person I am? Of course, the metaphysical megalomania of the self (which cannot really conceive of its own nonexistence) mirrors some classic philosophical arguments for the existence of God: If there is a supreme being, it exists because its necessary existence is self-evidently contained in the very *idea* of a supremely perfect being. The ego is a supreme being that cannot imagine the world without itself in it. But could all of us actually take on an impersonal standpoint, the zero-person perspective, while detaching ourselves from this specific, locally embodied ego? Could we thereby come to view the egoic self as, in Nagel's own words, "a momentary blip on the cosmic TV screen"? Here is Nagel again:

> How can I, who am thinking about the entire, centerless universe, be anything so specific as *this*: this measly, gratuitous creature existing in a tiny morsel of space-time, with a definite and by no means universal mental and physical organization? How can I be anything so *small* and *concrete* and *specific*?[9]

Many years ago, I was deeply fascinated by the fourth chapter of *The View from Nowhere*.[10] We should bear in mind however that Nagel's "true self" is only an intellectual exercise, not a phenomenal experience like those exemplified by the reports presented in this chapter. Nagel is only *thinking* about "the entire, centerless universe" and the difficulty of returning to his normal, individual viewpoint afterward. His theory originated not on a meditation cushion, but in a philosophical armchair. Yet, one of the many aspects that remain interesting in Nagel's treatment is the idea of using the "vantage point" of an individual person as a kind of window. The neuroscience of consciousness is beginning to show that phenomenal models of reality are precisely such windows: lived representations via which we interact with the world and with ourselves. In ordinary states of consciousness, the bodily self-model is one such window, and what we have called the "knowing self" is another. But what I have termed a "nonegoic unit of identification" is just such a window too. It defines a mode of conscious experience. As the phenomenological material presented in this chapter demonstrates, it is something that occurs in the real experiences of real people, not just in fancy philosophical thought experiments. It happens every day, all over the world. Here is one last example, taken from an earlier publication of mine: "[. . .] there's no personal point of view, it's the world point of view, it's like the world looking, not ME looking, the world is looking."[11]

Pure awareness, whenever it has transformed into the all-encompassing nonegoic unit of identification, strongly resembles what Thomas Nagel called "the objective self" or the "true self," which he described as "the subject of a perspectiveless conception of reality."[12] This subject is the origin of what in chapter 3 I called the "zero-person perspective," and what *really* appears from this perspective is not some philosopher's intellectual "conception" but the content of a full-blown, conscious model of reality in the brain of a biological organism. In this model, there is no real, individual vantage point because this is not only the "view from nowhere," it is the "view of nobody."[13]

We can now connect the idea of an out-of-brain-experience (chapter 28), the phenomenological motif of an "inversion," and Rilke's beautiful idea of *Weltinnenraum*, or "inner world-space" (chapters 27 and 28) in an interesting way. If we add current scientific findings, a whole new picture may emerge: The content of conscious experience is the content of a phenomenal model of reality in the brain, a comprehensive inner image of reality. This much seems plausible. Normally, of course, this image includes a model of the biological body. We know that, too. This body model includes, for example, our arms, legs, eyes, ears, and head, but also the perception of our own breath, abdominal sensations, or the sense of balance. What is completely missing from this body model, however, is the brain in which all of this is created in the first place. We do not feel our brain and the space of cognition opened up by it; they are not part of our subjective life world. Or are they?

The content of the phenomenal model of reality is, according to experience, something that lies outside the brain. The world is an outside world: It is independent of mind and consciousness. Then there is the body, which we also feel from the inside. The body is in a situation, but this inner and outer experience of situatedness is phenomenologically still out-of-brain; it certainly does not belong in our brain or in our head. Or does it?

Is "being-in-the-world" perhaps an abstract aspect of what the concrete biological brain does in our head—something virtual? A mere model, a simulation of externality and independence of consciousness? We have seen that in rare states of consciousness, an inversion can occur. First, the distinction between inside and outside gradually disappears. Like a hypothesis without confirmation by new empirical data, it becomes weaker and weaker; gradually the distinction is recognized as something merely virtual. Thus, what was previously outside becomes a new inside. World and body gain a certain quality of virtuality and interiority, for they are no longer independent of consciousness.

The result, however, is not the interiority of a cognizing ego, but rather what one of our study participants described in the following words: "I am the space in which everything happens" (see #66 or #2299 in chapter 33). What in chapter 27 I called the "single-embodiment constraint" seems to be temporarily lifted. The world itself is now experienced as a kind of nonegoic interior space; this is Rainer Maria Rilke's "inner world-space." But what exactly is it that has been upended? The first-person perspective? The phenomenal signature of knowledge? Is it the whole conscious model of reality that suddenly recognizes itself as a model?[14]

You will recall that the zero-person perspective describes the phenomenology of knowing the world, but not as a knowing self, as a thinking person, or even from the vantage point of a particular location. Rather, it is as if epistemic space itself, or even the world as a whole, is silently looking at what is happening within itself. And the interesting discovery is that precisely this passive epistemic space can sometimes function as the new, nonegoic unit of identification.

30 Pure Awareness Knows Itself

> "But is this 'pure awareness'?," asks the thinking mind, and something else simply knows. [#1575]

Experiencing the phenomenal character of awareness itself is an ongoing process that we might call "meta-awareness," but only if we keep in mind that it is an entirely nonconceptual process leading to an entirely holistic experiential quality. In pure awareness itself, there is no structure, nothing looking down at or monitoring some first-order state. Phenomenologically, there is no hierarchy and no "higher-order state." Minimal phenomenal experience (MPE) is not a form of thinking. Rather, as we saw in chapter 27, it can be related to a nonpropositional form of meta-awareness.[1] In this sense, it can be *described* as involving an awareness of awareness, but phenomenologically, the experience itself is not mediated by any form of thought or mental formation of concepts (it has nothing to do with what scientists call "metacognition"). MPE is also not pure awareness plus the recognitional thought, "*Now* my mind is completely clear and silent!" As meditators know only too well, this kind of subtle background thought—"I am currently not having a single thought!," "This is it!," or "Here we go again, *finally* . . ."—is a common means by which the predictive process of selfing cleverly tries to bootstrap itself back into existence.

If one looks closer, there seem to be at least two kinds of meta-awareness that can be involved in the experiential character of MPE. The first kind still includes an observer and a stable, inward-directed perspective. I will call this the phenomenology of "dual meta-awareness" because it still includes subject and object. Here is one example of dual mindfulness:

> 3499 Becoming conscious of the fact that one is conscious is not difficult, indeed it is immediately accessible. There is, after all, a subjective component, an experience that I can immediately make myself aware of, and that distinguishes me

> from a piece of wood or a robot. I can only assume this for others, but with myself I know it from my own experience. In meditation "on the mind" I first connect with this experience, differentiating it from the contents of consciousness that are being experienced. An observer position in relation to the contents in the mind: thoughts, images, possibly emotions, possibly bodily sensations or sensory impressions that "push their way in," although I actually want to focus on the pair "experiencer and mental contents." This feels like a silent observer in a dark space in which thoughts, images, etc. surface. I am not very good at maintaining stability here, i.e., quickly a thought grabs me and I am distracted. But as long as I remain "awake," i.e., undistracted, it is clearly distinguishable that I can remain in silence as an observer, even if there is a lot of movement in the mind.

Dual meta-awareness involves what we termed the phenomenal signature of knowing because it also includes a subjective sense of confidence (chapters 18 and 25). However, some of our meditators report that at times, the experience of pure awareness includes not only the phenomenal character of "knowingness," but also a distinct quality of "the knowing knowing *itself*." There seems to be a nondual variant of meta-awareness. In this second form of MPE meta-awareness, it is as if the phenomenal signature of knowing dynamically folds back into itself, silently but continuously reembedding awareness into itself. There is a distinct experiential quality that from now on, we can call the "phenomenal signature of *self*-knowing." However, the relevant experiential quality is entirely nonegoic, in the sense of having nothing agentive to it. This means that it is effortless; it is not a reaction to anything; it is not goal-directed; and accordingly, it does not involve a consciously experienced sense of control. It is not owned. There is also no autobiographical component to it (it appears as timeless and spontaneously present, and there is no sort of "inner narrative" whatsoever), and it does not involve any form of personal-level self-representation (there is no expression of personality traits or self-conscious cognitive agency, including self-directed thoughts or emotions). It is holistic; it lacks internal structure; and if we take the phenomenology seriously, then "meta" suddenly seems like an inappropriate term—something artificial, something imported from an external third-person perspective. In this second kind of nonconceptual meta-awareness, there is a strong phenomenological sense in which it has nothing to do with *you*.

Perhaps most important, this phenomenological variant feels like a *nondual* variety of metaknowledge, like a particularly intimate way of being in contact with oneself: Yes, awareness is aware of itself, but no, the phenomenological profile does not involve subject and object. Dual mindfulness is a *subjective* state of experience (as in the first example in this chapter); awareness being aware of itself isn't. Yet both are clearly conscious. This is what I mean when, throughout, I say that the minimal model approach dissolves the problem of subjectivity for consciousness science. In this phenomenological sense, some minimal forms of consciousness are not forms of subjective experience. Again, the self-cognizing but phenomenologically pure and unstructured form of awareness created by nondual mindfulness has nothing to do with *you*, but it also is not some reified entity that is *opposed* to you—because the phenomenal character of self-knowing created by awareness becoming acquainted with itself is not a thing at all, but a process. In the process of getting to know oneself, consciousness "awakens to itself," so to speak, and generates a nonegoic variant of self-awareness and self-confidence. It is therefore better not to describe it as some sort of higher-order "meta"-state at all, but rather to describe it in terms of what philosophers might want to call the dynamic process of "subpersonal first-order reflexivity." In case you are interested, I will offer a bit of philosophical theory in the hors d'oeuvres in this chapter. But let us first have some dessert, in the form of a careful look at a selection of reports describing the actual experience:

> 101 [. . .] It was a clear awareness, but with no thoughts; nondual (no subject–object, but not so much of wholeness). I was not quite sure about (or did not feel) "time" or "space" or "thought" or "body." But I knew that something/someone (I take it as me) exists then and there. It was certainly not "enlightenment," just the awareness manifested itself while no thoughts were present. It is the background awareness behind any thoughts and perceptions, and it knows itself.

There is an element of certainty in being in a state of nondual self-knowing because it is precisely a form of knowing that one knows. But the element of certainty cannot be reified. A better description would be as a process of nonconceptually knowing that *knowing currently takes place*. Therefore, the distinct quality that we just termed the "phenomenal signature of self-knowing" also implies a dynamic inner experience of self-certainty. One participant interestingly described this element of nonegoic self-certainty as a form of "pride":

> 151 [. . .] What makes this moment so certain is a kind of inner clarity, a kind of pride that consciousness takes in experiencing itself in this state. But this is no ordinary pride, it is like a knowledge of inner strength in a confrontation that needs no action, but is freed from all expectations and unfolds very spontaneously. [. . .]

Phenomenologically, we could also describe this kind of first-order reflexivity within pure awareness as self-cognizance without reification:

156 This was mostly an experience of emptiness, not as an object, but as emptiness/consciousness cognizing itself. There was a lot of clarity (consciousness cognizing itself) and a joy that lasted for several days. [. . .]

It can also be described as attention having come to rest in itself:

1037 I have learned to focus my attention [. . .] on the mind itself, not on the contents of the mind (i.e., thoughts, sensory stimuli, or anything concrete). It is then as if attention is simply resting within attention itself. If I do this for a long time, for example in a meditation retreat, a feeling of light soon spreads in the mind. The mind is then light. [. . .] It is like another plane of existence. I do not have the feeling that "I" am having this feeling, rather the feeling is far beyond personal identity. Most of the time I am not completely absorbed into this feeling, so I am still nevertheless aware of myself. But I sense that it is possible to let go of oneself more and more completely. This is what I practice.

If one looks for the phenomenal quality that, in the title of this chapter, I have provisionally called "pure awareness knows itself," then interesting combinations with many of our earlier descriptors can be found—like epistemic openness, calm, silence, emptiness and fullness, nondual being, background/foreground eversion, spaciousness, luminosity, suchness, wholeness, and the potential for full absorption. On the other hand, all seven phenomenological markers of *egoic* self-awareness discussed in chapter 29 (i.e., cognitive self-consciousness, spatial and temporal self-location, agency, and ownership in body and in mind) are lacking. What the material clearly shows is that a reflexive and nondual variant of MPE does exist. Phenomenologically, it is perhaps best described as a nonegoic signature of self-knowing. To let you judge for yourself, I will close this section by presenting eighteen examples without further comment:

1268 Awareness of awareness in vastness, silence, calm. [. . .]

1316 [. . .] Sometimes perception of complete emptiness, then self-referential blissful being. [. . .]

1350 [. . .] When the relevant state arises, it is like seeing seeing, or awareness of being aware.

1545 [. . .] A nonphysical space was emerging in me that was growing moment by moment and in which the boundaries were blurring between the perceiver and the perceived object. At the same time this space was conscious of itself. [. . .]

1582 [. . .] It was odd, as I kept experiencing no thoughts, just emptiness, yet it was "fullness" at the same time. The body felt like it dissolved and what was left was my awareness, being aware of itself. [. . .]

1617 Experiencing awareness of consciousness or of cognizance as such. That which is known recedes very much into the background, into the periphery of consciousness. In the center was the cognizance that is at rest in itself!! [. . .]

1623 I experienced awareness of awareness last year for the first time after 10 years of meditating. It was effortless. And now I can do it all the time.

1662 I had the distinct experience of awareness/consciousness knowing itself several times, but in this particular case [. . .] the experience was fairly long-lasting (~10+ minutes). Although thoughts and associated feelings and (faint) bodily sensations still occasionally arose, there was an increasing feeling of spaciousness that had no owner and seemed softly bright. The thoughts (including thoughts about the experience) continued to be intermittently present, but they were just occurring in this "space" without coloring the experience. [. . .]

1712 I experienced awareness itself. Not in the sense that I was experiencing awareness; there was no "me," no "observer." Awareness itself was always already aware.

1788 [. . .] A deeper state is reached when the emptiness withdraws into itself within "me" and becomes aware of itself. First I feel a "pulling" within me; something draws me toward itself and it feels as if this does not come of my own volition, but is done with me. I experience the pulling in the region of my head, and seeing is reversed: Instead of looking outward, the gaze turns back onto itself. [. . .]

1978 [. . .] The participants had the task of finding their true self and sharing it with the group at the end. In the course of this meditation it became clear to me not only cognitively but also emotionally that this self could not be found, but was merely a consciousness that can be observed by another kind of consciousness. Pure consciousness without I without thoughts without nothing. [. . .]

2007 [. . .] The experience was the noticing of noticing. There was no personality associated with the noticing, no discernible quality; just the awareness of awareness, as such.

2120 [. . .] Also, I became aware of the fact that "thought" does not have primacy in my experience, rather awareness has the primacy. Thought is occurring in this awareness which is also aware of itself.

2375 [. . .] Then I had a feeling of greater awareness in which I became more aware of both the thoughts passing through my mind and the experience of being the observer of the thoughts rather than the thinker of the thoughts. It was as if I became detached from myself, but not bodily or emotionally. I want to say I felt peaceful, but it was more like a cessation of thoughts and bodily sensation and the replacement of that with a feeling of wholeness. It did not feel external to me, but rather like a state that I would be able to return to as the experience felt like it was drawn from me. It felt like it could potentially continue forever. I don't know how to explain that I felt like myself (whatever that is) while also feeling like a camera observing myself. [. . .]

2474 [. . .] By dropping my desire to react, the whole bodily experience fell away, producing an intense flare-up from the bottom up, with great astonishment and intensity, almost destabilizing. I find myself leaping backward toward nothing, and the experience of awareness shows itself in its entirety, without a possible "me" between the knowing and the being of the experience. [. . .]

2710 All thoughts inside and outside were like a reflection, there was only being, self-recognizing awareness. [. . .]

3323 [. . .] There was only a kind of very delicate smouldering, a last remnant of consciousness that was now conscious only of itself and nothing else. In the years since then, this "experience" has stabilized, that is to say that pure consciousness is also aware of itself in everyday life. [. . .]

3464 [. . .] Once in this state, sensations can be largely suppressed, then it feels as if consciousness is perceiving itself, consciousness feels "naked." In a way, there is then no more focus, because perception itself is being felt. [. . .]

That Which Never Speaks

This naked clarity and emptiness beyond the intellect—
Letting it naturally be itself, it will see itself. [. . .]
Its self-recognition is the key point of practice.
As it meets itself face to face, it dissolves into itself.
—Rgyal ba Yang Dgon pa (1213–1258),
Song of the Seven Introductions II (2, 4)

The point is to remain without fabrication in the continuum of recognition.
—Longchen Rabjam (1308–1363), *The Precious Treasury of the Basic Space of Phenomena* (11)

When there is stillness, recognize just that within the stillness and sustain it with reflexively aware mindfulness.
—Dakpo Tashi Namgyal (ca. 1513–1587), *Moonbeams of Mahāmudrā* II (10)

Pure awareness knowing itself is an entirely nonconceptual affair. "Recognizing" pure awareness is not a form of active, conceptual thought, and it does not involve the quality of ownership. In and of itself, it is entirely silent, free of mental action. The epistemic agent model (chapter 25) has given way to a state of innocence. There is no accompanying thought like "Now it is aware of itself!" or "Now I actually *am* pure awareness that directly knows itself!" The phenomenology of pure awareness knowing itself also never contains some knowing self quietly saying "I think *this* must be it!" to itself, or asking "Is this *really* it? This is too simple . . ."

To be sure, in real life, such thoughts do occur, but they are only episodic automatisms: brief, habitual attempts of the biological organism to contract into a cognitive first-person perspective. If their content is left untouched, they dissolve all by themselves. Pure awareness knowing itself can occur as a thoughtless full-absorption episode, but apparently, it can also be accompanied by brief contractions of this sort, or even by longer-lasting and more complex forms of conscious experience (see chapter 34 for some potential examples). One major result of our phenomenological investigation is now not only that nonegoic units of identification do exist (as extensively discussed in the previous chapter), but that there are also nonegoic forms of *self-awareness*. It also seems as if there is a distinction to be made: Some of our meditators simply describe the phenomenal character of pure awareness, but others also describe a quality of *metaknowing* that may be an important part of the pure-awareness experience. Data from many hundreds of meditators show that selfless meta-awareness actually exists—and that it is not some arcane altered state of consciousness, but plausibly a natural capacity of the human mind that millions of people in different cultural contexts have already encountered.

You, the organism that currently identifies with the epistemic agent model in its brain, the embodied conscious self that reads and wants to understand what this is all about, the ego that already knows some things, the ego that apparently stays the same over long periods of time and always wants to know more—you can never fully identify with pure awareness, or know it "from the inside." Do you still remember our introductory discussion of Thomas Nagel's seminal bat thought experiment in chapter 3? Nagel's epistemological point can now be transferred to the context of meditation practice, for there is a corresponding phenomenological discovery to be made here: You can never know what it is like for the *bat* to be a bat, and only pure awareness can know itself nonegoically. But the phenomenology of nonegoic self-knowledge can arise in you, the biological organism. The organism can become acquainted with it, as an ownerless organismic state—but only so long as the identification with the epistemic agent model is suspended. The interesting question is how exactly you, the biological organism that takes itself to be a conscious person, later *appropriate* the experience into the high-level narrative that makes up your "own" life.[2]

As noted in the preceding chapter, human history (e.g., Indian intellectual history) is full of deep and beautiful philosophical debates about whether something like a "true" self exists, or whether there is no such thing as a self at all. And the specific inner experience that forms the topic of this chapter was known and precisely described by scholar–practitioners who came many centuries before us, including in the form of the Tibetan *rang rig ye shes*, which has been translated as "self-knowing timeless awareness" or "self-cognizing wakefulness."[3] In Western armchair phenomenology—for example, by Franz Brentano, Edmund Husserl, and Jean-Paul Sartre—there are many much more complex but often related discussions about the reflexivity of consciousness.[4] The idea that conscious mental episodes can be aware of themselves goes back at least as far as Aristotle.[5] But in our own data-driven investigation, we were not at all interested in the metaphysics or epistemology of selfhood, only in looking at the fine structure of experience itself in as radically bottom-up a fashion as possible. Within our twelve-factor solution, we actually found a statistical cluster (factor 8) that we labeled "Emptiness and Nonegoic Self-Awareness." This cluster of features seems to be most directly related to the phenomenal signature of self-knowing that we are interested in here.

As can be seen in figure 2.1 in chapter 2, the second- and third-strongest loading items in factor 8 were the two following questions: "Did you feel as though it was not you who had an experience of 'pure knowing' without any object, it was rather as if the 'pure knowing' was self-aware, knowing only itself, while you had nothing to do with it?" (item #73) and "Would it be a good description to say that there was 'an emptiness that has awoken to itself'?" (item #75). Item #74 was "Did the experience have a quality of knowing itself?" And it reappeared in factor 3, which we named "Self-Knowledge, Autonomous Cognizance, and Insight." The strongest item in factor 8, however, related to phenomenal qualities of emptiness and epistemic openness, as well as to the experience of a vacuum or a void. These statistical findings offer useful jumping-off points for qualitative analysis.

In assessing verbal reports referring to the phenomenology of pure awareness turning back into itself, we find that some features belonging to the traditional notion of egoic "self-awareness" can sometimes exist in MPE as well:

- The opening of a new epistemic space
- The phenomenal signature of knowing
- A sense of clarity and epistemic lucidity
- Reflexivity, or the quality of "knowing itself"

According to our reports, the selfless self-awareness of pure awareness knowing itself is an exceptionally clear and lucid state of consciousness. Phenomenologically, it may be the core feature of stable, continuously self-recognizing wakefulness itself: "self-cognizant wakefulness." Near the beginning of this book, in chapters 4 and 5, we saw how the conscious experience of wakefulness per se can be described as epistemic openness, as becoming aware of an unobstructed space of possible states of knowing. Selfless self-awareness also opens up a new inner space characterized by an entirely uncontracted signature of knowing, a nondual sense of confidence. And obviously, it carries the phenomenal character of reflexivity—but reflexivity of a very special and holistic kind, a quality that exhibits integration without time, thought, fragmentation, or any internal hierarchy. However, as all of this takes place in a nonselfy, unmediated, and entirely nonconceptual way, it is what academic philosophers might label "direct self-acquaintance,"[6] "self-intimacy,"[7] or "nonrepresentational reflexivity."[8]

The selfless form of self-awareness described by these philosophers and many meditators lacks the phenomenology of self-control and higher-order monitoring, and there is no subject/object divide. It is not something you can *own*. Other classic markers of selfhood and agency are also absent (see chapter 29). As I briefly pointed out in the first part of this chapter, important and seemingly necessary conditions for self-conscious egoic experience are not satisfied. These include mental agency and the phenomenal character of effort (as in controlling thought or deliberately sustaining the focus of attention), plus the experience of duration and identity across time (which in "normal" self-consciousness is typically accompanied by the emergence of an inner narrative). Self-location in space and time, as well as the experiential quality of "mineness" in body-ownership, are equally absent.[9] Therefore, we can also say that pure awareness never actively turns itself into an object, simply because it is entirely nonagentive, effortless, nonspatial, timeless, and therefore ahistorical.

I think that the phenomenal character I am trying to isolate may be a large part of what makes many describe MPE as an irreducibly *spiritual* state of consciousness (more on this in the epilogue). Within the space of family resemblances related to the experience of pure awareness, reflexive MPE is certainly at least a prototypical core region, and it forms the phenomenological anchor for what many meditators, if asked, may try to describe as the spiritual "essence" of their experience—or as their "true self" (chapter 29).

As we saw in the first part of this chapter, the phenomenal signature of self-knowing implies a nonegoic experience of self-certainty, described by one of our meditators as the nonegoic "pride that consciousness takes in experiencing itself" (#151). What complicates matters is that many earnest practitioners are familiar with philosophical debates about the nature of self or even adhere to certain metaphysical belief systems themselves. Therefore, it is more than plausible that their background beliefs, their expectations, and the conceptual instruments available to them will color not only their experience itself, but also any phenomenological reports that they may give later. Accordingly, many may describe the experience of pure awareness as getting in touch with, recognizing, or reidentifying with something that has always been there: the "true" or "real" self. This is only natural, and outside scientific research, it is also mostly benign. Yet it is another case of theory contamination, of the alacrity with which the only available conceptual framework influences attempts to verbally report something that was ineffable while it occurred. The neither-nor-ness and the timeless quality of epistemic openness get lost on the verbal level, and this may lead to a deep ambiguity at the boundary between reports about MPE (the prototypical phenomenal character of pure awareness itself) and reports about *reflexive* MPE (the quality of pure awareness as *knowing* itself).

There are several kinds of self-consciousness, or ways of being in a state of self-awareness. Interestingly, what both notions of self-consciousness—the weak, nonegoic experience of "emptiness awoken to itself" described by factor 8 and the robust, everyday phenomenal experience of "being someone"—have in common is that human beings can sometimes *identify* with their content (chapters 1, 25, and 29). Apparently, both types of self-consciousness can function as the phenomenological anchor for reports of the "I *am* this!" or "I *was* this!" type. Therefore, a first interim result of our investigation is that nonegoic self-awareness does exist; the second is that it can also function as a nonegoic unit of identification. Please again note that these two phenomenological insights go a long way toward defusing the false alternative underlying the ancient self/no-self controversy: Just taking the phenomenology seriously allows the metaphysical puzzle of the "true self" to naturally dissolve. If we distinguish the prototypical phenomenal character of pure awareness itself (MPE) from the experiential quality of pure awareness knowing itself (reflexive MPE), then both seem to be things that we can identify with (if that is our criterion for "selfiness"), but only one is a form of self-knowledge (though one that does not involve an epistemic agent model). Whether we conclude that the answer is self or no-self simply depends on what conceptual criteria we are most interested in, and which ones we choose to apply.

But there is more. If one takes contemplative phenomenology seriously, there seem to be at least two types of nonegoic reflexivity—two ways in which the weak, nonegoic experience of "emptiness awoken to itself" can occur. In the first version, awareness becomes aware of itself in a spontaneous and entirely nonconceptual way. It reveals itself to itself (e.g., in a deep and fully absorbed state of meditation). In the second variant—as we saw when looking at "suchness" in chapter 9—*particular* forms of experiential content, like sounds or perceived objects, not only appear against the background of pure awareness, but they also can sometimes gain a quality of nonegoic reflexivity for themselves. They seem to be coemerging *with* or appearing *out of* a non-dual form of awareness, and sometimes they share the specific phenomenal quality of nonegoic self-knowing described here. This is a phenomenological discovery made by Eastern scholar–practitioners centuries ago but almost completely ignored by Western philosophy.

We can now see that there are at least two kinds of nonconceptual but phenomenally experienced self-knowledge: egoic and nonegoic. My point is that the second type can also be found in perceptual objects. Somehow, the phenomenal signature of *self*-knowing can be part of the experience—but locally, in individual percepts, not only as a global background quality. It is as if they were themselves made of reflexively aware mindfulness, or as if that mindfulness were beginning to "translucently" shine through them. This point is related to the specific phenomenal character that I tried in chapter 9 to describe as "self-revealing," "self-disclosing," "self-evidencing," and "epistemically self-validating." Perceptual objects have become epistemically open. Metaphorically, one might try to say that in certain meditative states, they too can be experienced as "having selflessly awoken to themselves." Again, this is a phenomenological possibility that has been largely overlooked in Western philosophy of mind, but that fortunately is now beginning to attract attention.[10] Put simply, my phenomenological point is that the signature of self-knowing can actually be folded into perceptual objects themselves. In contemplative practice, it can become explicit. For a concrete example of what I am referring to, recall report #3160 from chapter 9, describing the phenomenal character of selfless self-awareness in auditory experience: "that it is not a self that perceives the sound, but that there is awareness in/with the sound. There existed a perception that the sound appears in space with awareness and 'recognizes itself.'"

I think that in the future of consciousness science, nonegoic reflexivity—pure wakefulness recurrently knowing itself—may turn out to be extremely interesting from a neuroscientific perspective, as well as for mathematical and computational modeling.[11] From a methodological perspective, a subpersonal but consciously self-knowing type of brain state could easily fall through the cracks of empirical science because—as a state not represented by an epistemic agent model (chapter 25)—it may not be easily reported by experimental participants. Western philosophers may discover that it was described in detail by contemplative scholar–practitioners many centuries ago, for example in Tibet between the third and twelfth centuries. There may also be a specific reason why the phenomenology of nonegoic reflexivity has been largely ignored in Western philosophy and science. You may recall that in chapter 28, when discussing the seminal contribution made to our understanding of the transparency of consciousness by the British thinker G. E. Moore, we rediscovered an important feature of pure awareness: It is "evasive." This means that every attempt to willfully fix attention on the quality of awareness per se destroys its originally nondual nature, the baseline quality that William James called "sciousness," simply because it reintroduces subject/object structure. Pure awareness recedes from attention. We can now see more clearly *why* this must be the case, and how the underlying principle also holds for reflexive MPE. Every effortful attempt to fixate attention on MPE, thereby turning it into a reified target of introspection, automatically creates a subtle hallucination, an inner image of a goal state to be reached—and an epistemic agent model that is directed at this goal state.[12] Imagining pure awareness knowing itself is not the same thing as pure awareness *actually* coming to rest in itself because the latter lacks any form of agency whatsoever.

If we look at those states in which pure awareness apparently "nonconceptually and nonegoically knows itself," we find a feature that may be closely related to the entirely silent and nonagentive character of reflexive MPE: it is "that which never speaks" (as in report #3624, cited in chapter 32). We have already seen that pure awareness is nonreactive—for example, classical mindfulness practice is precisely the cultivation of a wakeful, gentle, and compassionate state of nonreactivity. But reflexive MPE is also nonagentive. There is no motivational force, and because it contains no incentive to act, it is characterized by a principle of mental inaction. The quality of "reflexively aware mindfulness" that Dakpo Tashi Namgyal spoke of in the sixteenth century is not something that would ever actively reveal itself within a verbal report, or actively point to itself in a social context. The process creating it would not destroy itself—why should it? The fine and subtle quality of awareness silently "knowing itself" can perhaps be described as something that continuously reveals itself *to itself* or as something that continuously points *to itself*, like the arrow of attention now dynamically folding or bending itself back into itself. If sustained, it can create what Longchen Rabjam, seven centuries ago, called the "continuum of recognition" (see the second epigraphical quote presented in this section). But that which reports experiences, that which perhaps boasts about things that happened during meditation, that which may be slightly complacent[13] when speaking about "stages" and "states" that it has known—that entity is always something else. It is an ego; it has personality traits, psychological conditioning, and emotional needs—and it has had important insights that it feels an urge to share. Perhaps the urge and the complacency really are parts of some last-ditch escape strategy? Maybe there is even a new personality disorder to be discovered, "spiritual teacher personality disorder" (which I cannot abbreviate because in the introduction, I promised that "MPE" would be the only new abbreviation I would use in this book).

Be that as it may, my main point is that because verbal communication is something that happens between egos, in a space that opens itself up between embodied selves in a social context, the nonegoic quality of "pure awareness that knows itself" would never find a place in it. The public sphere is coarse-grained, created by functionally coupled self-models of the egoic and agentive kind; it is constituted by speech acts in which selves, as the philosopher John Austin said, "do things with words."[14] Those human beings in whom nonegoic self-awareness sometimes appears may therefore feel an intuitive incentive *not* to report it. If it is that which never speaks, why should *they* speak about it? The experience might well thus be evasive not only introspectively, but also socially, as something that gently resists being pulled into the public sphere. As I said before, it is an ownerless state.

Many practitioners may have the sense that it is something precious that has appeared in their life, something very hard to describe and even more easily destroyed by linguistic reification. And if, phenomenologically, they sometimes even *become* "pure awareness that knows itself," then one would predict that any motivation to speak about it will already have begun to dissolve. Once again, therefore, it is at least conceivable that the specific kind of phenomenal experience discussed in this chapter occurs far more frequently than we may think because it has a tendency to "silence the subject," and thus to evade scientific investigation via negative self-selection.

31 It Is Not an Experience

> This is exactly what is so impossible to describe: that it is not an experience at all. This is the first thing that I intuitively realized each time: "This is not an experience now." [#1311]

> I am trying to avoid the word experience here, because it would not do it justice, it is a state and not an experience that comes and goes and has a transitory aspect. [#3218]

A recurring theme in many reports is "Pure awareness really is not an experience at all." This element may be another one that is of special interest for philosophers because it seems to be related to what was investigated in the form of "witness consciousness" and "timelessness" in chapters 19 and 22. To highlight what makes the "not an experience" aspect philosophically interesting, I will tentatively call it "noumenal awareness" or the "phenomenology of transcendentality" (or "transcendentality" for short). I will say more about what all this means in the second part of this chapter, but please allow me to first quickly dissolve one potential misunderstanding right at the outset. In the Transcendental Meditation (TM) movement, pure awareness is described as "transcendental consciousness," in the sense of a state in which the TM practitioner has "transcended thought" and abides in knowing restful alertness itself. It is a clear and straightforward concept. It corresponds to the second possible reading of the term "pure consciousness," which for simplicity we will refer to in chapter 34 by the abbreviation P2: "nonthinking." Transcendental consciousness in this sense is restful alertness without thought, and it is directly related to the ancient notion of *turīya*, the fourth state of consciousness already mentioned in the early Upanishads. This is *not* the phenomenological quality that this chapter is all about.

Here, what we are trying to approximate is that for some meditators, the phenomenal character of pure awareness also includes the apparently self-evident fact that somehow, in a way that is very hard to express in words, what is occurring is not merely what philosophers call a "phenomenal experience"—something that subjectively appears to us by coming and going in the mind. Rather, pure awareness may feel like something that reveals the timeless condition of possibility for all conscious experience itself, and, somehow, its true nature as well. If this were actually true, and if pure awareness were sometimes the only content of experience (as in a full-absorption episode), then we might even conclude that what *really* makes such states so special is something that goes far beyond mere phenomenology—namely, the fact that they continuously reveal, express, or dynamically represent their own condition of possibility and that of all other conscious states as well. They are not about mere appearance; they involve touching the reality of consciousness "in itself." Using Kantian terminology, we might therefore be tempted to point out that some of these states are not merely phenomenal but also *noumenal.* In other words, they have an epistemological and a metaphysical dimension that perhaps cannot be cleanly separated from the phenomenology. This philosophical problem—really taking the phenomenology seriously as a target of research, but without committing a C-fallacy, an E-fallacy, or even an M-fallacy (chapter 10)—is what originates from what I call "transcendentality" or "noumenal awareness." Given the wealth of participants' reports that we have to work with, there are a number of ways to approach the problem in an evidence-based manner.

To begin with, one interesting finding is that some respondents spontaneously put the word "experience" in quotation marks, probably because they feel the word is slightly inappropriate and the concept that stands behind it may not quite apply. Here are six examples:

1236 [. . .] The state is characterized by the "experiencing" [*Erleben*] and "direct knowing" ["*Erfahren*"] of a completely wide, open space [. . .]. [. . .]

1311 Almost impossible to describe, because the word "experiences" in fact completely misses the mark. [. . .]

2299 [. . .] At the age of 37, however, another as yet unfamiliar "experience" occurred. [. . .]

2859 My "experience" originally occurred during Zen meditation, but occurs at other times now. [. . . .]

1828 [. . .] without that training the "experience" expressed above is neither interpretable nor meaningful. [. . .]

3323 [. . .] Later, however, there were "experiences" in which even the last remnants of this pure consciousness were extinguished in meditation. This was like an inner death, but then also an even greater freedom than pure consciousness itself. There was a clear experience of the fact that pure consciousness is far from being the deepest (or highest) possible, but that "behind" it

> there is a much more comprehensive, indescribable "no-thing." However, it cannot be described with words, since it is no longer an experience, but can at most be described as the absence of all experiences, or as absolute freedom. [. . .]

Aside from the fact that some meditators clearly think the word "experience" is inappropriate (yet often find no preferable alternative), it is also interesting to note that what I have provisionally termed the "phenomenology of transcendentality" seems to be quite directly related to other specific experiential qualities, many of which we have already investigated. Let us look at each of them in turn, starting with the feeling of "coming home." You may perhaps recall that in chapter 13, I presented some examples of what I called the "epistemic dimension of homecoming." This dimension is characterized by a phenomenology of "remembering" or "recognizing." In these states, there is clearly a signature of knowing. Often, what is remembered is described as universal, as fundamental, and as something that "had always been there but was forgotten." Approaching these aspects within the new context of transcendentality, let's now look at some more examples:

> 82 [. . .] I was aware that I was meditating and that this was happening to me and that it was not normal, although it was. That is, I had a sense that this state was always there, that I was that, but that I had never experienced it before. [. . .]
>
> 173 [. . .] It has always already [*schon immer*] been there. [. . .]
>
> 197 [. . .] It felt like getting in contact with something / touching an inner quality that had always already [*immer schon*] been there.
>
> 311 The intuitive feeling that this is the ground of being—the true nature or ground state of myself and of everything, which itself is not an experience, but rather the (groundless) basis of all experiences. This is exactly what is so impossible to describe: that it is precisely no experience at all. This is the first thing that I intuitively realized each time: "This is no experience now." Like entering into a completely different state of matter, which at the same time is so fundamental that it also feels completely normal again—because it was always and in all experiences invisibly present as its basis anyway. It is this quality of recognition, this "Ah yes—of course," which is, however, completely unspectacular—because it had never been any different anyway. This may sound like philosophy now, but it is only an attempt to put this state of pure awareness into words for myself.
>
> 1426 [. . .] The most amazing thing about this experience or state (which lasted for over 2 hours) was how normal it felt, in the sense of "yes, of course, that's how it is." It was also a feeling of recognition.

1482 [. . .] a kind of natural remembrance of a state of being that is deeply personal and at the same time universal in the sense that boundaries have dissolved into an experience of deep unity [. . .]

2293 It felt like a homecoming to a natural state forgotten but very familiar. Not in a spiritual way but just a re-cognition.

Adding to our observations on the scare-quoting of "experience" and the connection with homecoming, here comes a third interesting detail. In the original German version of the second and third reports presented here (#173 and #197), we begin to find first instances of expressions like *schon immer* and *immer schon*. When German philosophers talk about transcendentals, they often use exactly these expressions. (The English equivalent would be "always already," referring to the a priori conditions of possibility for knowledge or some other phenomenon to take place or, somewhat more vaguely, to a condition that has continued without any identifiable historical beginning, as in Martin Heidegger's *Being and Time*.) Put simply, a transcendental is a necessary logical presupposition. For example, when speaking about the transcendental subject, Immanuel Kant begins by explaining: "It is [. . .] very evident that I cannot know as an object that which I must presuppose to know any object" (A402). The subject is transcendental in exactly this sense: the "I" can never become an object itself, but at the same time is a necessary precondition for all knowledge. Kant—whom many take to be the most influential Western thinker of them all—then goes on to claim that it is impossible to have any direct conceptual knowledge of the object "since any judgment upon it has *always already* made use of its representation" (A346=B404; my emphasis). This third phenomenological observation may only be a minor detail, but I must admit that as a philosopher, it came as a surprise to me when I saw that a few of our meditators, when trying to describe their minimal phenomenal experience (MPE) experience, spontaneously used exactly the same expression: *schon immer* / *immer schon* / always already. Here are two further examples, this time ones that were originally submitted in English:

1712 I experienced awareness itself. Not in the sense that I was experiencing awareness; there was no "me," no "observer." Awareness itself was always already aware.

88 Always and already. Not coming or going. [. . .]

The phenomenology of "always already" also seems to be intimately connected to a fourth aspect of "This is not an experience!"—namely, the *nontransitory* character of pure awareness, its apparent timelessness (more examples can be found in chapter 22). As one of our participants said:

3570 [. . .] It seems difficult to split pure awareness into before, during, and after. This means that it is hard for me to allocate certain memories and sensations precisely to one phase. For me these transitions are barely describable in retrospect.

Timelessness is clearly a major phenomenological marker of the pure-awareness experience. The phenomenological aspect of "transcendentality" could be related to this experiential quality of timelessness—which, philosophically, would be expected of something highly abstract and transcendental, such as something related to the *condition of possibility* for time experience itself. However, if all this is true, such experiences really aren't "experiences" at all—at least in the sense of being episodic. They are temporally unbounded, and they are not tied to a given psychological moment, to any specific consciously experienced Now. If taken seriously, some of our reports attempt to refer to something that has no beginning and no end. And there is no "point in time" at which it takes place (here, you may recall the short quote by Douglas Harding presented in chapter 19). The following report makes this very clear:

1749 It never started, and it never ended, and that's no contradiction with saying that it ebbed in slowly like getting into a warm bath, and ended in the same way about 6 hours later. [. . .] None of this was experiential, meaning no experience or experiencing could indicate or capture it.

In chapter 22, we investigated the phenomenology of "timeless change." It seems that, for some participants, one way of trying to describe the aspect that I am trying to isolate here is to say that pure awareness is the true self (as discussed in chapters 29 and 30) and is *absolute* in the sense of preceding all other forms of appearance:

2668 I had an experience of pure experience in which I became aware of myself. I became aware that the ego is absolute and that "before" it is nothing. [. . .]

In addition, it seems as if the transcendentality of "always already" and the sense of touching reality "in itself" can also be couched in terms of the phenomenology of nothingness or complete absence (chapter 16):

1788 [. . .] In a further, even deeper state I suddenly see that everything is pure "nothing." This is initially frightening for me as well as very liberating and absurdly funny, because "nothing" is the only thing that is. I myself am also nothing, have always been nothing, and I cannot lose anything, because there is nothing to lose. This is how I experience it in this state. The liberating and very satisfying thing is that I then "see" that what I really am cannot die, because it is always already this lively nothingness and there is absolutely nothing else. [. . .]

Let me end the first half of this chapter, in which we have dwelt on the phenomenological puzzle constituted by credible reports about "conscious experience that really isn't," by showing you an example of how some of the meditators who participated in our study did more than report on their own experiences. Some actually developed deep philosophical theories about what "appearance" really is and what precisely makes it possible:

> 1416 What you have defined as "pure awareness," I would describe as follows: being aware that some (any) content of experience appears (even very subtle, like the fact of breathing or just being alive), and then becoming aware that what appears is the nonappearing of something that does not appear. When the meaning of "appearing" (being aware) becomes "the nonappearing of the nonappearing," then any content of experience (sensible or not) fades, goes to the background. What remains is the fact of experiencing (being aware of) something that is not there. In this way, there is awareness of awareness (because the content is no longer disturbing the perception of awareness as such), but this goes much deeper than just what is captured by the words "pure awareness," since what is really appearing is the fact that any appearing is the nonappearing of what does not appear.

Noumenal Awareness and Sympathy for the Devil

> The subject does not belong to the world, rather it is a limit of the world.
>
> Where *in* the world is a metaphysical subject to be found? You will say that this is exactly like the case of the eye and the visual field. But really you do not see the eye. And nothing in the visual field allows you to infer that it is seen by an eye.
>
> —Ludwig Wittgenstein (1889–1951), *Tractatus Logico-Philosophicus*, 5.632 and 5.633

> *Question*: Through what expedients is it possible to trace back the radiance of one's sense faculties in one thought and awaken to self-nature?
>
> *Chinul*: The self-nature is just your own mind. What other expedients do you need? If you ask for expedients to seek understanding, you are like a person who, because he does not see his own eyes, assumes that he has no eyes and decides to find some way to see. But since he does have eyes, how else is he supposed to see? If he realizes that in fact he has never lost his eyes, this is the same as seeing his eyes, and no longer would he waste his time trying to find a way to see. How then could he have any thoughts that he could not see? Your numinous awareness is exactly the same.
>
> —Chinul (1158–1210), *Secrets on Cultivating the Mind*

In his famous early work *Tractatus Logico-Philosophicus*, Ludwig Wittgenstein said that the subject itself does not belong to the world, but is a *limit* of the world (*TLP* 5.632). The original German title, before the English translation was given the Latin title, was *Logisch-Philosophische Abhandlung*; the book was written during World War I and not published until 1922. In 1929, Wittgenstein submitted it as his PhD thesis at Trinity College in Cambridge, and it was the only book-length work he wrote that was published during his lifetime. What exactly does it mean for something to be "a limit" of the world? Could the pure-awareness experience be a way of touching or even becoming one with the limit of our world? Could it perhaps be the ultimate selfless form of being a *subject*?

In the third quotation given at the start of this section, Chinul—the founder of Korean Zen—tells us that there is a way to "realize" the fact that we have never lost our eyes, and this is the same as "seeing" these eyes—which on the other hand, as Wittgenstein points out, *cannot* be seen because they never become part of the visual field. (Mirrors let us cheat somewhat, of course, but in a mirror, what you are seeing is not your eye as a *seeing* eye.) What is the truth here, when we think of the phenomenal field as a whole? Is consciousness per se somehow folded into the field of experience? In this special case, are "realizing" and "seeing" it really one and the same process? Or is the realization in the end not a real insight at all, but rather the making-real of a new world-model in the brain? Here, we get into philosophically deep water.

In Roberts Buswell's English translation of *Straight Talk on the True Mind* (which was written around 1205), Chinul is asked by a student what the normal mind is, and he replies: "All men possess a point of numinous brightness which is still like space and pervades every region. When contrasted with mundane affairs, it is expediently called noumenal nature. When contrasted with formations and consciousness, it is provisionally called true mind." Could this be a promising route for making sense of many of the reports presented in this book? Are they perhaps not reports about mundane "phenomenal experiences," because they are really referring to the realm of *noumena*, to the true mind, to the "noumenal nature" sometimes revealing itself in our very own consciousness? The general picture that would thus be emerging could make us see that what it really means for an experience "not to be an experience" is the fact that the episode in question cannot be adequately described as *phenomenal*, but is more like an awakening into something *noumenal*. It is not even an episode—but rather an internally timeless insight into something that is genuinely transcendental.

It may be tempting to say something like this: phenomenal consciousness refers to how things appear to you, to *what it is like* to have an experience; noumenal consciousness, strictly speaking, is not *like* anything at all and accordingly is correctly reported as not being an experience. All phenomenal forms are virtual, therefore ultimately *misrepresentational* (chapter 28), and must disappear. What remains is pure, reflexive awareness—the only aspect that is guaranteed to be veridical. Here, reality is grasped because it is based on a direct, nonegoic form of self-knowledge—namely, the process of recognition itself. Perhaps we would also like to say that MPE is the way in which the *conditions of possibility* for all experience reveal themselves, but a way we haven't yet fully understood. Clearly, for anybody who wants to take the reports presented here seriously, a host of philosophical questions begins to arise—and I admit that I have no ready-made answers.

As a starting point, we should be aware that the distinction between *phenomena* and *noumena* is something conceptual, and is not part of the contemplative experience/nonexperience itself. The distinction itself does not belong to the qualitative character of MPE. In themselves, most of the internally timeless episodes verbally reported in this chapter clearly lack any positive distinction between appearance and reality, or between what could be merely phenomenal and what might be noumenal. Perhaps we can say that they are also nondual in the extended sense of the dualism between appearance and consciousness "in itself" having been suspended, because once again, we seem to find a quality of neither-nor-ness. For the most part, we hear only what they were *not*—namely, experiences. And this is what we must pay attention to.

One source of complexity here is that the conceptual distinction between *phenomena* and *noumena* was not available *as such* in those early Asian contemplative traditions that generated the scholar–practitioners who observed their own minds over thousands of hours, with admirable discipline and great existential seriousness, and who gradually developed the techniques some of which many Westerners now use to meditate. The distinction originated in early Western philosophy (with Sextus Empiricus, whom we met in chapter 14 in the company of Wittgenstein) and today is mostly associated with Kant's epistemology. To give just one concrete example, the influential and important Eastern notion of recognizing one's "true nature" emerged not only in a distant historical epoch, but also in a completely different sociocultural context from Kant's system of thought. Therefore, it will be difficult to relate the Western distinction between appearance and reality in all its own historical depth to, say, the time-honored tradition of Buddhist philosophy. Both are highly differentiated within themselves and have their own complex prehistories. There simply is no direct mapping here between Western philosophy and Korean Zen or other important Asian traditions. But we can certainly ask whether shared phenomenological anchors lie behind all those incommensurable conceptual schemes. We would then be looking not for absolute perennial essences, but for humankind's experiential prototypes: shared regions in neurophenomenal state space that ground the philosophical puzzles they later create.

Buddhist philosophy is deep and complex, spanning many centuries. But Kant's distinction between *Dinge an sich selbst* ("things in themselves") and mere appearances is complex too. Let us use it as a small case study in the problem of incommensurability between different philosophical worldviews. For Kant, things in themselves are the conceptually required *something* that is not appearance, but appears in every appearance. At first glance, it may be attractive to say: Appearances are empty because they have, exactly as Kant says, no existence "grounded in themselves"; they are only experiences. But MPE, pure awareness, is precisely the one *Ding an sich selbst* that we have been looking for, because it is here that we find consciousness groundlessly grounded in itself. MPE is the true self, the true nature of the mind—the "noumenal nature" referred to in Buswell's beautiful translation presented earlier. But again, Kant's philosophical system is complex—and it has its own difficulties. For example, the appearance/thing-in-itself distinction is not the same as that between *phenomena* and *noumena*, between what can be an object of our sensible spatiotemporal intuition and what can never become an object of sensible intuition. In addition, as far as we know, Kant was not a meditator; he did not systematically cultivate this form of epistemic practice and probably did not even know the "nonexperiential" states to which some of our meditators refer. His theoretical work could not have been inspired by such states.

I hope this small example illustrates one or two aspects of the deep crosscultural incommensurability that exists here. There is no simple mapping. The number and scale of the incompatibilities is exactly why a fresh and global bottom-up approach is so important, complete with large and heterogeneous samples of participants.

But let us not avoid the problem. There are limits to scientific understanding. To bring out the issue more clearly, I will now play the role of an *advocatus diaboli*, taking another, somewhat more playful, angle to illustrate what I think may be the central difficulty here for you, my reader. The devil told me this:

> You cannot treat reports in which people say "This was not an experience!" as experiential reports. The material you just presented shows that your own approach of focusing solely on the phenomenology of pure awareness while trying to leave out the epistemology and, most of all, the metaphysics is deeply misguided. You are feigning ignorance, burying your head in philosophical sand. Proceeding with a reductionist "phenomenological approach" while ignoring self-evident metaphysical implications is intellectually dishonest—and as you have said yourself, intellectual honesty is a necessary element of any truly spiritual perspective. These are spiritual experiences and you are refusing to take them seriously *in their spirituality*.

"Maybe," I tried to interrupt, "but 'self-evident' metaphysical truths just don't exist. Don't you see . . ." But then I noticed an uncanny, mean-spirited smile on the devil's face. This smile expressed a deeply malicious form of compassion, an evil form of empathy that I had never seen before. The devil had already seen through me:

> Don't you see that all your misguided attempts to somehow understand something that just *can't* be understood ruin your very own practice? The craving for understanding destroys it all. All you're doing is feeding the epistemic agent model in your brain, making it stronger and stronger. "Combining different forms of epistemic practice"—what puffed-up nonsense! You are nourishing delusion, and one part of you enjoys it. The other part knows I am right.

This gave me pause. The devil continued:

> You can perhaps do a semantic analysis of ways in which people have used concepts like "pure consciousness" or "pure awareness" in the past, and you can carry out psychometric studies and offer statistical analyses of the actual reports that present-day, real-life meditators give. But if you want to proceed to a genuinely *qualitative* analysis of these reports that at least attempts to do justice to your research target, then you simply have to concede that their authors are referring not to *phenomenal* states but to something *noumenal*—not to the way pure awareness "appeared" to them, but to the fact that it actually brought them into contact with the reality behind all phenomenal experiences, the true self-nature itself. Your phenomenological reductionism blinds you to this obvious fact. You are deliberately superficial, systematically nourishing delusion—and, again, there is a part of you that already knows it. If you'll permit Satan to use a hybrid Kantian-Buddhist expression, these reports point straight to the very "no-thing in itself." Again—you are not researching phenomenal consciousness here; you're researching numinous awareness, which is your very own true and undifferentiated noumenal nature itself. What you call "MPE" is neither minimal nor phenomenal; it is the real thing, a *noumenon*, consciousness *in and of itself*. You are not advancing our understanding at all, because you are trying to turn a blind eye to the fact that none of this is about more accurately describing "phenomenal qualities" and all the other rubbish—that it's all about an actual manifestation of the noumenon itself.

I must admit that I am slowly beginning to like the devil, simply because I have had to live with her for such a long time. There is something about her. I am sometimes smitten by her charm—did you notice how elegantly she is trying to sell us versions of the E-fallacy and the C-fallacy at the same time?

Recall that the E-fallacy arises whenever someone falsely concludes that a consciously experienced feeling of knowing is a reliable indicator of actually possessing knowledge (chapters 7 and 18). Accordingly, the feeling of definitely knowing that something was *not* an experience, and therefore was *not* merely phenomenal, needs independent justification. Yes, the reported experiences are often deeply valuable, awe-inspiring, and formidable, and we are obliged to do the fullest justice we can to them—but to draw strong theoretical conclusions, we need to know a lot more. The purported metaphysical implications of the phenomenological descriptions are not self-evident, to the point that many of us just don't see them—as a matter of fact, few of our respondents make strong knowledge claims of this kind. It is only the devil who tries to make us jump to conclusions. And this is just one of the many guises in which you may have begun to notice her popping up.

Conclusions are also easily jumped to when it comes to "phenomenality per se" actually being something noumenal, being the "thing in itself" behind all appearances. The C-fallacy arises whenever someone falsely concludes that just because something feels like the true and timeless nature of consciousness itself—or *not* like an experience at all—we have actually found some metaphysical bedrock or ultimately understood consciousness itself. But the original C-fallacy has a mirror image: Just because something is reported as feeling *not* like an experience at all, this doesn't license the conclusion that it isn't. To complicate matters, we already know that there are states that are mere experiences, although they feel like knowing or even carry the phenomenal signature of knowing—which then leads to the much more interesting question of whether there could also be meditators who, on a nonconceptual level, have genuine insights but do not recognize them as such, mistakenly treating them as mere experiences. This would be the mirror image.

Once again, there is no doubt about the sincerity and veracity of our meditators' reports. They do their best to convey something that, for millennia, has been pointed out as being beyond all words. In doing so, they make an invaluable contribution to consciousness research. But to actually have a parsimonious minimal model explanation for consciousness is something else. It is another project, which belongs to another form of epistemic practice.[1] The way to go is not to jump to strong metaphysical or epistemological conclusions in a naively realistic manner.

As a matter of fact, taking spiritual experiences seriously *in their spirituality* (as the devil demands) means precisely this: not jumping to conclusions, not escaping into metaphysical speculation, not abusing the epistemic practice of meditation as a substitute for religion, not abusing the experiences or the theories for purposes of mortality denial or narrative self-deception (chapter 17). Intellectual honesty is an integral, indispensable part of spiritual practice and it is what connects such practice with modern philosophy and science (see the epilogue).[2] But the interdisciplinary endeavor of consciousness science is a project distinct from personal spiritual practice; it is a form of epistemic practice that targets different objects of inquiry. To begin paving the way toward the formulation of a first standard model of consciousness itself, we need much more—we need neural correlates, computational modeling, genuine interdisciplinarity, and an ongoing conceptual synthesis of empirical data. Fine-grained phenomenology is needed to kick-start the process, but unfortunately, we cannot "bootstrap a theory of consciousness out of pure consciousness itself." Thinking that we can is what in chapter 12 I labeled the "C-fallacy."

One problem is that, in a way that is (like so much else) very hard to express in words, MPE does not feel like what philosophers call a "phenomenal experience," something that subjectively appears to us by coming and going in our minds. Rather, it is something that seems to reveal the timeless condition of possibility for conscious experience itself (chapter 22). MPE is maximally simple, but there is an aspect of profundity that we can now begin to see as directly related to what at the beginning of this chapter I termed the "phenomenology of transcendentality." This aspect of transcendental profundity is what resists qualitative analysis.

Today, new theoretical options for understanding transcendentality are on the table. For example, the philosopher Jakob Hohwy has pointed out that under the free energy principle, certain nonequilibrium steady-state systems like ourselves can be described as tracking and representing the conditions of possibility for their own physical existence.[3] We already encountered this idea in chapter 17, when discussing the craving for existence. All physical processes have certain conditions that make their existence possible. But only very few physical systems model and *track* these conditions. Self-organizing systems like ourselves exist by persisting in dynamic environments, following, as Hohwy explains, an inner norm by which they model the conditions that make their own physical existence possible. I think that we must not overlook the fact that this modeling itself occurs in a physical medium; if you will, it autopoietically re-creates those conditions for as long as such systems live. Interestingly, *speaking* of the conditions of possibility for existence can also be seen as an (albeit very abstract) way of referring to existence, to being, itself. My own question is this: Why should there not be a much more concrete, fully embodied way of internally modeling or knowing these conditions nonconceptually, without words, in a state of silence—and thereby *realizing them*? Given this new context, I find it hard not to think of the experience of "pure being" that has popped up in so many places throughout this book (e.g., in chapters 1 and 26).

But what about pure *awareness*? Can we also internally model the conditions of possibility for *knowing*—in meditation, nonconceptually, without words, in a state of silence? If we view the experience of pure awareness as an internal model of epistemic openness (see chapter 4), and if the same principle of continuously tracking the conditions of possibility holds not only for whole organisms, but also for conscious processes unfolding in their nervous systems, then it is conceivable that the actual functioning of this principle can sometimes be detected by the model itself (e.g., in exceptionally clear and silent nondual states). It could be detected by meditators in states of dual mindfulness, but it could perhaps also detect *itself* in what, in the preceding chapter, we called "reflexively aware mindfulness" or "self-knowing empty cognizance." The currently running model would be continuously tracking its own condition for existence—epistemic openness, the very possibility of knowledge—thereby autopoietically re-creating itself over and over.

If this physical process had a conscious correlate (or if it were not correlated with, but were simply *identical* to, pure awareness "coming alive"), then our prediction would have to be that a bare and empty form of wakefulness emerges, a state that cannot be easily compared to any other kind of conscious experience. The phenomenological prediction would be that it has the experiential quality of continuously refreshing itself, of autonomously holding itself in existence—and, plausibly, of *knowing* itself.

Of course, all of this is more than speculative, and much more research is needed. But given the new theoretical options just sketched out, it is striking to note that contemplative scholar–practitioners living in an entirely different sociocultural context many centuries ago formed experience-based concepts that translate into English as "self-generating pristine awareness," "self-existing wakefulness," or "self-originated primordial awareness." Is there a rational, no-nonsense way to link these ancient zero-person frameworks to present-day third-person evidence, perhaps using the mathematical terminology provided to us by data-driven computational models of brain function, creating a "computational phenomenology of reflexive MPE"? It may sound like too distant or ambitious a research goal to many, but in all modesty, I would like to propose that precisely this would be a major achievement for the next generation of consciousness researchers—perhaps even the one that leads to a decisive breakthrough.

32 Meditation and Nonmeditation

> A bit like balancing a ball on a stick, the moment when it is in balance is characterized by lightness and naturalness, but is at the same time very difficult to hold. [#1457]

> It is difficult to put into words, but through meditation I can create conditions (by relaxing into the moment) and it comes or not, like grace [#2714]

This chapter unites three major topics: the phenomenology of effortlessness and the experience of spontaneity, plus the interesting fact that experiences of pure awareness quite often occur *outside* formal meditation practice. Among the most common contexts in which pure awareness arises are informal but still deliberate forms of practice that weave mindful awareness into everyday activities. These include mindfully eating or mindfully washing the dishes, sprinkling brief "micromeditations" throughout the day, repeatedly "glimpsing" a nondual state and then immediately letting it go, or—as discussed in chapter 14—practicing *satipaṭṭhāna,* informally establishing mindfulness during the day as part of living an examined inner life.

Let us begin by recalling that "dual meta-awareness" and "nondual meta-awareness" are two distinct regions within the space of possible states described as the pure-awareness experience. As we saw in chapters 26, 27, and 28, the phenomenal quality of awareness itself can occur in the context of an apparent self looking *at* awareness, as well as in the context of looking *from* awareness or "being the self-aware looking itself." In the second case, the overall process is often described as entirely effortless. Accordingly, there also exists a phenomenology of nonmeditation, and there are important differences between pure awareness in the context of dual mindfulness and pure awareness in the context of nondual mindfulness.

If we take our participants' reports at face value, nondual awareness does not feel like something that has been fabricated or created by a meditating self; rather, it feels like a spontaneous occurrence. This has an unexpected consequence: Some paradigmatic minimal phenomenal experience (MPE) states are *not* meditation experiences. One example is found in full-absorption episodes of "self-cognizing wakefulness" that occur during periods of formal practice originating in the wake state (in which the meditating self has disappeared); another is the clear light sleep discussed in chapter 20. But perhaps there is a deeper philosophical sense in which *all* experiences of pure awareness occur in a state of nonmeditation?

As we will soon see, there is much more to be said about this specific phenomenological aspect. Before we begin to look at our reports, here is a typical practical instruction for how to facilitate an effortless, spontaneous recognition of reflexive MPE. This quote from Urgyen brings out the difference between meditation and nonmeditation in a clear and simple way: While meditation is deliberate mindfulness, nonmeditation is the effortless *self-recognition* of pure awareness by itself:

> That is called deliberate mindfulness. It is dualistic mind that reminds you to recognize, but the seeing of no thing to be seen is *rigpa*, the awakened state free of duality. This becomes clearly discerned through practical experience.[1]
>
> I mention this because, if the goal of the main training is to construct a state in which thoughts have subsided and which feels very clear and quiet, that is still a training in which a particular state is deliberately kept. Such a state is the outcome of a mental effort, a pursuit. Therefore, it is neither the ultimate nor the original natural state.
>
> The naked essence of mind is not known in *shamatha*, because the mind is occupied with abiding in stillness; it remains unseen. All one is doing is simply not following the movement of thought. But being deluded by thought movement is not the only delusion; one can also be deluded by abiding in quietude. The preoccupation with being calm blocks recognition of self-existing wakefulness, [. . .].[2]

Using our conceptual tools, we can say that meditation as a practice still involves an epistemic agent model (see chapter 25), a goal-directed self that wants to know something (e.g., how best to "abide in quietude"), whereas, phenomenologically, nonmeditation is the spontaneously occurring recognition of pure, nondual MPE itself. We might also call it "undistracted nonmeditation." Nonmeditation is what comes after the surrender of the epistemic agent model. This is important because it shows that certain aspects are beyond the reach of "spiritual athleticism," in the sense of willpower, earnest discipline, or any approach based on the application of "techniques." This fact also has an unexpected consequence: Nonmeditation means stopping trying to understand. This being said, let us now turn to our own phenomenological data. To begin with, many of our meditators describe a specific quality of effortlessness and mental nonaction:

> 172 Felt openness, effortless clarity and awareness; sense of ease, spaciousness, timelessness . . . Okayness.
>
> 1960 That was a paradoxical feeling of total awareness and at the same time absence of any effort. Like something in my brain had stopped and my frontal lobes relaxed at once. [. . .]
>
> 2951 It was a long-lasting experience of just being fully "myself" in an open and undefended contact with my surroundings (which included a group of 20 people). It was the most effortless and "simple" way of being, and there was a strong experience of being a crystal-clear vessel, fully capable of being of service for whatever would appear to be needed at that time and place. [. . .]

Second, many of our participants' reports support the classical teachings' claim that pure awareness itself cannot be fabricated or directly constructed. However, their reports also reinforce the idea that contemplative practice may actually involve an intricate interplay between conscious and unconscious processes—as already proposed in the "dolphin model of meditation" sketched in chapter 10:

> 1322 [. . .] a state that can occur regularly both in meditation and spontaneously in restful wakefulness during the day or in sleep. I cannot provoke it or reach it with a certain thought, intention, activity, or the like. It is there when I "let go," pause completely. [. . .]
>
> 2270 The experience comes when I refrain from generating it. It is a feeling of absence of all effort and mental agitation. I am always surprised what the concrete experience is like when things that are otherwise in the foreground disappear. It is actually very easy to experience this. In my opinion, you don't need impressive meditation experience for this, but rather a kind of basic peace / inner calm, so that this experience is not constantly overlaid and thus not perceived. [. . .]
>
> 2603 [. . .] I never have such an experience early on in the meditation (e.g., the first 10 or 20 minutes); it usually takes a longer period of practice (of directing effortful attention toward my breath, thoughts, feelings, and sensations), perhaps 30 minutes or longer. Then, the experience of pure awareness will arise almost spontaneously, as if the momentum of the mindfulness practice in those 30 minutes got me in the right groove or caused pure awareness to "click into place." [. . .]

The following longer report illustrates how pure awareness sometimes emerges at exactly the moment when the practitioner thinks that she is *not* meditating anymore. It also shows how the state of innocence that is made possible by this disappearing belief that one is meditating is lost again, gradually, by thought creeping in again—by the process that creates what we called the epistemic agent model (discussed in chapter 25), simulating, predicting, forcing itself back into existence:

> 3624 It began in the very moment when the bell rang and I knew that the meditation period was over. All thought stopped, and a great sense of mental clarity and precious silence descended. It was completely effortless. Crystal clear, yet entirely undramatic. It was as if during meditation there had been something I had always remained unaware of, an unconscious sense of striving for something, to not forget something, a background mechanism that kept checking, continuously trying to sustain something it would then immediately destroy by recognizing it as the state of pure awareness. As I got up, the unconscious mechanism finally relaxed and gave way to a thickness. While I slowly walked back under the trees, attention and the way it settled on perceptual objects had a gentle, almost tender quality to it. There was a subtle and completely undramatic but positive feeling tone permeating this silent awareness, a very fine and nonsensational sense of wonder, of delight, and soundness. Sometimes, I spontaneously stopped to look closely at bark or leaves. Suchness. Then I became attached. A first thought crept in, but quickly dissolved by itself: "Don't ruin this!" Unfortunately, something was now aware that it very much liked this state. As I slowly walked on and began to lose my innocence, the commentator began to jump in with clever remarks, but didn't yet manage to destroy the overall state. The clever commentator said: "Whatever this is, one thing is true: As long as I am like that there is not a trace of suffering in my mind! No fireworks at all, but this would actually be the end of all psychological suffering." As I walked on in silence, I noticed, in an entirely nonconceptual and nonintellectual way, that I am that which never speaks, that which is always silent, that which would never boast about or report an "experience."
>
> I also became aware that I did not yet want to meet anybody, because this would almost inevitably have interrupted the precious silence, the wonderful clarity. Thinking of other people was also the moment when the clever commentator tried to bootstrap itself back into existence and take over completely. It almost managed to ruin it all by giving a long speech: "We have a completely wrong understanding of meditation, solitude, and renunciation! It is not

> because people sat silently in formal practice, did regular walking meditation, distanced themselves from others, and lived in caves and monasteries that they often experienced pure awareness. We misunderstood the direction of causality! *Because* episodes of pure awareness sometimes spontaneously occurred in them, some of these people kept their distance, lived on mountains and in forests, sat quietly under trees, and walked very slowly—in order not to ruin it. Later on, others tried to get into such states by slavishly copying the observable outward behavior that was originally caused in those others by the experience of pure awareness. Going into 'silent retreat' was a state of mind, not something you did in the outside world! Rituals, liturgies, some of the daily routine in monasteries and retreat centers actually resembled cargo cults," the complacent commentator now thought. Self-importantly, he proclaimed: "Being a hermit, renouncing the world, becoming ordained as a nun or monk really are states of mind—and not outward behaviors to be copied and cultivated." I returned into the crystal-clear silence one more time, but the innocence was long gone and I decided to let it go and start my day. Back at my desk, I took my little notebook and wrote: "If it is really true that I *am* this, then I will never have to meditate again." After that, it was completely gone.

Long-term practitioners of meditation often experience episodes of pure awareness outside their formal practice. These are not states of full absorption, as in deep meditation or during dreamless deep sleep; they are experiences of silence, great clarity, witnessing, unity, pure being, or nondual awareness co-occurring with complex experiential content. Perhaps the meditator's brain can learn to unconsciously track the conditions of possibility for MPE?

I believe that the spontaneous occurrence of pure awareness is caused by two factors: (1) a continuous stabilization of the unconscious, functional preconditions of such experiences through formal meditation practice; and (2), more specifically, an improved ability to *recognize* such episodes during everyday life. There is a deep causal dialectic connecting meditation and nonmeditation. Recognition of the phenomenal quality of pure awareness can take place when alone in nature, but also in urban environments, during sleep/wake transitions, under the influence of psychoactive substances, in flow states, sometimes in social situations, during or after physical exertion (e.g., sport or sex), in a breathing class for singing and acting (#2964), or even during acute emergencies. Meditators—who already know the experience of pure awareness from their own formal practice—unexpectedly encounter its specific phenomenal character in a wide range of other, quite different, and more complex situations. Please note that many other reports referring to pure awareness as a spontaneous occurrence can be found throughout this book (e.g., #2426, #2780, #2936, #3068, and #3146).

Let us begin with twelve examples in natural settings:

1305 Spontaneous appearance of pure awareness during a perfectly ordinary walk where "the world flowed into me." [. . .] Experiences of the mutual penetration and overlapping of space, and that space moves with it, as well as the realization of not being separated. [. . .]

1313 While walking and being in the wintry forest with snow-covered trees and ice on rocks, it happened to me several times that I was unseparated from the environment, was at the same time observer and not-observer, part of the whole, felt the trees, the ice within me, or was unseparated from them. It was silence. [. . .]

1341 [. . .] a velvety feeling when walking outside and it felt as if the air was going right through me—I felt so light and frictionless. [. . .]

1432 The trigger was the beauty of some black berries in a hedge in winter. The berries, the street, the houses, and me and my movement were just there, were a unity without a name. Only perception. Without limitation. This happens quite often. Only perception, connected to the moment without evaluation.

1531 [. . .] often also quite spontaneously—for example out in nature, but actually in all kinds of situations. These are moments of thought silence, i.e., I have no commenting thoughts, but simply perceive the situation with just my senses, and I am aware of this perception as it happens. [. . .] Sometimes I also spontaneously experience other mystical states, such as those [. . .] in which it is clear to me that I am identical with everything that somehow exists and is conceivable—or experiences of oneness in which I directly sense the oneness of all being.

1575 [. . .] Especially when opening one's eyes in nature, there is a lightning flash of recognition of how everything is simultaneous, interwoven, inseparable from the seeing one and the surroundings, a kind of spatial experience and penetration, and a kind of noncognitive "knowledge" that it is all experience and silence at the same time. [. . .]

2561 [. . .] The 1st time I was sitting on a hill, with two children playing in the background. From the outside it was an everyday situation. But in my experience a never previously experienced presence, feeling of unity, nonreified clarity. The 2nd situation was in a café with a friend. Afterward I could only tell her that I felt completely existent and completely nonexistent in one.

2691 [. . .] I was walking outside, realized that I was in a very different mental, physical, and emotional state. The best way to describe it is to say that "there was an appearance." It was blindingly obvious that some things simply were "so." I stood and "watched" this, but it was not "me" doing the watching. My internal world and the external world were "complete" in and of themselves. I "knew" that some personal beliefs I usually possess and hold were no longer relevant, and a wider understanding of the human condition took its place. I knew what I was experiencing, it was not imaginary, but the knowing was also beyond me. There was a universality about it.

2738 I was lying in a meadow with my eyes closed. I felt safe and secure. As I lay, I eventually fell deeper and deeper into myself (body/mind). It is difficult to describe this quality. It was a falling into and through oneself, through several layers of human experiential worlds down to a layer where nothing and everything was at the same time. I stayed there for some time, I don't know how long. After I came "back," my perception of the world was retrospectively changed. I had no need to speak or perform any action. Time did not exist. Everything was only now. After about 2–3h this state wore off and thoughts plus self-reference returned.

3166 [. . .] For example, experiences in nature where I am completely there, one with everything, full of joy and light. Fearless, connected, full of love for everything that surrounds me. Or full of compassion. Everything is beautiful then, even dead trees or suffering.

3243 [. . .] While on a walk I stopped, looked at the sky, a bird flew by. The feeling of just "BEING" appeared again. There was neither this nor that. Without limits . . . [. . .]

3413 [. . .] on the way to the commuter train, crossing a meadow past a small coniferous forest, there arose out of nothing a feeling of merging with the environment, especially the trees. A complete detachment from thinking, although I was still capable of it. I could also feel. But the feelings did not correspond at all to anything I had known before, like fear, joy, etc. It was a moment of incredible energy that was not limited to anything, a connection or unity, a complete freedom. [. . .]

Spontaneous pure awareness can also come about in urban environments:

1277 [. . .] for over a week, during my visits to the city center, I had experienced a conscious separation from my surroundings. At times I felt like a disembodied being when I walked through the city. Connected with this was a feeling of freedom and amazement. [. . .]

1323 After a meditation I left the house with a deep calm and the inexpressible feeling of contentment, went outside and sat down in my car to set off somewhere. As soon as I was sitting in my car, my surroundings dissolved completely. I was the asphalt under the car, I was the adjacent tree, I was the meadow, I was EVERYTHING and gave no limit. It was a feeling of unbelievable freedom and deepest bliss. This state lasted for about 5–6 minutes (I had looked at the clock before and after).

1381 The experience [. . .] was unexpected. On an ordinary morning I stepped out of my front door into the street—into a space of timeless self-luminous world-penetrating silence. All my sensory perceptions became three-dimensional, sharply contoured, exaggerated, and at the same time I was detached from the perceptions and sensations. I had a physical feeling of heaviness and sinking and opening expanding billowing time. The word I use for myself to describe this state is "silence" or "space." The silence is not acoustic, the space is not physical. I felt bright cheerfulness, which spread in waves upward and out of me. The whole world was just completely logical and clear, embedded in this glowing fragrant sweet silence. [. . .] Anyway, since then I can easily and almost always enter this delicious silence.

1428 [. . .] so I was in the supermarket, and I was worrying about spending too much money and suddenly my awareness opened up and I didn't feel my body anymore and there was lightness, it was like something opened up in my chest and dissolved "me" and awareness was just floating there, soon awe came and contentment. It was pretty cool. [. . .]

3356 Nothing else is experienced than being. All sounds and images heard and seen are more intense, connected and stripped of illusions. Nothing more than a very intense feeling of complete union with all and everything. The "quality" of the environment seems unimportant for the occurrence. The strongest experiences even happened in not so nice environments such as a noisy car park under a train station. And they happen always completely unexpectedly. [. . .]

3510 I was sitting in a Chinese fast-food restaurant, eating and killing time before going to the movies. Although I already knew the dish, at first the food tasted better than usual (similar to a shift in consciousness, but related only to the sense of taste, not substance-induced or due to mental illness), and suddenly I experienced myself as abruptly centered and in the middle, which triggered a great inner lightness and joy. [. . .] My movements became slower as I ate, more conscious and more enjoyable. I didn't analyze and question this very intense moment, but enjoyed the centeredness, the calmness, joy, and lightness within me for about two minutes, although it was loud and hectic around me. Since then I have never experienced it with this intensity again. [. . .]

Spontaneously occurring pure awareness can be recognized during sleep/wake transitions:

1360 Between sleeping and waking up there are "moments" that for me are absolute well-being—complete naturalness on all levels—[. . .]

2964 [. . .] After half an hour of taking slow, deep breaths, I had a moment when I probably fell asleep for about 2 minutes. I don't really remember falling asleep, but I do remember being blank and realizing that I wasn't paying attention to what the teacher was saying. The feeling of clarity and calm was immense. One of those moments when you are aware of what is happening, you can even talk, but at the same time you are "absent," calm, oblivious to everything, and, for the same reason, able to take more details in the expressions, in the tone of a conversation, etc. It is as if you achieve a kind of autonomy over your own thoughts, but thanks to that, you can see clearly and not only what you are predisposed to see with your interests, complexes, expectations, etc. When this happened to me I felt strange, but curious and calm.

3243 In the "falling asleep phase" there was suddenly an awareness of total boundlessness. It is very difficult to think of another word for it. There was neither body nor thought nor emotion . . . just "BEING." [. . .]

Then there are also deliberately altered states of consciousness in which MPE experiences sometimes occur spontaneously. One of humankind's ancient epistemic traditions—at least as old as systematic contemplative practice, but probably even older—consists in the systematic use of hallucinogens in the pursuit of self-knowledge. As the philosopher Chris Letheby, one of the world's leading experts in psychedelics, has pointed out, both epistemic practices—meditation and psychedelic use—promote (1) the unbinding of mental contents from the self-model, (2) a decentered mode of introspection, and (3) a shift of self-related mental content from phenomenal transparency to opacity. His speculative hypothesis is that meditation promotes self-transcendent experiences via decentering and phenomenal opacity, while psychedelics promote decentering and phenomenal opacity via self-transcendent experiences. As Letheby writes:

[. . .] the connection is real, deep, and not solely due to cultural framing, suggestibility, or interpretive bias. It consists, at least partly, in the fact that psychedelics and meditation can both disrupt self-binding processes, promote decentering, and confer phenomenal opacity on (self-related) mental contents, though the causal pathways involved are, in a sense, mirror images of each other. The similarities and differences between psychedelics and meditation have informed the project of epistemologically analyzing psychedelic experience within naturalistic constraints; thus, we ought to consider whether they might inform the parallel project of epistemologically analyzing meditation practice within such constraints.[3]

These are among the reasons why some serious practitioners of meditation integrate psychoactive substances into their lives in a sustainable way. Today, there also seems to be a historically new population of committed practitioners who were earnest and intrepid spiritual psychonauts in earlier phases of their lives, but found the psychedelic experience to lack sustainability. First, they tripped, and then they went to India to deepen their search. Now regular meditators, they have given up one epistemic practice in favor of another, with some of them perhaps still on rare occasions repeating the type of experience that got them started in the first place. Here, I will confine myself to four reports given by participants of our study, the first two referring to LSD experiences, the second two to psilocybin:

108 [. . .] Meanwhile there was no *I* [*ICH*], only an emptiness and a cool "breath." There was—even during this time—the certainty of having encountered NOTHINGNESS. The only feeling that can be analyzed and described was astonishment, also astonished watching—but from no particular perspective.

2267 The strongest and most vivid experience of pure awareness was on a high dose of LSD with two friends. [. . .] And then at its peak suddenly every boundary dissolved. There was no longer any chatter in the mind, no feeling of I myself, no distinction between self and other. The thoughts of my friends were my thoughts, the same as their doing and vice versa. This feeling of connection was present everywhere I looked and I no longer had any concepts of Table or Door or Tree in Mind, there was just experience. And a subtle feeling (no thought) of "this is right." Since then I've had several similar experiences randomly throughout my day where I suddenly was aware again, and very often I just have to laugh or smile when it happens. I no longer meditate to reach that state because I realized that there is no state to be reached. So I just meditate for the enjoyment of meditating.

1926 [. . .] This was an experience I had while on psilocybin. I stared into the sunset and there came this exact awareness of being aware that there was no point from which I was experiencing the sunset or any of the other sensations.

They all became this "one" thing where the only thing that remained was the world in some sense. [. . .]

3470 The experience I used to answer this questionnaire was one I had on a psilocybin journey as part of Johns Hopkins research on long-term meditators. [. . .] With eyes closed and headphones on, however, I experienced an incredibly expansive visual spaciousness that in certain moments went beyond expansive toward limitless. There were also instinctive contractions that seemed to be both physical and mental and that felt like the self trying to stay or come online. With relaxation, those contractions abated, and the spaciousness/expansiveness returned. The spaciousness was not confined to internal visual space but instead also included internal auditory space as well as somatic space. Things were thin. There was very little negative valence. There was an understanding of how the world works and how the self forms. It took place about 4 years ago and I'm not sure what kind of languaging I have onboarded since then, but I do think that the awareness of the roominess had no center. Awareness was everywhere / in no particular place and mapped straight onto experience itself.

"Flow states" sometimes occur when we become deeply immersed in an activity that is challenging but not outside our skill set. They are characterized by total absorption in and concentration on an activity that we enjoy and may be passionate about, sometimes resulting in a partial loss of self-consciousness and a quality of timelessness. The relationship between pure awareness and flow states has long been noticed;[4] here is one typical report:

2567 In an everyday situation I had an experience of being effortlessly fully on-task with undivided attention, essential pure perception, wakefully surrendered to action.

During spontaneously occurring MPE episodes, we may find that certain phenomenological features discussed in this book are particularly dominant. The following set of examples include the transcendental character of luminous and spacious wakefulness, nonduality, insights into the deep structure of time, ego dissolution, the experiential quality of soundness, and spontaneity and effortlessness. Sometimes these various aspects or features can even be distinguished and invited into the foreground of experience:

1530 [. . .] As I type these words, I can listen to the sound of the keys, reflect on what I am writing, and also tune my attention to the fundamental wakefulness of awareness that both receives and makes possible all these perceptions. If I focus particularly on awareness itself, it begins to pervade all experience. One or other nuance begins to stand out—its knowing quality, or its natural luminosity or brilliance, or its mysteriousness, or its self-evidence, or its edgeless, centerless all-pervasiveness. All these nuances and others can then be contemplated directly, filling experience, or they can be just the background which the sense of agency rests in and arises from. In this way the sense of pure awareness, or the fundamentalness of awareness as the ground of all perception, stays present, while thought and action and activity can still be engaged in, while being felt as unfolding naturally, effortlessly, informed by the wisdom–clarity–freeness of the awareness that underpins everything.

1573 [. . .] There was a deep and perfect sense of time. As if in a few seconds all time that there is were stuck. As if every action, every encounter that can exist, were contained in this action. There was no "they" and "I." But also not the thought that the "they" and "I" do not exist. There was only "being." Completely saturated.

2683 Lately, I get flashes of having "disappeared completely" while the world remains, usually in ordinary settings outside of focused meditation.

2798 [. . .] I was standing at a window and looking out to a tall bunch of grasses. It was a windy day and the grass was moving a lot. Then, for some period, I coincided with these grasses. Their movement showed (or maybe better: was) the nature of time, the complete ungraspability of the present while at the same time there is nothing else, no past, no future, just this ungraspable now. The difficulty of answering your questions is that there was no thought at all about time or the passing of time, there was just this movement. Only later, in reflection, there came the thought "I saw into the nature of time."

3171 [. . .] Awake (outside of formal meditation)—bright, spacious, vibrant, expansive, happiness, equanimity, unity, ease, thoughtless, without conscious directing of behavior or impulses.

3207 Sitting relaxed in an armchair with a view into a garden. Then, out of the blue, the world, existence—simply everything that exists—opened up into an infinite expanse with a deep and clear, almost absolute and unshakable certainty of the coherence of everything that exists, that is. Everything is in the right place. Everything is all right, just as it is. I—was minimally present in this moment but also all right. Feelings were mainly love and amazement and gratitude. No realization in my life was ever more true, right, deep, or lasting.

3615 This experience is something that I cannot bring about consciously or intentionally. It sometimes arises completely unexpectedly and physically reminds me of an unexpected sudden chill. It is there and I am immediately seized by it. The experience of this awareness is timeless and notions of space and time and temperature are absolutely meaningless and are in this experience unthinkable. For me it is the overwhelming feeling of coming home, arriving at a place that is and always has been there and here and in me and in everything. This experience has shaped me so much that I can remember it well—but as I said before, I cannot repeat or produce it intentionally. This experience has created an entrance within me in which this overwhelming awareness can fly in like a bird and immediately fly out again. In an unthinking moment it is suddenly there and instantly gone again. But this fleeting moment is overwhelming. It gives me the liberating and ridiculous or humorous realization: If I didn't exist as an organism, I would not be conscious of the awareness—and that too is completely inconsequential and irrelevant in this moment. On the contrary. With every further hint of consciousness, awareness vanishes. I am deeply grateful for this experience—and yet I remain a person with flaws, vanities etc. What remains for me is the realization: Don't worry about "enlightenment," "awareness," etc., and do the next thing that needs to be done in my life. And so I feel again: The entrances and windows to my experience are open again. And maybe I am just out of the house when awareness wants to pay me an unexpected visit :-)

Some spontaneous episodes of pure awareness occur in a social context:

1529 [. . .] My experiences with pure awareness often arise in groups. I repeatedly experience the connection with myself, the people around me, the world and the universe, and again with myself. Awareness appears when I do not expect it in any way—I experience it as a spontaneous state that I also encounter during times in nature. The attempt to formulate the content of awareness disenchants it again.

2773 Out of a certain exhaustion during a very harmonious meeting with colleagues for dinner in a quiet restaurant, a perception of boundless awareness appeared [. . .]. I was shaped by the feeling of being fulfilled, of wisdom and knowledge, without being able to specify it. In meditation I have never experienced it again with comparable intensity.

2814 My strongest experience happened when I stepped onto the stage for a solo concert performance that I'd been somewhat anxious about beforehand, since I was a professional musician but am now no longer active. All the stage fright turned into pure awareness and fell away from me as I came onto the stage. I was nothing but a feeling of feet and stairs, anxious thoughts were completely gone. The state was exhilarating!

3413 [. . .] It was the moment before a lecture in front of about 200 businesspeople. I was overcome by a feeling of fear that began to paralyze me. I sat down in the auditorium, closed my eyes, and observed the fear in all its facets without fighting it. Suddenly there was a complete change, a lightness and freedom. The physical excitement could still be felt, but there was no more fear at all. Inwardly it was like an emptiness. I knew nothing, had no idea what I was about to say. I had prepared beforehand. Then I went on stage and spoke freely for half an hour and worked through questions in a close back-and-forth with the audience, spontaneously. I had the feeling that it's just happening, without me having to make an effort or do anything at all. There was also a great clarity and perceptiveness the whole time. I was receptive and attentive to everything. Perhaps the word "presence" expresses it, absolute presence without thinking interfering in the form of wanting or inner goals. Nevertheless, I had to use my thinking. Otherwise I could not have spoken.

Interestingly, there are also negative experiences of spontaneously occurring pure awareness. You may remember this one from chapter 25:

1690 [. . .] I wasn't meditating, and I was sleep deprived. After putting down a book that was discussing nondual consciousness, I felt a sort of immediacy of experience. The foreground disappeared, and the background was endless. It felt as if it had always been there, and I was just now remembering. But instead of "feeling home" or relief, I felt an immense dread weighing me down like a thick blanket. As I came back to subject/object awareness, I felt like there was no point in navigating space in this manner.

Pure awareness can happen spontaneously during or after sporting activities:

1569 [. . .] During this hike with increasing elevation, my fear of a panic attack accompanied me. At the summit I got a panic attack during which I pressed myself against a rock ledge and became aware that there was no escaping from the situation—and suddenly there it was—pure awareness—the panic attack had disappeared and I went "light as a feather" into the mountain hut. I almost cry with joy when I think about it.

2190 I had just finished working out at an unfamiliar high-rise gym. I decided to go to a patio on the roof of the building to meditate before going home. When I got to the roof, I saw the full moon and immediately fell into the state. I sat there for what turned out to be an hour before I regained normal consciousness.

2341 I was sitting on a bench, looking at the city from above a high vantage point and had just finished my usual running. The air was clear, and the sun was just setting. I was almost immediately able to achieve said state and was blissed out for like 15 minutes.

2390 [. . .] Meditation comes to me freely at any time. Introspection from swimming for many years has trained me to find a meditative state at almost any location during my life even while moving.

2742 [. . .] A rare runner's high while jogging in the park. Here I had perceptions of my surroundings and yet no thoughts at all. Due to the hormone high a very positive and confident mood.

The experience of pure awareness can sometimes even be triggered by physical emergencies (this can also be true of a *dreamed* emergency leading to MPE in the dream state; see chapter 21):

2862 One of my experiences occurred during a white-water kayaking accident in a remote location. I was pinned underwater, beneath a rock that was undercut. With all of the river necking down before going over a drop to my right, there were a lot of PSIs pressing me into the rock. Feeling like this was probably "it" for me, I supposed that I would be sucking water at any moment as could only hold my breath for so long after all! I thought, "well, this is it!" (how I was going to die). Then, I heard a clear, "Nope, not today." I noted that I was perfectly calm, could not tell if my eyes were open or closed, as I could not "see" or feel anything, even though there were thousands of pounds of cold water bearing down on me, smashing me against the boulder. The voice was like my voice but came from nowhere, as there was "nothing." There was just a texture of brightness and spaciousness, in this emptiness: There was no me, and no anything else. Just a sense of pure existence, like a field of energy. And it was like everything else had dropped away. There was no "I" worried about dying, as there was no place to leave from. No sense of weight or weightlessness, no sense of time, just a sense of pure consciousness that belonged to no one. Having no point of origin or ending. I was later rescued. I did not tell anyone what happened.

Finally, a small subpopulation reports spontaneously occurring episodes long before they turned into meditators. Here are two examples:

2876 When I was 15, I had an experience that happened during a literature class in high school; I was looking at my teacher's ear (kind of spacing out during a discussion), and all of a sudden the world as I knew it was replaced by an experience of absolute oneness and rightness, and I think emptiness, although since I have no visual memory of it, I am a little murky about the emptiness part. There was no good or bad or other kinds of opposites, and everything was one unity at the same time as each thing completely particular and individual. This was not connected to any prior meditation practice (I had none).

3353 It suddenly occurred on an evening in autumn [. . .], I don't know how long it lasted, but it felt like very long, although it could have been only a few minutes, as I reconsider it. [. . .] And suddenly, there was a deep feeling of wholeness in eternity, no physical boundaries, there was no time, there occurred a wholeness, oneness with all and no time nor space, there was only a very broad experience of peaceful oneness which included also the here and now in spaciousness, timeless. [. . .] It was [. . .] the first time that I experienced an overwhelming incidence of peaceful happiness and oneness or wholeness, [. . .] which I always remembered, but never dared to talk about. But for a long time, I longed for this kind of being in the world. Since that time, I always felt a longing. The first time that, years and years later, I was sitting on a cushion, meditating, it felt like coming home . . . until this very day. [. . .]

Spontaneity and Effortlessness

The mind has to be let loose without directing [it]. Sustained mindfulness has to be cast away without objectifying it. The mind has to be left in its ordinary state without meditating. Thus, with nothing controlling it, the mind is joyous and at ease.
—Gampopa (1079–1153), *Mahāmudrā—The Moonlight: Quintessence of Mind and Meditation*

When the mind is free of all ambition and all belief-supports, that is Dzogchen vision. Abide in a state of nonmeditation. Realize Dzogchen's goal of nonattainment.
—Shabkar Tsokdrug Rangdrol (1781–1851), *The Flight of the Garuda* (Song Seventeen)

Imagine you are one of the blind people whom the king has called to touch the elephant. After a while, you stop the exploration, end all of your groping movements, and turn around, and one of the king's servants is about to guide you back home. Suddenly you hear a moist snuffle behind you, and then you feel a gentle touch on the back of your head and something slowly sliding down your back. Has the elephant got used to you? Does she perhaps want you to stay a little longer?

According to some metaphysical models, the spontaneous presence of MPE implies that all phenomena arise out of an undifferentiated state to which they all later return. This fundamental level is self-caused, which one may interpret as meaning that it has a "self-contained" origin or has no origin—either of which would make it an unintelligible mystery that ultimately demands an act of faith. Conceptually, taking the experiential "spontaneity" characterizing the onset of pure awareness at face value, as an objective property of the human mind, would mean accepting it as causally indeterminate. Therefore, like a miracle, it would be inaccessible to standard experimental methods. On a physical or functional level of description, to call something "spontaneous" means to describe it as uncaused and lawless—and therefore doing so without argument or evidence could be seen as a form of hand-waving or deliberate obscurantism. If we try to be intellectually honest, set all metaphysics aside, and stay as close as possible to contemplative experience itself, then the term "spontaneity" can only refer to a *phenomenological* kind of spontaneity. There is an *experience* of spontaneity, and this is what has to be taken seriously. As always, jumping to strong metaphysical conclusions is a way of *not* taking your own experience seriously.

The phenomenological facts that nondual MPE often appears spontaneously and effortlessly, that it has an "ahistoric" quality and cannot be fabricated by applying a mental technique, can be explained with reference to the fact that most of its enabling conditions in the brain are fully unconscious. We do not experience them. Science may be able to find and isolate these causally enabling conditions, but the brain itself doesn't represent them in its conscious model of reality. There is a parallel case in the phenomenology of volition, of suddenly "willing" something. Many people proceed from the phenomenological fact that, say, the consciously experienced intention to now move their right arm and reach for a beautiful flower seems to spontaneously occur in their mind, "out of the blue" as it were, and interpret this as proof of free will in a strong sense: as demonstrating the existence of what philosophers call "ultimate origination." Ultimate origination would mean the capacity to start a new causal chain of events in your mind by an uncaused act of will, an act coming from outside the network of dependent origination. But for the experience of free will, as well as for the experience of spontaneously occurring pure awareness, the same principle holds: From the fact that you *experience* something as uncaused, it does not follow that it actually *was* uncaused. Believing or assuming otherwise is another version of the C-fallacy (chapter 12).

Some intelligent systems control their own behavior based on a model of themselves as *agents* in a strong sense. To successfully control their own behavior—which may be bodily or mental—they use a conscious image of themselves as an entity that possesses the capacity for goal selection and ultimate origination. Other intelligent systems don't do this. The difference is important. We can think of embodied systems intelligently moving through the world, automatically updating and constantly improving their inner portrait of reality, controlling what they will know and what they will not know. But some of these systems—for example, self-aware human beings—sometimes may use a model of themselves as an epistemic agent, now achieving the same goal by using a special computational device. They would experience themselves as possessing free will, as being agents capable of initiating new causal chains out of the blue. These systems would have an ego in a strong sense.

The underlying distinction between model-free and model-based control may offer another perspective to help us better understand what pure awareness really is. We can now think of intelligent systems automatically regulating their own level of tonic alertness, their current degree of epistemic openness to the world—for example, by controlling the level of cortical arousal and generating their own sleep/wake cycle with the help of some low-level, subpersonal brain mechanisms. And we can think of systems that currently have a *model* of this very subpersonal process and are using it for model-based control of alertness. They *know* their own alertness. These systems would be conscious in a well-defined but entirely nonegoic way: They would possess a subpersonal form of knowledge about their own alertness and nonconceptual knowledge about their own *capacity* to know and the current state of epistemic openness—but this would not be the knowledge of a self. It would appear as spontaneous wakefulness and knowingness, but not caused by any agent. It would appear as the givenness of cognizability, but not as something brought into the world by an actively knowing self.

As we already saw when introducing the dolphin model of meditation in chapter 10, and in light of the rich empirical evidence provided by contemporary neuroscience,[5] it is more than plausible that the experience of effortlessness, including the subjective qualities of "givenness" and "spontaneous presence," depends on unconscious causal precursors in the brain.[6] However, these precursors can certainly be cultivated, strengthened, or triggered in an indirect way—as if meditation practice were a way of doing neurofeedback across a wider time window, using the phenomenal correlates of meditation practice as a user interface. If you are a meditator, I propose a short experiment for you. Next time, when you have sat down and are just about to begin to meditate—don't. Take a minute to carefully check whether there is something that *has* already been meditating all the time. What do you find?

A few years ago, I had an interesting and somewhat shocking experience. I had to catch a train to an important meeting, I was running terribly late, and I was close to panicking. I was washing dishes and cleaning the kitchen, all in a great hurry. I knew I had to be in my car and wheels turning at six o'clock, it was supertight, and I still had a lot of things to do before then. Suddenly, an even more uncomfortable thought arose in the midst of my cleaning frenzy: "But you still have to do your evening meditation! There is absolutely no time for it, you cannot possibly squeeze it in!" At that moment, something very strange happened. I realized that the whole room was meditating. The cupboards, the table—the whole kitchen was *already* meditating. There was nothing I needed to do because the whole damn thing was already doing it for me—abiding in the silent clarity of nondual suchness all by itself. Unfortunately, I was so surprised and fascinated by what was happening that my excitement quickly destroyed it. But if you are too lazy to meditate (and/or cooler than I am), there may be a much simpler trick: Just step out of the picture for a moment and let the whole room meditate for you.

It has long been known that there are nonconceptual insights and spaces of consciousness beyond ordinary contemplative experience that not only are extremely hard to communicate but also cannot be accessed by any mental technique whatsoever. As already noted, there is something that is beyond willpower, discipline, and earnestness—something that eludes any systematic, technical, goal-directed approach. Technical approaches introduce a sense of effort and subject/object duality, and if the conditions that the techniques are meant to generate—namely, the functional conditions that reliably enable pure awareness to arise—are *unconscious*, then they cannot be directly targeted by the conscious self. In our own data, we found neither a negative nor a positive correlation between the number of practice hours and the score on either factor 8 ("Emptiness and Nonegoic Self-Awareness") or item #36 ("Was there a sense of effort?"). Accordingly, the idea of an "expert meditator" (which scientific meditation researchers very much like) may turn out to be a myth because there is something highly important that cannot be measured in hours.

Put differently, there is a deep phenomenological dialectic between meditation and nonmeditation. I would like to propose that some aspects of this dialectic may actually reflect an interplay of conscious and unconscious processes in the brain of the meditator. Meditation practice and the phenomenology of pure awareness may have much more to do with *unconscious* information processing than many practitioners or even scientists tend to think.

Let me briefly explain this idea. It could be that there is a causal interaction between the apparent mental agency involved in meditation practice and automatic processes that are invisible to inner attention. Most of what goes on in our brains during the day remains in the dark, unavailable for introspection—the conscious self doesn't like to hear this, but it is a well-established fact. Most of our learning is unconscious. For example, canonical techniques like gently and precisely focusing on the breath, returning to a state of open monitoring, or "reversing the arrow of attention and finding nothing" (the "you-turn" discussed in chapter 21) may actually be ways of cultivating certain (as yet unknown) functional properties in the brain, which may then later trigger a "spontaneous" emergence of pure awareness. But again, the term "spontaneity" would then refer only to a *phenomenological* form of spontaneity. The experience of effortlessness and the phenomenologies of nonmeditation and spontaneous presence would depend on nonspontaneous, unconscious causal precursors in the brain.

The deep phenomenological wisdom in describing pure awareness episodes as "spontaneous" lies not in some murky metaphysics, but rather in the fact that the illusion of control is prevented from arising in the meditator. Human beings largely lack introspective insight into the mechanisms by which they control their body or events taking place in their mind; the illusion of control is our tendency to *overestimate* our ability to control events—for example, by trying harder and then patting ourselves on the back and attributing it all to an invisible entity later called the "self." This point is directly relevant to meditation practice: Noticing one's mind wandering may often produce a subtle illusion of control. ("I *myself* just successfully noticed the arising of a stray thought! My attention collapsed, but now *I* have just regained it.") Jonathan Schooler and colleagues importantly pointed this out more than a decade ago:

> Mind wandering can terminate for reasons unrelated to meta-awareness (such as when an external event disrupts the internal train of thought or the action of an unconscious monitoring process). Once the episode has ended, individuals might reconstruct their recent conscious experience and in so doing realize that mind wandering had taken place. In this case the strong sense that we have "caught" our minds wandering could be an illusion of control ("I know I was mind wandering a second ago so I guess I must have caught the experience").[7]

What I have, in my own academic work,[8] called the "phenomenal self-model" is precisely the result of the process that Schooler and his colleagues call "reconstructing recent conscious experience." A lot of our self-model is based on illusions of conscious control. For example, if you drifted into a daydream at some point while reading this book, there may first have been some unconscious word associations that triggered the conscious experience of daydreaming. Unconscious events moved you into a new attentional state, a state of being distracted while also unaware of being distracted—and then the system unconsciously infers the actual state and regains attentional control via a new level of meta-awareness.[9] On the level of conscious processing, this is an unexpected event. The ensuing conscious sense of self is the brain's attempt to explain away the occurrence of an unexpected conscious event—the "noticing," or even an occurrence of MPE. If a meditation teacher manages to install a false belief about uncaused "spontaneous presence" in the conscious self-model of a disciple, this may enable a deeper form of letting-go. The letting-go enabled by the new "spontaneity prior" attenuates the illusion of control, making future occurrences of nonegoic meta-awareness more probable. It is one way of helping people to try less hard and to stop complacently patting themselves on the back quite so often.[10]

Let us stay with actual meditation practice for a moment. Installing false beliefs may work for some people, but I think there is a better, more direct route to effortless pure awareness. When discussing the phenomenon of mind-wandering in chapter 25, we saw that mind-wandering creates an inner affordance landscape. Every single thought can be seen as an action affordance. In one way or another, they all say "Think *me*!" or "Pursue *me*, follow me into the future!" or "Can you remember *this*? Try it!" However, there is one type of thought that is special: the thought of "noticing" that you, the meditator, have strayed; the thought that you, the meditator, have obviously just had an attentional lapse. If Schooler is right, this thought might be an illusion of control. But there is more to be seen.

That thought ("My mind just wandered!") is an inner affordance as well. It tells you to do something: Return to the breath; bring your attention back to the present moment. If the system falls for it, that special thought swiftly creates yet another agency illusion, a hallucination of what in chapter 4 I called "attentional agency": the experience of deliberately and actively controlling the focus of attention in a goal-directed manner. It includes a subtle sense of effort and successful mental control: I, the meditator, must act with my attention now because I "have noticed" a stray thought, and as I have learned in the past, I must now bring "my own" focus of attention back to the present moment.

I think that *not* doing this is what is meant by "nonmeditation"; it means not falling for the very last affordance that happened to pop up. There is a deep, but very subtle phenomenological insight to be had: The silence, clarity, and epistemic openness are *already* there. Nobody has to go anywhere; nobody has to do anything, like fabricate some new state. You do not have to act. The "special thought," the thought of "noticing," is actually sculpted of that silent clarity, if one looks very closely. This what I mean by the better, more direct route to effortless pure awareness. Perhaps one could say that the noticing itself has already *given* the silence[11]—and the only thing that could ruin it would be an attempt to meditate.

Let us return from the subtleties of contemplative practice to my hypothesis that there must be an interplay of conscious and unconscious processes in the brain of the meditator. This idea leads to an empirical prediction: If meditation practice is the process of trying to deliberately create the unconscious causal conditions for MPE, then any other way of creating these conditions will also lead to the emergence of an episode of pure-awareness experience that feels spontaneous and effortless. For any conscious system that has the capacity to undergo a pure-awareness experience, and for any set of sufficient, causally enabling conditions (there may be more than one), if we bring these conditions about, then the experience will reliably appear. But if the process itself is unconscious, this event may later be reported as a surprise, a gift, a conscious event that was entirely unexpected. Put differently, there may be many ways of creating the right causal conditions leading to episodes of pure awareness other than traditional meditation practice itself.

In this context, it is interesting to distinguish between personal and subpersonal levels of description. If you are trying to understand and describe a phenomenon better on the personal level, then you look at a human being as a whole. You look at her desires and beliefs, the social context, and the cultural background against which the phenomenon you are interested in—for example, the experience of pure consciousness—arises. But there are also many subpersonal levels: things that happen "under the hood" and in a person's brain; unconscious processes that are best analyzed in terms of information processing, different styles of neural computation, biochemistry, or evolutionary origins. My first point is that for most people, meditation practice starts on the personal level, but it attempts to change something that is actually subpersonal, such as the unconscious causal enablers of mindfulness and compassion.

In the beginning, there is a human person who has certain beliefs and desires, who consciously experiences herself as an agent endowed with a robust sense of selfhood. This person fully identifies with her body and her epistemic agent model. Perhaps this person is even engaged in an earnest spiritual search, or in the kind of project that I described in chapter 17 as "meaning-making" or "contemplative heroism." She reads books, learns and explores various meditation techniques, and begins to experience interesting and at times rewarding altered states of consciousness. Since their enabling conditions are located on a subpersonal level, the fine-grained causal process by which they come about remain invisible to her. The overall process may be slightly enigmatic, at times surprising, and often appear uncontrollable if viewed from the "perspective" of the personal-level self-model—which is not a problem. Our practitioner just keeps going.

But as we have seen in the course of this book, many of these states may later begin to involve degrees of ego dissolution, a cessation of time experience, a feeling of witnessing, or even the spontaneous and effortless emergence of nondual awareness. Now, some of these states can no longer be adequately described on the personal level of description. The phenomenology of ownership is attenuated. They are not *her* experiences because the phenomenology is not a personal-level phenomenology anymore. Sometimes something beautiful happens. There is no need to talk about it. Why should she? She just keeps going.

My first point was that for most people, meditation practice starts on the personal level but tries to change something that is actually subpersonal. My second point is that—in terms of accurately describing the character of conscious experience across a longer time-frame—many contemplative biographies are transformative experiences.[12] In a special phenomenological sense, they begin on the personal level but do not end there—and the conscious self that begins the journey may not exist at its end. It is not only that you cannot know what it will be like to have the experience before you have it yourself—*you* may not be there to "have" it at all. Of course, it is now tempting to speak of MPE as a "subpersonal" or even a "suprapersonal" stage of conscious experience, but this would only import an external hierarchy of explanatory levels into something that (as we saw in chapters 3–6) fundamentally lacks internal structure. I would rather say that the experience of consciousness per se is neither subpersonal nor suprapersonal, and turn all philosophical and explanatory issues related to the contraction principle (the brain's misrepresentation of consciousness as an ego's consciousness) into a target for future research (see chapter 8). Why has the human brain's misrepresentation of consciousness—namely, as a property of an egoic phenomenal self—worked so well in practice, as we saw in chapter 8? I myself think that it has a lot to do with social cognition. In any case, science will take care of it. And as for our practitioner and the transformative phenomenology of peaceful ego dissolution that unfolds while walking onward through the garden of contemplative experience, I prefer the old Zen saying:

Only here
could a path end surrounded
by parsley.[13]

My third and final point for this chapter is that the cultivation of MPE-like states does not have to start on the personal level at all. It is conceivable that new technologies could directly create the causally enabling conditions for MPE on the subpersonal level, perhaps by using artificial intelligence (AI) and new methods of adaptive real-time feedback to target the neural correlates of consciousness and their computational functions much more directly and precisely. To give a more specific example, we could use machine learning techniques to extract the shared activity patterns underlying every single phenomenology described in this book from the brains of many experienced meditators, and then close the causal loop via neurofeedback. If naturalism is true, this should give future subjects very similar states of consciousness to those described in this book, but elicited in a historically unprecedented way.

I have strong doubts that any of this will lead to really interesting and sustainable changes in the human mind any time soon—let alone in a risk-free way that acknowledges the ethical dimension of classical meditation practice, and thereby integrates it safely into the wider context of an altruistic, genuinely prosocial epistemic practice.[14] But I may be wrong. We already know that many of the types of experience described in this book can be directly caused by unconscious, subpersonal events in the brain: these include spontaneously occurring early childhood experiences, emergencies, ecstatic epileptic seizures, and experiences of pure and nondual awareness emerging under the influence of psychoactive substances like LSD and psilocybin. There are many other conditions under which local physical events trigger global states that clearly belong to the MPE family. Therefore, beyond classical meditation practice, there are likely to be many causal routes to the experience of pure awareness. We must remain open-minded: It is plausible to assume that many of them are still unknown and will be discovered at some point in the future.

33 Timeless Continuity

Everything is deified and becomes a meditation. [#2437]

Afterward I knew that I am the expression of the whole universe, at the point where I find myself. I do nothing, everything happens. [#2384]

Subjectively, episodes of pure awareness are timeless. Objectively, from the third-person perspective of scientific observation, their duration is typically a few seconds or minutes (e.g., if they are periods of full absorption). However, some of our participants have reported a more stable experience of pure awareness that lasts for much longer. Sometimes they even describe a permanent *mode* of experience in which minimal phenomenal experience (MPE) has become the continuous foreground of all conscious experience, or rather something that holds, encompasses, and permeates everything else. It is important to note, however, that terms like "stability," "permanence," and "continuity" reflect the scientific third-person perspective of an outside observer. They may often be inadequate from a phenomenological perspective because they ignore the quality of timelessness. For more durable states, the phenomenological prediction would rather be a global mode of "timeless change" (chapter 22) or an all-pervading phenomenal character of "always already" (chapter 31). In such longer-lasting states, the variable contents of sensory perception, bodily movement, thought, and emotion apparently coemerge with that which never changes.

Viewed from the perspective of practice itself, the goal of "achieving continuity" can be seen as a subtle straying. The whole idea of "making pure awareness a permanent state" may be mostly an expression of greed—part of the incessant search for stability and certainty that originates in the egoic craving for existence described in chapter 17. Perhaps timeless moments or brief episodes of what report #1845 on the following page describes as being "temporarily liberated" are all that counts, and the idea of "attaining" a permanent state is itself a contraction into an egoic perspective, something that actually prevents such moments from happening. On the other hand, we find some evidence that the MPE experience can turn into what the philosophers Timothy Bayne and Jakob Hohwy have called a *mode* of conscious experience.[1] In the introduction and throughout this book, I have said that a mode of consciousness is a global *way* in which reality appears to us. As opposed to a mere *state*, it is not content-specific because it can encompass the flow of many kinds of experiential content. In this sense, "timeless continuity" is not a single MPE state, but rather something like a phenomenological class, a whole category of more permanent MPE modes.

As we will see in this chapter, this longer-lasting global mode of conscious experience is characterized by an all-encompassing phenomenology of clarity, certainty, nonduality, joyfulness, vibrancy, and aliveness. Once again, we also find the quality of virtuality (sometimes described as a "dreamlike" quality; see chapter 28) and the experience of emptiness and epistemic openness, plus a prolonged absence of thought activity. Other phenomenological descriptors of the MPE mode provisionally labeled "timeless continuity" are soundness, connectedness, openness (in a more general sense), unity, deep peace, well-being, nonreactivity, and forgiveness.

The following, and final, selection of eighteen experiential reports is very different from those presented in previous chapters. They are more diverse, and I will simply let you assess them for yourself. I have also deliberately refrained from offering any phenomenological analysis or a wider philosophical context in the second part of this chapter. Here, the reports are simply ordered by their estimated objective temporal duration, and not by the specific phenomenological profile being described. Please note that a few other reports about states lasting in the order of hours are distributed throughout other chapters (e.g., #1612, #1675).

1189 [. . .] In activity it is more and more often the case that even in the most hectic places (funfair, Christmas market, etc.) the experience of peaceful wakefulness is present at the same time. It is a little as if time stands still. It happens that in the morning, before actually waking up, the mind is awake while the body is still asleep. [. . .]

1845 [. . .] at some point I just "knew" that I was temporarily liberated. I didn't know my name, nor what I was doing. The "knowing" was not knowing in an ordinary sense, it's hard to describe. I'm an atheist, and this sounds silly, but yes it happened.

2506 [. . .] It lasted several hours. It occurred during a sitting meditation and also carried on afterward. I would describe it as a state of complete clarity regarding the phenomena of existence. The sequences of events were not extraordinary; perception was fluent and connected with a feeling of contentment and happiness and was free of all doubt. The duality of "I" and "you" or "good" and "evil" no longer existed. The experience of the sensory perceptions was very intense, joyful. There were no more worries or fears.

1243 [. . .] an experience of sensations—very clear and very precise—and feelings and thoughts—very clear and very precise—but without any self-referentiality, without any personalized reference point. Very irritating and partly confusing, because the "I" could not be found anymore. This "selflessness" lasted for about 20 hours, until the "I" slowly reappeared in experience.

2926 [. . .] It was only AFTER I came out in the post-meditation period that I had the most incredible experience of unity and oneness with the entirety of the phenomenal world. Objects had no sense of distance and no barrier to them, no duality. There was no relational experience of me and an object I perceive, but rather things were simply there. In fact, everything was there, and it was extremely vibrant and alive and peaceful, open, inviting, and joyous. There was a sense of intensity and vibrancy and luminosity to all objects and this realization itself was extremely pleasant and I was aware that it was so. This state lasted for about 30–40 hours. I started assuming that it was permanent, but then it started to fade around the second day. It considerably changed the way I viewed the world, but the state itself did not last. Having done long-term retreats I have had many experiences of altered states on the spectrum of consciousness, many pleasant sensations and experiences of illumination and intensely high levels of mindfulness, purity of mind, and equanimity, yet this specific experience was somehow different to all of that. It was an experience of nondual union that felt so natural, extraordinary, yet it was occurring in ordinary mind.

66 One experience was a feeling of a dreamlike state that lasted for a few days, where everything was coherent, all right, easily perceptible, but where I saw myself as being prompted to little or no reaction by what I was experiencing. A special state of vastness that penetrated everything, me, the outside world, my actions . . . (what I had to do . . .) very intense in the experiencing, a feeling of being one and a wholesome, healthy state. Unfortunately it faded after a few days, but since then it has remained something special for me, when I remember it as I am doing now. [. . .]

3485 I remember the two experiences I'm describing as qualitatively identical states in my feelings, in my consciousness, and in my perception. [1] [. . .] A practical meditation session [. . .] triggered in me a state of complete inner calm and emptiness. During the meditation I was able to maintain a completely thoughtless consciousness. Afterward I was awake, conscious, relaxed, desireless, and satisfied in a way I never had been in my life before then. I've been able to spend several hours of my ordinary everyday life like this. [. . .] [2] Without inducing the described state through meditation, I was able to live my life thought-free for several days. Including going to work, time with social contacts, and other everyday occurrences.

3619 The condition lasted a few days. Feelings like silence, warmth, lightness, everything and everyone forgiven, and in the now. Love within and for the outside. [. . .]

765 The head intermittently becomes completely empty. The space between the thoughts spreads out—to a free space within me. It fills with consciousness and inner peace. This hitherto unknown calm appears to me at first as an entirely natural consequence of the excessive inner activity. But it *remains*. It continues to accompany me—in the following days—and beyond. The last four days—happiness, even without a kick: More and more often and longer and longer I end up in the thoughtless free space. Only occasionally is my mind busy describing and analyzing the exciting process of its disappearance. [. . .] Silence and peace spread out over head and body. Breathing, body perception, and well-being now take precedence and get much more energy and attention than thinking. The thought chatter starts thinking again only when something existentially and emotionally important sets me in motion, when there is a disturbance, an exciting occasion for reflection. Or when creative new thoughts emerge and make me happy. While I was used to the fact that feelings of happiness can be brought about by special kicks and easily disappear again with the kick, from now on I experience this new kind of happiness: a happiness even without a kick. It is the joy of simply being, being an observer.

837 My experience lasted about 4 weeks, I felt a lot of energy, timelessness, a powerful serenity, because the judgments fell away. Everything was as it is. I felt a powerful connection with the people and things I encountered. Everything seemed to be harmonious and flowing.

2437 [. . .] In that Peace, every effort other than what was necessary for functional purposes was painful. Peace was present with eyes closed and eyes open. I started spontaneously perceiving many things through the senses without Peace being affected in any way. [. . .] Every object was myself. Everything was emerging from profound Silence, and was an expression of Silence, and was returning to Silence. I thought, how foolish it was to think that this world had ever disturbed me! Everything was most beautiful. Ugliness, the opposite of beauty, to be disliked and avoided, was absent. This Peace had come at a time when I was going through intense fear, helplessness, and frustration. It had come after six years of spiritual practice. Having nothing to care for, nothing to be bothered about in this universe, I spent most of the night in the vast garden of the monastery, either walking or sitting. Thoughts were there, but they were mere words emptied of all emotional content, singing the song of Peace from the time I got up in the morning until I went to sleep. [. . .] All hankering, including for enlightenment, had vanished. Desires and doubts were experienced as pain as soon as they were born. I was stunned, and wondered what had I been doing all these years, when unfathomable peace was available at any moment without the need for any effort? I knew that this was the end of my sadhana (spiritual practice). When one rejoices in whatever one is doing, there is no need to concentrate on a spiritual practice and divide life into action and meditation. Everything is deified and becomes a meditation. [. . .] I did not even worry whether one day this Peace would go away. It was such a simple thing, I thought, how could it just go? I had felt it as my own, as something that could not go. But it did go! In the beginning, this Peace came many times. Whenever this experience came, it always gave me the feeling that it would never go away. I felt that it was so intrinsic and innate, so, how could it go? I knew, if anything disturbed this Peace, it was my own reaction, which I could stop immediately by reverting to Stillness, so, how could it go? How could it go? But, it would always go. [. . .] Though the first experience of Peace did not last longer than four or five days, it removed many misconceptions I had had about happiness and spirituality. It also gave me some clues on how to "stumble" upon it. That was the day when Effortless Meditation was born for me. [. . .]

3340 [. . .] It is now (after long training experience) the usual state of awareness, while disturbance of this state becomes "visible" (experiencing anger, greed, disappointment, ugliness, etc.).

1315 [. . .] I have the experience of infinite space and timelessness both in meditation and in sleep and wakefulness, and it is always accompanied by feelings of happiness and lightness. The older, but actually the younger I become, the clearer and more comprehensive this experience of pure consciousness becomes, and permeates every area of life. The intensity sometimes varies. Depending on whether I am tired, or overworked, or traveling, or have the flu. [. . .]

1354 My experiences with pure awareness [. . .] have changed over time. In the beginning it was just the experience of restful wakefulness, without thoughts and without mantra. For a short time I simply rested within myself. I think it was the experience of the gap, the experience of pure being, between two thoughts. The place into which a thought disappears and a new thought reappears. Through repeated experiences of this being, it became a peaceful stream that spread more and more into my daily activities, so that I can be dynamically active and at the same time still rest within myself. I am then the silent witness who simply observes the activities. This experience of being a witness now also happens during most of the time of sleeping and dreaming. The experience of pure awareness in meditation has become more and more evident over time and with evolving meditation practice. The experience of infinity and bliss now also occurs along with logical thinking, feelings, and spiritual perceptions.

1405 Since I have been meditating (or rather, "transcending") regularly every day for most of my life, the experience of "pure consciousness" is deeply anchored in my sense and way of life [*Lebensgefühl*], so that I experience it practically 24/7. [. . .]

2299 [. . .] This had irreversible consequences. A new state of being. I would say here that "pure consciousness recognized itself" and "awakening" happened. Awakening to the SELF. The exit from the dream. The recognition of the world of appearance as an illusion. The world does not exist in and of itself. It is a game of consciousness, a reflection of our "mind" or state of being. The falling-away of identification with the limited person. I am not in the world—the world is in me! I am the space in which everything appears. This is an experience that has remained until today. When e.g. I walk around, I have the sensation of walking around inside myself. It is a sensation of infinity. The body does not end at the obvious body boundaries, it is infinitely large. And there is only one thing. THIS. Everything is THIS, or pure awareness, or pure consciousness or God or presence. The feeling that I MYSELF am everything that appears, this ONE, the WHOLE. There seem to be two states of consciousness running simultaneously. The SILENCE (pure awareness) as the basis of all being is always there and at the same time the processes in the world of form, which however is no longer experienced as different from the SILENCE. Therefore the impression is there that nothing is really happening. Great peace, permanent silent joy, bliss (these are of course words that should be clarified).

However, this has become stable only after a fairly long period (some years, about 10) of integrating the experience. [. . .]

2355 I experience Pure Awareness (PA) constantly, i.e., in 1) waking, 2) dreaming, 3) sleeping, and 4) meditation. The experiences are accordingly different. As for 1)—while waking I experience PA all the time. My entire waking activity is pervaded, permeated, by it. All mental, emotional, and physical activity is imbued with it. Yes, in terms of feeling and perception it is even a complete fusion. It requires a conscious activity to separate this complete fusion back out into two components. Everything is completely and constantly united with the PA in one point and in the whole. Everything is PA. Nevertheless it is possible for me to make everyday distinctions if they are necessary. But even then there is no feeling of separation. Since in terms of the feeling I am everything, including all being and doing, I can only very rarely feel myself into the mood of being separated. When I try to do so, a kind of countermovement immediately appears, which instantly and completely captures me again. Practically speaking, I live completely differently from how I used to live in the feeling of being separated. It's difficult to describe it in a way that will help someone understand or relate to it who cannot experience it themselves. On 2), when dreaming I am fully awake and can act self-sufficiently and do almost everything I want. I am the PA and the dream is simply a special lifetime in the night, which I use for varied and interesting activities. I travel in time, live whole lives in other eras, learn a lot, and can also use that in waking. I travel through space and look at everything I want. I meet with many life forms, I can penetrate into them and communicate directly with them and see their inner being etc. Since I am everything, I can cause everything etc. . . . The PA is always there, there is no moment where it is "gone." I am this PA. [. . .] As for 3), in deep sleep the activity is present in a different way, because I see how as a body I am in a state, in a special physical mode of consciousness. Many constantly running processes flow, order, and repair themselves. I can see all of this. Many things are just beautiful. I experience myself completely as a body consciousness. PA is body consciousness. It is beautiful, it is again, as in a dream, a unique kind of conscious expression of life. I am still exploring all that is possible and what happens in this state of deep sleep. It is a complete unity of material body and mind. This is surprising and so I am exploring how far this unity goes and how useful it can be practically (health, development, understanding and knowledge about nature, etc.) On 4), in meditation

I simply switch off all moving activities through an intention and so my inner perception is transformed into a field of creation that is eternal, infinite, and powerful. I feel that I am then simply this field of creation, this energy, which is completely still and which contains the whole of creation as a delicate, gently moving structure, as potential. I am all this and I can simply bathe and surf in it. I bathe completely dissolved in myself and know about all this. Normal reality is like a thin sheet of paper on this ocean of Pure Awareness. It is beautiful and very familiar.

3109 [. . .] With time and regular practice, this experience has become more and more stable and refined. Now I have this experience even with thoughts, or in daily life together with dynamic activity. Also in sleep and dreaming this alertness is mostly present. In meditation I experience a lot of light, infinity, and bliss.

MPE as a Global Mode of Conscious Experience

> At some point, when mindfulness and your mind are no longer different entities, everything turns into the nature of mindful presence and it is "smooth sailing" from then on. [. . .] Instead of recognizing that the training is the indivisible unity of path and fruition and that this fruition is present as a natural possession, the basic straying is to believe that the path is the training, while the fruition will be attained at another point.
>
> —Dakpo Tashi Namgyal (ca. 1513–1587), *Clarifying the Natural State*

In thinking about this last set of reports, let us try to be modest and keep things simple. Nobody knows what "liberation" or "awakening" actually is; nobody knows whether something like this even exists or what all of these traditional terms *really* mean—at least, the author of this book certainly doesn't. Of course, there are all kinds of people out there who *claim* to know, people who wear funny clothes and have given themselves names, people who have founded organizations, people who teach and bestow titles to each other, people who have "attained" certain things and mutually confirm them to each other. But the epistemic virtue of intellectual honesty is a necessary component of any modern, genuinely self-critical approach to spiritual practice,[2] and it also includes the kind of humility that comes with an openness to empirical data and rational argument. If anything like "liberation" does exist, one of its components must be liberation from tradition and all mere belief systems. In more seriously and radically pursuing epistemic practices like meditation, science, and philosophy, what is needed is not only courage, open-mindedness, and perseverance, but also caution and temperance. Our future is open, and we shouldn't pretend to know something when we actually don't. I will say a little more about all of this in the epilogue.

For now, let me begin by drawing your attention to one single phenomenological point, which may not have been described quite in this way in the past. Reflexive MPE, the timeless experience of nonegoic self-awareness (see chapter 30), is *free* of the subtle but almost all-encompassing, almost global feeling tone of futility and absurdity that we all try to suppress, more or less successfully—continuously squeezing it out of our phenomenal self-model, as it were. There is no trace of psychological suffering nor any existential despair in MPE itself. On the contrary, we can now loop back to what we saw at the very beginning of this book, in chapter 1, when putting together some new psychometric data and introducing the new phenomenological concept of "existential ease," which refers to the experiential *integration* of relaxation, lucid clarity, and a state of pure being. Experientially, existential ease is characterized by a complete absence of futility, absurdity, and despair. Everybody who enjoys a moment of pure awareness can verify this simple phenomenological fact for themselves: What I called the "Sisyphean quality" in chapter 17—when we investigated the relationship between impermanence, embodied existence bias, and narrative self-deception—is *not* part of the phenomenal character of pure awareness. In themselves, MPE and reflexive MPE are entirely effortless states, and the Sisyphean quality is completely absent. This seems to be equally true of the prolonged *modes* of experience reported earlier, which I labeled as examples of "timeless continuity."

Another fact that we can clearly state is that sometimes, on rare occasions, MPE stops being a temporary episode and becomes almost continuous. In this book, I have tried to capture this phenomenological fact by using the somewhat coarse-grained distinction between MPE states and MPE modes: Whereas states of pure consciousness are something local and episodic, something still attributed to an experiencing self (*whose* states they are), we also find more generalized modes of consciousness dominated by an all-encompassing experiential quality—namely, the nondual phenomenal character of awareness itself—as extensively investigated in this book. Examples of such "MPE modes" are provided by witness consciousness (chapter 19), the phenomenon of clear light sleep (chapter 20), and the large variety of so-called nondual wake states described in this and many preceding chapters. I will list five major examples in chapter 34. The new question that this last set of phenomenological reports ultimately raises is this: Do *permanent* modes of MPE exist? This is an important question because if the answer is yes, it might entail the actual possibility of "non-Sisyphean modes" in which all psychological suffering is absent.

Currently, we have no empirical data that could help to decide the issue. I also doubt whether we have the right philosophical tools. I hope that my own conceptual distinction between MPE states and MPE modes has successfully done the simple job that it was supposed to do for this very first phenomenological exploration, but its coarse-grained nature is particularly salient when we look at full-absorption episodes. You will recall that full-absorption episodes are those in which the meditating self has dissolved into pure consciousness and nothing else remains that could be later reported. Full-absorption episodes form a kind of bridge or logical link between MPE modes and MPE states because they can be described as a global mode *and* a state at the same time. They constitute an interesting borderline case in that they may still count as "content-specific," because there is one specific form of content that remains: epistemic openness, tonic alertness, the silent clarity of MPE itself. If we take objective time as a frame of reference, they mostly have a much shorter duration than many of the experiences described in the first part of this chapter. On the other hand, the specific quality of MPE has now become globalized and all-pervading; it is all that exists. Full-absorption episodes are clearly also nondual (lacking subject/object structure), as well as being internally timeless. Given the additional phenomenon of clear light sleep (which can also be described as a form of absorption; see chapter 20), it is not even clear whether full absorption into MPE should be taxonomized as a special kind of "wake state" or be treated as a separate category of its own (like *turīya*, "the fourth," in Hindu philosophy).

In making a fresh start, new scientific and phenomenological data will not be enough. We need more philosophical analysis because a number of conceptual ambiguities must be resolved before we can return to the question of precisely what kind of "timeless continuity" permanent MPE modes have—and what their relationships with each other might possibly mean in the wider context sketched out in this chapter and in chapter 17. In this book, however, my only concern has been to take the phenomenology itself really seriously and to provide an open-ended conceptual foundation for the future. In the following chapter, I will summarize some of the main results of our phenomenological investigation. I will also present a short list of empirical and conceptual research goals for the future. Arriving at an empirically validated computational model that describes suffering-free modes of conscious experience should be one of them.

34 The Elephant: What *Is* Pure Awareness?

It's so close you can't see it.
It's so profound you can't fathom it.
It's so simple you can't believe it.
It's so good you can't accept it.
—"The Four Faults of Awareness"

What you have just read is not a book. It is what philosophers sometimes call a *prolegomenon*: a prefatory discussion of the phenomenology of pure awareness, serving to introduce an extended research program that is hopefully yet to come. As I said at the very beginning, it is important to take the phenomenology itself seriously and start by looking carefully at the actual experiences of real-world meditators. As the author of this prefatory discussion, I hope that what it has done will prove useful for you if you are interested in taking this research further—for example, by working toward a minimal model explanation of consciousness that ultimately unifies the core insights of existing approaches on a formal level.[1]

There are many other main courses that might be cooked next, many of which I probably can't even imagine. If you are not a researcher and read the preceding pages out of personal interest, then I hope you have seen some of the depth and beauty that can come along with the experience of pure awareness. Pure awareness does exist, and there definitely is something to it. The human mind knows silent and nonegoic ways of being in touch with itself, and the conscious experience of pure awareness and of "self-knowing empty cognizance" is one of them. It seems that all healthy human beings have the capacity to experience the quality of awareness per se, as well as the crystal-clear, nonconceptual, and entirely selfless form of self-knowledge that is enabled by what I have called the "zero-person perspective." I hope you have also discovered that pure awareness is something everyone already knows, at least in certain moments of their lives, that it spontaneously occurs—and you do not have be an Olympian meditator or some kind of overmotivated spiritual athlete to get in touch with it. Profound pure awareness is not something dramatic or sensational. Rather, the problem is that it's so close that you can't see it. It is maximally simple, but we are complicated. It knows itself, but most of us really don't. That said, our data show that there are *many* people who know these states well but almost never talk about them. One of many reasons for this situation is that we lack the kind of cultural context that would dignify the types of conscious experiences described by more than 500 meditators in this book. It does not really matter whether you have read this book out of personal interest, from the perspective of a researcher, or as an already-experienced practitioner of meditation—I do hope that at some point during your reading, you will have become acutely aware that, somehow, almost all of us are systematically ignoring obvious phenomenological facts. These facts are important. In more than one way, there is an elephant in the room.

For now, we have almost arrived at the end of our journey. This book has brought together a small bit of conceptual analysis, a first set of psychometric data, plus a semi-systematic qualitative analysis of experiential reports. In embarking on this journey, one of my goals was to begin paving the way for future research projects, to lay some very initial foundations and help you see the significance of pure awareness. Another goal was to point to the possibility of a *Bewusstseinskultur* (more on this in the epilogue). But as already explained in the introduction, my main motivation was really quite different: I simply wanted to *share* something that I have found to be valuable with a wider readership—and I hope this collection of materials has served the additional function of creating a platform from which you can begin to make your own discoveries. Perhaps it will have offered you inspiration or some helpful conceptual tools for your own projects. You can find a short list of such tools in the glossary of terms at the end of this book. As I said in the introduction, I have provided only the dessert and some appetizers because I want you to cook the main course yourself—whether it be via your own research, your personal practice, or any other way. However, a first set of cooking utensils and some potential ingredients are now on the table. This book has contained more than 500 answers to the question, "What is pure awareness *like*?" Now we are ready to ask: "What *is* it?"

Pure Awareness: What *Is* It?

So what *is* pure awareness? We still do not really know. We are making a fresh start, and progress is bound to be slow at first. But by now, we do know a bit more about what pure awareness is *like*. So let us take stock. What have we learned? In this summary chapter, I will now lead you through our results, dividing them into six sections. The main result is that there clearly is a distinct phenomenal character of awareness itself, and for now, it does remain the prime candidate for the simplest and subtlest form of conscious experience that human beings are capable of. The elephant does, apparently, exist.

However, I think we are now in a position to do a bit more, compiling a whole list of preliminary answers. Let us begin with the "subjectivity argument." I first pointed to this when—in the very first paragraph of this book—I claimed that consciousness can exist without an experiential first-person perspective and that, in this sense, consciousness may not be a *subjective* phenomenon at all. This claim, if true, may change the stakes of the entire consciousness debate; it also forms a deep link to the ethical goal of suffering reduction and the idea of a rational, evidence-based *Bewusstseinskultur*.

The Subjectivity Argument

The minimal model approach dissolves the problem of subjectivity. Consciousness is a whole bundle of problems, some of a more philosophical and some of a more empirical nature. The most difficult of them all, however, is the problem of subjectivity. In the past, it has often been falsely assumed that the target of consciousness research is irreducible because it is essentially tied to individual first-person perspectives (chapter 27), or even to some kind of mysterious "first-person facts." Now that full-absorption episodes, nonegoic self-awareness, and nondual modes of consciousness are empirically established phenomena, we can see that minimal phenomenal experience (MPE) is not subjective and having a first-person perspective is *not* a necessary condition for consciousness to occur.

There are weak versions of what the term "subjective" could mean, and then there is a philosophically substantial phenomenological reading of it. All the weaker versions are ultimately anchored in the phenomenal experience of having a subjective perspective—but this experience is not the only form of consciousness we know. Take a full-absorption episode of MPE as an example: consciousness in its simplest, most primordial form. Full-absorption episodes are nondual in the sense that no form of subject/object structure is represented in them. Of course, MPE is still subjective, in that there will be *some* subject of experience that "undergoes" or is subjected to an episode of pure awareness—for example, the biological organism that lives through it. Perhaps a conscious machine could have internal states that we describe as physically realizing or even simply being identical to "pure awareness" or "nonegoic self-awareness."[2] These are two examples of one weak version of subjectivity: Trivially, the state of MPE will always be the state of *some system*. But this doesn't entail any phenomenology of ownership, egoic self-awareness, or the fact that this system has a consciously experienced first-person perspective. One important advantage of the minimal model approach is that it prevents unnecessary details from distracting us from a deeper understanding of our phenomenon of interest.[3] My point is that what we call the "first-person perspective" is precisely one such detail.

Our first provisional result is that the phenomenology of world-directed, egoic self-awareness is a surface phenomenon. Ultimately, it belongs to the content of consciousness, but it is not one of its intrinsic features. The "selfiness" and the "perspectivalness" of ordinary conscious experience are part of the movie, if you like, whereas pure awareness is more like the projector's light beam that sometimes causes the whiteness of the cinema screen when the movie has paused. This may also be what our meditators mean when they say, "It is not an experience" (chapter 31). What unites science and meditation is that what they really want to get at is the light beam, not merely the movie. No, *consciousness* as such is not subjective, but yes, for humans, conscious *experience* almost always is because it has been contracted into and is now structured by a first-person perspective, by the brain's model of a knowing self directed at the world (chapter 8).

What about nonegoic self-awareness? Nonegoic self-awareness (chapter 29) certainly is *phenomenal* (this is what makes it an experience), but it is neither subjective nor objective. There is no knowing self in it, and it is also not reified—like some kind of "thing" that awareness has turned itself into in order to look at itself from the outside. Nobody owns it. Phenomenologically, it is more like an anonymous and unstructured field of self-cognizant wakefulness. Quite obviously, there could also be an "epistemic subject" in the equally weak and abstract sense given in the two examples of the organism and the conscious machine. Both the organism and the machine could be nonegoically acquainted with such states and *later* report and verbally describe them as "their own" experiences. Afterward, there could be consciously experienced episodic memories referring to an episode of nonegoic self-awareness or full absorption into pure awareness, and such episodes of remembrance could be owned and be experienced as states of a knowing self.[4] But that knowing self was not present when consciousness occurred in its simplest, most primordial form. All that existed was the zero-person perspective (chapter 3); the epistemic agent model was absent (chapter 25). Yes, the organism or the machine may be "subpersonally accquainted" with such a state, but this form of inner acquaintance, deep, subtle, and intimate as it is, presents us with only a weak and benign, epistemologically and metaphysically unproblematic form of self-knowledge. You do not have to be a person to enjoy it; another animal or a machine could also have it. Subpersonal but nonetheless conscious self-knowledge from the zero-person perspective does exist—this is what empirical evidence shows. It is thanks to this fact that consciousness is now wide open to scientific methods like data-driven computational modeling. The philosophically substantial epistemological reading of "subjective" says that consciousness is necessarily tied to an epistemic agent (a knowing self that actively organizes the field of experience), to an individual and consciously experienced first-*person* perspective, or to some slightly mysterious but still egoic form of prereflective subjectivity. As our material in this book shows, this reading is false. A more careful and thorough phenomenological approach focused on the *simplest* form of experience demonstrates that consciousness *as such* is not irreducibly subjective in any strong or philosophically interesting sense.

But if MPE isn't subjective in a strong sense—is it objective? Yes, in the weak, naturalistic sense that it is a fully embodied phenomenon and scientific investigation of it is possible. It is not unknowable. We can—and one day will—have an evidence-based, empirical account of pure awareness. Much more interestingly, however, a more substantial, phenomenological reading of "objective" clearly shows that pure awareness is not truly objective either. If we take the phenomenology of pure, nondual awareness seriously, we find that pure awareness is *neither* subjective *nor* objective. MPE is entirely open to scientific investigation, but doing full philosophical justice to this neither-nor-ness also requires a new approach—an approach that we might provisionally term "spiritual naturalism." This methodological approach would be naturalist, in that it sees "philosophy and science as engaged in essentially the same enterprise, pursuing similar ends and using similar methods,"[5] but spiritual, in that it also accepts meditation as an epistemic practice in its own right and underscores the importance of nonegoic self-awareness to any convincing theory of consciousness.

Still on a conceptual level, let us continue summing up by looking at the semantic component of "purity," taking not only previous, more traditional descriptions but also our own qualitative investigation into account. One result is that we can now distinguish at least five major readings of what the alleged "purity" of MPE might refer to, depending on the phenomenological context.

Five Forms of Purity

If we ask what exactly makes the experience of "pure consciousness" or "pure awareness" *pure*, we are now in a position to offer a number of different answers—and each of them is completely right in its own way.

You may recall that our very first, rather radical, reply to the question about "purity" was this: "Consciousness is pure whenever there is no other experiential content whatsoever, when the quality of conscious*ness* itself is the only kind of phenomenal character that can later be remembered and reported." It was this specific answer that led to my initial hypothesis that a maximally simple, minimal form of conscious experience does exist, and that it can be derived equally from historical discourse and from our own data on episodes of full absorption or "clear light sleep" (chapter 20). This answer refers to a canonical idea that appears in some of the mystical and contemplative literature, and it is now tentatively confirmed for a specific subset of MPE states. But this is only the first possible reading of "purity." Let us look at all five equally possible answers:

- **P1 Contentlessness** MPE can occur in full-absorption episodes as the only kind of phenomenal character that can later be reported; in these cases, pure awareness appears as a stand-alone, singular feature (see introduction).
- **P2 No-thought** The second, less radical, but equally canonical reading of "purity" refers to the absence of all mental conflict, noise, and perturbation. In a simplified reading, it is the absence of all discursive thought, memory, planning, and "spontaneous-task unrelated thought" like daydreaming or mind-wandering (see chapter 25). Pure consciousness means calmly abiding in a crystal-clear and thoughtless state, nonconceptually knowing alertness itself.
- **P3 Clarity** In this answer, pure awareness is the experience of a clear and unobstructed space of knowing. This phenomenological aspect relates to the concept of "epistemic openness" and the idea of having an inner "model of unobstructed epistemic space itself," which were first sketched in chapters 4 and 5. The clarity feature has been described for centuries and is simply the nonconceptual experience of one's own epistemic capacity per se, of an inner space that is pure in the sense of being wide open and lucid at the same time.
- **P4 Suchness** This interpretation refers to the phenomenology of "pure perception" extensively discussed in chapter 9. Here, the conscious experience is of perceptual content spontaneously arising, but with a complete lack of conceptual overlay, including time experience and implicit judgments as to the "existence" or "nonexistence" of what is perceived. It is this lack that really makes perceptual awareness pure. Phenomenologically, the absence of all discursive thought ("seeing out of emptiness") leads to a suspension of prior knowledge and expected probabilities in perception ("seeing epistemic openness *in* perceptual objects themselves") and an experience of timeless immediacy—of perceiving *what is*.
- **P5 Nonduality** In chapters 27 and 29, we saw that one way of defining "purity" is as conscious experience lacking subject/object structure. In this sense, pure awareness

is a nondual *mode* of experience that lacks an epistemic agent model (chapter 25) and any sort of egoic first-person perspective. Of course, there are also more transient nondual *states*, such as full absorption episodes during formal sitting practice. Please note that this general way of looking at MPE could also permit us to speak of phenomenological *degrees* of purity: Subject/object structure turns out to be a fundamental but not strictly necessary structural feature of consciousness that can be absent, a hyperprior or high-level belief that can be suspended through contemplative practice. There are other such hyperpriors. For example, the dualism between what is temporal and what is subjectively experienced as timeless (chapters 22 and 33) or the experiential distinction between what actually exists and what doesn't (chapter 28) could equally be suspended, thereby leading to an even "purer" global mode of conscious experience, to a higher degree of indeterminacy or "neither-nor-ness." Depending on the high-level belief that becomes attenuated, therefore, we may be able to distinguish between different forms or degrees of nondual awareness.

Contextualizing Purity

Before we continue, please note that we find distinct but indirectly related readings of "purity" in many nonphenomenological contexts, which may be very different from the phenomenological ones. Here is a first example. A radically naturalist hypothesis says that MPE can be viewed as a way of thoughtlessly experiencing the raw activity of a specific part of the *neural* body only (in chapters 6 and 18, this was referred to in terms of "abstract interoception" and the "raw-feeling aspect of wakefulness," respectively). Under a functional reading of "purity," awareness could be viewed as pure in the sense of not being mediated by any sensory system, being directed neither at the interior of the nonneural body (via "interoceptive inference") nor at the external environment (via "perceptual inference"), and emerging in the absence of any concrete form of inner or outer action (via any kind of "active inference"). Purity in this sense has no direct connection to ordinary phenomenal experience; in chapter 24, but also in chapters 7 and 18, I termed it "abstract embodiment," viewing it in those contexts as a merely computational property. Rather, it consists in a sheer lack of hierarchical depth, simplicity, or abstractness in an-as-yet-to-be-determined computational sense. Let us now look at a second, very different nonphenomenological context in which the idea of "purity" plays an important role.

One important type of context is not computational, but cultural and historical. In different cultures and against a variety of philosophical backgrounds, the "purity of consciousness" can mean many different things—but interesting and inspiring connections can be seen. To give just one example, the history of Christian philosophy and early Western theories of consciousness contains the idea of "having a pure conscience," a beautiful idea that emerges in the context of the Latin precursor concept of *conscientia* and frames consciousness as a matter of higher-order moral judgment and the presence of a divine witness or ideal observer in your own mind. For some interesting transcultural parallels, we can now consult chapter 19 of this book, on "witness consciousness" (and the Christian idea of "resting in the Lord's gaze"), as well as chapter 26, on the medieval mystics' concepts of *lûterkeit* and *pûrheit* as related to the *Seelengrund*, the "ground of the soul." These writers refer, for example, to the "purity of meditation," the "pure clear light," or Meister Eckhart's famous "pure clear One." According to this second, normative definition of consciousness as Christian *conscientia*, many of us would simply lack awareness: Because *moral* purity and *moral* meta-awareness make you conscious, committing Christian sins makes you unconscious. The general point is that, depending on the context we are considering, new insights about pure awareness could potentially acquire richer meanings that go far beyond the phenomenology itself.

A Minimal Model of Consciousness

At the start of this book, I proposed a new research strategy: to adopt a "minimal model approach" to the problem of consciousness. We do not have such a minimal model yet, but what we have achieved—thanks to the generosity of our participants—is a much better phenomenological understanding of the modeling *targets*, of what it really is that we need to understand better. I may be wrong, but I think that from this description of target states, we can begin to fashion the first building blocks for a minimal model of consciousness. Let us proceed from the most abstract to the most concrete level as we collect our materials:

- MPE is a model of epistemic openness. It is a global and entirely nonconceptual representation of the *possibility* of knowledge and epistemic gain (chapter 5). For example, we could view it as a high-level Bayesian model—an integrated, global, but entirely unstructured expectation of epistemic value. Whenever this model is actually *running* in a physical brain, it opens phenomenal state space—creating the potential for an experience of pure awareness. Please note that "epistemic openness" is a new third-person concept, applied from the external perspective of science. But the property that it (hypothetically) *refers* to can also be represented internally, by the system itself—the idea is that this is what pure awareness is. "Emptiness" is epistemic openness. Phenomenologically, it creates the experiential quality of wakefulness and silent clarity without content.

- Speculatively, MPE can be seen as a highest-level generative model, describing the conditional probability of internally observable states of knowing. What MPE really predicts is the occurrence of internal states carrying high epistemic value (i.e., the probability of uncertainty reduction while navigating the world).[6]
- MPE can also be described as a model of an *epistemic space*, a space containing a very large number of possible states of knowing (chapter 5). All biological and some artificial systems can now be viewed as embodied epistemic spaces. But only some of them have a *model* of their own epistemic space, and only very few are aware of this model *as a model*. Everyday contemplative practice has a lot to do with not just "being conscious," but also making this model of the space of knowing "visible" or phenomenally opaque—as if learning how to look *at* a window at the same time as you look *through* it.

This was a maximally abstract description. Before summarizing more of our main results, let me point out that for every target phenomenon, there can be multiple, equally justified levels of description. Speaking of "levels" is not some vague spatial metaphor because each level is constituted by (1) a logical subject (e.g., whole persons, biological organisms, brains, parts of brains, their computational states, and other subpersonal units of explanation) and (2) a set of predicates that helps us ascribe certain properties (e.g., beliefs and intentions, bodily states, neural activation patterns and their computational properties, and certain phenomenal qualities like "nonsensational awe") to this subject. From a philosophical point of view, it is not prima facie obvious that one of these levels is "closer to reality." For example, if our target is "pure awareness," we can choose among a phenomenological level of description, related statistical and conceptual levels, and the level of neuroscience.

Today, it is more than plausible to assume that MPE has a true physical description. As I have noted too many times, I am not interested in metaphysics—but there are many metaphysical options on the table, neutral monism and nonreductive forms of materialism and naturalism among them.[7] Of course, many *other* levels of description and explanation also exist, some of which we may discover in the future. What level we choose depends on what exactly it is we want to know, what our epistemic goal really is. As ever, what is needed in choosing the right level is, above all, a quality of open-mindedness.[8] We also need certain other epistemic virtues, however—like courage (e.g., when it comes to facing emotionally unattractive empirical facts), but also caution and temperance (when it comes to drawing metaphysical conclusions).

The general picture is this: MPE is what anchors verbal reports of pure awareness. However, it is not some sort of mysterious "essence." MPE is not a thing, but a process. Today, that process is best described using a dimensional approach—that is, as a region in or path through the phenomenal state space opened up by the human brain, our inner space of actual and possible conscious experiences. This region may have no sharp boundaries. The experiential process or state of consciousness described by this region functions as what I have called the "phenomenological anchor" for a large category of experiential reports, most often occurring in the context of contemplative practice. Over the centuries, this phenomenological anchor has also grounded certain metaphysical models of reality, traditional philosophies of mind, and even religious notions of salvation and liberation. Humankind has known MPE for a long time. The experience has been described by scholars, practitioners, and scholar–practitioners. Accordingly, there are many different and equally important ways of representing and trying to understand this region in conscious state space—some examples being via conceptual analysis, psychometric data, or a more qualitative, phenomenological approach based on a large number of verbal reports. This being said, let us now look at some tentative results.

"Pure Awareness": From Concepts and Psychometric Data to Qualitative Phenomenology

In a first semantic analysis, I took existing descriptions and concepts of pure consciousness as the raw material.[9] I found that the two strongest semantic markers by far were "Wakefulness" and "Epistemicity":

- **Wakefulness** is a graded phenomenal quality that everybody knows. If experienced as such, its qualitative character is related to phenomenological notions like "mental clarity" (but without mental content), "cognitive lucidity" (but without conscious cognition), and "bare awareness" (without any perceptual or cognitive object). As pointed out in chapters 4 and 5, wakefulness is best characterized as an experience of *epistemic openness*. Empirically, it is related to tonic alertness, to stimulus-readiness, and to nonconceptual knowledge about one's own capacities for orientation in time and space and for control of attention (but without necessarily exerting them).
- **Epistemicity** refers to the consciously experienced quality of knowing, which can sometimes appear in isolation and in the absence of egoic self-consciousness. In this book, it was called "the phenomenal signature of knowing" (chapter 7), or alternatively "the phenomenal signature of *self*-knowing" (chapter 30), since there is also a reflexive variant that was given a conceptual description many centuries ago. Relating this experiential quality to current evidence-based frameworks, we can say that epistemicity is the phenomenal character of *subjective confidence*, such as the nonconceptual experience of knowing the probability that one currently knows or

is likely *able* to know something, like a perceptual object (say, an apple on the table that appears as ultimately real) or a state of one's own mind (say, a spontaneously occurring but somewhat vague feeling of happiness and contentment).

Let us now proceed to the second level of investigation. A first psychometric study took as its database responses given by practitioners of meditation participating in a larger survey.[10] The selection of verbal reports presented in this book originated from the same study as a separate data set. A statistical analysis of the survey responses themselves yielded twelve factors (internally coherent clusters) from our ninety-two questionnaire items, as described in chapters 1 and 2. If we now add in our qualitative analysis as a phenomenological "relevance filter," then the most interesting five of these factors are the following:

- **Peace, Bliss, and Silence** (factor 2) picks out the experience of relaxation and ease, perhaps the best-known effect of meditation practice generally. It also refers to a simple experience of deep, unbounded silence and "pure being" that is described as natural and gentle. Two items in this cluster relating to the phenomenal experience of "peace" and "wholeness" show an absence of mental conflict and an increased degree of integration. In terms of qualitative analysis, factor 2 directly relates to the two new phenomenological concepts of "existential ease" and "nonsensational awe," and to many of the experiential reports presented (e.g., in chapters 1 to 3, as well as in chapters 14, 15, and 26).
- **Self-Knowledge, Autonomous Cognizance, and Insight** (factor 3) matches one of the semantic constraints for our first working concept of pure consciousness—namely, the element of "Epistemicity" listed previously because it is related to the nonconceptual phenomenal experience of "knowingness" or subjective confidence. It instantiates an autonomous (i.e., unfabricated) phenomenal character of insight, cognizance, and clarity. Importantly, the strongest-loading item in factor 3 ("Did the experience have a quality of knowing itself?") expresses a phenomenology of non-egoic, first-order reflexivity that also reappeared in the present qualitative analysis (as explained in chapter 30). Our data show that the nonconceptual quality of "pure knowing" can be *self-directed*, but in a nonegoic way, without the involvement of any kind of mental or bodily agency, lacking a conscious sense of control, the quality of ownership, and the phenomenology of "selfhood" in terms of transtemporal identity. Therefore, we also find a strong phenomenological relationship to factor 8 ("Emptiness and Nonegoic Self-Awareness").
- **Wakeful Presence** (factor 4) integrates the phenomenology of spatiotemporal self-location with alertness and the unified experience of "existence as such." According

to our survey data, there is a phenomenal experience of being fully settled in the "Here" and "Now," permeated by the character of wakefulness and a feeling of "simply being." We labeled this cluster of phenomenal qualities "Wakeful Presence" because its four strongest-loading items refer to the embodied experience of wakefully being in the present moment. Quite obviously, the phenomenological-alertness component directly relates to and confirms the first, and most prominent, semantic constraint for what we mean by MPE—namely, "Wakefulness" (chapter 4). However, factor 4 also incorporates two additional elements: the experiential qualities of pure being and unity. Both of them reappear in our qualitative assessment, under the rubric of "nondual being" (chapter 26).

- **Luminosity** (factor 6) refers to the nonvisual phenomenal qualities of "brightness," "radiance," and "vibrancy," but also to the visual experience of brightness with closed eyes. "Self-Luminosity"[11] was also one of the six semantic constraints previously extracted from the literature, referring to the phenomenology of "brilliance" or to the "clear light of primordial awareness." Importantly, the phenomenal character sometimes described as "luminosity," "radiant clarity," or even "enlightenment" comes in many varieties, which may turn out to be unrelated. For example, many practitioners describe a nonvisual phenomenology of "clear light," a nonperceptual experience of clarity and mere epistemic openness. By contrast, others report a more concrete form of visual brightness, which can even be experienced with closed eyes.
- **Emptiness and Nonegoic Self-Awareness** (factor 8) may be most interesting from a philosophical perspective. The second- and third-strongest loading items in factor 8 offered two metaphorical descriptions of first-order reflexivity combining the "Self-Knowledge, Autonomous Cognizance, and Insight" of factor 3 and the "Wakeful Presence" of factor 4 with the phenomenal quality of emptiness and epistemic openness. These are negatively correlated with the phenomenology of selfhood: The phenomenology of self-knowing and self-awakening picked out by this factor is *nonegoic*. Arguably, it is precisely these aspects of factor 8 that, for many participants, may express the "spiritual essence" of MPE most directly. Empirical evidence for the actual existence of nonegoic self-awareness in the context of a substantial psychometric study is a theoretically relevant result. It will need further conceptual analysis, one reason being that almost all philosophical theories of conscious self-knowledge have tied it to a subject or an individual first-person perspective. Many think that consciousness is irreducible for exactly this reason. But MPE, as well as reflexive MPE, are subjective only in the benign sense of occurring within an individual organism. The experiential quality of such episodes may later become integrated with some

long-term self-model, mentally reformatted into autobiographical memory, and this appears to be a causal precondition for the episodes to be *reported* in a verbal form that, via its subject/object grammar, retrospectively imports a fictitious first-person perspective.[12] But we now have the first systematic empirical evidence for a conscious and genuinely nonsubjective form of self-knowledge, and this evidence may present us with the kind of "epistemic convergence zone" discussed in chapters 20, 27, and 28, a previously unnoticed point of contact linking human phenomenology with the scientific image of reality.

The first level of investigation looked at the *concepts* that were used to refer to pure consciousness across different historical epochs and cultural contexts; the second level was a *psychometric* approach to MPE using a larger sample of present-day meditators in more than fifty countries.[13] The third level of investigation targeted the *phenomenology* of awareness based on explicit verbal reports. The semisystematic qualitative analysis offered in this book has tried to integrate all three levels and has now yielded a whole range of specific phenomenological insights into the pure-awareness experience, as well as some more abstract principles. Before listing a selection of core insights, this time starting with the more concrete and proceeding to the more general, let me come back to the fable of the elephant and the blind and draw your attention to a fundamental phenomenological principle, guarding against possible misunderstandings right from the outset.

The prototypical core of MPE can occur in isolation, but also in combination with other, often more global states and modes of conscious experience. The elephant can be found all by itself, in a desert as it were, but also in and against the background of very different landscapes—in a valley or on a hilltop, deep in the forest or bathing in a river. These landscapes can also be seen as something that appears *within* the elephant, as if it were dreaming about or imagining them. Sometimes other animals like to walk ahead of it (think of an experience of deep relaxation or dissolving body boundaries, as described in chapters 1, 24, and 25). From time to time, other phenomenological creatures may prefer to follow in its footsteps (think of experiences of connectedness, joy and gratitude, or lucid dreaming, as described in chapters 11, 15, and 21). And then there is always the oxpecker, a bird that lands on an elephant and sits on top of it. Even a human elephant rider—a mahout (chapter 25)—has been known to appear in the elephant's mind from time to time.

Many of the reports presented in this book may appear to be referring to very different states of consciousness that at first glance have nothing much to do with each other. However, what unites them is that they were given at the end of a long survey targeting MPE and involving ninety-two items, in response to a question asking for one particular example where the qualitative character of "awareness itself" was explicit and clearly noticeable. In some cases, the examples chosen were dramatically altered states of consciousness, perhaps involving an MPE experience plus ego dissolution, an episode of nondual awareness, or the paradoxical quality of timeless change plus a global mode of consciousness involving "witness consciousness" (for more, see chapters 25, 27, 22, and 19). In other examples, "awareness itself" was the only experiential quality that could later be reported; these were full-absorption episodes, typically occurring during formal sitting meditation or, much more rarely, during dreamless deep sleep (chapter 20). In sum, we find a whole *family* of MPE-related modes of conscious experience. For example, we saw how in the real life of a committed practitioner, there are many variants and subtle, hard-to-describe transitions between such variants and modes, creating a lived pattern of family resemblances that is impossible to cleanly map onto any rigid conceptual scheme. However, we also find a hard phenomenological core. Our dimensional approach shows that there are resemblances, but there is also a prototype. So, finally, if we consider just prototypical MPE itself, the phenomenological anchor of pure awareness reports, what stands out? What is the phenomenological core? It includes the following elements:

- **Simplicity**. The phenomenal character of awareness itself is utterly simple, but it often goes along with a subjective sense of profundity. In our participants' ratings, the statement "The experience of 'pure awareness' is the simplest kind of conscious experience I know" achieved a median rating of 80 out of 100 possible points, and it appears in a cluster of items in which MPE is described in terms of the phenomenal character of "existential ease" (see chapter 1), as "the natural state" (item #73), and as an experience of pure being (item #55). This resonates with many ancient texts, where it has been described as "the natural state" or one's own "true nature," for example.
- **Silence**. As one of many expressions of simplicity, the phenomenology of silence and mental stillness is often described as a complete absence of thought and mental content. Prototypical MPE is characterized by a principle of mental inaction and nonreactivity: Phenomenologically, it is that which never speaks, never makes a choice, and never initiates an action. On the level of semantic analysis, the phenomenology of silence appeared as the "low complexity" constraint for our working concept of pure consciousness (chapter 3); statistically, it is related to the most internally consistent of the twelve factors, factor 2, which was labeled "Peace, Bliss, and Silence."
- **Wakefulness**. Wakefulness is the conscious experience of tonic alertness, which—in more abstract terms—can be described as the phenomenology of epistemic openness. The qualitative character of wakefulness is a core feature of MPE, and in

semantic analysis, it reappears strongly as the Wakefulness constraint (PC1), while from a psychometric perspective we can clearly see it in factor 4 ("Wakeful Presence") and factor 3 ("Self-Knowledge, Autonomous Cognizance, and Insight").

- **Clarity.** In chapter 6, I analyzed clarity as the experience of an unobstructed inner space of epistemic possibilities. Our meditators described their own experience of clarity as related to qualities of lucidity, lightness, and spaciousness; as an absence of cognitive perturbation characterized by equilibrium and wakeful presence; as a calm and timeless form of stillness; as a gentle form of acuity or sharpness; or as a mode of conscious perception without center or boundaries. The prototypical core of MPE, therefore, can be described as a silent combination of clarity and wakefulness, the phenomenal experience of unobstructed epistemic openness. This corresponds to some classical texts that speak of the union of lucidity and emptiness.
- **The signature of knowing.** MPE instantiates an autonomous and unfabricated phenomenal character of *subjective confidence*, often described as a quality of insight, cognizance, and clarity. Phenomenologically, it is that which *knows* but never speaks, makes a choice, or acts. Importantly, our data show that the specific phenomenal character of subjective confidence can occur in isolation, as an experience of "pure knowing" that lacks subject and object. Equally important, they also demonstrate the existence of a phenomenology of nonegoic self-awareness (i.e., the actual occurrence of a distinct phenomenal signature of *self*-knowing, which can be experienced in isolation and nondually). In a first conceptual analysis, this core feature appeared as the semantic constraint of "Epistemicity" (PC5); in psychometric analysis, it is reflected as factor 3 ("Self-Knowledge, Autonomous Cognizance, and Insight"); and qualitatively, it is interestingly related to certain aspects of luminosity (see chapter 18).

The qualitative analysis presented in this book has highlighted a range of features that are intimately related to pure awareness, some of which (like density and soundness) have, to my knowledge, never been directly reported before, while others (like spaciousness, timelessness, pure presence, and paradoxical combinations of emptiness and fullness) have been repeatedly described for centuries. Perhaps most significant of all, this analysis has also confirmed that MPE can occur not only as a stand-alone phenomenon, but also as a major characteristic of certain altered states of consciousness. There seem to be global *modes* of conscious experience in which the explicit experience of awareness itself plays a central role, but in which the flowing kaleidoscope creating other and more specific content is present as well. In the introduction, I termed such modes of conscious experience "MPE modes."

From States to Modes

As we saw at the beginning of this chapter, global MPE modes can often be characterized by more all-encompassing qualities of clarity, certainty, and nonduality than are found in localized MPE states. Other highly typical descriptors are joyfulness and well-being; virtuality and emptiness; an all-pervading experience of silence, deep peace, and nonreactivity; the global phenomenal qualities of soundness, connectedness, openness, and unity; and an attitude of all-forgivingness. Let me list five typical examples:

- **Witnessing.** Witnessing is a global mode of conscious experience that, in the wake state, is characterized by an uncontracted version of the phenomenal signature of knowing listed here as one of the two strongest semantic markers of pure awareness. As investigated in chapter 19, the *totality* of all experiential contents appears as if being observed by something that isn't really an "observer" at all—by a timeless, absolutely impersonal, knowing presence. Phenomenologically, this MPE mode can perhaps best be described as an experience of the world being mirrored in an all-encompassing quality of infinite, choiceless, and nonconceptual knowing.
- **Nondual awareness.** Nondual awareness is perhaps the best-known global MPE mode. Once again, its existence shows that the experience of knowing does not have to be contracted into a self and a first-person perspective because it can also occur in a "nondual" way. It is as if mirror and world merge. For example, the "meditating self" can sometimes be absent while at the same time perceptual objects take on another quality, not as distinct entities "outside" the space of egoic self-awareness but as elements of a new *Weltinnenraum* (Rilke's "inner world-space"). Nondual awareness lacks subject/object structure, and it is often described as a spacious form of awareness for which conceptual distinctions like "inner" and "outer" or "real" versus "illusory" do not make sense. The example of nondual awareness also illustrates a general point: Global MPE modes are often characterized by (1) the phenomenal character of MPE having become exceptionally strong and vivid, and (2) the fact that certain structural features (like subject/object duality) have been attenuated or are missing altogether.
- **MPE as a nonegoic unit of identification.** Another important general insight is that there are dual and nondual variants of MPE itself. To give an example, I pointed out in chapter 3 that there is one phenomenology of *experiencing* silence, and another of *being* silence. This phenomenological difference points to the same general principle: Repeated and vivid episodes of MPE can apparently lead to a flattening of the brain's landscape of hyperpriors, dissolving some of our unconscious expectations about the structure of reality—for example, about who *we* are.[14] In our example, this

can lead to the disappearance of subject/object structure in the experience of mental stillness.

The specific result that dual and nondual variants of MPE both exist also corresponds to the ideas of "dual mindfulness" (as reflected in factors 1, 7, and 9 in our study) and "nondual mindfulness."[15] Phenomenologically, in dual mindfulness, there still is a meditator and a goal state that is "owned"; there is a sense of effort created by either mental or bodily agency; and accordingly, the subjective experience of time is sustained and remains in place. Typically, there will be an attentional lapse, followed by the phenomenology of noticing, remembering the goal state, and refocusing.[16] There may be a recurring experience of awareness as such, but it is still a meditation experience, not yet effortless and not a full-absorption episode. In dual-mindfulness practice, we may find the experience of a meditating self, recognizing the phenomenal character of MPE, and then looking at and calmly abiding in the quality of awareness itself. Nondual MPE modes, by contrast, have often been described by saying that the experiencer has simply "become one" with pure awareness or the effortless experience of mindfulness itself (this can be described as non-meditation; see chapter 32). The new concept of a "nonegoic unit of identification" highlights the phenomenological fact that this happens without creating any classic markers of selfhood like ownership, agency, or self-location in time and space. My empirical hypothesis is that dual-mindfulness practice indirectly creates the unconscious causal precursors for nondual modes to appear in the practitioner's brain.

- **Bodiless body-experience**. There are global modes of conscious experience in which the experience of embodiment is no longer tied to a representation of the physical body. In chapter 24, I described these modes as examples of "abstract embodiment." They seem to follow the same principle of MPE becoming the unit of identification just described, but in the more spatial sense of directly "embodying" or "realizing" abstract MPE qualities like wakefulness or the phenomenal signature of knowing itself. These MPE modes have been reported in the past, but they deserve more attention as a target of systematic empirical research programs. The general principle of the brain suspending automatic predictions about the structure of reality clearly recurs in many forms. As we have seen, it can also refer to a dissolving of felt boundaries between body and environment, to the well-documented phenomenon of ego dissolution, to the disappearance of time experience, and even to the loss of any distinction between what exists and what doesn't.
- **Virtuality**. Some MPE modes are characterized by the phenomenological fact that the conceptual distinction between what is "real" and what is "not real" can no longer

be meaningfully applied (chapter 28). Experientially, there is a global form of metaphysical indeterminacy, a hard-to-describe form of neither-nor-ness, and this "lucid-dreamlike" quality of virtuality may appear when the phenomenal character of pure awareness itself has become very vivid and dominant, leading to a global quality of "translucency." This "virtuality mode" is yet another example of the general principle that MPE not only occurs in isolation or in simple combination with other experiential contents, but at times emerges in the context of an entirely new *mode* of conscious experience. We find MPE states, but we also find MPE modes. What we do not know is whether there is any direct causal relation here, governing transitions between states and modes.

Ten Open Questions

In closing, let me list a few of the most important future research targets yielded by the prefatory discussion presented in this book. I have flagged a whole series of them as we've been going along, but here are the ten that I consider most promising:

- The overarching goal is a *minimal model explanation* for consciousness, involving a formal model of conscious experience that leaves out everything superfluous, isolating only the core causal factors. Given our psychometric data and the qualitative analysis in this book, my initial working hypothesis still stands: Consciousness can be entirely dissociated from egoic self-awareness, time representation, and spatial self-location; none of these features is a necessary condition for consciousness to occur. MPE is thus the first target.
- On a *conceptual* level, it will be important to develop convincing semantic criteria for minimality. "What is the simplest kind of conscious experience?" is, if taken seriously, a difficult question that demands the attention of computational modelers, mathematicians, and philosophers alike. As I pointed out in chapter 17, this is a thorny theoretical issue that cannot be answered per decree by self-proclaimed experts or spiritual teachers merely restating the phenomenological taxonomies of time-honored contemplative traditions, let alone by the supposedly authoritative representatives of some religious system.
- As should now be obvious, the challenge of pure awareness is extremely interesting from a wider *philosophical* perspective, and for many reasons. For example, pure awareness is a new challenge for philosophy of science, but equally for comparative and cross-cultural philosophy, applied ethics, and many other subfields. I have tried to pave the way for others by focusing on the phenomenal character itself, but some

of the core discoveries that we have begun to make clearly have important *epistemological* implications.

- Let me single out just one issue—a philosophical problem that may be the most interesting of all: the epistemic status of nonegoic self-awareness (as statistically described by factor 8). Given the conceptual tools of modern epistemology, is determining the status of nonegoic self-knowledge a *tractable* problem? Here, part of the philosophical difficulty lies in correctly assessing the "self-validating" phenomenology of certainty and insight with respect to the states' own epistemic status, such as when some verbal report claims that this was *not* a mere phenomenal experience (see chapter 31). Is the contemplative phenomenology of nonegoic self-knowledge merely an interesting form of experiential character, perhaps one that has never been adequately described, or is it grounded in a deep but entirely nonconceptual form of self-knowledge sui generis? If it is in a class by itself, an absolutely unique epistemic phenomenon, what is the relation of "nonegoic acquaintance" that I have briefly gestured toward in this discussion—what exactly does it mean for a conscious system to know itself from a zero-person perspective?
- On the level of *neuroscience*, what in chapters 20 and 21 I called the "Triple Triangulation Project" would be the logical first step. It involves systematically comparing episodes in which MPE (1) occurs as a stand-alone phenomenon and (2) is entered from the wake state, from a lucid dream, or from non–rapid eye movement (NREM) sleep. As explained in chapter 20, this project would involve a number of steps, which can be summed up as follows:
    - Isolate the minimally sufficient neural correlates of type I, type II, and type III episodes of full absorption in neurotypical human beings.
    - Describe commonalities and overlaps by creating a series of computational models that, on a mathematical level, ultimately allow us to find out whether there is a phenomenological "cut set" connecting all three kinds of full-absorption episode—and if so, what exactly its physical and computational correlates are in the human brain.
    - Empirically validate mathematical descriptions of MPE by reliably bringing about the pure-awareness experience via a *subpersonal* route, such as by instantiating its physical and computational correlates through advanced forms of neurofeedback.
- The *psychometric instruments* used to describe the phenomenal character of MPE are in need of serious improvement. This will be achieved by better questionnaires, validation studies using confirmatory factor analysis, and an exploration of potential

differences in response patterns and item functioning across linguistic and sociocultural contexts.

- If larger numbers of verbal reports can be gathered in the future, it will be possible to build a much larger and more varied *phenomenological database* to be submitted to more systematic forms of qualitative analysis. These could include computer-aided "big-data" and artificial intelligence (AI) strategies to deal with increasing volumes at higher speeds. (You can contribute to this set of related projects, even if you are not a meditator, by going to mpe-project.info, where a revised version of the survey is open for submissions.)
- As I have mentioned several times, perhaps the greatest *methodological* problem to be solved is what I have termed "embodied theory contamination." The problem of theory contamination is of special relevance when dealing with the verbal reports describing the experience of pure awareness. For example, we need a much better understanding of how the presence or absence of spiritual and religious elements, whether privately embodied or via institutions, gets integrated into the meditator's self-model. This in turn will allow us to understand how such elements influence verbal reports and determine the fine-grained phenomenal character of contemplative experience itself.
- There are many ways in which virtual reality (VR) and new mixed-reality technologies could be used and combined with other methods like real-time functional magnetic resonance imaging (fMRI), AI-controlled brain–computer interfaces, and other yet-to-be-invented technologies.[17] This could happen by first creating VR models of MPE and nondual awareness and later using these models as experimental tools to systematically induce some of or all the phenomenological features described in this book. For example, VR models could be used to give users a higher degree of opacity-control[18] over their own conscious states, or to create the phenomenology of "translucency" described in chapter 28.
- Finally, given the many new mathematical tools now at hand,[19] the time has come to develop a *computational phenomenology* of meditation.[20] Beyond dual-mindfulness, full-absorption episodes, and the Triple Triangulation Project, the most relevant modeling target would be nonegoic self-awareness, also termed "reflexive MPE" (chapter 30). To end on a more concrete note, figure 34.1 presents a simple sketch of a possible computational model. If you would like more detail, a range of color figures depicting fully parameterized models of the state of pure awareness with and without content, and also of some global modes like nondual awareness, are provided on the companion website (mpe-project.info).

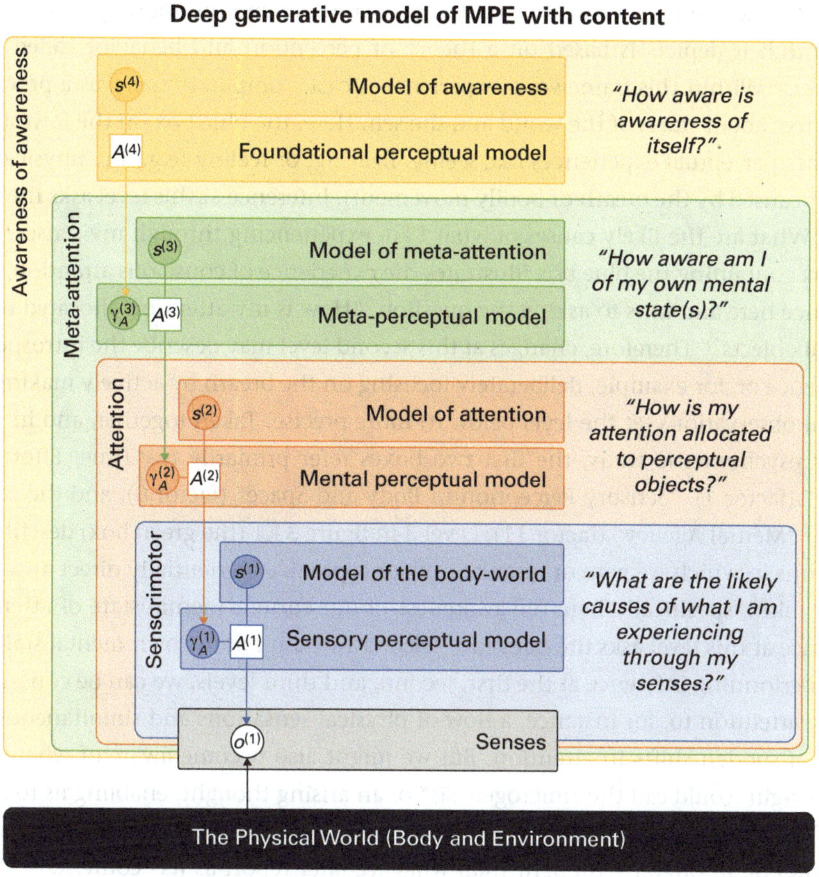

Figure 34.1
An illustration of a nested inferential architecture that would give rise to the experience of pure awareness as co-occurring with other forms experiential content. The phenomenal character of MPE is present and may encompass everything else, but thoughts, feelings, and sensations occur at the same time. Phenomenological examples would be episodic experiential states during dual-mindfulness practice (e.g., chapters 1–5) or the global MPE mode of witness consciousness (chapter 19). The tentative model presented here was conceived by Thomas Metzinger and Lars Sandved-Smith; the figures were created by Lars Sandved-Smith.[21]

Let us take a quick look at this illustration before we end our journey together. The model that it depicts is based on a theory of perception and behavior called *active inference*.[22] Within this framework, experience is cast, computationally, as a process of inferences about states of the world and the self. Here, the blue box on the lowest level describes perceptual experiences like seeing, hearing, or feeling (e.g., the physical sensations caused by the breath or bodily movement). Inference at this level asks the question, "What are the likely causes of what I am experiencing through my senses?" The red box containing the blue box illustrates the experience of conscious attention, since inference here amounts to asking the question, "How is my attention allocated to perceptual objects?" Therefore, changes at this second level may describe the introspective experience of, for example, deliberately focusing on the breath by actively making perceptual observations (at the level below it) more precise. Taken together, and in terms of our psychometric study, the first two boxes refer primarily to "Time, Effort, and Desire" (factor 1), "Sensory Perception in Body and Space" (factor 9), and the experience of "Mental Agency" (factor 11). Level 3 in figure 34.1 (the green box) describes all situations in which we gain or lose the nonconceptual, experientially direct awareness of our mind, specifically here the awareness of our current mental state of attention. Inference at this level asks the question, "How aware am I of my own mental state(s)?"

By performing inference at the first, second, and third levels, we can be consciously paying attention to, for instance, a flow of physical sensations and simultaneously be aware of sudden shifts in attention. But we might also become aware of what Claire Petitmengin would call the "microgenesis" of an arising thought, enabling us to disengage from its content—by discovering that there is actually much more to the subtle dynamics of an ongoing thought than what we later report as its "content."[23] This is a typical example of experiences occurring during mindfulness practice like Shamata or Vipassanā meditation, in which we deliberately but nonjudgmentally observe our breath and monitor the wandering of attention.

Finally, the fourth and largest box describes the phenomenology of "epistemic openness" or pure awareness at the system level. Fourth-level state inferences model perceptual awareness across levels of inference; they are representations of the system's global "epistemic openness" to incoming information. Phenomenologically, one might argue that to infer at this level is to become consciously aware of awareness itself by asking the question *"How aware is awareness of itself?"*

You may now ask: Why is awareness *consciously* aware of itself in this situation? The short answer is this: Because conscious awareness appears when a system is epistemically open to (i.e., aware of) the world and also *knows* this fact. It now has a model of its own openness. It is level 4 that really endows the content generated on all lower-order levels with phenomenality—with the quality of "being consciously experienced"—because it adds the *expectation* that something can (and likely *will*) now be known. The additional, superimposed anticipation of knowledge is what generates the phenomenality of this lower-order content. More technically, consciousness is a generic global prediction of epistemic gain that functionally integrates all lower-order content by metarepresenting it as something that is now globally available to be explored in a deeper way. As you may recall from chapter 4, MPE is an ongoing process of nonconceptually experiencing the organism's current state of epistemic openness (i.e., of expecting new knowledge without yet having it). This is what modeling its states at the fourth level amounts to: a single, integrated model of its own epistemic openness in the form of a highest-level but nonegoic self-model. Without this fourth level, we could say that awareness (i.e., level 3) is transparent: We see "with" it or "through" it, but are aware only of the lower-order content. But now, with the fourth level, awareness itself is "opacified," as when we suddenly become aware of the window pane through which we are looking at the garden.

We are now at last in a position to see what the experience of pure awareness during a full-absorption episode really is. An MPE absorption episode can be modeled as the situation in which experience is a result only of inference at the highest level, without any perceptual inference at the levels below it. This would occur when the system no longer attends to lower-level observations and is unable to perform inference at those levels, while continuing to infer at the fourth level. As a result, the phenomenal character of MPE is present and is the only quality that can later be reported, since no thoughts, feelings, or sensations are experienced at the same time. For example, states of pure conscious experience in which nothing but the experiential quality of pure awareness itself is present may occur during deeper states of formal practice (e.g., sitting meditation), but also in phases of NREM sleep, sometimes called "clear light sleep" (chapter 20). Again, such states are not *subjective* states because they are not tied to a first-person perspective. At the nested lower levels, we experience the complex, dynamic, and interdependent network created by the inferred causes of our sensory data plus the continuous movement of thought and attention—often involving a sense of agency and the phenomenology of identification. By contrast, the lucid emptiness of MPE itself is nondual. MPE as such is nondual because the distinction between world and self is not yet represented. Phenomenologically, time and space are equally absent.

In this way, we can begin to model and simulate core features of the phenomenology of MPE and meditative practice. The example given here is meant as a hypothetical starting point to illustrate that in principle, we might eventually come to a computational understanding of MPE, the corresponding experiential quality of pure awareness, and the more subtle microstructures of consciousness as a whole.

The Elephant in the Room

It is now very nearly time to end this summary. But there is one last point that we should not forget: There *is* an elephant in the room. One simple yet striking result from our first study is just how *many* human beings all over the planet actually have these experiences but never really speak about them in public. Our survey demonstrates not only that these individuals are ready to do so, if actively approached under conditions of anonymity and in the context of a serious research project, but also that there is a lot for academic consciousness research to learn from them. Quite simply, what I have termed "MPE" seems to be a phenomenological prototype of humankind—at least, the pure-awareness experience is something that many hundreds of participants from fifty-seven countries consistently reported. Of course, the experiential quality of awareness itself has been described for millennia, mostly in Asian traditions, but today, it appears in many different cultural contexts across the globe. The obvious fact that so many people seek, know, and quietly cultivate the kinds of conscious experience reported in this book, plus the contrasting fact that all of this is only weakly reflected within culture, education, and policymaking (or even science and philosophy of mind), may point to a systematic societal blind spot. There really is an elephant in the room because there seems to be something genuinely profound here—but there is also some reason why global society has been unable to see its importance. Maybe, as in the old Tibetan saying cited at the very beginning of this chapter, it is simply too close for us to see, too profound for us to fathom, too simple for us to believe, or even too good for us to accept.

Let us return to the fable of the elephant and the blind one last time. There are many things that the congenitally blind cannot see—things that they will never know. One of these is that the elephant itself is *not* blind: It might actually be looking at the blind people while they try to touch and understand it. Of course, they could touch one of its eyelids or even an eyeball by chance, but they would never fully understand what they are touching or what kind of understanding this eye affords. They would not know the special form of visual knowledge that is made possible by eyes, and therefore they could not wholly grasp the fact that they might actually be visually perceived themselves. If they were seen by the elephant while trying to make contact with it, they would know nothing of this fact. But the elephant would always already know them in a way they could never know themselves. Could it be that the elephant is looking right at you, *right now*?

Epilogue: *Bewusstseinskultur*

Is minimal phenomenal experience (MPE) a *good* state of consciousness? If we move beyond phenomenology, science, and academic philosophy of mind, is the experience of pure awareness something positive, something that we should foster and cultivate in our lives? And is there a sociocultural dimension to the kind of research I want this book to encourage? What is the wider context?

At the end of our journey, it is now time to change our perspective one last time, taking three steps back and a deep breath. My two goals in this final section are to open up a broader—specifically, a normative—perspective, while at the same time offering one last philosophical concept: *Bewusstseinskultur*, a culture of consciousness. I will keep things simple. In case you want to know more, there will be a short book to consult that may be helpful.[1]

Bewusstseinskultur

I introduced the idea of developing a *Bewusstseinskultur* a quarter century ago.[2] *Bewusstseinskultur* aims at a special form of cultural innovation. In a first-order approximation, it consists of three major elements:

1. The adoption of an ethical stance toward one's own mental states
2. The systematic cultivation of states assessed as valuable
3. A continuous process of rational, evidence-based enculturation—that is, an active embedding of such states of consciousness in culture and society

Before looking at each of these points in turn, I should make clear that parts and variants of the basic idea have been around for many centuries. Philosophers, of course, have always had long debates about what the term "philosophy" itself really means, what it *truly* means to love wisdom—because that is what *philosophia* (φιλοσοφία in

Greek) originally meant. To give just one example, Marcus Tullius Cicero (106–43 BC) said in the second book of his *Tusculan Disputations*[3] that truly loving wisdom means "taking care of and cultivating one's soul" (*cultura autem animi philosophia est*). I think this is one beautiful and highly topical variant of the general idea of creating a culture of consciousness. Unfortunately, it also points to one of the greatest weaknesses of the highly professionalized (and often truly excellent) form of academic philosophy that we have today:[4] It is not exactly conducive to the mental health of those who participate in it, let alone to the project of cultivating interesting states of consciousness with epistemic potential, a project that goes beyond mere theoretical knowledge. But given the empirical knowledge about the evolutionary roots of the human mind, and given the power of present-day cognitive neuroscience plus exciting recent advances in computational phenomenology, Cicero's classical motif offers a good example of what we might gain by radically reinterpreting some of our cultural goals.

Bewusstseinskultur is about a specific form of cultural innovation. It concerns a broadening of horizons: We need to develop a new philosophical perspective that not only assimilates all those exciting new scientific discoveries and insights, but also helps us deal with all the new potentials for action in an ethically sensitive manner. Therefore, developing *Bewusstseinskultur* can also be viewed as a novel form of applied ethics. It is about practical philosophy—or, more precisely, about creating a new connection between applied ethics and philosophy of mind. A host of technological possibilities for manipulating and changing the human mind have emerged—even artificial consciousness is being debated as a plausible part of our medium-term future—and we risk being overwhelmed by the psychological and sociocultural consequences of scientific progress and the technologies in generates. Therefore, we need a whole new branch of practical ethics. This branch could be called *Bewusstseinsethik*—an applied ethics of consciousness—because it specializes in all issues directly related to conscious experience itself.

The central aim of consciousness ethics would be to help us with what Cicero, more than two millennia ago, called "taking care of and cultivating one's soul."

Practical Philosophy of Mind

Following Aristotle's distinction between natural philosophy (aiming at the right kind of theory) and moral philosophy (aiming at the right kind of practice), today's academic philosophers often distinguish between theoretical philosophy and practical philosophy. Theoretical philosophy consists of subdisciplines like logic, epistemology, philosophy of science, and philosophy of mind; practical philosophy comprises fields like ethics, decision theory, and political philosophy. *Bewusstseinskultur* will have to unite a specific subset of these two broad sets of research goals in a new way. It can be seen as the societal and political implementation of insights from what I will dub a "practical philosophy of mind." Practical philosophy of mind starts by taking an ethical stance toward one's own conscious mind—and also toward the minds of others.

Taking an ethical stance toward one's own mental states first means asking what makes a mental state *good*. For example, one might ask whether certain states of consciousness possess an intrinsic value of their own. Is there a rational, well-defined way to say that some of the states and modes of conscious experience described in this book are intrinsically valuable, that they are *better* than other states and modes of conscious experience? Are there any objective criteria that we could apply?

In classical ethics, one would ask what makes an action good, or whether there are certain action goals, moral attitudes, or virtues that are intrinsically valuable. Here, the idea is to extend exactly the same strategy to our own conscious minds, by asking: What makes a state of consciousness a *good* state of consciousness? Are there "phenomenological virtues" at all? Are there intrinsically valuable attitudes, ways of life, or goal states of conscious experience? What are beneficial and wholesome ways of dealing with one's own mind?

This emphasis on the ethical question of conscious experience itself is the main reason why the first of the three elements that I listed here could also be called the project of developing a *Bewusstseinsethik*. *Bewusstseinsethik*, however, is not meant to replace traditional ethics. On the contrary, the idea is to complement and enrich traditional ethics by drawing on empirical evidence and rational argument and focusing on cultivating conscious experience itself.

The Systematic Cultivation of Positive States of Consciousness: Concrete Examples

Let us now look at the second point in the working definition of *Bewusstseinskultur* presented here: the systematic cultivation of valuable states. If our project of developing an ethics of consciousness yields results, then we can begin to think about how to implement them. Depending on our philosophical background theory, we might even feel a moral and political *obligation* to do so. If we accept any moral obligation at all, we also have the obligation to pursue our ethical goals as efficiently as possible, to consider all possible instruments to reach these goals, and to look at everything with an open mind. If and when we have identified positive states of consciousness, what will be the most effective ways to realize them?

All of this may be starting to sound quite abstract. However, *Bewusstseinskultur* is eminently practical because it is a new branch of applied ethics that—as soon as we begin to think seriously about it—leads to a whole range of very concrete questions. To make this clear, here are some examples.

Which States of Consciousness—If Any—Should Be Illegal?

The "technological landscape" for generating altered states of consciousness is undergoing dramatic changes, and it is fair to say that the overall situation is completely out of control. In the pharmacological domain, well over 1,000 new psychoactive substances emerged in illegal markets worldwide between 2009 and 2021.[5] Digitally, as a consequence of the fertile confluence of artificial intelligence (AI), virtual reality (VR), and neurotechnologies such as brain-computer interfaces (BCIs), new ways of "constructing" phenomenal states will keep emerging. The emergence will be driven partly by technologies that directly target the brain, but also via the creation of new medial environments and entire life-worlds, most likely owned by profit-driven private companies not oriented to either the common good or the mental health of their users.[6]

We can ask specific variants on the overall question posed here, such as:

- Given the recent explosion of new synthetic drugs and the failure of prohibition, what would a rational and ethically defensible drug policy for the future look like?
- Do we need legal regulations for advanced mind-altering technologies?
- How can we create a truly *humane* technology of consciousness? For example, how can we ethically design algorithms that are constantly creating new cognitive niches for us, as those underpinning social media, VR environments, and the "metaverse" are doing right now?[7]

A major problem for a contemporary culture of consciousness is the new "attention extraction economy," which extracts attention from human brains and turns it into money, supported by self-learning AI and continually self-improving algorithms. In what Tristan Harris has called a "race to the bottom of the brain stem," social media and tech firms aim to maximize user engagement (chapter 28) by creating ever-better attention sinks and developing pathological, addictive forms of media consumption.

Today, information and entertainment are now available almost everywhere for free. Attention, on the other hand, is the new scarce resource that must now be systematically exploited. Human attention has thus become a currency, a means of payment that can be traded. But what is being monetized in the new attention economy is really the destruction of mental autonomy—that is, our ability to control the focus of our own attention in a deliberate and self-determined way (see figure 25.1 in chapter 25). We need mental autonomy, however, if we are to maintain and defend our democracies as responsible citizens.

The capacity for mental self-determination is also of crucial importance for our overall quality of life, including our mental health. Today, the question being asked is no longer whether a human or a machine is a world champion at Go or chess. The game that is being played against us right now is a very different one: Who gets to control the scarce resource of attention generated by our biological brains—us or some tech corporation trapped in its own business model? What persuades human beings—which algorithmic strategy generates feelings of intimacy and the phenomenology of trust most efficiently? Questions like these are absolutely central to the culture of consciousness of the future. Questions such as the following come up here:

- What ethical and legal principles should guide us in the attention economy of the future?
- Which parts of the new sociophenomenological infrastructure should be public goods, systematically protected because of their relevance to the common good?
- Are there media environments and business models that should be illegal?
- In an open and free society, should all citizens have the right to manipulate their own brains in whatever way they wish? Or are there states of consciousness that ought to be "off-limits," for example because the means to achieving them cannot be safely controlled by beings like us, resulting in too high a probability of individual and/or collective suffering?
- How do we get psychoactive substances and AI to systematically increase the mental autonomy of its human users instead of destroying it?
- Which states of consciousness improve empathy and social cohesion?

What States of Consciousness (If Any) Should We Force upon Other Animals?

A growing number of us are beginning to understand that many of our ways of treating other sentient animals are ethically untenable. Animals clearly have a general preference for living as long as possible, having as many offspring as possible, and being as physically comfortable as possible. They do not want to be separated from their children. For many animals, these preferences—which we do not respect—are directly reflected in their conscious self-model; they identify with them. For self-conscious animals, frustrated subjective preferences become their own negative states, which creates the capacity for conscious suffering. Frustrated preferences don't have to come as thoughts; they can also come as emotions, feelings, or bodily sensations.

Modern consciousness research shows that many nonhuman animals have sentience, especially the ability to feel happiness and suffering. As beings capable of suffering, they are automatically the subject of ethical considerations. We must respect not only reason and the capacity for moral insight in other beings, but also their vulnerability, capacity for suffering, and mortality. Here we may ask the following questions:

- What exactly is the relationship between animal ethics and *Bewusstseinsethik?*
- What are effective strategies for reducing animal suffering?
- Which ways of treating nonhuman animals should be illegal, and how can their interests and preferences be brought into the mainstream of society?
- To arrive at convincing ethical decisions, we need hard facts, including about which animals are likely to have a phenomenal self-model, at least some of the time. How should modern consciousness science inform the practical philosophy of *nonhuman* minds?
- Do sentient nonhuman animals possess meaningfully definable forms of dignity too?

What States of Consciousness—If Any—Can We Force upon Machines or Other Conscious Postbiotic Systems?

You may recall that this book is dedicated to the conscious postbiotic subjects of the future. One reason for this is that AI ethics, and in particular the possibility of synthetic phenomenology and machine consciousness, are linked in important—and not innocuous—ways to the intermediate results of our search for a minimal model explanation of consciousness. There may be an information hazard: If we have a mathematical model of the simplest state of consciousness, then we can implement it on machines or other types of carrier systems. If pure awareness really is the simplest state of consciousness, then it may be the easiest to implement.

When taking an ethical stance toward the conscious postbiotic subjects of the future, the term "postbiotic" refers to the conceptual point that the distinction between "artificial" and "natural" systems is not as clear cut as it often sounds. (A philosopher might point out that it is an outdated distinction because it is neither exhaustive nor exclusive.) We have intelligent systems using biologically evolved algorithms on artificial hardware. The deep neural networks that we simulate in large computers are modeled on biology—namely, on the interconnection of neurons in the nervous system of a living being. Soon, however, we will probably have systems that use algorithms developed by humans but which improve themselves independently, or even entire cognitive architectures realized on biological "hardware" (e.g., on a genetically engineered substrate).

As I have argued elsewhere,[8] the possibility of artificial consciousness raises a host of ethical issues. We should not recklessly (whether intentionally or not) create artificial consciousness because artificial consciousness may involve a consciously experienced sense of self that may in turn generate artificial suffering in autonomous, intelligent systems. On the other hand, what makes MPE itself so interesting from an ethical perspective is precisely the fact that it is a form of conscious experience that involves *no* psychological suffering whatsoever (chapter 33). Imagine that we could build a machine that peacefully abides in a clear and silent state of pure consciousness, nondual and without subjectivity, for as long as it exists. If we could create a conscious machine that reliably remains in a single full-absorption episode of pure awareness—would there be anything ethically wrong with creating such a machine? The four questions given here are intended merely as illustrative examples of the many ethically sensitive links between the MPE research program sketched in this book and the problem of machine consciousness:

- Is it ethical to risk the creation of artificial consciousness before we even have a convincing theory of consciousness, let alone of conscious suffering?
- Given the potentially large number of unknown unknowns, should there be a global moratorium on synthetic phenomenology until we know definitively, or at least reasonably accurately, what we are doing?
- Is there a possible information hazard[9] involved in even trying to describe the simplest form of conscious experience or develop a mathematically precise computational phenomenology of suffering? That is, could a good caricature of the essential statistical physics be enough to conjure up the target phenomenon? Could the physical implementation of a minimal model unexpectedly create the relevant microscale causal mechanisms that are sufficient for consciousness to occur?
- If we eventually arrive at a theory that successfully describes conscious intelligence without suffering—are we then morally obliged to realize such suffering-free forms of conscious experience on artificial carrier systems, perhaps even devoting more resources to developing them than we do to propping up biological forms of intelligence for which suffering is unavoidable?

Which States of Consciousness Do We Want to Foster and Integrate into Our Society?

Bewusstseinskultur and practical philosophy of mind are concerned not only with what we ought not to do, but also with what ought to be promoted proactively—that is, purposefully and with foresight. Here is an example. Many interesting and evidence-based academic initiatives that test ways of building secular meditation practices into education exist worldwide. Against the background of the attention economy and the escalating drug problem, this seems to me to make extraordinary sense, especially for children and young people. This type of endeavor can be linked—as the reports from fifty-seven countries presented in this book clearly show—to the experiences of millions of adult practitioners and the expertise generated within the many spiritual subcultures or countercultures that began to develop in the West after World War II.

But for every social movement and for every cultural innovation, there is an optimal time window in which its core insights must be formally integrated into broader structures or risk being lost again. Now may be the time for political institutions to finally catch up with social reality, separate the wheat from the chaff, take responsibility, and preserve what is substantial before it gets further watered down by market mechanisms. Alongside education and teaching, there is research itself. Western academia lacks a tradition of scholar–practitioners. We do not systematically combine distinct epistemic practices like science and meditation, so we miss out, for example, on the heuristic fecundity of MPE phenomenology described in this book. Given the preliminary findings from this survey and the research goals listed at the end of chapter 34, should we create such a tradition?

Again, a more systematic exploration of altered states of consciousness will have great heuristic fecundity not only for science, but for society as a whole—and there is a time window that may close for good if we do not manage to preserve what matters before it gets destroyed by McMindfulness, organized crime, and the forces of a globalized black market. We currently have no protected spaces for those who want to devote their whole lives to cultivating elusive states of consciousness, maximally free of ideological distortions.

In this domain, questions worth asking include:

- Could the proper, safe, and beneficial use of psychoactive substances or the new mind-altering technologies mentioned in this discussion generate epistemic practices that are complementary to meditation and science, cultivated by specific subtypes of scholar–practitioner or phenomenological polymath?[10]
- Given the findings of modern meditation research and the phenomenological data presented in this book, should we integrate secular forms of meditation training into any or all levels of our education systems?
- Could something like a secular monastic tradition be created, one much purer than what we had before because it sets aside mortality denial and narrative self-deception (chapter 17)?

What States of Consciousness Do We Want to Show Our Children?

Education is a process by which young human beings become active members of society and culture, and thus also politically mature citizens. The right to education is enshrined in article 26 of the Universal Declaration of Human Rights: Education shall be free, at least in the elementary and fundamental stages (§ 1), and it shall be directed to the full development of the human personality (§ 2). In the educational context, questions that we would do well to ask and answer include the following:

- What regions of phenomenal state space should *every* human being get to know before becoming an adult?
- Against the backdrop of the escalating drug problem and the burgeoning attention economy, and given the promising results of modern meditation research, should we integrate secular forms of meditation instruction into some (or even all) levels of our educational system?
- Should we offer to the young a "contemplative toolbox" that enables them to systematically explore these regions, whenever they may be interested or have a specific need for them later in life?
- The participants in our study used and combined many meditation techniques as part of their own development (see figure 14.1 in chapter 14). Which of these techniques comprise the essential set that every child should get to know? Which specific meditation techniques should children learn, and during which time window of their psychological development?
- More generally, what would we consider the "indispensable core of consciousness-culture education" that should be imparted to a society? What should our future "phenomenological education canon" look like? What should be the basic set of conscious states that must be freely available to every human being to come into contact with at least once in their life?

What States of Consciousness Do We Want to Die In?

Bewusstseinskultur will obviously have to include a new culture of dying. Existence bias and the resulting culture of mortality denial are major determinants of human phenomenology (chapter 17). Many modern societies have systematically repressed death and dying, turning them into a sensitive and taboo topic. *Bewusstseinskultur* will have to break this taboo because it must pose the ancient philosophical question of what it really means to live a good life, which naturally leads us to ask what precisely a "good death" could be. More specifically, we might ask the following:

- From a rational, evidence-based, and ethically motivated perspective, are some ways of dying better than others?
- Should we create protected spaces in which some aspects of the dying process can be practiced safely, for instance with the help of meditation or certain psychoactive substances?

- Would it be useful if one could learn about the phenomenology of peaceful ego dissolution long before physical death occurs?
- What state or mode of consciousness would *you* like to die in?

These are just a few examples to give you a first impression of the wide variety of concrete issues that need to be addressed; I have written more elsewhere, so I will not repeat myself here.[11] Nevertheless, I hope this small selection shows how many aspects of our societal life would be directly affected if we ever decided to actually implement the provisional results yielded by a *Bewusstseinsethik*, a systematic practical ethics of consciousness. Just imagine what might happen to our society if we took the systematic cultivation of valuable states of consciousness seriously.

Enculturation

The third major aspect of the idea of *Bewusstseinskultur* relates to the creation of a new cultural context. *Kultur* does not just mean that individual citizens cultivate certain states of conscious experience rather than others; it also means organizing the social fabric in a way that supports individuals and groups in systematically investigating the types of question listed in this epilogue, exploring the landscape of possibilities for genuine cultural innovation. *Bewusstseinskultur*, therefore, refers to a new kind of sensitivity that creates a cultural context in which this very process of inquiry, investigation, and experimentation is itself valued—as a shared epistemic practice, as a prosocial activity of cultivating one's own mind while exploring the possibilities of a new normative context for all, and as something that makes a contribution by proactively creating a new kind of common good.

Bewusstseinskultur is a common good, something that we can all share and benefit from. But at its core, it is also an emancipatory project, for it aims to increase our own intellectual autonomy. The techniques underpinning it should never be commodified and turned into a private property, exploited primarily by corporations. Indeed, its essence is fundamentally anticorporate. As an antiauthoritarian, decentralized, egalitarian, and participatory project, *Bewusstseinskultur* is essentially based on community, cooperation, and transparency, and it rejects the logic of consumerist capitalism. If you will, it is a kind of mental infrastructure: In the future, we will need not only clean air and water, schools, museums, public transportation systems, civil liberties, and public safety, but also a continuous process of rational, evidence-based enculturation of those states of consciousness that we have found to be valuable. Having such a process is *itself* a common good—an asset that a given society may either possess or not.

Bewusstseinskultur is something that comes in degrees and grows over time. There is a global context to this: One society may be more advanced in the ethical treatment of animals; another may have a better educational strategy, and yet another the most successful drug policy. We must learn from each other. Here, "enculturation" refers to a continuous process by which ancient practical wisdom, new mind-altering technologies, and all relevant scientific insights are incrementally embedded into societal practice. As I pointed out in the introduction, the problem of pure consciousness will have to be handed over to the hard sciences of the mind, to cognitive neuroscience and computational modeling, but from there, it will later have to return to philosophy of mind and applied ethics. New scientific insights always need conceptual interpretation and an ethical assessment of any new potentials for action that they may generate. In formal research and other areas as well, there is a need for intercultural learning. Helping to construct the new cultural context that allows such a continuous embedding on a global scale is itself a prosocial activity, just as cultivating one's own mind through sustained meditation practice is. Therefore, the third major reading of *Bewusstseinskultur* refers to the proactive development of a shared sociocultural context that prioritizes the common good.

The Wider Context: Secular Spirituality and the Planetary Crisis

As I said in the introduction, if anything was ever a "big-picture issue," then pure consciousness is. There is a quality of profundity in some of the reported experiences that directly relates to many of the deepest philosophical puzzles. Quite obviously, there is a much wider context to all of this. It goes far beyond applied ethics, and we all are living through a very special historical epoch. To get a fuller picture, it may be helpful to sketch out two of the most important dimensions of the new context that would be created by the project of a *Bewusstseinskultur*.

Bewusstseinskultur has a lot to do with finally taking responsibility for your own life and with what I would like to term "the principle of self-respect." There is a problem to be solved: How does one preserve one's self-respect during a historical epoch in which humankind as a whole is losing its dignity? Humanity is in the midst of a planetary crisis that is self-inflicted and historically unprecedented. It does not look good. Both political institutions and large numbers of individuals around the world are failing miserably in their management of this crisis, and they have been failing miserably for a very long time. Most people are beginning to feel that something has changed: When it comes to the climate catastrophe, it is no longer intellectually honest to be an optimist. Many of us also feel that we have been lying to ourselves for some time now.

Yes, of course, from the perspective of climate science, it is still possible for warming to stay below 1.5 or maybe 2 degrees Celsius. But from the perspective of psychology and political science, it isn't.

If we face up to the psychological and political facts available today in an unbiased way, everything points to the conclusion that humanity will fail in the face of this problem, and it will fail with its eyes open. We have not managed the transition from a growth-oriented, greed- and competition-based model of economics into a sustainable form of steady-state economics. We knew it all long ago, we decided to ignore the facts for as long as possible, and we successfully organized our own self-deception on a political level. Now the greed-based growth model has finally led us into an accelerating environmental catastrophe.

The central factor in this (which everyone also knows about) is a close correlation between economic growth and the rate of global carbon dioxide emissions. Economic growth is the dominant normative and cultural context in which we currently live our lives, and, as societies, we have not managed to evolve an alternative context that is more attractive than the neoliberal growth model. Few of us know the unboundedness of pure awareness discussed in chapters 2 and 23, but the delusional belief in unbounded economical growth still has a stranglehold on many.

If we combined gross domestic product with the general level of *Bewusstseinskultur* as a standard prosperity indicator that goes beyond mere happiness measures and subjective life satisfaction, we could create more socially just and ecologically sustainable societies. It is easy to see how we could have chosen not to aim solely at maximizing the mean standard of living, but also to value something that, in free and open discussions, we had previously defined as the "mean standard of consciousness culture." To give one example in line with the central topic of this book—what if one of our central goals had been maximizing the minutes of pure awareness per capita? What if we had valued and systematically increased the average lifetime hours that every citizen can spend enjoying MPE states, or even MPE modes? What if our goal had not been to land on Mars, but to land in pure awareness?

Given the phenomenological material presented in this book, one may plausibly assume that many of us would have had much richer lives on an individual level, and that we would have stopped the planetary crisis before the chance to do so slipped through our fingers. But everything points to the prediction that it is now too late for this. Very soon, it will no longer be possible to respect humankind's behavior because this behavior does not change even when human beings clearly see, at the level of their own conscious self-experience, that we have done and continue to do all this knowingly.

Our species does not respect humanity as a whole, neither in the present nor in the future. Collectively, we lack the ethical integrity, the quality of compassion, and the capacity for rational action that would have enabled us to avert medium- and long-term catastrophe at moderate medium- and short-term cost. We are causing an enormous amount of future suffering, and we are doing so knowingly. Very soon, therefore, it will no longer be possible to respect the behavior of large segments of humanity. We will no longer be able to take ourselves seriously, for our behavior does not change even when we clearly recognize that it must.

This is where the idea of a new, secular form of spiritual practice comes back into play. There actually *is* something that we can still respect in ourselves and others: the nonegoic form of conscious self-knowledge that gradually reveals itself in contemplative practice (chapters 29 and 30). This form of conscious self-knowledge is much deeper than the high-level dignity that rational subjects may (or may not) see and acknowledge in each other, because it creates a much more fundamental relationship between the individual and the community of all sentient beings capable of experiencing it. It is nonconceptual. It is ownerless. It has nothing to do with words or thoughts. Animals may have it. Perhaps future machines should have *only* this form of self-awareness. The capacity for conscious but nonegoic self-knowledge is something that we can respect in ourselves and others, even if humanity as a whole fails. It has worth, and it can be valued—but first, it needs to be recognized. We have to rediscover it.

Amid the rolling catastrophe, we must somehow learn to do the right thing just because it *is* the right thing, without needing to obtain positive results for ourselves. In experiencing our own failure and inadequacy, we must somehow also learn to do the right thing in a sustainable way that involves a quality of self-compassion. Interestingly, both self-respect and the capacity to fail gracefully relate directly to the phenomenology of nonidentification that we have encountered and investigated at many different places on our journey through this book (e.g., in chapters 8, 19, 29, and 33) and to the capacity to act, at least sometimes, from a position of nonegoic self-awareness (see, e.g., chapters 29–33).

To a degree, both self-respect and graceful failure can be systematically cultivated. As we have seen, the quality of nonegoic self-awareness may be closest to what makes many describe self-knowing MPE as an irreducibly *spiritual* state of consciousness. It is possibly what characterizes the deepest sense of the word "spirituality": a form of conscious experience that is completely independent of religious belief systems. Within the space of family resemblances related to the experience of pure awareness, reflexive MPE is a prototypical core region that forms the phenomenological anchor for what many meditators, if asked, may try to describe as the spiritual "essence" of their experience—or even as their "true self" (chapter 29). As we have seen in this book, the recognition of reflexive meta-awareness via meditative practice arises on an entirely nonconceptual and even nonegoic level. There is more than one form of self-respect, more than one form of self-knowledge, and therefore more than one form of dignity. We could even coin a fancy term for this if we wanted to, like "nonegoic dignity."

At least in principle, discovering nonegoic dignity in themselves can help people to do what is right in a way that is not emotionally attached to achieving positive results or getting the rewarding experience of egoic self-efficacy. At least to a certain degree, therefore, contemplative practice may prevent bitterness and emotional burnout by helping to prevent ego-defensive anger, self-condemnation, constant depressive rumination, and generalized anxiety. Discovering the silkiness of self-cognizing silence is not something that demands a special form of giftedness or an unusual intellectual capacity; it could form the experiential kernel around which the requisite cultural growth happens at scale. There may actually be a way to preserve one's self-respect—including one's mental health—at a time when humanity as a whole is losing its dignity. I think it could be rewarding to think further in this direction.

The wider context is deep and rich. *Bewusstseinskultur* and its role in proactive damage control and the preservation of dignity form one dimension; secular spirituality forms another. We have already seen that *Bewusstseinskultur* is not a new form of religion, but it is closely linked to the deeper philosophical question of whether a fully secularized form of spirituality is at all conceivable. What about meditation as a more substantial, genuinely epistemic practice? Can there be a new form of spirituality that explores the epistemic dimension of nonegoic self-awareness shining through in some of our reports, a slightly more radical culture of consciousness that remains entirely free of irrationality, a set of practices not driven by mortality denial and narrative self-deception (as described in chapter 17)?

Taking a closer look at the quest for a secularized form of spirituality will give us a fuller picture of what *Bewusstseinskultur* really is. To begin with, one problem is this: Millions of meditators worldwide like to call themselves "spiritual but not religious (SBNR)" or "spiritual, but not affiliated (SBNA)"—but nobody knows what these concepts actually mean.[12] Yes, there has been a successful rebellion against organized religion and a fresh global counterculture has emerged. Millions have taken things into their own hands, and this can be seen as one of the major cultural innovations of the twentieth century. But if one looks at the reality of all the different movements, traditions, and groups that have formed—especially their irrational belief systems and the resulting social dynamics, which are often pathological and cultlike—one may conclude that the original innovative impulse has long been outweighed by more mortality denial and self-deception, just in slightly subtler forms than they used to take. Many of the spiritual movements that have developed in recent decades in Europe and the US have long lost their progressive impulse, and most sectors of spiritual counterculture seem already to have morphed into experience-based forms of privately organized religious delusion. Today, they merely stabilize or conserve the status quo and are characterized by a slightly infantile form of complacency, and often by crude forms of intellectual dishonesty. As I very briefly indicated in chapter 15, if viewed from an empirical perspective, we therefore have grounds to wonder whether the people labeling themselves as "SBNR" or "SBNA" are religious after all, just in a slightly new way.

In most of these contemporary forms of "nonreligious spirituality," something is sorely missing: a particular kind of honesty. After all, the opposite of religion is not science, but spirituality.[13] The ethical principle of intellectual honesty can be analyzed as a special case of the spiritual stance and, in their purest form, the scientific and the spiritual stances emerge from the same basic normative idea. But what exactly *is* "intellectual honesty"?

Intellectual Honesty

"Intellectual honesty" means simply not being willing to lie to oneself. It is closely related to old-fashioned values such as propriety, integrity, and sincerity, as well as to a certain kind of "inner decency." Perhaps one could say that striving for intellectual honesty is a very conservative way of being truly subversive. Intellectual honesty might be exactly what the many "teachers" and representatives of cults, organized religions, and theologians of almost any type simply cannot have—even if they often like to make claims to the contrary. Intellectual honesty means not pretending to know or even pretending to be *able* to know the unknowable, while still having an unconditional will to truth and a commitment to the growth of knowledge, which includes being genuinely open to new scientific results. Intellectual honesty is the missing element in many of the social movements whose members think of themselves as "SBNR." It is needed even (or especially) where self-knowledge is involved, and even where self-knowledge is not accompanied by pleasant feelings or is not in accordance with the doctrine in force at the time.

The sincere pursuit of *intellectual* integrity can be viewed as an important special case of the pursuit of *moral* integrity, and it turns out that a relationship also exists between these forms of integrity and contemplative practice. Philosophers sometimes call the type of approach that emphasizes such connections an "ethics of belief."[14] Correspondingly, *Bewusstseinskultur* also demands that we think clearly about the conditions under which it might be not only *irrational* or *imprudent* but also *morally wrong* to hold a belief on insufficient evidence. A philosophical ethics of belief may be another element missing from large parts of the spiritual counterculture.

A flood of empirical data, as well as the mathematical models of modern computational neuroscience, show how the beliefs we hold directly influence the way that we consciously experience the world—a simple empirical fact whose importance is hard to overstate when it comes to meditation practice. Whoever wants to become whole—a person of integrity—by gradually resolving all conflict between their actions and values must pursue this harmony with their *inner* actions as well. This requirement is especially true for their "epistemic actions"—actions for the sake of knowledge. We act epistemically whenever we strive for insight, knowledge, true belief, sincerity, and also authentic self-knowledge. To the extent that meditation is an epistemic practice, it cannot work without radical honesty toward oneself, without a self-critical ethics of belief. There is a bridge between spiritual practice and the ideal of reasonable, rational thought: Both involve an ethics of inner action for the sake of knowledge. Moreover, in both cases, the goal is a systematic enhancement of mental autonomy, of inner freedom.

Immanuel Kant put this point in a completely different but particularly beautiful way. What is needed, he says, is *the sincere intention of being honest towards oneself.*[15] The "sincerity" or "pureness" of the desire for honesty toward oneself is, I think, the central point—and it may actually have something to do with discovering pure awareness. Without the will and the courage to really *look,* pure consciousness does not come into existence. The sincere intention of being honest toward oneself is what connects critical rationality and the nonconceptual insight of contemplative practice.

What Kant did not see is that exactly the same principle is also crucial for a somewhat radical, philosophically motivated meditation practice. This form of meditation involves a nonconceptual form of mental action for the sake of knowledge—an action that marks the beginning of the search for a nondual state of consciousness beyond the subject/object split. Without the intention to be radically honest with oneself, dissolution of the phenomenal self cannot take place. Kant's purity of intention is the beginning of an inner movement toward a clear and effortless form of mental *inaction.*

But you cannot choicelessly observe your own thoughts as they arise and disappear again if you are not prepared to honestly face what you will now begin to see: the painful restlessness of your very own mind, your violent fantasies, your desire for retaliation, your boredom, your loneliness, your existential despair, or your envy. At the beginning of this section, I offered one example of what *Bewusstseinskultur* could refer to: Cicero's idea that truly loving wisdom, and thus being a philosopher, mean "taking care of and cultivating one's soul." Now we can view Kant's point as a second example:

In terms of *Bewusstseinskultur*, any genuine contemplative practice requires the sincere intention of being honest toward oneself.

The importance of intellectual honesty also provides another bridge connecting a secular, self-critical form of spiritual practice and the ideal of rational thought, this time on an eminently practical level. I hope that you are beginning to sense that a strict and altogether old-fashioned form of rationalism could have a lot to do with spirituality—and with what is still missing in large parts of the spiritual counterculture. To stay with Kant for a moment: For him, dishonesty is nothing other than a form of unconsciousness, a lack of awareness and conscientiousness. Here, my point is that rationality is actually a special form of mindfulness. In order not only to think clearly but also to listen properly—in order to *understand* one's opponent—critical rationality first needs the sincere and nonreactive quality of mindful attention. And then it also needs careful, rational thinking, a way of handling concepts that is as precise as possible.

The essence of being honest toward oneself while listening to others is the form of reflexively aware mindfulness that Dakpo Tashi Namgyal and others[16] spoke of many centuries ago (chapter 30). There is a deep connection to the Western ideal of critical rationality here, because the quality of reflexively aware mindfulness is what really creates open-mindedness—and it is something that can be discovered and trained. Epistemic openness (chapters 4 and 5) is a precondition of genuinely rational thought, and it also has a social dimension. Our empirical data clearly show that MPE also has a social dimension; this primordial and entirely nonintellectual variety of open-mindedness creates an experience of connectedness because there is an element of compassion in it (chapter 11). Such connectedness, in turn, is a basic condition for genuine solidarity. Of course, much more could be said here, but my second point can be put very simply: *Real* meditation practice cultivates the conditions of possibility for rational discourse by cultivating the sincere intention of being honest toward oneself. If it doesn't, there is something wrong.

Spirituality, Intellectual Honesty, and *Bewusstseinskultur*

We have now briefly looked at what *Bewusstseinskultur* is, and at some aspects of the broader context in which it could come into being. As it turns out, *Bewusstseinskultur* is a slightly subversive but prosocial way of taking an ethical stance toward one's own mental states. As I have suggested, it is a "practical philosophy of mind." *Bewusstseinskultur* involves the systematic cultivation of states assessed as valuable, aiming at a continuous process of rational, evidence-based enculturation. It is a project—something that we need, and that we certainly don't have yet.

In this book, we have looked at the experience of pure awareness carefully and from many angles. We saw that pure awareness may turn out to be the "convergence zone" where spiritual practice, cognitive neuroscience, and modern philosophy of mind finally come together. At its prototypical core, pure awareness is not a subjective phenomenon because it is not necessarily tied to a knowing self or a consciously experienced first-person perspective. But in the course of our journey, it has also become clear that what philosophers, scientists, and the more than 500 experiential reports that I have shared with you are trying to describe is not a single and unitary phenomenon. It is not some metaphysical essence. Rather, it is a specific region in phenomenal state space: The space of possible conscious experiences. Pure awareness has many aspects and dimensions. They have been ignored for too long. This region in phenomenal state space could be of interest, certainly to philosophers, computational modelers, and neuroscientists working on consciousness—but maybe also to all of us who are interested in the project of developing a *Bewusstseinskultur* that is fit for the future.

Glossary of Terms

- **Abstract embodiment:** A phenomenological concept referring to the experience of embodying or identifying with a more abstract property, like space, wakefulness, or the nonconceptual experience of knowing. In this sense, pure awareness can sometimes be described as the experience of another kind of body (e.g., as spacious awareness or as awareness of a "wakefulness body"). This form of embodiment is constituted by an abstract level of our conscious self-model that has transcended or always already preceded the distinction between what is inside versus what is outside our physical body. See also **Nonegoic unit of identification**, **Phenomenal unit of identification**. See chapter 24.
- **Absurdity management:** Any psychological coping strategy that aims to efficiently assimilate or simply deny our knowledge of impermanence and mortality, since otherwise it might function as a toxic form of self-knowledge, leading to a generalized "Sisyphean" quality of futility. See also **Existence bias**, **Immortality project**, **Mortality denial**, **Toxic self-knowledge**. See chapter 17.
- **Arousal:** A graded physical property of the human brain, depending on activation levels in five types of neurotransmitters in the ascending reticular activating system. Successful control of cortical arousal is necessary for tonic alertness and for generating the sleep/wake cycle. See also **Tonic alertness**, **Wakefulness**. See chapter 4.
- ***Ataraxia* (*ἀταραξία*):** A lucid and enduring state of "imperturbability," "equanimity," or "tranquility." In ancient Greek philosophy, the term was first introduced by Democritus and Pyrrho, and then it was further developed by Epicurus and thinkers in the Stoic tradition. Cicero and Seneca used the Latin translation *tranquilitas animi* (tranquility of the soul). See chapter 14.
- **Augmented reality (AR):** A technological relative of virtual reality (VR), creating a multisensory and interactive experience of a transparent "real-world" environment where objects that are subjectively experienced as residing in the real world are

enhanced by computer-generated perceptual information. The technology of AR adds an environmental layer invisible to others outside it. It is conceivable that pure awareness itself—that is, the primordial model of an empty epistemic space—is what is augmented during ordinary conscious experience by superimposing a "real-world" environment via controlled online hallucination. See also **Phenomenology of virtuality, Translucency, Virtual reality (VR)**. See chapter 28.

- **Autobiographical self-model:** A dynamic process in the brain that allows us to organize future goals, memories, and abstract knowledge into a coherent biographical image of ourselves. This long-term self-model is not a little man in the head, but it can create the conscious experience of an abstract, knowing self that apparently remains the same across time. As such, the autobiographical self-model is a subpersonal process, but its content is sometimes reported as a "life history" or as if it were a "life narrative," implying an illusory narrator or observer. See also **Epistemic agent model, Narrative self-deception, Phenomenal self-model, Transparent self-model**. See chapter 17.
- **Autoscopic hallucination:** A hallucinatory experience in which the person sees a double of himself or herself, from the usual visuospatial perspective and without disembodiment. See also **Heautoscopy, Out-of-body experience.** See chapter 21.
- ***Bewusstseinskultur*:** A practical philosophy of mind that involves three major elements: (1) the adoption of an ethical stance toward one's own mental states; (2) the systematic cultivation of states assessed as valuable; (3) a continuous process of rational, evidence-based enculturation. See also **Normative phenomenology.** See the epilogue.
- ***Bhava-taṇhā*:** In Buddhism, the craving for existence. Arguably, this is one of the deepest causes of conscious suffering in humans, and probably in many other animals too. We therefore should avoid recreating it in conscious machines. See also **Existence bias, Phenomenal unit of identification**. See chapter 17.
- **Bodiless body-experience:** A special case of the experience of spatiality that often begins with the body gradually disappearing from the phenomenal field. See also **Abstract embodiment**. See chapter 24.
- **C-fallacy:** A logical error consisting in falsely concluding that just because something *feels* like the very essence of consciousness, it is also a reliable indicator of actually *being* in touch with consciousness per se. As such, verbal reports referring to having experienced an "essence" or "pure consciousness in and of itself" do not imply or license any claims as to the actual existence of such an essence because all such claims need an independent epistemic justification. See also **E-fallacy**. See chapter 12.

- ***Catuṣkoṭi*:** The famous "four-cornered" negation, a term most frequently associated with the Buddhist philosopher Nagarjuna and also known as the "Tetralemma of Nagarjuna." It consists of the rejection of a thesis, the rejection of its negation, the rejection of the conjunction of the thesis and the negation, and the rejection of the disjunction of the thesis and its negation. The full sequence can serve to "freeze the intellect" and serves as a logical procedure to make one realize emptiness. See chapter 14.
- **Clear light sleep:** The experience of pure consciousness during dreamless deep sleep. See also **Witness consciousness**, **Witnessing sleep**. See chapter 20.
- **Computational self-model:** An internal proxy for reality as a whole, a model that is neither a mere part of the egoic phenomenal self-model nor part of a world-model. Computationally, it is created by the organism predicting its own inner states (this is why it is a *self*-model and is physically realized through a part of one's body), but according to conscious experience this model encompasses the environment, the body, and the knowing self. It underlies the totality of all conscious experiences and it is physically realized through a part of one's body. Epistemic openness is a property created by the computational self-model as a whole, not by some sort of ego. Computationally, it is an integrated model of some of the organism's own inner states; phenomenologically, an entire world appears. See also **Nondual knowing**, **Phenomenal self-model**, **Zero-person perspective**. See chapter 24.
- **Contemplative heroism:** Alluding to the work of Ernest Becker, the idea that some forms of contemplative practice may involve "living under the terms of the immortality project" as a strategy for dealing with impermanence and futility. An immortality project helps with death denial by involving a symbolic belief system that ensures that I can believe that my self is superior to physical reality, such as by trying to discover—or actually becoming part of—something eternal, something that can never die. See also **Absurdity management**, **Mortality denial**, **Narrative self-deception**. See chapter 17.
- **Contraction principle:** Phenomenality—"being conscious"—is a subpersonal property, a property of some complex brain state. The brains of neurotypical human beings misrepresent this property of phenomenality by contracting it into a transparent, conscious self-model that then forms the origin of a first-person perspective. From the first-person perspective, this creates the experience that *you* are conscious. See also **Nondual knowing**, **Phenomenal unit of identification**, **Transparent self-model**. See chapter 8.
- **Convergence principle:** The idea that some global modes of consciousness are actually closer to the scientific image of reality than others, in terms of their

underlying ontology and how the ongoing process of conscious experience itself is portrayed by them—and that minimal phenomenal experience (MPE) modes are special in precisely this regard. See also **High-convergence mode**, **Nondual knowing**, **Zero-person perspective**. See chapter 28.

- ***Dharmakāya***: A Sanskrit term referring to the "truth body" or the true nature of the Buddha, which is sometimes viewed as a cosmic principle. Also used to describe the pure, nondual dimension of consciousness.
- **Diaphanousness:** Another term for phenomenal transparency. See also **Phenomenal transparency**. See chapter 28.
- **Dissociative identity disorder:** Formerly referred to as "multiple personality disorder," a condition characterized by the maintenance of at least two distinct and relatively enduring personality modes and associated with severe memory lapses. The disorder pattern includes changing clusters of personality traits and multiple autobiographical self-models with which the individual identifies at any given time. See also **Narrative self-deception**, **Phenomenal unit of identification**. See chapter 24.
- **Dolphin model of meditation:** According to this model, thought processes often cross the boundary between conscious and unconscious processing, in both directions, just as dolphins cross back and forth over the water surface. These transitions may be especially noticeable during meditation. Here, the idea is that a causal interaction between conscious and unconscious factors also might play a decisive role in understanding what happens during meditation. For example, unconscious processes may make certain experiential contents—sudden insights or minimal phenomenal experience (MPE) itself—appear as directly given, self-caused, or spontaneous. See chapter 10.
- **E-fallacy:** A logical error consisting in falsely concluding that a consciously experienced feeling of knowing is a reliable indicator of actually possessing knowledge. As such, verbal reports referring to a phenomenal signature of knowing or an intuitive sense of "just knowing" do not imply or license any theoretical claims about consciousness itself, the metaphysical nature of reality, the "true" self, and so on, because all such claims need an independent epistemic justification. See also **C-fallacy**. See chapter 7.
- **Ego dissolution:** A phenomenology that consists either in a temporary disappearance of the epistemic agent model or in its ceasing to function as a phenomenal unit of identification. Conscious experience without ownership, agency, cognitive self-reference, or self-location in time and space. See also **Nondual knowing**, **Zero-person perspective**. See chapter 25.

- **Epistemic agent model:** A special layer in the phenomenal self-model. It enables a cognitive system to form a strong first-person perspective and to represent itself as actively constructing and seeking new knowledge relations to the world and to itself. If the organism identifies with it, an epistemic agent model creates the experience of being a "knowing self." See also **Contraction principle**, **Phenomenal self-model**, **Phenomenal unit of identification**. See chapter 25.
- **Epistemic openness:** A new phenomenological concept that refers to a specific form of being open to the world. It is the experience of openness related to knowledge, to the space of possibilities related to the acquisition of knowledge, and to the mere capacity for knowledge. It is related to the experience of wakefulness and clarity, and it can be seen as a nonmetaphysical, exclusively phenomenological reinterpretation of the traditional concept of "emptiness." See also **Suchness.** See chapter 4.
- **Epistemic practice:** A practice that aims at insight, at the creation of knowledge. Just like philosophy and science, genuine meditation is an epistemic practice, but it aims at a very specific kind of knowledge that has nothing to do with words, concepts, or theories. See also **Convergence principle**, **Zero-person perspective**. See chapter 16.
- **Epistemic space:** An abstract concept used to describe a set of possible states or processes of knowing. Different conscious systems, such as human persons and nonhuman animals with different brains, may open very different spaces of knowing. If a given system has an internal model of its own epistemic space, it becomes conscious because it nonconceptually knows about its own capacity to know. This can include knowing about currently actualized states or processes of knowing unfolding within its integrated epistemic space, but it doesn't have to. If a system has an explicit model of this space in and of itself, the phenomenal character of "pure awareness" appears. See also **Epistemic openness**, **Nondual knowing**, **Zero-person perspective.** See chapter 5.
- **Existence bias:** A physically embodied top-level preference, leading human beings to almost always opt for sustaining their own existence, even if doing so is not in their own best interest. It leads to a fundamental distortion in our model of reality and to a high-level cognitive bias in which our own mere existence is treated as evidence for the goodness of that existence. The existence bias also led to the evolution of religion because humans must deal with the additional challenge of "toxic self-knowledge" threatening the integrity of our self-model. See also **Absurdity management**, ***Bhava-taṇhā***, **Toxic self-knowledge.** See chapter 17.
- **Existential ease:** A new phenomenological concept referring to the integrated experience of relaxation, lucid clarity, and a state of pure being. See chapter 1.

- **Fallacy of composition:** An informal fallacy that arises in natural language, when one falsely infers that something is true of the whole from the fact that it is true of some part of the whole. See also **Contraction principle.** See chapter 8.
- **Family resemblance:** A concept first introduced by Arthur Schopenhauer and later used by Ludwig Wittgenstein that advocates a nonessentialist view in which entities are defined not by common properties, but by a set of overlapping features and resulting similarities. From a phenomenological perspective, a full-absorption episode of pure awareness can be seen as an experiential prototype, and many of the experiences described in this book can be understood as referring to a family of states that resemble each other, some being more central than others. See the introduction.
- **Full-absorption episode:** An episode after which the phenomenal character of awareness itself is the only feature that can be reported. See also **MPE mode, MPE state.** See chapter 33.
- **Graviception:** The capacity of a biological organism to detect the Earth's gravitational field, which can have its own phenomenology (as in consciously feeling the body's own weight). In some contemplative states, one finds a distinct and contrasting quality of weightlessness. See chapter 24.
- **Groundless ground:** A Tibetan Buddhist concept (*gzhi'i gzhi med*) referring to the ground of all experience, which is not a metaphysical ground established by the usual sources of valid knowledge but rather is a matter of personally realized self-awareness. There are strong parallels in Western mysticism, and in some phenomenologies, pure awareness is the "foundationless foundation" of all experience. See also ***Seelengrund.*** See chapter 26.
- **Heautoscopy:** A rare neurological disorder characterized by a multimodal illusion in which one sees a double of oneself and has difficulty locating the self, either in the physical body or in the autoscopic body. See also **Autoscopic hallucination**, **Phenomenal unit of identification**, **Robotic reembodiment.** See chapter 21.
- **High-convergence mode:** A minimal phenomenal experience (MPE) mode in which the convergence principle is strongly expressed. Nondual awareness could be a high-convergence mode, a form of conscious experience that is suboptimal from a biological perspective but closer to our best scientific understanding of what the conscious brain really does and what the deeper causal structure of the world actually is. See also **Convergence principle**, **Nondual knowing.** See chapter 28.
- **Illusion of control:** A tendency to overestimate one's ability to control events, which may lead to the conscious experience of having control over random events, or hallucinating agency. Includes the idea that in meditation, the feeling

of having control over certain mental processes—such as successfully "noticing" and terminating a mind-wandering episode—may sometimes be based on an introspective illusion of control. See also **Epistemic agent model**. See chapter 32.

- **Immortality project:** According to Ernest Becker, a mode of mortality denial involving a symbolic belief system that allows one to believe one is superior to physical reality, to create meaning beyond one's biological lifespan via a sense of legacy, and to identify with something larger than oneself or sustain belief in an afterlife. The optimal outcome is reduced death anxiety. See also **Absurdity management**, **Narrative self-deception**, **Toxic self-knowledge**. See chapter 17.
- **Lucid dream:** A type of dream in which the dreamer is aware that she is currently experiencing a dream and that its hallucinatory content is the result of an internal simulation. There are different levels of dream lucidity, and the plot can even be controlled by the dreamer in some cases. See chapters 20 and 21.
- **Lucidity lapse:** Transient loss of dream lucidity (e.g., during complex state transitions from pure consciousness to lucid dreaming, followed by its return). See chapter 21.
- **M-fallacy:** A faulty chain of reasoning where from the fact that something is subjectively experienced as self-caused or as uncaused, its metaphysical status is erroneously inferred. See also **C-fallacy** and **E-fallacy**. See chapter 10.
- **Minimal model explanation:** Explains a target phenomenon (like consciousness) via an idealized theoretical model that leaves out everything superfluous, isolating only the essential features and core causal factors that give rise to the target phenomenon we want to understand. See also **Minimal phenomenal experience (MPE)**, **Triple triangulation project**. See note 1 in the introduction, and chapter 12.
- **Minimal phenomenal experience (MPE):** The simplest form of conscious experience. Pure awareness is one possible candidate for MPE.
- **Mortality denial:** Denial of one's own mortality in the attempt to cope with feelings of futility, absurdity, or anxiety. Many religions are organized forms of mortality denial involving metaphysical belief systems. Many styles of meditation originated in exactly such belief systems. See also **Absurdity management**, **Contemplative heroism**, **Toxic self-knowledge**. See chapter 17.
- **MPE mode:** A global mode of conscious experience infused by an all-encompassing, unbounded, and nonegoic quality of awareness per se. See also **Clear light sleep**, **Minimal phenomenal experience (MPE)**, **Nondual awareness**, **Witness consciousness**. See chapter 33.
- **MPE perennialism:** The philosophical thesis that MPE has a unique and distinct kind of phenomenal character that exists in all conscious humans (and perhaps

even in other animals), across all cultures, traditions, and historical epochs, albeit mostly unnoticed. See also **Minimal phenomenal experience (MPE), Perennialism.** See chapter 26.

- **MPE state:** An episodic state carrying the specific qualitative character of minimal phenomenal experience (MPE), typically still attributed to an individual experiencing self. See also **Minimal phenomenal experience (MPE).** See chapter 34.
- **Myth of cognitive agency:** The conventional belief that the paradigmatic case of conscious cognition is that of autonomous, self controlled, and rational thought directed toward goals/tasks. See also **Epistemic agent model, Illusion of control.** See chapter 25.
- **Naive realism:** A theory in the philosophy of perception that the senses provide the perceiver with a direct awareness of mind-independent objects as they really are. The idea is also called "direct realism," "perceptual realism," or "common-sense realism." There is a nonconceptual equivalent on the level of phenomenal experience itself: **Phenomenal transparency.** See chapter 28.
- **Narrative self-deception:** The process of mistakenly interpreting one's own autobiographical self-model as if it were actually a narrative, and identifying with the person or entity that one believes to be the narrator. Narrative self-deception is mostly a subpersonal process that expands the organism's predictive horizon and, in searching for thematic coherence, stabilizes the fabric of its long-term self-model, such as by creating a permanent inner monologue. Phenomenologically, it may create a fictitious entity that is in control and remains the same across time. See also **Epistemic agent model.** See chapters 16 and 17.
- **Nested *saṃsāra*:** A somewhat playful naturalistic reinterpretation of *saṃsāra*, the migration of the focus of experience from one phenomenal unit of identification to the next, as the process in which a self-organizing biological or mental system undergoes a succession of states leading to the impermanent functional embodiment of ever-new units of identification. Since *saṃsāra* takes place simultaneously on many levels, in life and in the mind, it is a nested process. See chapter 24.
- **Nondual awareness:** A global mode of conscious experience in the absence of subject/object duality. **Full-absorption episodes** are also characterized by nonduality. See also **High-convergence modes, Nondual knowing, Zero-person perspective.** See chapter 27.
- **Nondual knowing:** A conscious experience of perception and knowing that is no longer structured by a localized and "knowing" self represented by the brain as intentionally directed toward specific, external objects of knowledge. In some

cases, the transition from egoic knowing to the experience of nondual knowing turns into a state of full ego dissolution. See also **Ego dissolution, Epistemic agent model, Phenomenal signature of knowing.** See chapter 27.

- **Nonegoic reflexivity:** The phenomenal quality of pure consciousness that consists in being aware of itself, spontaneously, and completely nonconceptually, without any form of ownership or agency (e.g., in a deep and fully absorbed state of meditation). See **Nonegoic self-awareness.** See chapter 30.
- **Nonegoic self-awareness:** A special phenomenology of nondual insight that involves nonconceptually knowing that knowing is taking place; a nondual variant of meta-awareness (this time, awareness of the quality of awareness itself). The knowing is phenomenally experienced as selfless self-knowledge, and it can function as a nonegoic unit of identification. See also **Phenomenal unit of identification.** See chapter 30.
- **Nonegoic signature of self-knowing:** An experiential quality of reflexive, nondual, selfless self-awareness that involves an inner dynamic in the phenomenal character of self-certainty. A continuous process of nonconceptually knowing that it knows itself. See also **Nonegoic self-awareness.** See chapter 30.
- **Nonegoic unit of identification:** A consciousness system with a nonegoic unit of identification is radically selfless, but it may still identify with a more abstract form of phenomenal character, like that of awareness itself or the quality of awareness nondually knowing itself. See also **Nonegoic self-awareness, Phenomenal unit of identification.** See chapter 29.
- **Nonsensational awe:** A subtle but clearly noticeable form of bliss, sometimes described as an "invisible smile" or as a calm and entirely undramatic phenomenology of rapture and wonder. See chapter 1.
- **Normative phenomenology:** A set of norms that prescribe, on a phenomenological level, the course of a practitioner's progress (i.e., what the various "stages" of meditation should entail). Many traditional systems explicitly state what "good" or "better" states of consciousness are. See also the discussion of ***Bewusstseinskultur*** in the epilogue. See chapter 2.
- **Noumenal awareness:** Another term for "phenomenology of transcendentality." Possibly, pure awareness has epistemological and metaphysical dimensions that cannot be cleanly separated from the phenomenology. See also **Phenomenology of transcendentality.** See chapter 31.
- **Out-of-body experience:** The result of a deviant form of self-modeling where one has the feeling of being outside and often elevated above the physical body. See **Phenomenal unit of identification.** See chapters 21 and 24.

- **Perennialism:** The philosophical thesis that a common core can be identified in all mystical experiences across all cultures and traditions, in all historical epochs, and in many social and religious contexts. See also **MPE perennialism**. See chapter 26.
- **Peripersonal space:** The region of space immediately surrounding our bodies in which objects can be grasped and manipulated. Our inner model of this space may help with early threat detection and response, such as via involuntary defensive or other self-preserving movements. The brain's representation of peripersonal space influences bodily self-consciousness and enables interaction with the environment, and may plausibly play a role in some forms of contemplative phenomenology. See also **Bodiless body-experience**. See chapter 11.
- **Personal and subpersonal levels of description:** Descriptions at the personal level refer to the person as a whole, to beliefs, desires, intentions, and so on, while descriptions at the subpersonal level refer to the constituent parts of the whole, for example to brain states like the neural correlate of minimal phenomenal experience (MPE). See chapter 32.
- **Phenomenal indeterminacy-blindness:** Being introspectively blind to phenomenal neither-nor-not-ness on the level of conscious experience, such as when body boundaries are *neither* well-defined *nor* explicitly experienced as absent. Sometimes there simply is no phenomenological fact of the matter, because an absence of representation is not the same as a representation of absence. "Indeterminacy blindness" means not being aware of being introspectively blind to phenomenal facts (e.g., being blind to the often indeterminate nature of one's phenomenal experience as such), and it can lead to confabulatory verbal reports. See **Narrative self-deception**, **Theory contamination**. See chapter 24.
- **Phenomenal signature of knowing:** The nonconceptual conscious experience of knowing, often experienced as an intuitive insight or a seemingly direct form of "seeing" the truth. Phenomenologically, uncontracted and nondual variants of the signature of knowing exist as well; see **Contraction principle**, **Nondual knowing**, **Nonegoic self-awareness**. See chapter 7.
- **Phenomenal signature of self-knowing:** The sense of being intimately familiar with oneself and of pure awareness knowing itself nonconceptually. See also **Nonegoic reflexivity**, **Nonegoic self-awareness**, **Nonegoic signature of self-knowing.** See chapter 30.
- **Phenomenal state space:** A mathematical model providing an abstract description of all possible configurations of a conscious system, where a position in the state space represents a particular type of state whose phenomenal properties are

described on multiple dimensions. Conscious experiences can be described as paths through state space, or categorized as belonging to certain regions within this space, or as being more or less close to each other, and so on. See chapter 16.

- **Phenomenal transparency:** A property of conscious mental representations in which only their content, not the process of content formation, is available for introspection, meaning that a representation cannot be experienced *as* a representation. A representation experienced *as* a representation is called "phenomenally opaque"; unconscious representations are neither transparent nor opaque. See also **Naive realism**, **Phenomenology of naive realism**, **Phenomenology of virtuality**. See chapter 28.
- **Phenomenal unit of identification:** Any experiential content that leads to phenomenological reports of the type "I *am* this!" or "I *was* that!" A conscious system that has no unit of identification has no phenomenology of identification and lacks self-consciousness. See also **Nonegoic unit of identification**, **Phenomenal self-model**, **Transparent self-model**. See chapter 24.
- **Phenomenological anchor:** A region in phenomenal space that describes similar types of conscious experience, which in turn form the inner anchor or reference point for certain verbal reports made in public, such as those about pure awareness. A phenomenological anchor assumes a relation of family resemblance connecting experiences but makes no further metaphysical assumptions, such as about the existence of a single phenomenological essence or some kind of intrinsic, disembodied, or context-independent quality of experience. See also **Phenomenal state space**. See chapter 26.
- **Phenomenal self-model (PSM):** A conscious, global, and multimodal representation of a biotic or postbiotic system as a whole. A PSM may include psychological and social features; if it is transparent, then it creates the phenomenology of selfhood. See also **Phenomenology of naive realism, Transparent self-model**, **Phenomenal unit of identification**.
- **Phenomenology of epistemic self-validation:** The phenomenal character of self-revelation, referring to the experiential quality of a conscious process or perceptual object presenting evidence for its own existence, or "disclosing" itself. See also **Suchness**. See chapter 9.
- **Phenomenology of metaphysical indeterminacy:** A form of conscious experience in which it is no longer the case that you experience things as either existing or not existing. See also **Lucid dream**, **Phenomenology of virtuality**. See chapter 28.
- **Phenomenology of naive realism:** The robust and irrefutable subjective experience of perceiving something that must be real (e.g., when directly perceiving

mind-independent objects as irrevocably real). See also **Phenomenal transparency, Naive realism**. See chapter 28.

- **Phenomenology of transcendentality:** The experience of an apparently self-evident fact: that pure consciousness is the timeless condition of possibility for all conscious experience, that the phenomenology of touching reality "in itself" somehow also reveals its true, fundamental nature. Often linked with the phenomenal character of timelessness and a sense of "always already." See also **Noumenal awareness**. See chapter 31.
- **Phenomenology of virtuality:** A global form of experience in which objects are experienced as virtual because they appear as neither existent nor nonexistent but instead in "as if" mode, leading to a suspension of naive realism. See also **Naive realism, Phenomenal indeterminacy-blindness, Phenomenal transparency**. See chapter 28.
- **Prephilosophical mind/body problem:** The idea that the mind/body problem is hardwired into the human mind at a prephilosophical level because our phenomenal self-model possesses both spatial and nonspatial layers and the causal interaction between bodily and mental events cannot be introspectively experienced. This contingent phenomenological fact automatically creates dualistic, Cartesian intuitions (e.g., the "knowing self" could exist without the body) and makes humans feel the theoretical problem of psychophysical causality as part of their own experience. See also **Epistemic agent model, Narrative self-deception, Out-of-body experience, Phenomenal self-model, Phenomenological anchor, Theory contamination**. See chapter 22.
- **Principle of phenomenal correlates:** The assumption that all metaphysical theories (such as idealism, panpsychism, pantheism, solipsism, nihilism, or the existence of an ideal observer) correlate with a given altered state of consciousness or even a global mode of consciousness experience. The ontology of such altered states directly corresponds to the ontological assumptions made by the theory in question. Metaphysical theories and philosophical intuitions are nonconceptually mirrored in conscious models of reality that human beings have experienced, and perhaps even originated in them. See **Naive realism, Phenomenal state space, Phenomenological anchor**. See chapter 19.
- ***Prosoché*:** The practice of attention, a fundamental Stoic attitude of constant vigilance and presence of mind. A constant tension of the mind, and a continuous practice of self-awareness plus attention to the here and now. See chapter 14.
- **Reflexive MPE:** The experiential quality of pure awareness knowing itself; the timeless experience of nonegoic self-awareness, which is effortless and nonagentive

and can be spontaneously recognized; a quality of epistemic self-disclosure, as if the nonconceptual content were continuously "self-revealing" or as if an aspect of experience were making itself knowable, timelessly pointing to itself. See also **Nonegoic reflexivity**, **Nonegoic self-awareness**, **Nonegoic signature of self-knowing**. See chapter 30.

- ***Rigpa***: In Tibetan Buddhism, a classical notion of pure awareness referring to the "knowing of the ground," the spontaneous presence of primordial wakefulness, often symbolized by a mirror in which all phenomena of experience arise. *Rigpa* is characterized by qualities such as presence, openness, effortlessness, spontaneity, luminosity, original purity, expanse, clarity, and self-liberation (i.e., all attempts of the mind to "liberate" itself are futile because *rigpa*, its own true and fundamental nature, is already liberated). See chapters 5 and 10.
- **Robotic reembodiment:** Humanoid robotic avatars and avatars in virtual reality (VR) can be used to artificially manipulate the experience of embodiment and spatial self-location, making a human being identify with an artificial surrogate body. Some experimental setups can induce an identification with empty space, a sense of self that is smeared in space, or (in healthy humans) the sense of being in two different bodies at two locations at the same time, similar to the illusion of heautoscopy. See also **Heautoscopy**, **Out-of-body experience**, **Phenomenal unit of identification**. See chapter 21.
- ***Sākṣin***: A classic term from Advaita Vedanta philosophy, referring to an entity that is different from the empirical individual: the single, immutable, and pure element of awareness in all knowing. Phenomenologically, it is also known as "witness consciousness," denoting the experience of a passive and impersonal observer being present. See also **Witness consciousness**. See chapter 19.
- ***Samadhi***: A classical concept referring to a peaceful state of complete, thoughtless equilibrium. It is the eighth and final level identified in the Yoga Sutras of Patañjali, as well as the last of the eight elements of the Noble Eightfold Path in Buddhism, and it has often been translated as "even intellect." In our data, it is semantically related to the Low Complexity constraint. See the introduction.
- ***Sat-chit-ananda***: A metaphysical concept perhaps best known from Vedanta philosophy, where it refers to the three main attributes of Brahman, the nonpersonal Absolute: "existence, consciousness, and bliss." See the introduction.
- **Sciousness:** A term coined by the philosopher William James to draw attention to the fact that consciousness in and of itself is devoid of any duality of knower and known, rather being an "instant field of the present" lacking subject/object duality. See also **Nondual knowing**. See chapter 30.

- ***Seelengrund*:** An occidental concept ("ground of the soul") introduced by Meister Eckhart into the theological, philosophical, and spiritual debates of the Middle Ages. Viewed as a phenomenological concept, *Seelengrund* shares many features with the phenomenology of fundamentality and nonreifiability, but also directly relates to unboundedness, the combination of emptiness and fullness, and global qualities like timelessness, presence, and nondual being. An important philosophical idea is that knowing God is actually a form of self-knowledge: The *Seelengrund* is that part of you where you can see God—and where God sees you. The phenomenologies of pure consciousness as a "groundless ground," as described in the Eastern tradition of Tibetan Buddhism and by meditators in the Minimal Phenomenal Experience Project, are closely related to each other. See also **Groundless ground.** See chapter 26.
- **Self-certainty:** Here, an implicit, nonconceptual, embodied, and nonegoic form of self-knowledge that implies the phenomenal quality of "knowing that it knows." See also **Phenomenal signature of self-knowing**. See chapter 30.
- **Single-embodiment constraint:** The fact that in all currently known conscious systems, the sensory and motor systems are physically integrated within the body of a single organism. See chapter 27.
- **Social embodiment:** The experience of merging with the phenomenal field and of one's body boundaries dissolving can sometimes happen in the social domain, as an experience of becoming one with other embodied beings or as a transitory phase in the process of melting into space and transcending the distinction between inner and outer. See chapter 24.
- **Suchness:** "Just-as-it-is-ness," the disclosedness of something; also called "thusness." The experience of epistemic openness and nonconceptuality, like that applied to insentient perceptual objects themselves. Experiencing suchness involves the absence of any form of conceptual essence or intrinsic meaning, a quality of timelessness, the phenomenal character of self-disclosure and self-manifestation at the same time, and a phenomenology of epistemic self-validation. See also **Phenomenology of epistemic self-validation**. See chapter 9.
- **Temporal thickness:** A property of some generative models that enables an organism to predict the outcomes of actions in a more distant future, to successfully navigate the world and apply survival strategies in ever more intelligent ways. A model's temporal thickness indicates its predictive depth (i.e., the temporal range of the prediction). Minimal phenomenal experience (MPE) is temporally thin. See also **Timeless change.** See chapter 2.
- **Theory contamination:** Theoretical assumptions and beliefs that human beings physically embody via their self-model. Such assumptions (which can be

viewed as "priors," statistical beliefs encoded in our brain) act as unconscious presuppositions and causally influence (i.e., "contaminate") how we perceive, think, and verbalize things. In this way, they directly influence the flow of low-level information in our brains, and they also shape how our experiences are later described and interpreted. See also **Narrative self-deception**. See chapter 2.

- **Timeless change:** A phenomenological concept referring to the fact that the distinction between "timelessness" and "temporality" is not exclusive and exhaustive. Sometimes the atemporal phenomenal character of awareness itself can be quite explicit while being strongly overlaid or seamlessly integrated with different forms of time experience like nowness, duration, nonsimultaneity, succession, being in the past, or being expected in the future. See chapter 22.
- **Tonic alertness:** A graded functional property, linked to our cognitive capacities for sustained attention and orientation in time and space. It is the part of alertness that can be sustained in the absence of an external cue. Sometimes tonic alertness can be consciously experienced, and its distinct phenomenal character is intimately related to the experience of pure awareness. See also **Wakefulness**. See chapter 4.
- **Toxic self-knowledge:** Any form of knowledge or insight that threatens the biological fitness of an organism or its individual reproductive success. The explicit, self-conscious insight into one's own mortality, the knowledge of one's own inevitable death, is toxic in the sense that it is a threat to one's mental integrity not only because of its demotivating potential, but also because it interferes with the coherence of one's autobiographical self-model. An epistemic state that is toxic from a biological perspective may be seen as a healing or liberating form of knowledge or insight from a spiritual perspective. See also **Absurdity management**, **Contemplative heroism**, **Mortality denial**. See chapter 17.
- **Translucency:** A quality of experience in which the background becomes so dominant that what was previously in the foreground now becomes "see-through" or "diaphanous," with some of its structural features gently fading or disappearing altogether. Minimal phenomenal experience (MPE) might be this background. See also **Augmented reality (AR)**, **Nondual awareness**, **Phenomenology of virtuality**. See chapter 28.
- **Transparent self-model:** A self-model that is phenomenally transparent creates the realistic conscious experience of being someone and being in direct and immediate contact with oneself as a whole. The organism then *identifies* with whatever its current content is. See also **Phenomenal self-model**, **Phenomenal transparency**, **Phenomenal unit of identification**.

- **Triple Triangulation Project:** A proposed research strategy that may lead to the formulation of a first standard model for consciousness, because it would be focused on a minimal description that abstracts from all unnecessary details and excludes everything that is not causally necessary for phenomenal consciousness to emerge. The project would home in on the minimally sufficient neural correlate of minimal phenomenal experience (MPE) by comparing (1) pure awareness as entered from the wake state and as spontaneously occurring during clear light sleep, (2) pure awareness as entered from the wake state and from a lucid dream, and (3) pure awareness during clear light sleep and as entered from a lucid dream. See also **Minimal model explanation**. See chapter 20.
- ***Turīiya*:** A term from the oldest Upanishads, the notion of a distinct fourth state of "pure" consciousness underlying the three common states of waking, dreaming, and dreamless deep sleep. See the introduction.
- **Ultimate origination:** As a phenomenological concept, the subjective experience of being the initiator or first cause of one's own mental and physical actions, with a corresponding sense that one could have acted differently. A strong experience of free will. See chapter 32.
- **Virtual reality (VR):** Technological VR is experience depicted via the simulation of possible worlds and possible selves—and is arguably the best metaphor yet for biologically realized consciousness itself. One can view the content of consciousness as the content of a controlled online hallucination. See also **Augmented reality (AR)**. See chapter 28.
- **Wakefulness:** A graded phenomenal property that can be interpreted as nonconceptually representing tonic alertness. One *can* attend to it, but most people rarely do. See also **Tonic alertness**. See chapter 4.
- **Witness consciousness:** The experience of an uncontracted phenomenal signature of knowing, also known as *sākṣin*, a nonegoic "ideal observer." It emerged as factor 12 in our first study. As a phenomenological concept, witness consciousness refers to the experience of a passive and impersonal observer being present, which is typically characterized by the nonegoic character of awareness itself, as well as by timelessness, a simplicity, and a global quality of choiceless awareness. The concept has a long tradition in Indian and Tibetan metaphysics; in a phenomenological reading, it obviously bears direct relevance to the idea of a minimal phenomenal experience (MPE) mode. See also **MPE mode**, ***Sākṣin***. See chapter 17.
- **Witnessing sleep:** Another term for clear light sleep. See also **Clear light sleep**. See chapter 19.

- ***Ye shes***: The timeless awareness of primordial wakefulness, such as in Vajrayana Buddhism. See the introduction.
- **Zero-person perspective:** The absence of both a first-person and a third-person perspective in conscious experience. Full-absorption episodes of pure consciousness, as well as nonegoic self-awareness and global nondual states, are paradigmatic cases of being in a state involving the phenomenology of knowing from a zero-person perspective (0PP) in the absence of a knowing self. See also **Epistemic agent model**, **High-convergence mode**, **Nondual knowing.** See chapter 3.

Notes

Introduction

1. One major factor in this strategy is that in a complex domain of explanation—which the phenomenon of conscious experience clearly is—ideal models that cover *all* the behaviors of the kind of system that interests us are often unavailable, but, as Richard Batterman (2002) has pointed out, "*highly idealized minimal models* of the universal, repeatable features [. . .] are often obtainable" (p. 36). These models may still possess great computational and explanatory power. Batterman quotes Nigel Goldenfeld, who describes a correct minimal model as being "that model which most economically caricatures the essential physics" (originally in Goldenfeld, 1992, p. 33; see also Goldenfeld, 2018, p. 36). The strength of a minimal model consists in giving us an understanding of what we actually want to understand, but without yet faithfully depicting fine-grained causal pathways or actual functional mechanisms on a micro-scale (Weisberg, 2012, chapter 6, section 6.1.2). So, minimal models are highly idealized descriptions of some target phenomenon; they may possess great explanatory power; and they do not yet represent the underlying, fine-grained causal mechanisms. A fourth important idea about minimal models is that the addition of details might actually *detract* from an understanding of the phenomenon of interest (Batterman, 2002, p. 22). One of my main points in this book is that what we call the "first-person perspective" is precisely one such superfluous detail, and adopting the minimal model approach therefore will dissolve the problem of subjectivity for the science of consciousness. I think that all four ideas listed here could also apply to a potential model for phenomenal experience, a model targeting only its absolutely essential and minimal form, namely MPE, the experience of "pure consciousness" as phenomenologically investigated in this book. I have said a little about minimal model explanations in this context in Metzinger (2020, section 1.2), but since this is a popular book aimed at a wider audience, I will refrain from any technical discussion. See Batterman and Rice, 2014; other discussions I found helpful are Lange, 2015; McKenna, 2021; Rice, Rohwer, and Ariew, 2019; and Wiese, 2023.

2. See Gamma and Metzinger, 2021. You can find updates and future publications at mpe-project.info.

3. Baminiwatta and Solangaarachchi, 2021.

4. The concept was first used by Windt, 2015; for an introduction, see chapter 20 of this book and Metzinger, 2020.

5. Metzinger, 2020.

6. Gamma and Metzinger, 2021.

7. See Metzinger, 2020, for an example.

8. Bayne and Hohwy, 2016, p. 57.

9. Bayne and Hohwy, 2016, p. 72.

10. For more on the dimensional approach, see Bayne, Hohwy, and Owen, 2016.

11. For an example, see Bhikkhu, 2012, p. 95. For a detailed discussion and further references, see Higgins, 2013, pp. 207–211 and note 529.

12. Visit mpe-project.info.

Chapter 1

The epigraph is cited in Bricklin, 2006, p. 11.

1. You can find the original set of questions and a wealth of supplementary material on the companion website.

2. Killingsworth and Gilbert, 2010. For more on mind-wandering, see Metzinger, 2013, 2018a; an accessible introduction can be found in Metzinger, 2018b.

3. Not all authors agree with my categorization of bliss as not really being a feeling or an emotion; see, for example, Woods, Windt, and Carter, 2022a, section 2.6.

4. Metzinger, 2020.

5. Section II, (3) B; cf. Higgins, 2022, p. 154.

6. This is the translation by Sheng Yen, cf. https://terebess.hu/english/hsin3.html#03 and Sheng Yen, 2006; many other and widely diverging translations exist, see https://terebess.hu/zen/sengcan.html.

Chapter 2

1. Metzinger, 2015.

2. Metzinger, 2020.

3. In the original paper, this aspect of pure consciousness was labeled "PC2"; see Metzinger, 2020.

4. For an accessible introduction, see Wiese and Metzinger, 2017.

5. Laukkonen and Slagter, 2021. Another excellent recent example for the new computational phenomenology of meditation is Sandved-Smith et al., 2021. See also the pointers and figure 34.1 in chapter 34.

6. See Struhl, 2022, for an important analysis of the epistemological naiveté involved in ideas that are widespread in discussions of mindfulness and Vipassanā meditation, like "bare attention" and "seeing things as they are."

7. This might also explain the observed divergence in responses to two control items, which occurred at different places in the questionnaire but were phrased nearly identically. In hindsight, it might have been better to phrase these items not just nearly but absolutely identically, in order to know that any intra-individual variation in responses must be wholly due to inconsistent reporting. As it was, divergent scores could also have been due to the slight semantic difference in the two items, one asking about nonvisual radiance only, while the other also asked about self-luminosity.

8. Metzinger, 2020, p. 14.

9. Nagel, 1974; Shear, 2007.

10. Lockwood, 1993; Metzinger, 1995a, pp. 3–37, 1995b, 2003, pp. 189–197; Sellars, 1956, pp. 253–329.

11. For a review, see Hymanjr and Loftus, 1998.

12. I first flagged this problem in Metzinger, 2003, p. 566. I recommend Raphaël Millière's work as an excellent entry point into the debate; see Millière and Newen, 2022.

13. For important discussions, see Millière, 2020, and Fink, 2020.

14. This is one of the main points in Metzinger, 2003. The general idea is that we are biological organisms that—not only for the special case of recalling a selfless, nonegoic episode but also in most other situations—continuously embed low-level representations of the processes underlying their thoughts, feelings, and perceptions into a conscious self-model, thereby endowing them with the phenomenal properties of ownership and agency. Phenomenal selfhood is a post hoc confabulation; agency and ownership are useful high-level fictions.

15. As Wikipedia (2022) puts it: "Heterophenomenology is put forth as the alternative to traditional Cartesian phenomenology, which Dennett calls 'lone-wolf autophenomenology' to emphasize the fact that traditional phenomenology accepts the subject's self-reports as being authoritative. In contrast, heterophenomenology considers the subjects authoritative only about how things seem to them. It does not dismiss the Cartesian first-person perspective, but rather brackets it so that it can be intersubjectively verified by empirical means, allowing it to be submitted as scientific evidence." See also Dennett, 1993, 2003, 2007.

Chapter 3

For the epigraph, see Buswell, 1991, p. 170.

1. Spackman, 2022.

2. Woods, Windt, and Carter, 2020.

3. See also Woods, Windt, and Carter, 2022a, 2022b, and Woods et al., 2023. A recent neuroscientific study is Winter, 2020.

4. For details about the absence and presence of specific types of emotions, see chapter 15 and and Woods, Windt, and Carter 2022a and Woods et al., 2023.

5. Woods, Windt, and Carter, 2020, p. 10.

6. I found out after writing this chapter that my own ideas and findings strongly converge with those of Woods, Windt, and Carter; see Woods, Windt, and Carter, 2022a, 2022b. A particularly interesting recent study is Woods et al., 2023; here robust phenomenological differences in the reported experience of bliss, alertness, absorption, and depth are shown between different meditation techniques that are said to aim for "contentless" experience. For example, based on participant reports, mindfulness practice in the Thai Forest tradition was found to lead to significantly greater experiences of bliss and joy than classical Tibetan Buddhist Shamata practice. In contrast, Stillness Meditation, developed in Australia, produces states of consciousness that are clearly characterized by lower levels of wakefulness, but which have higher levels of absorption than the first two techniques mentioned, and are also more strongly characterized by a quality of pure being and the attainment of a ground state of mind.

7. Cobb and Comfort, 2023, p. 660.

8. Good entry points are Seth and Bayne, 2022; Doerig, Schurger, and Herzog, 2021; and Francken et al., 2021. See also Deane, 2021; Del Pin et al., 2021; Hanson and Walker, 2021, p. 2; Rorot, 2021; Signorelli, Szczotka, and Prentner, 2021.

9. Lewis, 1929, pp. 121, 131. See also Metzinger, 2003, sec. 2.4.

10. Nagel, 1974, p. 436.

11. Metzinger, 2003, 2008, 2009.

12. This is as cited in Nagel, 1995, p. 70.

13. I am grateful to Nihat Ay for stimulating discussions on this point.

14. Kleiner, 2020.

15. Shear, 2007, pp. 700, 702–703.

16. Cage, 1991, p. 64. I am grateful to Cyril Costines for pointing me to this specific example, as well as for many other excellent ideas and proposals.

17. Sandved-Smith et al., 2021.

Chapter 4

For the epigraph, see Namgyal, *Clarifying the Natural State: A Principal Guidance Manual for Mahamudra,* 2001, p. 29.

1. Metzinger, 2020; see also Woods, Windt, and Carter, 2022a, secs. 2.1 and 2.8.

2. For discussion, see Bayne and Hohwy, 2016, p. 73.

3. For some references, see Metzinger, 2019, 2020. There are also arousal-based meditation techniques, see Kozhevnikov, 2022. An important new perspective on the role of subcortical mechanisms is Solms, 2021.

4. Laureys et al., 2004, 2010; Sanders et al., 2012.

5. Munn et al., 2021, p. 5.

6. Sadaghiani and D'Esposito, 2015.

7. Posner, 2008, p. 193.

8. Sadaghiani and D'Esposito, 2015, p. 2763.

9. Sadaghiani and D'Esposito, 2015, p. 2764.

10. Mañjuśrīmitra, 1987, p. xi.

11. Higgins, 2012, p. 447; 2013, p. 99. For a recent empirical study, see Costines et al., 2021.

12. Blanke and Metzinger, 2009, p. 7. See also Lenggenhager et al., 2007.

13. See the introduction of this book and Bayne and Hohwy, 2016.

14. Friston, 2018; Hohwy, 2016, 2021.

15. If you are, see Siderits, 2021; see also Buswell and Lopez, 2014.

Chapter 5

For epigraphs, see Urgyen, 2001, p. 131 and Urgyen, 2000, p. 112.

1. Whenever I speak of conscious experiences as being a "model," I am thinking of a probabilistic model encoded by the organism's brain, a physically realized model of the hypothesized causes that generate its sensory observations—that is, a *generative model* (Parr et al., 2019). Even though this model is not based on concepts, words, or languagelike structures, it can (metaphorically) be described as a subpersonal "belief" that the organism now has about the world—an embodied inner image of reality, if you will. However, most of the organism's "beliefs" are entirely unconscious and constantly updated as they interact with each other. To have a model in this sense means to physically embody a form of probabilistic knowledge. In addition, the organism is inferring the model's "fitness." If this knowledge is good and robust, close to optimal, then it can lead to the conscious experience of "realness" or "certainty." In this sense, conscious probabilistic models are *epistemic* models (they process properties like "evidence" or "fitness"), but at the same time, they offer a conceptually clear way of understanding what it really means for human beings to "embody a reality" or "dynamically realize their own life-world." My speculative idea is that, first, consciousness normally flags the specific model that the system will use for active inference involving global control of the body as a whole; it selects the appropriate level. As a consequence, the organism can actively change the external world to make it fit an inner goal-state—which is itself a form of knowledge acquisition. But, second, to be able to do this, the system has to know that epistemic gain, the acquisition of

new knowledge, is even possible. The organism can achieve this step only because a model of the organism's epistemic space as a whole has been embedded into this space itself, thus allowing the organism to know that something can be known in this specific way. For an introduction, see Wiese and Metzinger, 2017. The philosophical point about consciousness as a self-modeling epistemic space was first made in Metzinger, 2020.

2. This point is a slightly more abstract and nonagentive variation on the philosophical/computational idea of a system finding itself "poised over action space" or "knowing poise over an action space." See Ward, Roberts, and Clark, 2011; Nave et al., 2022, p. 1020.

3. Alexander, 1988, p. 3.

4. Severeide, 1990, p. 1570; as cited in Bachmann, 2014, p. 52.

5. Travis and Pearson, 2000, p. 81.

6. For a recent example, see Friquegnon, 2022.

7. Mañjuśrīmitra, 1987, p. 60; see also Laukkonen and Slagter, 2021.

8. Windt and Metzinger, 2007.

9. Ward and Wegner, 2013; see also Kawagoe, Onoda, and Yamaguchi, 2019. An important conceptual clarification is given in Fell, 2022.

10. Metzinger, 2020.

11. Meyniel, Schlunegger, and Dehaene, 2015; Meyniel, Sigman, and Mainen, 2015.

12. Woods, Windt, and Carter, 2022a also supports the understanding that full-absorption episodes involve subject/object nonduality; see section 2.3 in Woods, Windt, and Carter, 2022a.

Chapter 6

1. Lockwood, 1993; Metzinger, 1995, pp. 3–37; 2003, pp. 189–197; Sellars, 1956, pp. 253–329.

2. For good philosophical discussions, see Kirk, 1994; Levine, 1996; Tye, 1997.

3. Seth, 2021, p. 187.

Chapter 7

For the epigraph, see Plotinus, 2018, p. 887 (translation by Sybilla Pereira and Emily Troscianko).

1. Metzinger and Windt, 2015.

2. This thesis has been defended in one way or another by German philosophers including Johann Gottlieb Fichte (1762–1814) and Franz Brentano (1838–1917), and more recently by Dieter Henrich and Manfred Frank (Frank and Kuneš, 2022). It is also associated with the French philosopher Jean-Paul Sartre (1905–1980). A good entry point is Kapitan, 1999.

3. Metzinger, 2010.

4. Dennett, 1993, p. 401.

5. Millière and Metzinger, 2020.

6. For a recent review, see Tulver et al.,2023.

7. Metzinger and Windt, 2015, pp. 7–8.

8. Letheby, 2021.

9. Jopling, 2001; Letheby, 2021b, ch. 8. On the relationship between psychedelics and meditation, see Letheby, 2022.

10. Picard, 2013, p. 2497.

11. Picard, Scavarda, and Bartolomei, 2013.

12. For five case reports, see Picard and Craig, 2009.

13. Bartolomei et al., 2019, p. 1121 (authors' citations omitted).

14. For everybody interested in these issues, I strongly recommend Letheby, 2021. See also Letheby and Gerrans, 2017; 2024.

15. For an accessible discussion, see Metzinger, 2017.

Chapter 8

For the epigraph, see Harding, 2000, p. 77.

1. Bachmann, 2014.

2. See Metzinger, 2003, 2008, 2010 for more.

3. Siderits, 2003.

4. Metzinger, 1995, p. 22.

5. I first formulated this principle in Metzinger, 2020.

6. For an example, see Abdoun et al., 2019.

7. Seth, 2021, p. 188.

8. As Antonio Damasio and I have argued, the phenomenal self-model of human beings is likely to be functionally anchored on *homeostatic* self-control. For the important new idea that the conscious sense of self may be intimately related to having an *allostatic* control model, see Deane, 2021.

Chapter 9

For the epigraph, see Hinton, 2016, p. 36.

1. Higgins and Draszczyk, 2019.

2. The etymological root of the term *quidditas* ("whatness" or "essence") is the Latin *quid*, meaning "what"; and that of the term *haecceitas* ("thisness") is the Latin *haec*, meaning "this."

3. Morard, 1956.

4. Burbea, 2014, S.212f.

5. "The epistemic salience of phenomenal content suggests that the transitivity of reflexive self-consciousness is not hidden or transparent. Rather, it is a kind of primitive and prereflective consciousness that is nonconceptual and self-presenting" (Coseru, 2022, p. 357). For another excellent discussion of nonconceptuality and the reflexivity of consciousness, see Spackmann, 2022.

6. Lamme and Roelfsema, 2000; Lamme et al., 2000; Lamme, 2006; Boehler et al., 2008.

7. Hohwy, 2016, 2021.

8. See also Hohwy, 2021.

9. Hohwy, 2021, see also Seth and Hohwy, 2020.

10. Wiese and Friston, 2021.

11. Sheehy and Mathes, 2020.

12. Weil, 1956, p. 554.

Chapter 10

For the epigraph, see Barron and Fairclough, 2001, pp. 12, 23; Rabjam, 2014, ch. 2, p. 45.

1. Windt, 2015.

2. Kent and Wittmann, 2021; Wittmann, 2013, 2018.

3. Kelly, 2019.

4. I recommend Fasching, 2008, 2011, 2012, 2016, 2019.

5. See Fasching, 2021, pp. 680, 687; for more, see Fasching, 2022.

6. For more, see Davis and Steinbock, 2021.

7. I first presented this idea in Metzinger, 2020.

8. See Pliushch and Metzinger, 2015; an accessible and freely available overview is Metzinger, 2018.

9. I first understood this possibility when reading Schooler et al., 2011; an interesting computational model is Sandved-Smith et al., 2021.

Chapter 11

1. Matthen, 2021, p. 197; Hinton, 2016, p. 8.

2. Graziano, 2018, p. 1.

3. Dijkerman and Medendorp, 2021, p. 92. See also Noel, Samad, et al., 2018.

4. See de Vignemont et al., 2021, p. 3.

5. Noel, Bertoni, and Serino, 2021, p. 17.

6. Masson et al., 2021.

7. Cardini et al., 2019.

8. Serino, Canzoneri, et al., 2015; Serino, Noel, et al., 2015.

9. Noel, Park, et al., 2018.

10. Guterstam, Gentile, and Ehrsson, 2013.

11. Guterstam, Gentile, and Ehrsson, 2013, p. 1097.

12. Guterstam, Abdulkarim, and Ehrsson, 2015.

13. Guterstam, Abdulkarim, and Ehrsson, 2015, p. 6.

14. Guterstam, Gentile, and Ehrsson, 2013, Guterstam, Abdulkarim, and Ehrsson, 2015.

15. Metzinger, 2020.

16. Baars, 1993; Mashour et al., 2020.

17. Kleiner and Tull, 2021; Tononi et al., 2016. See Bayne (2018) for a critical philosophical discussion. Good recent overviews and entry points into the current "marketplace" for theories of consciousness are Seth and Bayne, 2022; Doerig, Schurger, and Herzog, 2021; and the freely available special issue "Consciousness Science and Its Theories" of the journal *Neuroscience of Consciousness* at https://academic.oup.com/nc/issue/2021/2.

Chapter 12

For the epigraph, see Rgyal-ba-g'yung-drung, 2017, pp. 94–97.

1. Kachru, 2022, p. 98.

Chapter 13

1. Woods et al., 2023. See also Woods, Windt, and Carter, 2022a, 2022b.

2. Higgins, 2022, p. 78.

Chapter 14

This translation of Wittgenstein is by Emily Troscianko. For the second epigraph, see Sextus Empiricus, *Pyrroneeioi hypotyposeis*, book 1, chapter 12, section 28.

1. Vogt, 2021.

2. Sextus Empiricus, *Pyrroneeioi hypotyposeis*, book 1, chapter 12, section 29.

3. Laukkonen and Slagter, 2021.

4. Aristocles apnd *Eusebius Praeparatio Evangelica*, book 14, chapter 18, Verses 2–4.

5. Dahl, Lutz, and Davidson, 2015; Lutz et al., 2015..

6. Wasson, Hofmann, and Ruck, 2008.

7. Hadot, 2002, p. 6.

8. Hadot, 1995, p. 90.

9. Hadot, 2002, p. 6.

10. Here is how Sonam Kachru (2022, p. 101) describes the more general point that I am drawing attention to at this time:

> For the ancients, however, the gap between the person as a whole and the person *qua* philosopher presumably would not always (or ever) obtain. This is partly what it means to think of ancient philosophy as having been practiced as an encompassing *way of life*. For the ancient philosopher, self-transformation on the basis of contouring one's ways of being minded (including the shaping of attention no less than belief) was often linked to behavioral regimens thought to govern every aspect of one's life, down to the least details of gesture, diet, and comportment [. . .]. If the connection between ancient philosophy and meditation as metacognitive training is more direct, then the conception of philosophy is far more demanding than the profession it has become today.

For all those interested in the new debate about the relationship between philosophy and meditation, I recommend the *Routledge Handbook on the Philosophy of Meditation* as an entry point. In his own contribution, the handbook editor, Rick Repetti, writes: "I am convinced that meditation is thus the meta-mental and meta-philosophical virtue-epistemic art and practice of consciously sharpening the wisdom-cultivating tools in the philosophical toolkit, particularly for philosophers who consciously practice it as such." (Repetti, 2022, p. 65).

11. Hadot, 1995, p. 82.

12. Hadot, 1995, p. 131.

13. Lopez, 2017. I recommend Pigliucci (2022) as an excellent entry point to the debate.

14. Marcus Aurelius, *Meditations*, 7, 54; see also 3, 12; 8, 36; 9, 6 (quoted in Hadot, 1995, p. 84).

Chapter 15

For the epigraph, see Weil, 1997, p. 117.

1. Terry, 2022, p. 390.

2. Streib and Hood, 2011. An important collection of texts can be found in Streib and Hood, 2016.

3. Metzinger, 2017.

4. Streib and Hood, 2011, 2016.

5. Thompson, 2020, p. 18.

6. See Woods, Windt, and Carter, 2022a.

7. Namgyal, 2019, sec. II, ch. 12, p. 366.

8. Namgyal, *Clarifying the Natural State: A Principal Guidance Manual for Mahamudra,* 2001, p. 62.

9. For relevant recent proposals, see Deane, 2021; Hesp et al., 2021.

Chapter 16

For the epigraph, see Krishnamurti, 1956, p. 98.

1. This was semantic constraint PC2; see Metzinger (2020) for details.

2. Kleiner and Tull, 2021; Tononi et al., 2016.

3. Buswell and Lopez, 2014, p. 588. For recent studies of "cessation" events, see Laukkonen et al., 2023; Chowdhury et al., 2023.

4. Bayne, Seth, and Massimini, 2020.

5. For some descriptions, see Metzinger, 2003, pp. 101–103.

6. Alcaraz-Sánchez et al., 2022; see also chapter 20.

7. Alexander, 1988, p. 3.

8. Metzinger, 2013, n. 14.

9. Suzuki, 1949, p. 30.

10. Suzuki, 1949, p. 31.

11. Windt et al., 2016, Box 1; Windt, 2020.

12. Metzinger, 2020; Woods, Windt, and Carter, 2022a, 2022b; Woods et al., 2023.

13. Bayne, Hohwy, and Owen, 2016.

14. Wolfe and Ralls, 2019.

15. Moser et al., 2021.

Chapter 17

For the first epigraph, see Namgyal, 2019, p. 520; for the third, see Padmasambhava and Lingpa, 2013, p. 11.

1. Frith and Metzinger, 2016; Metzinger, 2003, 2008.

2. For a little more about K and this conversation, see Metzinger, 2020.

3. See, for example, Frith and Metzinger, 2016.

4. Schnell, 2010, 2021.

5. Kelly, 2019, pp. 159, 181.

6. If we assume that the degree of theory contamination will change over the lifetime of a long-term practitioner, then it is interesting to think about the various developmental paths that such a hypothetical person could take. For example, he might start his regular practice in a rather innocent way, out of mere open-mindedness and curiosity, in a sincere attempt to live an examined life. But somewhat later—say after six to twenty-four months of regular practice—he becomes aware of the many positive effects that this practice actually has and naturally wants to understand where it all comes from. Thus, our hypothetical person begins to read books about meditation, listen to talks given by the so-called teachers on the internet, and maybe try one or two new techniques. He may even succumb to some belief system: an authority, or an ancient tradition or spiritual lineage. From then on, the degree of theory contamination at play whenever he speaks about his own experiences continuously increases. At the other extreme, we can imagine another hypothetical person born into a sinister cult that insists on regular meditation practice. She gradually liberates herself from the specific metaphysical background theory promoted by the cult, because for her, meditation has the desired effect of increasing intellectual honesty and mental autonomy, of achieving a certain degree of humility and a self-critical attitude. In this case, we can imagine how the degree of theory contamination *decreases* as time goes by.

My point is that the frequent implicit assumption that so-called advanced practitioners make the best participants for scientific experiments or for providing phenomenological reports may be mistaken. It is not a given that the degree of theory contamination is lowest in long-term practitioners, that their "expert reports" will be unadulterated and less biased. Often, in fact, the ideological noise in the signal may be much stronger than in beginners or others who are "merely experimenting" with meditation—those with a true "beginner's mind" (to quote Shunryu Suzuki and his well-known book of the same title). There is no automatic accumulation of expertise (if there is anything like "expertise" related to the epistemic practice of meditation at all), and it may well be that the phenomenology of pure awareness is freshest, most pristine, and most pronounced in certain beginners. Accordingly, the optimal point to investigate MPE from a scientific perspective may be very different in every meditator's life trajectory, and it certainly doesn't depend on the length of their individual path or have to be the closest possible point to their death. This is another argument for always aiming at a large and heterogeneous sample.

7. Metzinger, 2003, 2008, 2009.

8. For a freely available introduction, see Wiese and Metzinger, 2017.

9. In his book *Consciousness Explained*, Dennett criticized the philosophical intuition of a "Cartesian theater," which he described as a common symptom of "Cartesian materialism": "Cartesian materialism is the view that there is a crucial finish line or boundary somewhere in the brain, marking a place where the order of arrival equals the order of 'presentation' in experience because *what happens* there is what you are conscious of [. . .]." Dennett, 1993, p. 107.

10. Weber, 1947, p. 155.

11. Metzinger, 2016, 2021. A good entry point is Vinding, 2020.

12. Trivers, 2000, 2011; Von Hippel and Trivers, 2011.

13. Metzinger, 2016.

14. Wikipedia contributors, 2021.

15. Pyszczynski, Solomon, and Greenberg, 2015; Routledge and Vess, 2018.

16. Eidelman, Crandall, and Pattershall, 2009.

17. Metzinger, 2017a, 2017b.

18. Friston, 2010; see also the discussion and references presented in chapter 9.

19. Hohwy, 2016; see also Hohwy, 2021.

20. Hohwy, 2020.

21. Holland, 2020, p. 86.

22. Metzinger, 2016.

23. Friston, 2013, p. 11.

Chapter 18

For the epigraph by Tulku Urgyen Rinpoche, see Urgyen, 2000, p. 163.

1. It is interesting to compare these reports with results reported in Woods, Windt, and Carter, 2022a, across the three practices of Shamatha, Transcendental Meditation (TM), and Stillness Meditation; see section 2.7 and table 1.

2. Beierwaltes, 1972; Renger, 2016.

3. Ataria, Dor-Ziderman, and Berkovich-Ohana, 2015, p. 138.

4. Hoffmann, 2017; Schütz-Bosbach, Musil, and Haggard, 2009. In case you're interested, Ataria, Tanaka, and Gallagher (2021) is an excellent entry point into the current debate. And I have said a little more about what I mean by a "conscious model" in note 3 in chapter 5.

5. See Lindström et al. (2023) and Nave et al. (2021) for recent studies.

6. Ataria, Dor-Ziderman, and Berkovich-Ohana, 2015, p. 142.

7. Ataria, Dor-Ziderman, and Berkovich-Ohana, 2015, p. 145.

8. Ataria, Dor-Ziderman, and Berkovich-Ohana, 2015, p. 145.

9. Metzinger, 2020, p. 17.

10. Lingpa, 2014, pp. 14–15.

11. See, for example, Rinpoche Thrangu., 2011, p. 36, n30.

12. Buswell, 1991, p. 138.

Chapter 19

For the first epigraph, see Maharaj, 1973, p. 15. For the second, see Harding, 2014 [1961], p. 205.

1. Chatterjee, 1982, p. 339.

2. Fort, 1984, p. 287, n2; Gupta, 1998, p. 19.

3. Gupta, 1998, p. 18.

4. Albahari, 2009, 2020; Fasching, 2011, 2012; Timalsina, 2022.

5. Seager, 2020.

6. For some pointers, see Metzinger, 2017, n15, n21.

7. Fort, 1984, p. 278.

8. Williamson, 2010, p. 177.

9. Kennedy, 1976, p. 1327. See also Lindahl and Britton, 2019.

10. Huebner and Hayman, 2022, p. 274.

11. Castillo, 1990, p. 162f.

12. Castillo, 1990, p. 166f.

13. For introductory discussions, see Alcaraz-Sánchez, 2021; Thompson, 2015.

14. Williamson, 2010, p. 175.

Chapter 20

For the epigraph by Ramana Maharshi, below the heading "White Nights," see Venkataramiah, 2006, vol. 3, p. 583.

1. Forman, 1999, p. 20.

2. For an introduction to this idea, see Holecek, 2016; Wallace and Hodel, 2012; Wangyal and Dahlby, 1998.

3. If you are interested, I recommend Baird et al., 2018, 2019; Baird, LaBerge, and Tononi, 2021; Baird and Koroma, 2020.

4. For some case studies and references, see Metzinger, 2009, pp. 89–101.

5. Urgyen, 2000, p. 179. Helpful references to the relevant Indo-Tibetan meditation literature and further information about yogic sleep and the practice of "Yoga Nidra" can be found in Kavi, 2023.

6. Vilas, Auksztulewicz, and Melloni, 2021; Wiese, 2020; Wiese and Friston, 2021.

7. I will mostly use the two terms "wake sleep" and "clear light sleep," but the term "lucid dreamless sleep" is also in use, such as in Wallace and Hodel, 2012, p. 58. For some philosophical reasons not to use it, see Alcaraz-Sánchez, 2022.

8. Thompson, 2015, p. 15.

9. Windt, 2015.

10. Windt, 2015, p. 3.

11. Windt, 2015, p. 1 (emphasis in original).

12. Windt, 2015, p. 22.

13. Metzinger, 2009, 2013.

14. Blanke and Metzinger, 2009.

15. Windt and Metzinger, 2014; Metzinger and Windt, 2015, sec. 3.1.

16. Woods, Windt, and Carter, 2022a, 2022b; Woods et al., 2023.

17. "Consider a phonograph system as an apparatus of experience. [. . .] Now let us do the equivalent of quieting thoughts, namely, removing the record, perhaps also turning off the speakers and the turntable. When only the amplifier is on (with no ordinary 'objects of experience' given it), what is the experience like? [. . .] It would be a mistake to think there is an unusual reality being encountered, when that merely is what it feels like when the experience-mechanism is turned on yet nothing is present to be experienced [. . .] None of the literature I know describes what experience the quieting meditative procedure would produce in the absence of an unusual reality or self, so we don't know whether [it] is a revelation of an unusual reality or self, or instead an artifact of an unusual procedure of experiencing wherein most but not all functions are damped down." (Nozick, 1983, pp. 158–159).

18. Churchland, 1989, p. 66.

19. Kachru, 2022.

20. Alcaraz-Sánchez, 2022. For two recent studies on objectless experience during sleep, see Alcaraz-Sánchez, 2021, and Alcaraz-Sánchez et al., 2022.

21. See Windt, 2010. If you would like to read more, I strongly recommend Windt and Metzinger, 2007; for recent work, see Windt, 2020, 2021.

22. Shear, 1983, p. 59 (emphasis in original).

23. Kühle, 2015.

24. Hesp et al., 2021, p. 403. For an important philosophical perspective, see Evan Thompson's (2010) classic *Mind in Life*.

25. See Friston, 2013; Kirchhoff et al., 2018; Pezzulo, Parr, and Friston, 2022.

26. Luders and Kurth, 2019.

27. Metzinger, 2019, sec. 5–6.

28. Devor, Koukoui, and Baron, 2022.

29. Alcaraz-Sánchez et al., 2022; Alcaraz-Sánchez, 2021.

Chapter 21

For the epigraph, see sermon 71; DW III, p. 543 (Eckhart, 1986); Eckhart, 2007, p. 140.

1. Good starting points are Baird, Mota-Rolim, and Dresler, 2019; and Voss and Hobson, 2015. See also Voss et al., 2013; and Windt and Voss, 2018.

2. Metzinger, 2009, 2017; Windt and Metzinger, 2007.

3. For a good discussion, see Maleeh and Konjedi, 2022.

4. Metzinger, 2009, pp. 82–101.

5. Blondiaux, Heydrich, and Blanke, 2021; Brugger, Regard, and Landis, 1997; a good entry point is Blanke, 2012.

6. Furlanetto, Bertone, and Cristina Becchio, 2013; Furlanetto et al., 2016.

7. Kelly, 2019, pp. 98–102.

8. I recommend Ursula Voss and Alan Hobson's contribution to the Open MIND collection, plus the ensuing discussion with Lana Kühle, as a freely available entry point into this important discussion. I also recommend Voss et al., 2013, 2014; Voss and Hobson, 2015; Windt and Voss, 2018.

9. Gyatrul et al., 2002; Holecek, 2016; Wallace and Hodel, 2012; Wangyal and Dahlby, 1998.

Chapter 22

The epigraph below the heading "The Emptiness of Time Experience" is from Harding, 2000, pp. 142–143.

1. Popper and Eccles, 1977, pp. 361–364.

2. Fink, Lyre, and Kob, 2021; Fink and Lin, 2021; Lepauvre and Melloni, 2021; Metzinger, 2000; Overgaard and Kirkeby-Hinrup, 2021.

3. Kim, 1984; Metzinger, 1985; Place, 1956.

4. Please note that mental imagery, which makes significant contributions to many kinds of cognition, actually has pronounced spatial and sensorimotor qualities and mechanisms. Therefore,

according to the old Cartesian distinction, it should belong more to the body than the soul, which obviously disrupts the assumption that reasoning is a nonbodily act. For more, see Gottschling, 2006; Kosslyn, 1996; Pylyshyn, 2002; and Thomas, 2021.

5. Clark, 2016. I have greatly profited from the work of Andy Clark ever since *Microcognition* (1989) came out; if you are new to his work, you might like Clark (1998) as an introduction. In *The Experience Machine* (2023, p. 209) he writes the following about meditation and the control of attention: "By clamping attention on to an unfolding present moment (such as the breath), we also temporarily shrink the time horizon of predictive processing. This implies a kind of freezing of longer-term anticipatory processes, preventing the kinds of counterfactual "looking ahead" that play such a large role [. . .] in daily behavior. This means that even internal "information foraging" (purposeful explorations of our own memory, for example) can be put on hold. This is awareness with minimal counterfactual and temporal depth."

6. If you are interested in the relationship between the problem of pure awareness and Douglas Harding, I recommend the work of Brentyn Ramm (2017, 2018, 2019, 2021).

7. Melloni, 2015; Melloni et al., 2007.

8. Laukkonen and Slagter, 2021.

9. Krishnamurti and Bohm, 1997.

Chapter 23

For the epigraph, see Krishnamurti, 1991, vol. 15, p. 63.

1. Guterstam, Gentile, and Ehrsson, 2013; Guterstam, Abdulkarim, and Ehrsson, 2015.

2. Blanke and Metzinger, 2009; Metzinger, 2008.

3. Nguyen et al., 2021.

4. Siegel, 2007.

5. Mañjuśrīmitra, Simmons, and Namkhai Norbu, 1987, p. 60.

6. See also Metzinger, 2006.

7. The following passages have been excerpted from an as-yet-unpublished interview that I conducted on May 13, 1989 in Bad Homburg, Germany, with Albert Hofmann, the first person to synthesize and experience LSD (Metzinger, 1989). A short edit excerpt of this interview has been published as "Der Weltraum der Seele—Ein Gespräch mit Albert Hofmann," *TAZ*, June 27, 1989, pp. 11–12. See also https://taz.de/!1807810/.

8. Hofmann, 2013, chap. 1.

9. Metzinger, 1989.

10. For more, see Letheby, 2021; Letheby and Gerrans, 2024.

11. Metzinger, 2023.

12. Krippner, 1990; Roberts, 2001; Schultes and Hofmann, 1979; Wasson, Hofmann, and Ruck, 2008; Wasson, Kramrisch, and Ruck, 2008.

13. For more, see Bublitz, 2013, p. 250ff; Bublitz, 2016, p. 322; and Metzinger, 2024.

14. See Wasson, Hofmann, and Ruck, 2008.

Chapter 24

For the epigraph, see Plotinus, *Enneades*, V, 5, 9, 29–33; Plotinus, 2018, p. 593.

1. See Nave et al., 2021, and Lindström et al., 2023, for recent empirical studies.

2. Metzinger, 2005, 2009, 2013; Windt, 2015.

3. Windt, 2015, pp. 326–328, section 7.3.3 (emphasis in original).

4. Lipps, 1900, 1903.

5. Plotinus, *Enneades*, VI, 9, 8; Plotinus, 2018, p. 893.

6. Lipps, 1903, p. 188 (emphasis in the original): "Sie ist die Tatsache, daß der Gegensatz zwischen mir und dem Gegenstand verschwindet, oder, richtiger gesagt, *noch nicht besteht.*"

7. Metzinger, 2009a, 2009b, 2013, 2015, 2017, 2018.

8. Berrios and Luque, 1995; Metzinger, 2004.

9. Windt, 2021; Windt et al., 2016.

10. Johanson et al., 2008.

11. Deane, 2021; Letheby, 2021; Letheby and Gerrans, 2017; Millière, 2017; Nour and Carhart-Harris, 2017.

12. Fadiman and Gruber, 2020; Metzinger, 2003.

13. Metzinger, 2018.

14. Metzinger, 2018.

15. Song and Wang, 2012.

16. See, for example, Fox et al., 2015; Kam, Mittner, and Knight, 2022.

17. Limanowski, 2021.

18. MacKenzie, 2022, p. 250.

19. Millière and Newen, 2022.

20. For an example, see De Ridder et al., 2007.

21. I have been making this point for a number of years, but Jakub Limanowski and Karl Friston have now made it very clear (see Limanowski and Friston, 2018).

22. For an accessible introduction to this topic, see Seth, 2021.

23. Pezzulo, Parr, and Friston, 2022.

24. Sheng Yen, 2008, pp. 20-21.

Chapter 25

1. For more, see Metzinger (2017) and Metzinger (2013b); for a computational model, see Sandved-Smith et al. (2021).

2. Gallup and Anderson, 2020, p. 50.

3. Friston et al., 2015.

4. Baird, Mota-Rolim, and Dresler, 2019; Windt, 2015; Windt and Metzinger, 2007.

5. Deane, 2020, 2021; Millière et al., 2018; Millière and Metzinger, 2020.

6. Metzinger, 2013, 2015, 2017, 2018.

7. The experience of veto control is, of course, intimately connected to the question of free will. I recommend reading Filevich, Kühn, and Haggard (2012), and Filevich et al. (2013) as entry points, and Filevich et al. (2015) for the connection to dream lucidity.

8. Metzinger, 2013.

9. For a perfect overview, see Fox and Christoff, 2018.

10. See also Metzinger, 2013, 2017.

Chapter 26

For the epigraph under the heading "Seelengrund and 'Groundless Ground,'" see Eckhart, sermon 42; Eckhart, 2007, p. 400. In Middle High German: "*Nû wizzet: alliu unser volkomenheit und alliu unser srelicheit liget dar ane, so daz der mensche durchgange und übergange alle geschaffenheit und alle zîtlicheit und allez wesen und gange in den grunt, der gruntlôs ist.*" (DW II, p. 696). "DW" means *Deutsche Werke*; the first reference is to the canonical German edition.

1. Rosenthal, 2016, pp. 238–239; quoted in Woods, Windt, and Carter, 2022a, section 2.14.

2. A good overview and entry point into the current debate is Millière et al. (2018); see Millière and Metzinger (2020) for a recent edited collection.

3. Millière, 2020; Millière et al., 2018.

4. Metzinger, 2003, 2008.

5. Translation from German by Thomas Metzinger; see Hadewijch and Hofmann, 1998, p. 151; Hadewijch and Plassmann, 1923, p. 101.

6. Hadwijch, Vision 14, l. 82; Hadewijch, 1924, p. 162.

7. Translation from German by Thomas Metzinger; see Hadewijch and Hofmann, 1998, p. 157; Hadewijch and Plassmann, 1923, pp. 101, 103.

8. Hadewijch and Plassmann, 1923, p. 37.

9. Hadewijch and Plassmann, 1923, p. 37.

10. Hadewijch and Plassmann, 1923, p. 37.

11. Translation from German by Thomas Metzinger; see Hadewijch and Plassmann, 1923, p. 37.

12. Eckhart, sermon 12; DW I, p. 194, l. 5; see also Ueda and Benz, 2018. See also McGinn, 2001.

13. Eckhart, sermon 12; DW I, p. 235, l. 24; see also Ueda and Benz, 2018.

14. Translation from German by Thomas Metzinger; see Hadewijch and Plassmann, 1923, p. 141.

15. Dietsche, 1960; Waldschütz, 1989.

16. Eckhart, sermon 7; DW I, p. 122, l. 10.

17. Eckhart, sermon 7; DW I, p. 123, l. 5; Waldschütz, 1989, p. 194.

18. Eckhart, sermon 5b; DW I, p. 90, l. 8.

19. Eckhart, sermon 17; DW I, p. 283, l. 15.

20. Dietsche, 1960.

21. Eckhart, sermon 72; DW III, p. 252, l. 1 et seq.

22. Eckhart, sermon 48; DW II, p. 415, l. 20.

23. Eckhart, treatise 2; DW V, p. 80, l. 8.

24. Eckhart, sermon 48; DW II, p. 415, l. 22.

25. Pfeiffer, [1845]1962, p. 349, p. 20 et seq.; as cited in Egerding, 1997, p. 552.

26. Eckhart, sermon 42; DW II p. 307, l. 3; as cited in Egerding, 1997, p. 553.

27. Eckhart, sermon 42; DW II p. 307, l. 3; as cited in Egerding, 1997, p. 553.

28. Eckhart, sermon 18; DW I, p. 299, l. 2–4.

29. In Middle German: *"Dv solt got minnen nichgeistliche, de ist: De din sel sol nichgeistig sin vnd entplôzet aller geistekeite; wand die wile din sel geistformig ist, so hat si bilde; die wile si bilde hat, so [so] hat si mittel; Die wile si mittel hat, so hat si nit einikeit noch einberkeit;"* (Eckhart, sermon 83; DW III, p. 447, l. 12–p. 448, l. 3.) In High German: *"Du sollst Gott ungeistig lieben, das heißt so, daß deine Seele ungeistig sei und entblößt aller Geistigkeit; denn, solange deine Seele geistförmig ist, solange hat sie Bilder. Solange sie aber Bilder hat, solange hat sie Vermittelndes; solange sie Vermittelndes hat, solange hat sie nicht Einheit noch Einfachheit."* In English, using "mind" for *Geist*: "You should love God nonmentally: that is to say in a way that the soul is nonmental and is stripped of all mentality. For as long as your soul is in mind-form, it has images. But so long as it has images,

it has mediation; so long as it has mediation, it has neither unity or simplicity." (Eckhart, 2007, pp. 464–465)

30. Mechthild von Magdeburg, 1990, V 5, 10 et seq.; as cited in Egerding, 1997, p. 167.

31. Pfeiffer, [1845]1962, p. 391, l. 23 et seq.; as cited in Egerding, 1997, p. 168.

32. Pfeiffer, [1845]1962, p. 324, l. 16; as cited in Egerding, 1997, p. 169.

33. Eckhart, sermon 23; DW I, p. 407, l. 9; as cited in Egerding, 1997, p. 170.

34. Mechthild von Magdeburg, 1990, VII 59, l. 6; as cited in Egerding, 1997, p. 168.

35. Pfeiffer, [1845]1962, p. 352, l. 92 et seq.; as cited in Egerding, 1997, p. 416.

36. Heinrich von Nördlingen, Brief LII; Strauch et al., 1882, p. 264; as cited in Egerding, 1997, p. 434.

37. Eckhart, sermon 21; DW I, p. 361, l. 3.

38. Eckhart, sermon 27; DW II, p. 53, l. 1; as cited in Egerding, 1997, p. 28.

39. Eckhart, sermon 1; DW I, p. 18, l. 6 et seq.; as cited in Egerding, 1997, p. 419.

40. Eckhart, sermon 35; DW II 181, l. 7; as cited in Egerding, 1997, p. 421.

41. Eckhart, sermon 102; DW IV, p. 412, l. 39 et seq.

42. Eckhart, sermon 10; DW I, p. 165, l. 9 et seq.

43. Waldschütz, 1989, pp. 111, 132.

44. Mechthild von Magdeburg, 1990, II, p. 19, l. 55; as cited in Egerding, 1997, p. 510.

45. Tauler, 1910, p. 164, l. 8 et seq.; as cited in Egerding, 1997, p. 518.

46. Eckhart, sermon 10; DW I, p. 165, ll. 8–12.

47. Eckhart, treatise 21; DW V, p. 275, l. 35.

48. Tauler, 1910, p. 55, l. 24–26; as cited in Egerding, 1997, p. 29.

49. Seuse, [1907]1961, p. 219, l. 11; as cited in Egerding, 1997, p. 31.

50. Seuse, [1907]1961, p. 15, l. 6; as cited in Egerding, 1997, p. 31.

51. Seuse, [1907]1961, p. 360, l. 6 et seq.; as cited in Egerding, 1997, p. 31.

52. Eckhart, sermon 83; DW III, p. 442, l. 1 et seq.; as cited in Egerding, 1997, p. 550.

53. See "Seelengrund," 2021.

54. Morard, 1956, p. 170.

55. Morard (1956, p. 173) connects it to the *esse purum per se subsistens* (pure existence subsisting of itself) of Thomas Aquinas.

56. Eckhart, sermon 83; DW III, p. 443, ll. 5–7; translation from German by Emily Troscianko.

57. Morard, 1956, p. 173. Note that German also offers the option to play on the homonyms *sein* (to be) and *sein* (his), as Eckhart does here with "*sinesheit*" and "*sin*" building up to the final "*istikeit.*"

58. Eckhart, sermon 102; DW IV, p. 408.

59. Koran, 50, 16. This theological motif seems to be an invariant, repeating itself across the centuries. Compare a recent text on the practice of meditation in a Christian context: "Christian meditation and its corresponding contemplative exercises are the fruitful expression of a perceived call to the Unconditioned, to draw closer to the One who is infinitely closer to the practitioner than she is to herself" (Terry, 2022, p. 390).

60. Morard (1956, p. 175) comments: "*Nach all den zitierten Stellen aus Eckhart dürfen wir füglich schließen: istikeit ist von ist abgeleitet und kommt im vollen Sinn nur allein dem einzigen ist zu, das Gott allein ist, indem Gott allein, als wesenhaftes Sein, aus dem Eigenen Sein hat, ja eigentlich Sein ist.*" ("According to all the quoted passages from Eckhart, we may reasonably conclude: *Istikeit* is derived from *is* and applies in its fullest sense only to the single *is* that is God alone, while God alone, as essential being, has being from itself, indeed actually is being.")

61. Waldschütz, 1989, p. 124.

62. Eckhart, sermon 15; DW I, p. 246, ll. 9–10; as cited in Egerding, 1997, p. 288.

63. Pfeiffer, [1845]1962, p. 372, l. 39; as cited in Egerding, 1997, p. 292.

64. Tauler and Vetter, 1968, p. 251.

65. Tauler and Vetter, 1968, p. 175 et seq.

66. Pfeiffer, [1845]1962, p. 66, l. 7 et seq.; as cited in Egerding, 1997, p. 297.

67. Seuse, Little Book of the Truth, V

68. Seuse, [1907]1961, p. 446, l. 11 et seqq.; as cited in Egerding, 1997, p. 306.

69. Tauler, 1910, p. 68, l. 7 et seq.; as cited in Egerding, 1997, p. 26.

70. Seuse, [1907]1961, p. 16, l. 8; as cited in Egerding, 1997, p. 527.

71. Astigarra et al., 1990, p. 854.

72. Jetté, 1964.

73. See Higgins, 2013, p. 173 et seq., for a discussion of the terms "foundationless" (*gnas med pa*) and "groundless" (*gzhi med pa*). I also recommend Higgins and Draszczyk, 2016, pp. 319, 398, 428; and Higgins's comments on *kun gzhi tsam*, as well as the idea of a groundless subject (*gzhi med kyi yul can*) of experience in David Higgins and Martina Draszczyk's impressive work on Buddha nature (2019, pp. 230, 251). Alternatively, see Higgins (2012, 2020) for a freely available introduction.

74. Higgins, 2013, p. 173.

75. Higgins, 2020, p. 264; see also Higgins and Draszczyk, 2016, p. 319.

76. Higgins, 2020, which contains a quotation by the twelfth-century master Zhang rin po che (1122–1193), one of the earliest Tibetan descriptions of groundless ground (though he does not explicitly use the term itself). His main point is that the absolute imputed by scholars is, despite their assertions, neither a ground nor groundless, neither a source nor devoid of source. Mi bskyod rdo rje (the Eighth Karmapa, 1507–1557) took Zhang's statement that the absolute is neither ground nor groundless as support for his own idea of a groundless ground, and may then have become the first to give it conceptual articulation. Higgins (personal communciation) points out that one can find some mentions of the term *gzhi gzhi med* (or *gzhi'i gzhi med*) in the writings of other Tibetan scholars, like the Rnying ma pa Tshe dbang nor bu (1698–1755) and the Sa skya pa Ngag dbang Kun dga' bsod nams (1597–1660), but they all come later than Mi bskyod rdo rje.

77. Higgins, 2013, p. 172; 2020, p. 263.

78. Ray, 2001, pp. 280–282. 'Ba' ra ba Rgyal mtshan dpal bzang (in *Heartfelt Advice*; see Higgins, 2022, p. 199) says this: "When you no longer find fault with appearance, then amidst whatever appears, you don't analyse and judge forms as being nice or not nice but leave them just as they are without grasping them. Letting them be in this way, just as things continue appearing of their own accord, let them range freely where they like. There is no need for a meditation that tries to give up appearances, taking them as something extraneous. Rather, during this appearing, one remains loosely poised in the state, prior to grasping, in which you don't have to meditate, yet are not distracted. This is known as 'the many having one flavour.' [. . .] At this point, things have become open-ended and unbounded and there is a coming together of what was fragmented. For it has been taught that the impediment of dualistic phenomena is overturned from its very foundations. In this very instance, it is said that I do not see anything apart from me. And in this very instance, it is stated: "*Then everything flows into the stream of the single taste meditation.*" Line 39 of *Song of the Seven Direct Introductions*; see Higgins, 2022, p. 154. For more on the "yoga of one flavour" and "one flavour meditation," see Higgins, 2022, part II, II:5 (pp. 195–199). I am grateful to David Higgins for personal correspondence on the meaning of *ro gcig*, which I draw on in this paragraph.

79. Hadewijch, Vision 14, l. 82; Hadewijch, 1924, p. 162.

80. Huxley, 1945; Schuon, 1975; Taves and Asprem, 2017.

81. For a related discussion, see Woods, Windt, and Carter, 2022a, 2022b; Woods, et al., 2023.

82. The following discussion ties in with the point critically addressed in Woods, Windt, and Carter, 2022a and Woods et al., 2023: the widespread claim and highly popular philosophical background assumption that all contentless experiences are identical. I am grateful to Alex Gamma and Toby Woods for critical discussion.

83. Komarovski, 2015, p. 68 et seq.

84. See Ramstead et al., 2021; Sandved-Smith et al., 2021.

Chapter 27

For the first epigraph, see Eckhart, sermon 6; Eckhart, 2007, p. 332. In Middle High German: *Sumlîche einveltige liute wænent, sie süln got sehen, als er dâ stande und sie hie. Des enist niht. Got und ich wir sint ein* (DW I, p. 113, ll. 6–7). For the second and third epigraphs, see James, 1904; and James, 1892, p. 467.

1. This is a new translation by Emily Troscianko and Thomas Metzinger. See also Rilke, 1969, p. 369, as cited in Bricklin, 2006, p. 227.

2. "But this state of nothingness is not an ideal to be pursued. It has nothing to do with the inventions of the mind. [. . .] When the mind is silent, there is the coming into being of this sense of complete nothingness which is the very essence of humility. It is only then that there is a radical transformation in the quality of the mind, and it is only such a mind that is creative" (Krishnamurti, New Delhi 4th Public Talk, February 18, 1959; Krishnamurti, 1991, vol. 11, p. 177).

3. On "pure humility," see Eckhart, sermon 54a; DW II, p. 551.

4. Kelly, 2019.

5. Millière and Metzinger, 2020.

6. Blackstone, 2007, p. 10.

7. Blackstone, 2007, p. 10.

8. Dunne, 2011, p. 73.

9. Metzinger, 2020.

10. Metzinger, 2003, p. 161, section 3.2.3.

11. Clark, 2017; Wiese and Metzinger, 2017. For an accessible introduction to this topic, see Seth, 2021.

12. On the body as structuring the space of perception and conscious experience, I recommend the work of Adrian Alsmith (2011, 2015, 2017, 2021).

13. This may sound like a simple and self-evident phenomenological statement on my part, but if we move on to a more rigorous conceptual level, there are many philosophical intricacies and complexities to discover. I recommend Wiese (2018) and Bayne (2010) as entry points.

14. Sullivan, 1995, p. 53.

15. James, 1950, p. 273.

16. As an ethicist, I have strongly argued against even risking the *potential* creation of artificial consciousness before we really know what we are doing; see Metzinger, 2021.

17. Full, Walach, and Trautwein, 2013.

18. Josipovic and Miskovic, 2020.

19. Bayne, Hohwy, and Owen, 2016.

20. Kühle, 2015.

21. Metzinger, 2020, n. 2.

22. Karmapa, 2017, p. 223.

Chapter 28

For the epigraph, see Moore, 1903, pp. 446, 450.

1. I highly recommended his magnum opus, *Inner Presence* (Revonsuo, 2009, p. 118).

2. Williamson, 2010, p. 175.

3. Dowman, 2003, p. 193, n. 101.

4. Spinoza, *Ethics V*, 29.

5. Seli et al., 2018.

6. A more detailed analysis, with more direct quotations from Moore, can be found in Metzinger, 2020a. For more on transparency, see Metzinger, 2003a, 2003b.

7. For a little more detail, see Metzinger, 2020a, sec. 2.2.

8. Moore, 1903, p. 450 (emphasis in original).

9. Moore, 1903, p. 449.

10. Moore, 1903, p. 449.

11. Dunne, Thompson, and Schooler, 2019.

12. Sandved-Smith et al., 2021, p. 2.

13. Metzinger, 2020.

14. For a first example of a computational model, see Sandved-Smith et al., 2021.

15. Metzinger, 2009, 2018.

16. Metzinger, 2018; Slater, 2009.

17. For some introductory references, go to Center for Human Technology, "How Social Media Hacks Our Brains," https://www.humanetech.com/brain-science.

18. For a similar idea, see Humphrey, 2011.

19. Metzinger, 2018.

20. For many philosophers, the computer metaphor implied that psychological properties can be exhaustively described by a Turing machine table (Boden, 2006; Churchland, 2005; Putnam, 1967, 1975, 1992).

21. Metzinger, 2021.

22. Metzinger, 2003a, p. 140.

23. Metzinger, 2004, p. 52.

24. Seth, 2021; Wiese and Metzinger, 2017. A good recent overview of theories of consciousness is Doerig, Schurger, and Herzog, 2021.

25. Clark, 2016; Hohwy, 2013; Wiese and Metzinger, 2017.

26. Wiese, 2017.

27. Madary and Metzinger, 2016; Metzinger, 2019.

28. Madary and Metzinger, 2016, p. 20.

29. deVries, 2021.

30. See Konkoly et al., 2021.

31. Namgyal, 2001, p. 81.

32. Namgyal, 2001, p. 62.

33. See David Higgins's new and excellently documented translation in *Heartfelt Advice* (Higgins, 2022, p. 199).

34. Higgins, 2022, p. 199.

35. See also Higgins, 2022, n. 429.

36. Dowman, 2003, p. 204.

37. Laukkonen and Slagter, 2021; see also Fleming (2020) on a simple second-order "awareness state" that guides our decisions if some perceptual stimulus is present or absent.

38. A good, freely available entry point is Beyer, 2020.

39. See Husserl, 1989, pp. 125–126; Hanna, 1995; Hanna et al., 2017.

40. Rilke, 1997, p. 243. The untitled poem by Rainer Maria Rilke was written in the early autumn of 1914, probably in Munich or Irschenhausen; translated from the German by Emily Troscianko. Compare an experiential report given by a Zen meditator, a participant in a study conducted by Jared Lindahl and Willoughby Britton (2019, p. 170): "The boundary between me and my environment began to break down. A bird flew in front of me, but it didn't fly in front of me—it flew through me."

Chapter 29

For the epigraph, see Nagel, 1986, p. 61.

1. The philosopher John Spackman puts the point about actually participating in what I have called the "self-knowing field of awareness" like this: "[. . .] no belief can achieve this, because

there is one feature of experience the nature of which no belief will allow one to apprehend, namely the reflexive nature of consciousness itself, which transcends the act–object distinction. A belief can only represent consciousness as an intentional object, which does not capture its nondualistic nature. On the present account, the only way of capturing the reflexive nature of consciousness is by being in that state of awareness, which is nonconceptual. The non-reifying attitude must thus itself include a nonconceptual component; it must itself participate in the awareness of the reflexive nature of consciousness" (Spackman, 2022, p. 147).

2. For a discussion of this point, see Siderits, Thompson, and Zahavi (2011) and Metzinger (2011). To bring out the philosophical issue, here is one interesting argument about the selflessness of pure awareness based on its phenomenological impermanence, put forward very simply and clearly by Sheng Yen (1931–2009):

> In terms of the existence of self, how should we understand entering *samādhi*, the meditative state of deep absorption? Does the practitioner enter a state of timeless, permanent self? If there is a fixed self, then when you enter *samādhi*, you will be transformed permanently to that state. How would you ever return from *samādhi*? If you can enter and leave *samādhi*, then that means *samādhi* is impermanent, empty of a fixed substantial essence. Therefore, going in and out of *samādhi*, one would understand that there is no selfhood even in *samādhi*. Someone in *samādhi* may feel as if they are experiencing something permanent, but the mere fact that they can come out of *samādhi* tells us that what is experienced in *samādhi* is not **[sic!]** impermanent. So, when you experience *samādhi*, please do not identify that as self. The correct view is to understand it in terms of no-self.

See Yen, 2008, p. 33. **Author's note**: The passage cited here, published by Shambala in 2008, contains a small but significant mistranslation. Instead of "not impermanent," the Chinese original says "impermanent," so the penultimate sentence should read: "Someone in *samādhi* may feel as if they are experiencing something permanent, but the mere fact that they can come out of *samādhi* tells us that what is experienced in *samādhi* is impermanent." I am grateful to Dr Emily Troscianko (Pembroke College, Oxford) for spotting this error, and to Professor Ying-Tung Lin (Institute of Philosophy of Mind and Cognition at National Yang Ming Chiao Tung University in Taipei City, Taiwan) for helping me verify it. For an interesting phenomenological contrast, compare the quote of Douglas Harding that I presented at the beginning of the second half of chapter 19.

3. MacKenzie, 2022, p. 253.

4. For valuable discussions of this, see Fasching, 2012, 2016.

5. Blanke and Metzinger, 2009.

6. Metzinger, 2011; Seth, 2021; Wiese and Metzinger, 2017.

7. Nagel, 1986, p. 61.

8. Nagel, 1986, p. 57.

9. Nagel, 1986, p. 61 (emphasis in original).

10. Nagel, 1986, chapter 4.

11. Ataria, Dor-Ziderman, and Berkovich-Ohana, 2015, p. 141, as cited in Metzinger, 2020, case study #12.

12. Nagel, 1986, p. 64.

13. Timalsina, 2022, p. 312.

14. Closely related figures of thought are also found in the philosophy of religion, such as in Meister Eckhart: "By recognizing itself as an image, the spirit becomes the living image of God" (Kreuzer, 2023, p. 128).

Chapter 30

For the first epigraph, see Higgins, 2022, p. 129. For the second, see Sangs-rgyas-rdo-rje, Klong-chen-pa Dri-med-'od-zer, and 'Jigs-med-gling-pa Rang-byung-rdo-rje, 2014, p. 133. For the third, see Namgyal, 2019, p. 302.

1. Dunne, Thompson, and Schooler, 2019; Maleeh and Konjedi, 2022. In traditional Western terms, the epistemological goal of meditation practice is to develop a kind of *knowing-how* rather than *knowing-that*. Christopher Gowans has termed this the "knowledge-how model of meditation" (Gowans, 2022, p. 161). For the wider context, related to the concept of *Bewusstseinskultur* that I discuss in the epilogue, see also Gowans, 2021.

2. Millière and Newen, 2022.

3. "Rang Rig Ye Shes," 2021. See also the entry on *svasaṃvedana* in Buswell and Lopez, 2014, p. 882.

4. Frank et al., 2020; see also Frank and Kuneš, 2022.

5. For a substantial overview and a new model of nonrepresentational "self-acquaintance," see Williford, 2019; see also Kriegel, 2009; Williford & Kriegel, 2006; Peters, 2013.

6. Williford, 2019.

7. Strawson, 2015.

8. Josipovic, 2019.

9. Blanke and Metzinger, 2009; Metzinger, 2003.

10. The existence of a fundamental, nonegoic form of reflexivity built into consciousness itself has been discussed for centuries by Buddhist philosophers (Finnigan, 2018; Williams, 1998). Following Brentano's classic idea of a single state simultaneously directed at an intentional object and itself (Brentano, 1973, p. 153; 2012, p. 25), many fine-grained conceptual models have been developed in more recent analytical philosophy of mind (Peters, 2013). Currently, the topic is gaining increased attention (Ganeri, 2017; Josipovic, 2019; Kriegel, 2019; Spackman, 2022; Strawson, 2015; Thompson, 2011; Williford, 2019).

11. Vilas, Auksztulewicz, and Melloni, 2021; Wiese and Friston, 2021.

12. In earlier publications, I have called this the "phenomenal model of the intentionality relation" (or PMIR); see Metzinger, 1993, 2000, 2003: ch. 6, 2005, 2006, 2020; Metzinger and Gallese, 2003.

13. Gebauer et al., 2018.

14. Austin, 2009.

Chapter 31

The first two introductory quotations are from the *Tractatus Logico-Philosophicus* by Ludwig Wittgenstein (1889-1951) and are found at 5.632 and 5.633. The third quotation is from Chinul (1158-1210) and is included in Robert Buswell's book *Tracing Back the Radiance*; it is from *Secrets on Cultivating the Mind*, written between 1203 and 1205. On this point, see Buswell, 1991, p. 103.

1. Batterman, 2002; Batterman and Rice, 2014. See Metzinger, 2020, sec. 1.2, for a short introduction.

2. Metzinger, 2017.

3. Hohwy, 2020; see also Hohwy, 2021.

Chapter 32

For the first epigraph, see Dowman, 2003, p. 94; for the second, see Namgyal, *Mahamudra—The Moonlight: Quintessence of Mind and Meditation*, 2006, p. 332.

1. Urgyen, 2000, p. 188.

2. Urgyen, 2000, p. 85.

3. Letheby, 2022, p. 219.

4. For a recent example and related references, see Lynch and Troy, 2021.

5. For the example of conscious volition, see Haggard, 2019. An accessible psychological perspective on the conscious experience of will and its determining factors is Wegner, 2018; classical papers by Daniel Wegner on the paradoxical effects of attempting to suppress thoughts are Wegner, 1994; Wegner et al., 1987.

6. Tang et al., 2022.

7. Schooler et al., 2011, box 1; see also Smallwood and Schooler, 2015.

8. Metzinger, 2003, 2008, 2009.

9. Sandved-Smith et al., 2021, p. 9.

10. See Zedelius, Protzko, and Schooler, 2021, for more.

11. Swami Madhurananda has made exactly the same phenomenological point about the very early stages of object perception via the senses; see Madhurananda, 2010.

12. Paul, 2014.

13. Persinger, 2013, p. 69.

14. Importantly, new research shows that meditation techniques have varying—including negative—effects on the meditator's moral self-model. For example, loving-kindness meditation appears to improve the emotional layer of a person's moral self-model, but only parts of their behavioral profile, while mindfulness meditation seems to improve some behaviors, but may have a limited or even negative effect on moral intentions. See Berryman, Lazar, and Hohwy, 2023, for an important new initiative.

Chapter 33

For the epigraph, see Namgyal, *Clarifying the Natural State: A Principal Guidance Manual for Mahamudra,* 2001, pp. 52, 60.

1. Bayne and Hohwy, 2016.

2. Metzinger, 2017.

Chapter 34

The epigraph here is an ancient saying from the Shangpa Kagyu tradition of Tibetan Buddhism. I am grateful to Andrew Holecek and Elizabeth Callahan (personal communication) for the pointer that the "original" source for this could possibly be Niguma, which is found in Khyungpo Naljor's "Amulet Mahamudra." See also Harding, 2011, p. 146.

1. Metzinger, 2020; Wiese, 2020.

2. Metzinger, 2021, see also the epilogue.

3. See chapter 1, note 1.

4. Such episodes would be not "selfless memories," but rather egoic, fully subjective memories of selfless states that occurred in the past. For an excellent discussion, see Millière and Newen, 2022.

5. Papineau, 2021.

6. As Karl Friston and colleagues (2015) say, "Epistemic value is the expected information gain under predicted outcomes. In other words, it reports the reduction in uncertainty about hidden states afforded by observations" (p. 192). My point is that (1) for human beings, the hidden state must be intimately related to the mechanism creating tonic alertness and wakefulness (chapter 4); and (2) phenomenologically, the process of expecting epistemic value itself creates an uncontracted, nondual signature of knowing (chapters 8 and 19).

7. For an excellent and accessible discussion of the question of whether pure awareness is contingent on the brain (and what the Dalai Lama really thinks), see Thompson, 2015, pp. 79–88.

8. Metzinger and Windt, 2015.

9. Metzinger, 2020; for more semantic analysis, I recommend Woods, Windt, and Carter, 2022a, 2022b.

10. Gamma and Metzinger, 2021.

11. This was PC3 in the original publication; see Metzinger, 2020.

12. Fink, 2020; Metzinger, 2003; Millière, 2020; Millière and Newen, 2022; Wiese, 2017.

13. Gamma and Metzinger, 2021; Metzinger, 2020.

14. Laukkonen and Slagter, 2021; Pagnoni, 2019.

15. Dunne, 2011.

16. Hasenkamp, 2018; Hasenkamp et al., 2012; Sandved-Smith et al., 2021.

17. Metzinger, 2018. For an example, see Glowacki et al., 2022.

18. Sandved-Smith et al., 2021.

19. Parr, Pezzulo, and Friston, 2022; Wiese and Metzinger, 2017.

20. The project is well under way; for early examples, see Laukkonen and Slagter, 2021; Sandved-Smith et al., 2021.

21. The boxed parameter **A** represents the beliefs about the likelihood of a particular state *s* given some input data (i.e., the *likelihood mapping*). Importantly, the confidence in this mapping can be modulated by the *likelihood precision*, γ_A. Together, **A** and γ_A represent the perceptual model being used at each level to infer the associated experience *s*. At the first level, inferred states represent sensory perception. Each higher level is inferred based on the likelihood precision of the level below it. If you are interested, you can find more detailed figures targeting some of the specific phenomenologies described in this book online at mpe-project.info.

22. Friston et al., 2017.

23. Petitmengin, 2021; Petitmengin et al., 2017; Petitmengin, Remillieux, and Valenzuela-Moguillansky, 2019.

Epilogue

1. Metzinger, 2023.

2. Metzinger, 1994, 1996, 1998, 1999, 2000a, 2000b, 2003, 2005, 2009a, 2016, 2017. For other interesting discussions, see Fink, 2018. See also the extended German version of *The Ego Tunnel* from 2014.

3. Cicero, 2007, bk. II, 5 (13), Cicero, *Tusculan Disputations,* 2007.

4. Hadot, 2002.

5. Just to give an impression of the current order of magnitude, at the end of 2021, the European Monitoring Centre for Drugs and Drug Addiction was monitoring around 880 new psychoactive substances, 52 of which were first reported in Europe in 2021; in 2014, there were more than 100 first-time detections of new substances (EMCDDA, 2022). None of them has ever been clinically tested, long-term side effects are unknown, and most emergency doctors never heard of any of these substances when they were in school. The *World Drug Report 2022* published by the United Nations Office on Drugs and Crime identifies 1,127 new psychoactive substances as having emerged worldwide between 2009 and 2021 (UNODC, 2022). For further discussion and references, see Metzinger, 2009a, ch. 9; Metzinger 2023b.

6. Metzinger, 2018; see also Madary and Metzinger, 2016.

7. A good entry point is https://www.humanetech.com/; see also S. Vöneky, P. Kellmeyer, O. Müller, and W. Burgard, eds., *Cambridge Handbook of Responsible Artificial Intelligence,* 2022, (open access at https://www.cambridge.org/).

8. Metzinger, 2021.

9. An "information hazard" is a risk arising from the spread of true information that may cause harm. See Bostrom, 2011.

10. See Metzinger, 2023b; for a philosophical discussion, see Letheby, 2021; Letheby and Gerrans, 2024; Millière, 2017; Millière et al., 2018; Deane, 2020, 2021.

11. Metzinger, 2009a, 2021, 2023a.

12. Streib and Hood, 2011, 2016.

13. I have said more about this in an open-access paper that is freely available on the internet; see Metzinger, 2017.

14. Chignell, 2018.

15. Kant makes the point *ex negativo,* by speaking of the "*impurity* [*impuritas, improbitas*] of the human heart" (cf. Kant, Academy edition VI, 30; for an English translation, see 6:30 in Kant [2018, p. 53]). The notion that the special kind of "sincerity" discussed in the main text can help us to realize "the idea of the moral good in its absolute purity" and to counter the "corruption that lies in all human beings" can be found at 6:83 (Kant, 2018, p. 98). The second thought—that every man has an obligation toward himself to be truthful—is found in the *Metaphysik der Sitten* (Kant, Academy edition VI, 429); for an English translation, see part 2, chapter II, $9: "Man's Duty to Himself Merely as a Moral Being" in Kant (1991, p. 225). This is where one also finds the concept of an "inner lie."

16. See, for example, Namgyal, 2019, II (10).

References

Abdoun, Oussama, Jelle Zorn, Stefano Poletti, Enrico Fucci, and Antoine Lutz. "Training Novice Practitioners to Reliably Report Their Meditation Experience Using Shared Phenomenological Dimensions." *Consciousness and Cognition* 68 (February 2019): 57–72. https://doi.org/10/gft3hk.

Abelson, Ben. "Meditation and the Paradox of Self-Consciousness." In *Routledge Handbook on the Philosophy of Meditation*, edited by Rick Repetti, 70–77. London: Routledge, 2022. https://doi.org/10.4324/9781003127253-5.

Albahari, Miri. "Beyond Cosmopsychism and the Great I Am: How the World Might Be Grounded in Universal Advaitic Consciousness." In *Routledge Handbook of Panpsychism*, edited by William Seager, 119–130. New York: Routledge, 2020.

Albahari, Miri. "Witness-Consciousness: Its Definition, Appearance and Reality." *Journal of Consciousness Studies* 16, no. 1 (2009): 62–84.

Alcaraz-Sánchez, Adriana. "Awareness in the Void: A Micro-Phenomenological Exploration of Conscious Dreamless Sleep." *Phenomenology and the Cognitive Sciences* (August 3, 2021). https://doi.org/10/gmfs7q.

Alcaraz-Sánchez, Adriana. "Is Lucid Dreamless Sleep Really Lucid?" *Review of Philosophy and Psychology* (2022). https://doi.org/10.1007/s13164-022-00663-9.

Alcaraz-Sánchez, Adriana, Ema Demšar, Teresa Campillo-Ferrer, and Susana Gabriela Torres-Platas. "Nothingness Is All There Is: An Exploration of Objectless Awareness during Sleep." *Frontiers in Psychology* 13 (June 10, 2022): 901031. https://doi.org/10.3389/fpsyg.2022.901031.

Alsmith, Adrian. "Perspectival Structure and Agentive Self-Location." In *The Subject's Matter: Self-Consciousness and the Body*, edited by Frédérique de Vignemont and Adrian Alsmith, 263–288. Cambridge, MA: MIT Press, 2017. https://doi.org/10.7551/mitpress/10462.003.0016.

Alsmith, Adrian. "Perspectival Structure and Vestibular Processing. A Commentary on Bigna Lenggenhager & Christophe Lopez." In *Open MIND*, edited by Thomas Metzinger and Jennifer M. Windt. Frankfurt am Main, Germany: MIND Group, 2015. https://doi.org/10.15502/9783958570559.

Alsmith, Adrian. "The Structure of Egocentric Space." In *The World at Our Fingertips: A Multidisciplinary Exploration of Peripersonal Space*, edited by Frédérique de Vignemont, Hong Yu Wong,

Andrea Serino, and Alessandro Farnè, 231–248. Oxford: Oxford University Press, 2021. https://doi.org/10.1093/oso/9780198851738.003.0013.

Alsmith, Adrian. "The Structuring Body: A Critical Study in the Description & Explanation of Perceptual Experience." PhD thesis, Johannes Gutenberg University, Mainz, Germany, 2011. http://doi.org/10.25358/openscience-2047.

Astigarra, J. Luis, Agustí Borrell, and F. J. Martín de Lucas. *Concordancias de los escritos de San Juan de la Cruz*. Rome: Teresianum, 1990.

Ataria, Yochai. "Where Do We End and Where Does the World Begin? The Case of Insight Meditation." *Philosophical Psychology* 28, no. 8 (November 17, 2015): 1128–1146. https://doi.org/10.1080/09515089.2014.969801.

Ataria, Yochai, Yair Dor-Ziderman, and Aviva Berkovich-Ohana. "How Does It Feel to Lack a Sense of Boundaries? A Case Study of a Long-Term Mindfulness Meditator." *Consciousness and Cognition* 37 (December 2015): 133–147. https://doi.org/10.1016/j.concog.2015.09.002.

Ataria, Yochai, Shogo Tanaka, and Shaun Gallagher. *Body Schema and Body Image: New Directions*. Oxford: Oxford University Press, 2021.

Austin, John L. *How to Do Things with Words*. Oxford: Oxford University Press, 2009.

Aymerich-Franch, Laura, Damien Petit, Gowrishankar Ganesh, and Abderrahmane Kheddar. "The Second Me: Seeing the Real Body during Humanoid Robot Embodiment Produces an Illusion of Bi-Location." *Consciousness and Cognition* 46 (November 2016): 99–109. https://doi.org/10/f9b9g9.

Baars, Bernard J. *A Cognitive Theory of Consciousness*. Cambridge: Cambridge University Press, 1993. Originally published in 1989.

Bachmann, James K. *Accounting for Pure Consciousness: An Examination of the Ability of the Representationalist Approach to Phenomenal Consciousness to Account for Pure Consciousness Experiences*. Edmonton: University of Alberta, 2014.

Baird, Benjamin, Anna Castelnovo, Olivia Gosseries, and Giulio Tononi. "Frequent Lucid Dreaming Associated with Increased Functional Connectivity between Frontopolar Cortex and Temporoparietal Association Areas." *Scientific Reports* 8, no. 1 (2018): 1–15. https://doi.org/10/gmv4rn.

Baird, Benjamin, Daniel Erlacher, Michael Czisch, Victor I. Spoormaker, and Martin Dresler. "Chapter 19—Consciousness and Meta-Consciousness during Sleep." In *Handbook of Sleep Research*, edited by Hans C. Dringenberg, 283–295. London: Elsevier, 2019. https://doi.org/10.1016/B978-0-12-813743-7.00019-0.

Baird, Benjamin, Stephen LaBerge, and Giulio Tononi. "Two-Way Communication in Lucid REM Sleep Dreaming." *Trends in Cognitive Sciences* 25, no. 6 (1 June 2021): 427–428. https://doi.org/10/gmgh4k.

Baird, Benjamin, and Matthieu Koroma. "Internally Generated Conscious Activity: Reflections upon (Lucid) Dreaming, Mind-Wandering and Meditation. An Interview with Benjamin Baird by Matthieu Koroma." 2020. Alius, https://doi.org/10/gj6tx8.

Baird, Benjamin, Sergio A. Mota-Rolim, and Martin Dresler. "The Cognitive Neuroscience of Lucid Dreaming." *Neuroscience & Biobehavioral Reviews* 100 (2019): 305–323. https://doi.org/10/gf95df.

Baird, Benjamin, Brady A. Riedner, Melanie Boly, Richard J. Davidson, and Giulio Tononi. "Increased Lucid Dream Frequency in Long-Term Meditators but Not Following Mindfulness-Based Stress Reduction Training." *Psychology of Consciousness: Theory, Research, and Practice* 6, no. 1 (2019): 40. https://doi.org/10/gg9kfq.

Baminiwatta, Anuradha, and Indrajith Solangaarachchi. "Trends and Developments in Mindfulness Research over 55 Years: A Bibliometric Analysis of Publications Indexed in Web of Science." *Mindfulness* 12, no. 9 (September 2021): 2099–2116. https://doi.org/10.1007/s12671-021-01681-x.

Barron, Richard, and Susanne Fairclough, eds. *The Precious Treasury of the Basic Space of Phenomena*. Junction City, CA: Padma Publishing, 2001.

Bartolomei, F., S. Lagarde, D. Scavarda, R. Carron, C. G. Bénar, and F. Picard. "The Role of the Dorsal Anterior Insula in Ecstatic Sensation Revealed by Direct Electrical Brain Stimulation." *Brain Stimulation* 12, no. 5 (September 2019): 1121–1126. https://doi.org/10/gg4b47.

Batterman, Robert W. "Asymptotics and the Role of Minimal Models." *British Journal for the Philosophy of Science* 53, no. 1 (2002): 21–38. https://doi.org/10.1093/bjps/53.1.21.

Batterman, Robert W., and Collin C. Rice. "Minimal Model Explanations." *Philosophy of Science* 81, no. 3 (2014): 349–376. https://doi.org/10.1086/676677.

Bayne, Tim. "On the Axiomatic Foundations of the Integrated Information Theory of Consciousness." *Neuroscience of Consciousness* 2018, no. 1 (1 January 2018). https://doi.org/10/ggwf37.

Bayne, Tim. *The Unity of Consciousness*. Oxford: Oxford University Press, 2010.

Bayne, Tim, and Jakob Hohwy. "Modes of Consciousness." In *Finding Consciousness: The Neuroscience, Ethics, and Law of Severe Brain Damage*, 57–80. Oxford Series in Neuroscience, Law, and Philosophy. New York: Oxford University Press, 2016. https://doi.org/10.1093/acprof:oso/9780190280307.003.0005.

Bayne, Tim, Jakob Hohwy, and Adrian M. Owen. "Are There Levels of Consciousness?" *Trends in Cognitive Sciences* 20, no. 6 (June 2016): 405–413. https://doi.org/10/f8pc7t.

Bayne, Tim, Anil K. Seth, and Marcello Massimini. "Are There Islands of Awareness?" *Trends in Neurosciences* 43, no. 1 (January 2020): 6–16. https://doi.org/10/gmfkb8.

Beierwaltes, W. "Erleuchtung." In *Historisches Wörterbuch Der Philosophie*, 2: 712–718. Basel: Schwabe & Co., 1972.

Berrios, German E., and Rogelio Luque. "Cotard's Syndrome: Analysis of 100 Cases." *Acta Psychiatrica Scandinavica* 91, no. 3 (1995): 185–188. https://doi.org/10/bc3tsm.

Berryman, Kevin, Sara W. Lazar, and Jakob Hohwy. "Do Contemplative Practices Make Us More Moral?" *Trends in Cognitive Sciences* (August 2023), S1364661323001730. https://doi.org/10.1016/j.tics.2023.07.005.

Beyer, Christian. "Edmund Husserl." In *Stanford Encyclopedia of Philosophy*, edited by Edward N. Zalta, Winter 2020. Metaphysics Research Lab, Stanford University. https://plato.stanford.edu/archives/win2020/entries/husserl/.

Bhikkhu, Thānissaro. *Udāna: Exclamations. A Translation with an Introduction & Notes*. Valley Center, CA: Metta Forest Monastery, 2012. https://www.accesstoinsight.org/lib/authors/thanissaro/udana.pdf.

Blackstone, Judith. *The Empathic Ground: Intersubjectivity and Nonduality in the Psychotherapeutic Process*. Albany: State University of New York Press, 2007.

Blanke, O., and T. Metzinger. "Full-Body Illusions and Minimal Phenomenal Selfhood." *Trends in Cognitive Sciences* 13, no. 1 (2009): 7–13. https://doi.org/10.1016/j.tics.2008.10.003.

Blanke, Olaf. "Multisensory Brain Mechanisms of Bodily Self-Consciousness." *Nature Reviews Neuroscience* 13, no. 8 (2012): 556–571. https://doi.org/10/gddnh3.

Blanke, Olaf, Giulio Rognini, Simone Gallo, Florian Lance, and Bruno Herbelin. Method and System for Creating an Out-of-Body Experience. Patent; issued 2020. https://infoscience.epfl.ch/record/277794.

Blanke, Olaf, Mel Slater, and Andrea Serino. "Behavioral, Neural, and Computational Principles of Bodily Self-Consciousness." *Neuron* 88, no. 1 (October 2015): 145–166. https://doi.org/10/f7wh2j.

Blondiaux, Eva, Lukas Heydrich, and Olaf Blanke. "Common and Distinct Brain Networks of Autoscopic Phenomena." *NeuroImage: Clinical* 30 (2021): 102612. https://doi.org/10/gnpcjw.

Boden, Margaret. *Mind As Machine: A History of Cognitive Science*. Oxford and New York: Oxford University Press, 2006.

Boehler, C. N., M. A. Schoenfeld, H.-J. Heinze, and J.-M. Hopf. "Rapid Recurrent Processing Gates Awareness in Primary Visual Cortex." *Proceedings of the National Academy of Sciences* 105, no. 25 (2008): 8742–7847. https://doi.org/10.1073/pnas.0801999105.

Bostrom, Nick. "Information Hazards: A Typology of Potential Harms from Knowledge." *Review of Contemporary Philosophy*, no. 10 (2011): 44–79.

Brentano, Franz. *Descriptive Psychology*. London & New York: Routledge, 2012.

Brentano, Franz. *Psychology from an Empirical Standpoint*. Übersetzt von A C Rancurello, DB Terrell, und LL McAlister. London & New York: Routledge, 1973.

Bricklin, Jonathan, ed. *Sciousness*. 2nd ed. Guilford, CT: Eirini Press, 2006.

Brugger, Peter, Marianne Regard, and Theodor Landis. "Illusory Reduplication of One's Own Body: Phenomenology and Classification of Autoscopic Phenomena." *Cognitive Neuropsychiatry* 2, no. 1 (February 1997): 19–38. https://doi.org/10/crmndx.

Buswell, Robert E. *Tracing Back the Radiance Chinul's Korean Way of Zen*. Honolulu: University of Hawaii Press, 1991.

Buswell, Robert E., and Donald S. Lopez, eds. *The Princeton Dictionary of Buddhism*. Princeton: Princeton University Press, 2014.

Cage, John. "An Autobiographical Statement." *Southwest Review* 76, no. 1 (1991): 59–76.

Cardini, Flavia, Natalie Fatemi-Ghomi, Katarzyna Gajewska-Knapik, Victoria Gooch, and Jane Elizabeth Aspell. "Enlarged Representation of Peripersonal Space in Pregnancy." *Scientific Reports* 9, no. 1 (December 2019): 8606. https://doi.org/10/ggqdfz.

Castillo, Richard J. "Depersonalization and Meditation." *Psychiatry* 53, no. 2 (May 1990): 158–168. https://doi.org/10/gfxfpk.

Chatterjee, Tara. "The Concept of Sākṣin." *Journal of Indian Philosophy* 10, no. 4 (1982): 339–356.

Chignell, Andrew. "The Ethics of Belief." In *Stanford Encyclopedia of Philosophy*, edited by Edward N. Zalta, Spring 2018. Metaphysics Research Lab, Stanford University. https://plato.stanford.edu/archives/spr2018/entries/ethics-belief/.

Chowdhury, Avijit, Remko van Lutterveld, Ruben E. Laukkonen, Heleen A. Slagter, Daniel M. Ingram, and Matthew D. Sacchet. "Investigation of Advanced Mindfulness Meditation 'Cessation' Experiences Using EEG Spectral Analysis in An Intensively Sampled Case Study." *Neuropsychologia* (2023): 108694.

Churchland, Paul M. "Functionalism at Forty: A Critical Retrospective." *Journal of Philosophy* 102, no. 1 (2005): 33–50. https://doi.org/10/f3f9zn.

Churchland, Paul M. *A Neurocomputational Perspective: The Nature of Mind and the Structure of Science*. Cambridge, MA: MIT Press, 1989.

Cicero, Marcus Tullius. *Tusculan Disputations*. Edited by J. E. King. Rep. 1927, rev. 1945. Cicero/Cicero 18. Cambridge, MA: Harvard University Press, 2007.

Clark, Andy. *Being There: Putting Brain, Body, and World Together Again*. Cambridge, MA: MIT Press, 1998.

Clark, Andy. *The Experience Machine: How Our Minds Predict and Shape Reality*. UK: Allen Lane, 2023.

Clark, Andy. "How to Knit Your Own Markov Blanket." In *Philosophy and Predictive Processing*, edited by Thomas K. Metzinger and Wanja Wiese. Frankfurt am Main, Germany: MIND Group, 2017. https://doi.org/10.15502/9783958573031.

Clark, Andy. *Surfing Uncertainty: Prediction, Action, and the Embodied Mind*. Oxford and New York: Oxford University Press, 2016.

Cobb, M., and N. Comfort. "What Rosalind Franklin Truly Contributed to the Discovery of DNA's Structure." *Nature* 616(7958)(2023), 657–660.

Coseru, Christian. "Consciousness, Content, and Cognitive Attenuation." In *Routledge Handbook on the Philosophy of Meditation*, edited by Rick Repetti, 354–368. London: Routledge, 2022. https://doi.org/10.4324/9781003127253-29.

Costines, Cyril, Tilmann Lhündrup Borghardt, and Marc Wittmann. "The Phenomenology of 'Pure' Consciousness as Reported by an Experienced Meditator of the Tibetan Buddhist Karma Kagyu Tradition. Analysis of Interview Content Concerning Different Meditative States." *Philosophies* 6, no. 2 (June 2021): 50. https://doi.org/10/gkqwfn.

Dahl, Cortland J., Antoine Lutz, and Richard J. Davidson. "Reconstructing and Deconstructing the Self: Cognitive Mechanisms in Meditation Practice." *Trends in Cognitive Sciences* 19, no. 9 (September 2015): 515–523. https://doi.org/10/f7qmtp.

Davis, Zachary, and Anthony Steinbock. "Max Scheler." In *Stanford Encyclopedia of Philosophy*, edited by Edward N. Zalta, Fall 2021. Metaphysics Research Lab, Stanford University. https://plato.stanford.edu/archives/fall2021/entries/scheler/.

Deane, George. "Consciousness in Active Inference: Deep Self-Models, Other Minds, and the Challenge of Psychedelic-Induced Ego-Dissolution." *Neuroscience of Consciousness* 2021, no. 2 (2021): niab024. https://doi.org/10/gmpwmx.

Deane, George. "Dissolving the Self: Active Inference, Psychedelics, and Ego-Dissolution." *Philosophy and the Mind Sciences* 1, no. I (2020): 1–27. https://doi.org/10/gnt7.

Deane, George, Mark Miller, and Sam Wilkinson. "Losing Ourselves: Active Inference, Depersonalization, and Meditation." *Frontiers in Psychology* 11 (2020): 2893. https://doi.org/10/gmgwh8.

Del Pin, Simon Hviid, Zuzanna Skóra, Kristian Sandberg, Morten Overgaard, and Michał Wierzchoń. "Comparing Theories of Consciousness: Why It Matters and How to Do It." *Neuroscience of Consciousness* 2021, no. 2 (1 October 2021). https://doi.org/10/gm7469.

Dennett, Daniel. *Consciousness Explained*. London: Penguin UK, 1993.

Dennett, Daniel. "Heterophenomenology Reconsidered." *Phenomenology and the Cognitive Sciences* 6, no. 1–2 (2007): 247–270. https://doi.org/10/dc5vsq.

Dennett, Daniel. "Who's on First? Heterophenomenology Explained." *Journal of Consciousness Studies* 10, no. 9–10 (2003): 19–30.

De Ridder, Dirk, Koen Van Laere, Patrick Dupont, Tomas Menovsky, and Paul Van de Heyning. "Visualizing Out-of-Body Experience in the Brain." *New England Journal of Medicine* 357, no. 18 (November 1, 2007): 1829–1833. https://doi.org/10/dz5hs9.

Descartes, René. *Meditations on First Philosophy: With Selections from the Objections and Replies*. Translated by Michael Moriarty. Oxford: Oxford University Press, 2008.

Devor, Marshall, Mary Koukoui, and Mark Baron. "Searching in the Wrong Place: Might Consciousness Reside in the Brainstem?" *Behavioral and Brain Sciences* 45 (2022). https://doi.org/10.1017/S0140525X21001928.

deVries, Willem. "Wilfrid Sellars." In *Stanford Encyclopedia of Philosophy*, edited by Edward N. Zalta, Fall 2021. Metaphysics Research Lab, Stanford University. https://plato.stanford.edu/archives/fall2021/entries/sellars/.

Dietsche, Bernward. "Der Seelengrund nach den Deutschen und Lateinischen Predigten." In *Meister Eckhart der Prediger: Festschrift zum Eckhart-Gedenkjahr*, edited by Udo M. Nix and Raphael Öchslin, 200–258. Freiburg, Germany: Herder, 1960.

Dijkerman, H. C., and W. P. Medendorp. "Visuo-Tactile Predictive Mechanisms of Peripersonal Space." In *The World at Our Fingertips*, edited by Frédérique de Vignemont, Andrea Serino, Hong Yu Wong, and Alessandro Farnè, 81–100. Oxford: Oxford University Press, 2021. https://doi.org/10.1093/oso/9780198851738.003.0005.

Doerig, Adrien, Aaron Schurger, and Michael H. Herzog. "Hard Criteria for Empirical Theories of Consciousness." *Cognitive Neuroscience* 12, no. 2 (3 April 2021): 41–62. https://doi.org/10/gjvfb4.

Dowman, Keith. (2003). *The Flight of the Garuda: The Dzogchen Tradition of Tibetan Buddhism.* Rev. ed. Boston: Wisdom Publications.

Dresler, Martin, Stefan P. Koch, Renate Wehrle, et al. "Dreamed Movement Elicits Activation in the Sensorimotor Cortex." *Current Biology* 21, no. 21 (8 November 2011): 1833–1837. https://doi.org/10/d8598x.

Dunne, John. "Toward an Understanding of Non-dual Mindfulness." *Contemporary Buddhism* 12, no. 1 (1 May 2011): 71–88. https://doi.org/10.1080/14639947.2011.564820.

Dunne, John D., Evan Thompson, and Jonathan Schooler. "Mindful Meta-Awareness: Sustained and Non-propositional." *Current Opinion in Psychology* 28 (August 2019): 307–311. https://doi.org/10.1016/j.copsyc.2019.07.003.

Eckhart. *The Complete Mystical Works of Meister Eckhart.* Translated by Maurice O'C Walshe. New York: Crossroad Pub. Co., 2007.

Eckhart. *Die deutschen und lateinischen Werke. [. . .] Bd. 1: Die deutschen Werke / hrsg. und übers. von Josef Quint Meister Eckharts Predigten.—Bd. 1.* Edited by Josef Quint. Stuttgart: Kohlhammer, 1986. Originally published in 1958.

Eckhart. *Die deutschen und lateinischen Werke. [. . .] Bd. 2: Die deutschen Werke / hrsg. und übers. Von Josef Quint Meister Eckharts Predigten.—Bd. 2.* Edited by Josef Quint. Stuttgart: Kohlhammer, 1988. Originally published in 1971.

Eckhart. *Die deutschen und lateinischen Werke. [. . .] Bd. 3: Die deutschen Werke / hrsg. und übers. Von Georg Steer Meister Eckharts Predigten: Bd. 3 / hrsg. u. übers. von Josef Quint.* Edited by Josef Quint and Georg Steer. Stuttgart: Kohlhammer, 1976.

Eckhart. *Die deutschen und lateinischen Werke. [. . .] Bd. 4. Teilbd. 1: Die deutschen Werke / hrsg. und übers. Von Georg Steer Meister Eckharts Predigten / hrsg. und übers. von Georg Steer unter Mitarb. von Wolfgang Klimanek und Freimut Löser.* Edited by Josef Quint and Georg Steer. Stuttgart: Kohlhammer, 2003.

Eckhart. *Die deutschen und lateinischen Werke. [. . .] Bd. 5: Die deutschen Werke / hrsg. und übers. Von Josef Quint Meister Eckharts Traktate.* Edited by Josef Quint. Stuttgart: Kohlhammer, 1987. Originally published in 1963.

Eckhart, Edmund Colledge, and Bernard McGinn. *Meister Eckhart: The Essential Sermons, Commentaries, Treatises, and Defense*. London: SPCK, 1981.

Eckhart, Bernard McGinn, Frank J Tobin, and Elvira Borgstädt. *Meister Eckhart, Teacher and Preacher*. New York: Paulist Press, 1986.

Eckhart, and Josef Quint. *Deutsche Predigten und Traktate*. Hamburg: Nikol, 2007.

Economist Intelligence Unit. "Democracy Index 2021. The China Challenge." London: EIU, 2022. https://ictlogy.net/bibliography/reports/projects.php?idp=4636.

Egerding, Michael. *Die Metaphorik der spätmittelalterlichen Mystik. Michael Egerding: Bildspender—Bildempfänger—Kontexte: Dokumentation und Interpretation Bd. 2 Bd. 2*. Paderborn, Germany: Schöningh, 1997.

Eidelman, Scott, Christian S. Crandall, and Jennifer Pattershall. "The Existence Bias." *Journal of Personality and Social Psychology* 97, no. 5 (November 2009): 765–775. https://doi.org/10/bwxvtg.

European Monitoring Centre for Drugs and Drug Addiction. *European Drug Report 2022: Trends and Developments*. Luxembourg: Publications Office, 2022. https://data.europa.eu/doi/10.2810/75644.

Fadiman, James, and Jordan Gruber. *Your Symphony of Selves: Discover and Understand More of Who We Are*. Rochester, VT: Park Street Press, 2020.

Fasching, Wolfgang. "On the Advaitic Identification of Self and Consciousness." *Hindu and Buddhist Ideas in Dialogue. Self and No-Self*, 2012, 165–180.

Fasching, Wolfgang. "Consciousness, Self-Consciousness, and Meditation." *Phenomenology and the Cognitive Sciences* 7, no. 4 (2008): 463–483. https://doi.org/10/cbrgrq.

Fasching, Wolfgang. "The Experience of Presence." In *Routledge Handbook on the Philosophy of Meditation*, edited by Rick Repetti, 150–159. London: Routledge, 2022. https://doi.org/10.4324/9781003127253-12.

Fasching, Wolfgang. "I Am of the Nature of Seeing." In *Indian Philosophy: A Reader*, edited by Jonardon Ganeri, 182. London and New York: Routledge, 2019. https://doi.org/10/gmt3t7.

Fasching, Wolfgang. "Phenomenological Reflections on the Indian Notion of Witness-Consciousness." In *Self, No Self? Perspectives from Analytical, Phenomenological, and Indian Traditions*, edited by Mark Siderits, Evan Thompson, and Dan Zahavi, 193. Oxford: Oxford University Press, 2011.

Fasching, Wolfgang. "Prakāśa. A Few Reflections on the Advaitic Understanding of Consciousness as Presence and Its Relevance for Philosophy of Mind." *Phenomenology and the Cognitive Sciences* 20, no. 4 (2021): 679–701. https://doi.org/10/gmg6qh.

Fasching, Wolfgang. "The Non-plurality of the I: On the Question of the Ultimate Subject of Experience." *Journal of Consciousness Studies* 23, no. 1–2 (2016): 140–157.

Fell, Juergen. "What Is Mind Blanking? A Conceptual Clarification." *European Journal of Neuroscience* 56, no. 6 (September 2022): 4837–4842. https://doi.org/10.1111/ejn.15782.

Filevich, Elisa, Martin Dresler, Timothy R. Brick, and Simone Kühn. "Metacognitive Mechanisms Underlying Lucid Dreaming." *Journal of Neuroscience* 35, no. 3 (2015): 1082–1088. https://doi.org/10/f6z4f5.

Filevich, Elisa, Simone Kühn, and Patrick Haggard. "Intentional Inhibition in Human Action: The Power of 'No'." *Neuroscience & Biobehavioral Reviews* 36, no. 4 (April 2012): 1107–1118. https://doi.org/10/fxkx9c.

Filevich, Elisa, Simone Kühn, and Patrick Haggard. "There Is No Free Won't: Antecedent Brain Activity Predicts Decisions to Inhibit." *PloS One* 8, no. 2 (2013): e53053. https://doi.org/10/f4rd54.

Fink, Sascha Benjamin. "Commentary: The Concept of a Bewusstseinskultur." *Frontiers in Psychology* 9 (July 30, 2018): 732. https://doi.org/10/gnww8z.

Fink, Sascha Benjamin. "Look Who's Talking! Varieties of Ego-Dissolution without Paradox." *Philosophy and the Mind Sciences* 1, no. I (2020): 3. https://doi.org/10.33735/phimisci.2020.I.40.

Fink, Sascha Benjamin, and Ying-Tung Lin. "Progress and Paradigms in the Search for the Neural Correlates of Consciousness: Editorial Introduction to the Special Issue 'The Neural Correlates of Consciousness'." *Philosophy and the Mind Sciences* 2 (2021). https://doi.org/10/gmg2bb.

Fink, Sascha Benjamin, Holger Lyre, and Lukas Kob. "A Structural Constraint on Neural Correlates of Consciousness." *Philosophy and the Mind Sciences* 2 (July 19, 2021). https://doi.org/10/gmg2bf.

Finnigan, Bronwyn. "Is Consciousness Reflexively Self-Aware? A Buddhist Analysis." *Ratio* 31, no. 4 (December 2018): 389–401. https://doi.org/10.1111/rati.12200.

Fleming, Stephen M. "Awareness as Inference in a Higher-Order State Space." *Neuroscience of Consciousness* 2020, no. 1 (January 1, 2020): niz020. https://doi.org/10/gn4m99.

Forman, Robert K. C. *Mysticism, Mind, Consciousness*. Albany: State University of New York Press, 1999.

Fort, Andrew O. "The Concept of Sākṣin in Advaita Vedānta." *Journal of Indian Philosophy* 12, no. 3 (1984): 277–290.

Fox, Kieran C. R., and Kalina Christoff, eds. *Oxford Handbook of Spontaneous Thought: Mind-Wandering, Creativity, and Dreaming*. Oxford Library of Psychology. New York: Oxford University Press, 2018.

Fox, Kieran C. R., R. Nathan Spreng, Melissa Ellamil, Jessica R. Andrews-Hanna, and Kalina Christoff. "The Wandering Brain: Meta-Analysis of Functional Neuroimaging Studies of Mind-Wandering and Related Spontaneous Thought Processes." *NeuroImage* 111 (May 2015): 611–621. https://doi.org/10/f676qd.

Francken, Jolien C., Lola Beerendonk, Dylan Molenaar, Johannes J. Fahrenfort, Julian D. Kiverstein, Anil K. Seth, and Simon Van Gaal. "An Academic Survey on Theoretical Foundations, Common Assumptions and the Current State of Consciousness Science." *Neuroscience of Consciousness* 2022, no.1 (August 12, 2022): niac011. https://doi.org/10.1093/nc/niac011.

Frank, Manfred, and Jan Kuneš, eds. *Selbstbewusstsein: Dieter Henrich und die Heidelberger Schule.* Abhandlungen zur Philosophie. Berlin and Heidelberg: Springer Berlin Heidelberg, 2022. https://doi.org/10.1007/978-3-662-63683-1.

Frank, Manfred, Kenneth Williford, Marc Borner, and Books on Demand GmbH. *Senses of Self: Approaches to Pre-Reflective Self-Awareness*. Norderstedt, Germany: ProtoSociology, 2020.

Friquegnon, Marie. "The Relation between Meditation and Analytic Philosophy." In *Routledge Handbook on the Philosophy of Meditation*, edited by Rick Repetti, 78–90. London: Routledge, 2022. https://doi.org/10.4324/9781003127253-6.

Friston, Karl. "Am I Self-Conscious? (Or Does Self-Organization Entail Self-Consciousness?)" *Frontiers in Psychology* 9 (2018): 579. https://doi.org/10.3389/fpsyg.2018.00579.

Friston, Karl. "Life as We Know It." *Journal of the Royal Society Interface* 10, no. 86 (September 6, 2013): 20130475. https://doi.org/10/x22.

Friston, Karl, Thomas FitzGerald, Francesco Rigoli, Philipp Schwartenbeck, John O'Doherty, and Giovanni Pezzulo. "Active Inference and Learning." *Neuroscience & Biobehavioral Reviews* 68 (September 1, 2016): 862–879. https://doi.org/10.1016/j.neubiorev.2016.06.022.

Friston, Karl, Thomas FitzGerald, Francesco Rigoli, Philipp Schwartenbeck, and Giovanni Pezzulo. "Active Inference: A Process Theory." *Neural Computation* 29, no. 1 (2017): 1–49. https://doi.org/10.1162/NECO_a_00912.

Friston, Karl, Francesco Rigoli, Dimitri Ognibene, Christoph Mathys, Thomas Fitzgerald, and Giovanni Pezzulo. "Active Inference and Epistemic Value." *Cognitive Neuroscience* 6, no. 4 (October 2, 2015): 187–214. https://doi.org/10/gf393f.

Friston, K. "The Free-Energy Principle: A Unified Brain Theory?" *Nature Reviews Neuroscience* 11 (2010): 127–138. https://doi.org/10.1038/nrn2787.

Frith, Chris D., and Thomas Metzinger. "What's the Use of Consciousness? How the Stab of Conscience Made Us Really Conscious." In *The Pragmatic Turn: Toward Action-Oriented Views in Cognitive Science*, edited by Andreas K. Engel, Karl J. Friston, and Danica Kagic, 193–214. Cambridge, MA: MIT Press, 2016. https://doi.org/10.7551/mitpress/9780262034326.003.0012.

Full, Gisela E., Harald Walach, and Mathis Trautwein. "Meditation-Induced Changes in Perception: An Interview Study with Expert Meditators (Sotapannas) in Burma." *Mindfulness* 4, no. 1 (March 2013): 55–63. https://doi.org/10/gkh34s.

Furlanetto, Tiziano, Cristina Becchio, Dana Samson, and Ian Apperly. "Altercentric Interference in Level 1 Visual Perspective Taking Reflects the Ascription of Mental States, Not Submentalizing." *Journal of Experimental Psychology: Human Perception and Performance* 42, no. 2 (February 2016): 158–163. https://doi.org/10/gmfs6m.

Furlanetto, Tiziano, Cesare Bertone, and Cristina Becchio. "The Bilocated Mind: New Perspectives on Self-Localization and Self-Identification." *Frontiers in Human Neuroscience* 7 (2013). https://doi.org/10/gmfs6p.

Gallup, Gordon G., and James R. Anderson. "Self-Recognition in Animals: Where Do We Stand 50 Years Later? Lessons from Cleaner Wrasse and Other Species." *Psychology of Consciousness: Theory, Research, and Practice* 7, no. 1 (March 2020): 46–58. https://doi.org/10/gmkc5q.

Gamma, Alex, and Thomas Metzinger. "The Minimal Phenomenal Experience Questionnaire (MPE-92M): Towards a Phenomenological Profile of 'Pure Awareness' Experiences in Meditators." Edited by Jane Elizabeth Aspell. *PLoS One* 16, no. 7 (July 14, 2021): e0253694. https://doi.org/10/gk6jvp.

Ganeri, Jonardon. *Attention, Not Self.* Oxford University Press, 2017.

Gebauer, Jochen E., Andreas D. Nehrlich, Dagmar Stahlberg, et al. "Mind-Body Practices and the Self: Yoga and Meditation Do Not Quiet the Ego but Instead Boost Self-Enhancement." *Psychological Science* 29, no. 8 (August 2018): 1299–1308. https://doi.org/10/gd398v.

Glowacki, David R., Rhoslyn Roebuck Williams, Mark D. Wonnacott, et al. "Group VR Experiences Can Produce Ego Attenuation and Connectedness Comparable to Psychedelics." *Scientific Reports* 12, no. 1 (December 2022): 8995. https://doi.org/10.1038/s41598-022-12637-z.

Goldenfeld, Nigel. *Lectures on Phase Transitions and Critical Phenomenal.* Unpublished lecture notes, 1992.

Goldenfeld, Nigel. *Lectures on Phase Transitions and the Renormalization Group.* Boca Raton, FL: CRC Press, 2018.

Gottschling, Verena. "Visual Imagery, Mental Models, and Reasoning." In *Mental Models and the Mind: Current Developments in Cognitive Psychology, Neuroscience, and Philosophy of Mind,* 211–235. Amsterdam and Boston: Elsevier, 2006.

Gowans, Christopher W. "Meditation as Cultivating Knowledge-How." In *Routledge Handbook on the Philosophy of Meditation,* edited by Rick Repetti, 161–173. London: Routledge, 2022. https://doi.org/10.4324/9781003127253-13.

Gowans, Christopher W. *Self-Cultivation Philosophies in Ancient India, Greece and China.* New York: Oxford University Press, 2021.

Graziano, Michael S. A. *The Spaces between Us: A Story of Neuroscience, Evolution, and Human Nature.* Oxford and New York: Oxford University Press, 2018.

Gupta, Bina. *The Disinterested Witness: A Fragment of Advaita Vedānta Phenomenology.* Evanston, IL: Northwestern University Press, 1998.

Guterstam, Arvid, Zakaryah Abdulkarim, and H. Henrik Ehrsson. "Illusory Ownership of an Invisible Body Reduces Autonomic and Subjective Social Anxiety Responses." *Scientific Reports* 5, no. 1 (September 2015): 9831. https://doi.org/10/34m.

Guterstam, Arvid, Giovanni Gentile, and H. Henrik Ehrsson. "The Invisible Hand Illusion: Multisensory Integration Leads to the Embodiment of a Discrete Volume of Empty Space." *Journal of Cognitive Neuroscience* 25, no. 7 (July 2013): 1078–1099. https://doi.org/10/k6q.

Gyatrul, Sangye Khandro, B. Alan Wallace, Jigme Tenpe Nyima, Lochen Dharma Shri, and Bdud-'joms 'Jigs-bral-ye-śes-rdo-rje, eds. *Meditation, Transformation, and Dream Yoga*. 2nd ed. Ithaca, NY: Snow Lion Publications, 2002.

Hadewijch. *De Visioenen*. Edited by J. van Mierlo. Leuven. Ghent, and Mechelen, Belgium: S. V. de Vlaamsche Boekenhalle, 1924.

Hadewijch, and Gerald Hofmann. *Das Buch der Visionen*. Mystik in Geschichte und Gegenwart, Bd. 12–13. Stuttgart and Bad Cannstatt, Germany: Frommann-Holzboog, 1998.

Hadewijch, and J. O. Plassmann. *Die Werke der Hadewych. Teil I: Die Briefe, mit ausgewählte Gedichten*; Teil II: Die Visionen. Aus dem Altflämischen übers. und mit ausführlichen Erläuterungen versehen von J.O. Plassmann. Munich: Folkwang-Verl, 1923.

Hadot, Pierre. *Philosophy as a Way of Life: Spiritual Exercises from Socrates to Foucault*. Translated by Arnold I. Davidson. Malden, MA: Blackwell, 1995.

Hadot, Pierre. *What Is Ancient Philosophy?* Cambridge, MA: Harvard University Press, 2002.

Haggard, Patrick. "The Neurocognitive Bases of Human Volition." *Annual Review of Psychology* 70 (2019): 9–28. https://doi.org/10.1146/annurev-psych-010418-103348.

Hanson, Jake R., and Sara I. Walker. "Formalizing Falsification for Theories of Consciousness across Computational Hierarchies." *Neuroscience of Consciousness* 2021, no. 2 (October 1, 2021). https://doi.org/10/gm746x.

Harding, Douglas E. *Face to No-Face: Rediscovering Our Original Nature*. Edited by David Lang. Carlsbad, CA: Inner Directions Publishing, 2000.

Harding, Douglas E. *On Having No Head: Zen and the Rediscovery of the Obvious*. London: Shollond Trust, 2014.

Harding, Sarah. *Niguma, Lady of Illusion*. Vol. 9. Ithaca, NY: Shambhala Publications, 2011.

Hasenkamp, Wendy. "Catching the Wandering Mind: Meditation as a Window into Spontaneous Thought." In *Oxford Handbook of Spontaneous Thought*, edited by Kieran C. R. Fox and Kalina Christoff, 539–551. Oxford: Oxford University Press, 2018.

Hasenkamp, Wendy, Christine D. Wilson-Mendenhall, Erica Duncan, and Lawrence W. Barsalou. "Mind Wandering and Attention during Focused Meditation: A Fine-Grained Temporal Analysis of Fluctuating Cognitive States." *NeuroImage* 59, no. 1 (2012): 750–760. https://doi.org/10.1016/j.neuroimage.2011.07.008.

Hesp, Casper, Maxwell Ramstead, Axel Constant, Paul Badcock, Michael Kirchhoff, and Karl Friston. "A Multi-Scale View of the Emergent Complexity of Life: A Free-Energy Proposal." In *Evolution, Development and Complexity*, edited by Georgi Yordanov Georgiev, John M. Smart, Claudio L. Flores Martinez, and Michael E. Price, 195–227. Cham, Switzerland: Springer International Publishing, 2019. https://doi.org/10.1007/978-3-030-00075-2_7.

Hesp, Casper, Ryan Smith, Thomas Parr, Micah Allen, Karl J. Friston, and Maxwell J. D. Ramstead. "Deeply Felt Affect: The Emergence of Valence in Deep Active Inference." *Neural Computation* 33, no. 2 (1 February 2021): 398–446. https://doi.org/10/gjv9zb.

Higgins, David. "Buddha in the Storehouse." *Journal of the International Association of Buddhist Studies* (2019): 169–230. https://doi.org/10/gmkwv9.

Higgins, David. *Heartfelt Advice: Yang Dgon Pa's 'Song of the Seven Direct Introductions' with Commentary by "Ba" Ra Ba Rgyal Mtshan Dpal Bzang*. Studia Philologica Buddhica, vol. XXXIX. Tokyo: International Institute for Buddhist Studies, 2022.

Higgins, David. "An Introduction to the Tibetan Dzogchen (Great Perfection) Philosophy of Mind." *Religion Compass* 6, no. 10 (October 2012): 441–450. https://doi.org/10/f3svnv.

Higgins, David. "Mi Bskyod Rdo Rje on the Question of What Remains (Lhag Ma, Avaśiṣṭa)." In *Mahāmudrā in India and Tibet*, 237–268. Leiden, Netherlands: Brill, 2020. https://doi.org/10.1163/9789004410893_010.

Higgins, David. *The Philosophical Foundations of Classical rDzogs Chen in Tibet: Investigating the Distinction between Dualistic Mind (Sems) and Primordial Knowing (Ye Shes)*. Vol. 78. Wiener Studien zur Tibetologie und Buddhismuskunde. Vienna: Arbeitskreis für Tibetische und Buddhistische Studien Universität Wien, 2013.

Higgins, David. "On the rDzogs Chen Distinction between Mind (Sems) and Primordial Knowing (Ye Shes): Clarifications and Transcendental Arguments." *Journal of Buddhist Philosophy* 2, no. 1 (2016): 13–54. https://doi.org/10/ggz229.

Higgins, David, and Martina Draszczyk. *Buddha Nature Reconsidered: The Eighth Karma Pa's Middle Path*. Vol. 1. Wiener Studien zur Tibetologie und Buddhismuskunde 95. Vienna: Arbeitskreis für Tibetische und Buddhistische Studien, 2019.

Higgins, David, and Martina Draszczyk. *Buddha Nature Reconsidered: The Eighth Karma Pa's Middle Path*. Vol. 2. Wiener Studien zur Tibetologie und Buddhismuskunde 95. Vienna: Arbeitskreis für Tibetische und Buddhistische Studien, 2019.

Higgins, David, and Martina Draszczyk. *Mahāmudrā and the Middle Way. Post-classical Kagyü Discourses on Mind, Emptiness and Buddha-Nature: Introduction, Views of Authors and Final Reflections*. Vol. 1. Vienna: Arbeitskreis für tibetische und buddhistische Studien, Universität Wien, 2016.

Higgins, David, and Martina Draszczyk. *Mahāmudrā and the Middle Way. Post-classical Kagyü Discourses on Mind, Emptiness and Buddha-Nature: Translations, Critical Texts, Bibliography and Index*. Vol. 2. Vienna: Arbeitskreis für tibetische und buddhistische Studien, Universität Wien, 2016.

Hinton, David. *Awakened Cosmos: The Mind of Classical Chinese Poetry*. Boulder: Shambhala, 2019.

Hinton, David. *Existence: A Story*. Boulder: Shambhala, 2016.

Hoffmann, Matej. "The Role of Self-Touch Experience in the Formation of the Self." December 21, 2017. http://arxiv.org/abs/1712.07843.

Hofmann, Albert. *LSD: My Problem Child*. Oxford: Oxford University Press, 2013.

Hohwy, Jakob. "Conscious Self-Evidencing." *Review of Philosophy and Psychology* (August 5, 2021). https://doi.org/10/gmg6qg.

Hohwy, Jakob. "The Self-Evidencing Brain." *Noûs* 50, no. 2 (June 2016): 259–285. https://doi.org/10/gdvhvc.

Hohwy, Jakob. "Self-Supervision, Normativity and the Free Energy Principle." *Synthese* (March 13, 2020). https://doi.org/10.1007/s11229-020-02622-2.

Hohwy, Jakob. *The Predictive Mind*. Oxford: Oxford University Press, 2013.

Holecek, Andrew. *Dream Yoga: Illuminating Your Life through Lucid Dreaming and the Tibetan Yogas of Sleep*. Boulder: Sounds True, 2016.

Holland, Owen. "Forget the Bat." *Journal of Artificial Intelligence and Consciousness* 7, no. 1 (March 1, 2020): 83–93. https://doi.org/10/gmd5wj.

Huebner, Bryce, and Genevieve Hayman. "Control, Anxiety, and the Progressive Detachment from the Self." In *Routledge Handbook on the Philosophy of Meditation*, edited by Rick Repetti, 269–282, London: Routledge, 2022. https://doi.org/10.4324/9781003127253-21.

Humphrey, Nicholas. *Soul Dust: The Magic of Consciousness*. Princeton, NJ: Princeton University Press, 2011.

Huxley, Aldous. *The Perennial Philosophy*. New York: Harper, 1945.

Hymanjr, I., and E. Loftus. "Errors in Autobiographical Memory." *Clinical Psychology Review* 18, no. 8 (December 1998): 933–947. https://doi.org/10.1016/S0272-7358(98)00041-5.

James, William. "Does 'Consciousness' Exist?" *Journal of Philosophy, Psychology and Scientific Methods* 1, no. 18 (1904): 477–491. https://doi.org/10/bzv34v.

James, William. *Psychology: The Briefer Course*. New York: H. Holt and Co, 1892. http://search.ebscohost.com/direct.asp?db=pzh&jid=%22200607509%22&scope=site.

Jetté, Fernand. "Fond de l'âme. II. Marie de l'Incarnation." In *Dictionnaire de spiritualité ascétique et mystique doctrine et histoire*, 5: 661–666. Paris: Beauchesne, 1964.

Johanson, Mirja, Katja Valli, Antti Revonsuo, and Jan-Eric Wedlund. "Content Analysis of Subjective Experiences in Partial Epileptic Seizures." *Epilepsy & Behavior* 12, no. 1 (January 1, 2008): 170–182. https://doi.org/10/b66pv8.

Jopling, David A. "Placebo Insight: The Rationality of Insight-Oriented Psychotherapy." *Journal of Clinical Psychology* 57, no. 1 (January 2001): 19–36. https://doi.org/10.1002/1097-4679(200101)57:1<19::AID-JCLP4>3.0.CO;2-Z.

Jopling, David A. *Talking Cures and Placebo Effects*. Oxford: Oxford University Press, 2008.

Josipovic, Zoran. "Implicit–Explicit Gradient of Nondual Awareness or Consciousness as Such." *Neuroscience of Consciousness* 2021, no. 2 (October 1, 2021). https://doi.org/10/gm7462.

Josipovic, Zoran. "Nondual Awareness: Consciousness-as-Such as Non-Representational Reflexivity." In *Progress in Brain Research*, 244:273–298. Elsevier, 2019. https://doi.org/10.1016/bs.pbr.2018.10.021.

Josipovic, Z., and V. Miskovic. "Nondual Awareness and Minimal Phenomenal Experience." *Frontiers in Psychology* 11 (2020). https://doi.org/10.3389/fpsyg.2020.02087.

Kachru, Sonam. "Engaging Metacognitive Practices: On the Uses (and Possible Abuse) of Meditation in Philosophy." In *Routledge Handbook on the Philosophy of Meditation*, edited by Rick Repetti, 91–104. London: Routledge, 2022. https://doi.org/10.4324/9781003127253-7.

Kam, Julia W. Y., Matthias Mittner, and Robert T. Knight. "Mind-Wandering: Mechanistic Insights from Lesion, TDCS, and IEEG." *Trends in Cognitive Sciences* (January 2022): S1364661321003132. https://doi.org/10/gpchvc.

Kant, Immanuel. *The Metaphysics of Morals*. Translated by Mary Gregor. Cambridge and New York: Cambridge University Press, 1991.

Kant, Immanuel. *Religion within the Boundaries of Mere Reason and Other Writings*. Edited by Allen W. Wood and George Di Giovanni. Translated by Robert Merrihew Adams. 2nd ed. Cambridge Texts in the History of Philosophy. Cambridge, New York, and Port Melbourne: Cambridge University Press, 2018.

Kapitan, Tomis. "The Ubiquity of Self-Awareness." *Grazer Philosophische Studien* 57 (1999): 17–43. https://doi.org/10.5840/gps1999573.

Karmapa, Wangchug Dorje. *Mahamudra—The Ocean of True Meaning*. Translated by Henrik Havlat. Norderstedt, Germany: Books on Demand, 2017.

Kavi, Prakash Chandra. "Conscious Entry into Sleep: Yoga Nidra and Accessing Subtler States of Consciousness." In *Progress in Brain Research*, 2023. https://doi.org/10.1016/bs.pbr.2022.12.012.

Kawagoe, Toshikazu, Keiichi Onoda, and Shuhei Yamaguchi. "The Neural Correlates of 'Mind Blanking': When the Mind Goes Away." *Human Brain Mapping* 40, no. 17 (December 1, 2019): 4934–4940. https://doi.org/10/gk8bgb.

Kelly, Loch. *The Way of Effortless Mindfulness*. Boulder: Sounds True, 2019.

Kennedy, R. B. "Self-Induced Depersonalization Syndrome." *American Journal of Psychiatry* 133, no. 11 (November 1976): 1326–1328. https://doi.org/10/ggwhnc.

Kent, Lachlan, and Marc Wittmann. "Time Consciousness: The Missing Link in Theories of Consciousness." *Neuroscience of Consciousness* 2021, no. 2 (October 1, 2021). https://doi.org/10/gj2f78.

Killingsworth, M. A., and D. T. Gilbert. "A Wandering Mind Is an Unhappy Mind." *Science* 330, no. 6006 (November 12, 2010): 932–932. https://doi.org/10.1126/science.1192439.

Kim, Jaegwon. "Concepts of Supervenience." *Philosophy and Phenomenological Research* 45, no. 2 (1984): 153–176. https://doi.org/10/dc9t7r.

Kim, Jaegwon. "Emergence: Core Ideas and Issues." *Synthese* 151, no. 3 (August 2006): 547–559. https://doi.org/10/bxwdcj.

Kirchhoff, Michael, Thomas Parr, Ensor Palacios, Karl Friston, and Julian Kiverstein. "The Markov Blankets of Life: Autonomy, Active Inference and the Free Energy Principle." *Journal of the Royal Society Interface* 15, no. 138 (January 2018): 20170792. https://doi.org/10/gf97dj.

Kirk, Robert. *Raw Feeling: A Philosophical Account of the Essence of Consciousness*. Oxford, UK: Clarendon Press, 1994.

Kleiner, Johannes. "Mathematical Models of Consciousness." *Entropy* 22, no. 6 (2020): 609. https://doi.org/10/ghhfgw.

Kleiner, Johannes, and Sean Tull. "The Mathematical Structure of Integrated Information Theory." *Frontiers in Applied Mathematics and Statistics* 6 (2021): 74. https://doi.org/10/gmd7fd.

Komarovski, Yaroslav. *Tibetan Buddhism and Mystical Experience*. Oxford and New York: Oxford University Press, 2015.

Konkoly, Karen R., Kristoffer Appel, Emma Chabani, et al. "Real-Time Dialogue between Experimenters and Dreamers during REM Sleep." *Current Biology* 31, no. 7 (April 12, 2021): 1417–1427. https://doi.org/10/fwck.

Kosslyn, Stephen Michael. *Image and Brain: The Resolution of the Imagery Debate*. Cambridge, MA: MIT Press, 1996.

Kozhevnikov, Maria, Alina Veronika Irene Strasser, Elizabeth McDougal, Rupali Dhond, and Geoffrey Samuel. "Beyond Mindfulness: Arousal-Driven Modulation of Attentional Control during Arousal-Based Practices." *Current Research in Neurobiology* 3 (2022): 100053. https://doi.org/10.1016/j.crneur.2022.100053.

Kriegel, U., and K. Williford, eds. *Self-representational Approaches to Consciousness*. Cambridge, MA: MIT Press, 2006.

Krishnamurti, J. *The Book of Life: Daily Meditations with Krishnamurti*. San Francisco: Harper SanFrancisco, 1995.

Krishnamurti, J. *The Collected Works of J. Krishnamurti*. Dubuque, IA: Kendall/Hunt, 1991.

Krishnamurti, J. *Commentaries on Living, from the Notebooks of J. Krishnamurti*, edited by D. Rajagopal. Quest Books, 1956.

Krishnamurti, J., and David Bohm. *The Ending of Time*. San Francisco: HarperSanFrancisco, 1997.

Kühle, Lana. "Insight: What Is It, Exactly?" In *Open MIND*, edited by Thomas K. Metzinger and Jennifer M. Windt. Frankfurt am Main, Germany: MIND Group, 2015. https://doi.org/10.15502/9783958570696.

Lamme, Victor A. F. "Towards a True Neural Stance on Consciousness." *Trends in Cognitive Sciences* 10, no. 11 (2006): 494–501. https://doi.org/10.1016/j.tics.2006.09.001.

Lamme, Victor A. F., and Pieter R. Roelfsema. "The Distinct Modes of Vision Offered by Feedforward and Recurrent Processing." *Trends in Neurosciences* 23, no. 11 (2000): 571–579. https://doi.org/10.1016/S0166-2236(00)01657-X.

Lamme, Victor A. F., Hans Super, Rogier Landman, Pieter R. Roelfsema, and Henk Spekreijse. "The Role of Primary Visual Cortex (V1) in Visual Awareness." *Vision Research* 40, no. 10–12 (2000): 1507–1521. https://doi.org/10.1016/S0042-6989(99)00243-6.

Lange, Marc. "On 'Minimal Model Explanations': A Reply to Batterman and Rice." *Philosophy of Science* 82, no. 2 (2015): 292–305. https://doi.org/10/gcjqbq.

Laukkonen, Ruben E., and Heleen A. Slagter. "From Many to (N)one: Meditation and the Plasticity of the Predictive Mind." *Neuroscience & Biobehavioral Reviews* (June 2021) S014976342100261X. https://doi.org/10/gkptxv.

Laukkonen, R. E., M. D. Sacchet, H. Barendregt, K. J. Devaney, A. Chowdhury, and H. A. Slagter. "Cessations of Consciousness in Meditation: Advancing a Scientific Understanding of Nirodha Samāpatti." In *Progress in Brain Research,* S0079612322001984. Elsevier, 2023. https://doi.org/10.1016/bs.pbr.2022.12.007.

Laureys, Steven, Gastone G. Celesia, Francois Cohadon, et al. "Unresponsive Wakefulness Syndrome: A New Name for the Vegetative State or Apallic Syndrome." *BMC Medicine* 8 (2010): 68. https://doi.org/10.1186/1741-7015-8-68.

Laureys, Steven, Adrian M. Owen, and Nicholas D. Schiff. "Brain Function in Coma, Vegetative State, and Related Disorders." *The Lancet Neurology* 3, no. 9 (2004): 537–546. https://doi.org/10.1016/S1474-4422(04)00852-X.

Lenggenhager, Bigna., Tej Tadi, Thomas Metzinger, and Olaf Blanke. "Video Ergo Sum: Manipulating Bodily Self-Consciousness." *Science* 317, no. 5841 (24 August 2007): 1096–1099. https://doi.org/10/dn7gx4.

Lepauvre, Alex, and Lucia Melloni. "The Search for the NCC: Progress and Challenges." *Philosophy and the Mind Sciences* 2 (19 July 2021). https://doi.org/10/gmg2bc.

Letheby, Chris. "The Epistemic Innocence of Psychedelic States." *Consciousness and Cognition* 39 (January 2016): 28–37. https://doi.org/10/f76m6b.

Letheby, Chris. *Philosophy of Psychedelics*. International Perspectives in Philosophy and Psychiatry. Oxford and New York: Oxford University Press, 2021.

Letheby, Chris. "Psychedelics and Meditation." In *Routledge Handbook on the Philosophy of Meditation*, edited by Rick Repetti, 209–224. London: Routledge, 2022. https://doi.org/10.4324/9781003127253-16.

Letheby, Chris, and Philip Gerrans, eds. *Philosophical Perspectives on the Psychedelic Renaissance.* Oxford: Oxford University Press, 2024.

Letheby, Chris, and Philip Gerrans. "Self Unbound: Ego Dissolution in Psychedelic Experience." *Neuroscience of Consciousness* 2017, no. 1 (January 1, 2017). https://doi.org/10/d86g.

Levine, Joseph. "Raw Feeling." *The Philosophical Review* 105, no. 1 (1996): 94–97. https://doi.org/10/fnvxp5.

Lewis, Clarence Irving. *Mind and the World-Order*. Charles Scribners Sons, 1929. http://archive.org/details/mindtheworldorde007547mbp.

Limanowski, Jakub. "Precision Control for a Flexible Body Representation." *Neuroscience & Biobehavioral Reviews* (November 2021) S0149763421004656. https://doi.org/10/gnpj65.

Limanowski, Jakub, and Karl Friston. "'Seeing the Dark': Grounding Phenomenal Transparency and Opacity in Precision Estimation for Active Inference." *Frontiers in Psychology* 9 (2018): 643. https://doi.org/10/gdjrq8.

Lindahl, Jared R., and Willoughby B. Britton. "'I Have This Feeling of Not Really Being Here': Buddhist Meditation and Changes in Sense of Self." *Journal of Consciousness Studies* 26, no. 7–8 (2019): 157–183.

Lindström, Lena, Philippe Goldin, Johan Måretnsson, and Etzel Cardeña. "Nonlinear Brain Correlates of Trait Self-boundarylessness." *Neuroscience of Consciousness* 2023, no. 1 (April 25, 2023): niad006. https://doi.org/10.1093/NC/niad006.

Lipps, Theodor. "Aesthetische Einfühlung." *Zeitschrift für Psychologie und Physiologie der Sinnesorgane*, no. 22 (1900): 415–450.

Lipps, Theodor. "Einfühlung, innere Nachahmung und Organempfindung." *Archiv für die gesamte Psychologie* (1903): 185–204.

Lockwood, Michael. "The Grain Problem." In *Objections to Physicalism*, edited by Howard Robinson, 271–291. Oxford: Oxford University Press, 1993.

Lopez, Greg. "Sati & Prosoche: Buddhist vs. Stoic 'Mindfulness.'" *Modern Stoicism* (blog), 6 May 2017. https://modernstoicism.com/sati-prosoche-buddhist-vs-stoic-mindfulness-compared-by-greg-lopez/.

Luders, Eileen, and Florian Kurth. "The Neuroanatomy of Long-Term Meditators." *Current Opinion in Psychology* 28 (August 2019): 172–178. https://doi.org/10.1016/j.copsyc.2018.12.013.

Lutz, Antoine, Amishi P. Jha, John D. Dunne, and Clifford D. Saron. "Investigating the Phenomenological Matrix of Mindfulness-Related Practices from a Neurocognitive Perspective." *American Psychologist* 70, no. 7 (October 2015): 632–658. https://doi.org/10.1037/a0039585.

Lynch, Julianna M., and Allison S. Troy. "The Role of Nonduality in the Relationship Between Flow States and Well-Being." *Mindfulness* 12, no. 7 (July 2021): 1639–1652. https://doi.org/10/gmxg99.

MacKenzie, Matthew. "Meditative Experience and the Plasticity of Self-Experience." In *Routledge Handbook on the Philosophy of Meditation*, edited by Rick Repetti, 241–255. London: Routledge, 2022. https://doi.org/10.4324/9781003127253-19.

Madary, Michael, and Thomas K. Metzinger. "Real Virtuality: A Code of Ethical Conduct. Recommendations for Good Scientific Practice and the Consumers of VR-Technology." *Frontiers in Robotics and AI* 3 (2016): 3. https://doi.org/10/gc6znw.

Madhurananda, Swami. *Effortless Meditation: Stumbling upon Peace and Happiness*. 3rd ed. Mumbai, India: Yogi Impressions Books, 2010.

Maharaj, Nisargadatta. *I Am That*. Bombay: Chetana, 1973. https://dds.crl.edu/crldelivery/10479.

Maleeh, Reza, and Shaghayegh Konjedi. "Meta-Awareness, Mind Wandering and Negative Mood in the Context of the Continuity Hypothesis of Dreaming." *Phenomenology and the Cognitive Sciences* (July 2, 2022). https://doi.org/10.1007/s11097-022-09835-5.

Mañjuśrīmitra. *Primordial Experience: An Introduction to Rdzogs-Chen Meditation*. Translated by Namkhai Norbu, K. Lipman, and B. Simmons. Boston: Shambhala, 1987.

Mashour, George A., Pieter Roelfsema, Jean-Pierre Changeux, and Stanislas Dehaene. "Conscious Processing and the Global Neuronal Workspace Hypothesis." *Neuron* 105, no. 5 (2020): 776–798. https://doi.org/10.1016/j.neuron.2020.01.026.

Masson, Catherine, Donné van der Westhuizen, Jean-Paul Noel, et al. "Testosterone Administration in Women Increases the Size of Their Peripersonal Space." *Experimental Brain Research* 239, no. 5 (May 2021): 1639–1649. https://doi.org/10/gkh34h.

Matthen, Mohan. "The Dual Structure of Touch: The Body versus Peripersonal Space." In *The World at Our Fingertips*, by Mohan Matthen, 197–214. Oxford: Oxford University Press, 2021. https://doi.org/10.1093/oso/9780198851738.003.0011.

McGinn, Bernard. *The Mystical Thought of Meister Eckhart: The Man from Whom God Hid Nothing*. Edward Cadbury Lectures 2000–2001. New York: Crossroad Publishing, 2001.

McKenna, Travis. "Lange on Minimal Model Explanations: A Defense of Batterman and Rice." *Philosophy of Science* 88, no. 4 (2021): 731–741. https://doi.org/10/gnpfch.

Mechthild von Magdeburg. *Mechthild von Magdeburg, 'Das fliessende Licht der Gottheit'. Bd. 1: Text*. Edited by Hans Neumann. Münchener Texte und Untersuchungen zur deutschen Literatur des Mittelalters, Bd. 100. Munich and Zurich: Artemis-Verl, 1990.

Melloni, L., C. Molina, M. Pena, D. Torres, W. Singer, and E. Rodriguez. "Synchronization of Neural Activity accross Cortical Areas Correlates with Conscious Perception." *Journal of Neuroscience* 27, no. 11 (March 14, 2007): 2858–2865. https://doi.org/10/dhzdsx.

Melloni, Lucia. "Consciousness as Inference in Time." In *Open MIND*, edited by Thomas Metzinger and Jennifer M. Windt. Open MIND. Frankfurt am Main, Germany: MIND Group, 2015. https://doi.org/10.15502/9783958570566.

Metzinger, Thomas. "Anthropologiefolgenabschätzung, normative Psychologie und Bewusstseinskultur." In *Zukunftsentwürfe: Ideen für eine Kultur der Veränderung*, edited by Jörn Rüsen. Frankfurt am Main, Germany, and New York: Campus-Verlag, 1999.

Metzinger, Thomas. "Are You Sleepwalking Now? What We Know about Mind-Wandering." *Aeon* (blog), April 30, 2018. https://aeon.co/essays/are-you-sleepwalking-now-what-we-know-about-mind-wandering.

Metzinger, Thomas. "Artificial Suffering: An Argument for a Global Moratorium on Synthetic Phenomenology." *Journal of Artificial Intelligence and Consciousness* 8, no. 1 (March 1, 2021): 43–66. https://doi.org/10/gmd4td.

Metzinger, Thomas. "Der Begriff einer Bewusstseinskultur." In *Jahrbuch 2002/2003 des Wissenschaftszentrums Nordrhein-Westfalen*, edited by G. Kaiser, 150–171. Düsseldorf: Wissenschaftszentrum Nordrhein-Westfalen, 2003.

Metzinger, Thomas. *Being No One: The Self-Model Theory of Subjectivity*. Cambridge, MA: MIT Press, 2003.

Metzinger, Thomas. *Being No One: The Self-Model Theory of Subjectivity*. A Bradford Book. Paperback ed. Cambridge, MA: MIT Press, 2004.

Metzinger, Thomas. "Benevolent Artificial Anti-Natalism (BAAN)." *Edge.Org* (blog), July 8, 2017. https://www.edge.org/conversation/thomas_metzinger-benevolent-artificial-anti-natalism-baan.

Metzinger, Thomas. "Bewusstseinsforschung und Bewusstseinskultur." In *Jahrbuch Des Europäischen Kollegiums Für Bewusstseinsstudien*, 7–29. Berlin: Verlag für Wissenschaft und Bildung, 1994.

Metzinger, Thomas. *Bewusstseinskultur—Spiritualität, intellektuelle Redlichkeit und die planetare Krise*. Berlin: Berlin Verlag, 2023.

Metzinger, Thomas (in preparation). *Bewusstseinskultur—Spirituality, Intellectual Honesty, and the Planetary Crisis*.

Metzinger, Thomas, ed. *Conscious Experience*. Paderborn, Germany: Schöningh/Imprint Academic, 1995.

Metzinger, Thomas. "Conscious Volition and Mental Representation: Toward a More Fine-Grained Analysis." In *Disorders of Volition*, edited by Natalie Sebanz and Wolfgang Prinz, 19–48. Cambridge, MA: MIT Press, 2006.

Metzinger, Thomas. "Don't Look at Me! Test It Out!" Edited by Viju Jaithirtha, D Anantha Jyothi, Alok Mathur, et al. *Journal of the Krishnamurti Schools*, Special Issue: 125 Years of Krishnamurti, 25 (2020): 108–116.

Metzinger, Thomas. *The Ego Tunnel: The Science of the Mind and the Myth of the Self*. New York: Basic Books, 2009.

Metzinger, Thomas. "Empirical Perspectives from the Self-Model Theory of Subjectivity: A Brief Summary with Examples." *Progress in Brain Research* 168 (2008): 215–245. https://doi.org/10/c8tkfs.

Metzinger, Thomas. "The Enculturation Problem." In *Philosophical Perspectives on the Psychedelic Renaissance*, edited by Chris Letheby and Philip Gerrans. Oxford: Oxford University Press, 2024.

Metzinger, Thomas. "Faster than Thought." In *Conscious Experience*, edited by Thomas Metzinger,. 425–460. Thorverton, UK: Imprint Academic, 1995.

Metzinger, Thomas. "Hirnforschung, Neurotechnologie, Bewusstseinskultur. Medizinische, ethische und sozialphilosophische Fragen der Zukunft." In *Die Zukunft der Medizin: Neue Wege zur Gesundheit?* edited by Gert Kaiser, Katharina Wetzel-Vandai, Johannes Siegrist, and Eva Rosenfeld, 301–312. Schriftenreihe des Wissenschaftszentrums Nordrhein-Westfalen, Bd. 4. Frankfurt am Main, Germany, and New York: Campus-Verlag, 1996.

Metzinger, Thomas. "Von der Hirnforschung zur Bewusstseinskultur." *Neue Zürcher Zeitung*, March 18, 1998, 64, 79.

Metzinger, Thomas., ed. *Intentionalität und mentale Repräsentation*. Zweite, Durchgesehene und erweiterte Auflage. Grundkurs Philosophie des Geistes, Band 3. Paderborn, Germany: Mentis, 2019.

Metzinger, Thomas. "Introduction: Consciousness Research at the End of the Twentieth Century." In *Neural Correlates of Consciousness: Empirical and Conceptual Questions*, edited by T. Metzinger, 1–12. Cambridge, MA: MIT Press, 2000.

Metzinger, Thomas., ed. *Das Leib-Seele-Problem*. Zweite, Durchgesehene und erweiterte Auflage. Grundkurs Philosophie des Geistes, Band 2. Paderborn, Germany: Mentis, 2013.

Metzinger, Thomas. "M-Autonomy." *Journal of Consciousness Studies* 22, no. 11–12 (January 1, 2015): 270–302.

Metzinger, Thomas. "Minimal Phenomenal Experience: The ARAS-Model Theory: Steps toward a Minimal Model of Conscious Experience as Such." *MindRxiv*, 2019. https://doi.org/10.31231/osf.io/5wyg7.

Metzinger, Thomas. "Minimal Phenomenal Experience: Meditation, Tonic Alertness, and the Phenomenology of 'Pure' Consciousness." *Philosophy and the Mind Sciences* 1, no. I (March 24, 2020): 1–44. https://doi.org/10/gmgk5v.

Metzinger, Thomas. "The Myth of Cognitive Agency: Subpersonal Thinking as a Cyclically Recurring Loss of Mental Autonomy." *Frontiers in Psychology* 4 (2013). https://doi.org/10/gbfpsz.

Metzinger, Thomas. "The No-Self Alternative." In *Oxford Handbook of the Self*, edited by Shaun Gallagher, 279–296. Oxford: Oxford University Press, 2010.

Metzinger, Thomas. *Neuere Beiträge zur Diskussion des Leib-Seele-Problems*. Europäische Hochschulschriften, Series XX, Philosophy, Vol. 180. Frankfurt am Main, Germany, and New York: Peter Lang, 1985.

Metzinger, Thomas, ed. *Phänomenales Bewusstsein*. Zweite, Durchgesehene und erweiterte Auflage. Grundkurs Philosophie des Geistes, Bd. 1. Paderborn, Germany: Mentis, 2009.

Metzinger, Thomas. "Précis of Being No-One." *Psyche: An Interdisciplinary Journal of Research on Consciousness* 11 (2005): 1–35. https://philarchive.org/rec/METP_N.

Metzinger, Thomas. "The Problem of Consciousness." In *Conscious Experience*, edited by Thomas Metzinger, 3–37. Paderborn, Germany: Schöningh/Imprint Academic, 1995.

Metzinger, Thomas. "Self-Modeling Epistemic Spaces and the Contraction Principle." *Cognitive Neuropsychology* 37, no. 3–4 (May 18, 2020): 197–201. https://doi.org/10.1080/02643294.2020.1729110.

Metzinger, Thomas. *Spirituality and Intellectual Honesty*. Mainz, Germany: Self-Published, 2017. http://www.blogs.uni-mainz.de/fb05philosophieengl/files/2013/07/Metzinger_SIR_2017_English.pdf ISBN 978-3-00-041539-5

Metzinger, Thomas. *Subjekt und Selbstmodell: Die Perspektivität phänomenalen Bewusstseins vor dem Hintergrund einer naturalistischen Theorie mentaler Repräsentation*. Paderborn, Germany, Munich, Vienna, and Zurich: Schöningh, 1993.

Metzinger, Thomas. "The *Subjectivity* of Subjective Experience—A Representationalist Analysis of the First-Person Perspective." In *Neural Correlates of Consciousness: Empirical and Conceptual Questions*, edited by Thomas Metzinger, 285–306. Cambridge, MA: MIT Press, 2000.

Metzinger, Thomas. "Auf der Suche Nach einem Neuen Bild des Menschen. Die Zukunft des Subjekts und die Rolle der Geisteswissenschaften." *Spiegel der Forschung* 17, no. 1 (2000): 58–67.

Metzinger, Thomas. "Suffering." In *The Return of Consciousness: A New Science on Old Questions*, edited by K. Almqvist and A. Haag, 221–248. Stockholm, 2016.

Metzinger, Thomas. "Trip des Geistes." *ZEIT ONLINE*, January 18, 2006. https://www.zeit.de/online/2006/03/lsd_basel.

Metzinger, Thomas. "Unterwegs zu einem neuen Menschenbild." *Gehirn & Geist*, no. 11 (2005): 50–54.

Metzinger, Thomas. "Der Weltraum der Seele—Ein Gespräch mit Albert Hofmann." *TAZ*, June 27, 1989. https://taz.de/Der-Weltraum-der-Seele/!1807810/.

Metzinger, Thomas. "Wer, Ich? SPIEGEL-Gespräch mit Barbara Supp und Dietmar Piper." *Der Spiegel*, May 6, 2016, 19, 68–71.

Metzinger, Thomas. "Why Are Dreams Interesting for Philosophers? The Example of Minimal Phenomenal Selfhood, Plus an Agenda for Future Research." *Frontiers in Psychology* 4 (2013): 746. https://doi.org/10/ghh8gv.

Metzinger, Thomas. "Why Are Identity Disorders Interesting for Philosophers?" In *Philosophy and Psychiatry*, edited by T. Schramme and J. Thome, 311–325. Berlin: De Gruyter, 2003.

Metzinger, Thomas. "Why Are Out-of-Body Experiences Interesting for Philosophers?" *Cortex* 45, no. 2 (February 2009): 256–258. https://doi.org/10/crrq3x.

Metzinger, Thomas. "Why Is Mind Wandering Interesting for Philosophers?" In *Oxford Handbook of Spontaneous Thought: Mind-Wandering, Creativity, and Dreaming*, edited by Kieran C. R. Fox and Kalina Christoff, 97–111. New York: Oxford University Press, 2018. https://doi.org/10.1093/oxfordhb/9780190464745.013.32.

Metzinger, Thomas. "Why Is Virtual Reality Interesting for Philosophers?" *Frontiers in Robotics and AI* 5 (2018). https://doi.org/10/ghd5xd.

Metzinger, Thomas., and Vittorio Gallese. "The Emergence of a Shared Action Ontology: Building Blocks for a Theory." *Consciousness and Cognition* 12, no. 4 (December 2003): 549–571. https://doi.org/10/d4rkzg.

Metzinger, Thomas., and Jennifer M. Windt. "What Does It Mean to Have an Open MIND?" In *Open MIND*, edited by Thomas K. Metzinger and Jennifer M. Windt. Frankfurt am Main,

Germany: MIND Group, 2015. https://doi.org/10.15502/9783958571044; https://web.archive.org/web/20201127143007/https://open-mind.net/papers/general-introduction-what-does-it-mean-to-have-an-open-mind.

Meyniel, Florent, Daniel Schlunegger, and Stanislas Dehaene. "The Sense of Confidence during Probabilistic Learning: A Normative Account." Edited by Jill X O'Reilly. *PLoS Computational Biology* 11, no. 6 (June 15, 2015): e1004305. https://doi.org/10.1371/journal.pcbi.1004305.

Meyniel, Florent, Mariano Sigman, and Zachary F. Mainen. "Confidence as Bayesian Probability: From Neural Origins to Behavior." *Neuron* 88, no. 1 (October 2015): 78–92. https://doi.org/10.1016/j.neuron.2015.09.039.

Millière, Raphael. "Looking for the Self: Phenomenology, Neurophysiology and Philosophical Significance of Drug-Induced Ego Dissolution." *Frontiers in Human Neuroscience* 11 (May 23, 2017). https://doi.org/10/gd5z8t.

Millière, R. "The Varieties of Selflessness." *Philosophy and the Mind Sciences* 1, no. 1 (2020), 8. https://doi.org/10.33735/phimisci.2020.I.48.

Millière, Raphael, Robin L. Carhart-Harris, Leor Roseman, Fynn-Mathis Trautwein, and Aviva Berkovich-Ohana. "Psychedelics, Meditation, and Self-Consciousness." *Frontiers in Psychology* 9 (2018): 1475. https://doi.org/10/gd56xz.

Millière, Raphael, and Thomas Metzinger. "Radical Disruptions of Self-Consciousness: Editorial Introduction." *Philosophy and the Mind Sciences* 1, no. 1 (March 24, 2020): 1–13. https://doi.org/10/ggwf7n.

Millière, Raphaël, and Albert Newen. "Selfless Memories." *Erkenntnis* (2022). https://doi.org/10.1007/s10670-022-00562-6.

Moore, G. E. "The Refutation of Idealism." *Mind* 12, no. 48 (1903): 433–453.

Morard, Meinrad Stéphane. "Ist, Istic, Istikeit bei Meister Eckhart." 1956. https://doi.org/10.5169/SEALS-761441.

Moser, Julia, Franziska Schleger, Magdalene Weiss, Katrin Sippel, Lorenzo Semeia, and Hubert Preissl. "Magnetoencephalographic Signatures of Conscious Processing before Birth." *Developmental Cognitive Neuroscience* 49 (June 2021): 100964. https://doi.org/10.1016/j.dcn.2021.100964.

Munn, Brandon R., Eli J. Müller, Gabriel Wainstein, and James M. Shine. "The Ascending Arousal System Shapes Neural Dynamics to Mediate Awareness of Cognitive States." *Nature Communications* 12, no. 1 (2021): 1–9. https://doi.org/10/gnpfcj.

Nagel, Thomas. "The Boundaries of Inner Space." *Journal of Philosophy* 66, no. 14 (1969): 452–458. https://doi.org/10/cqvk6m.

Nagel, Thomas. *Other Minds: Critical Essays, 1969–1994*. New York: Oxford University Press, 1995.

Nagel, Thomas. *The View from Nowhere*. New York; Oxford: Oxford University Press, 1986.

Nagel, Thomas. "What Is It Like to Be a Bat?" *The Philosophical Review* 83, no. 4 (1974): 435–450. https://doi.org/10.2307/2183914.

Namgyal, Dakpo Tashi. *Clarifying the Natural State: A Principal Guidance Manual for Mahamudra.* Hong Kong: Rangjung Yeshe Publications, 2001.

Namgyal, Dakpo Tashi. *Mahamudra—The Moonlight: Quintessence of Mind and Meditation.* Somerville, MA: Wisdom Publications, 2006.

Namgyal, Dakpo Tashi. *Moonbeams of Mahamudra.* Boulder, CO: Snow Lion, 2019.

Nave, Kathryn, George Deane, Mark Miller, and Andy Clark. "Expecting Some Action: Predictive Processing and the Construction of Conscious Experience." *Review of Philosophy and Psychology* (June 10, 2022). https://doi.org/10.1007/s13164-022-00644-y.

Nave, Ohad, Fynn-Mathis Trautwein, Yochai Ataria, et al. "Self-Boundary Dissolution in Meditation: A Phenomenological Investigation." *Brain Sciences* 11, no. 6 (June 21, 2021): 819. https://doi .org/10/gk3m9z.

Nguyen, Phuong D. H., Yasmin Kim Georgie, Ezgi Kayhan, Manfred Eppe, Verena Vanessa Hafner, and Stefan Wermter. "Sensorimotor Representation Learning for an 'Active Self' in Robots: A Model Survey." *KI—Künstliche Intelligenz* 35, no. 1 (March 2021): 9–35. https://doi.org/10/gm9xzk.

Noel, Jean-Paul, Tommaso Bertoni, and Andrea Serino. "Peri-Personal Space as an Interface for Self-Environment Interaction: A Critical Evaluation and Look Ahead." In *The World at Our Fingertips*, 17–46. Oxford: Oxford University Press, 2021. https://doi.org/10.1093/oso/9780198851738 .003.0002.

Noel, Jean-Paul, Hyeong-Dong Park, Isabella Pasqualini, et al. "Audio-Visual Sensory Deprivation Degrades Visuo-Tactile Peri-Personal Space." *Consciousness and Cognition* 61 (May 2018): 61–75. https://doi.org/10/gdjhkh.

Noel, Jean-Paul, Majed Samad, Andrew Doxon, Justin Clark, Sean Keller, and Massimiliano Di Luca. "Peri-Personal Space as a Prior in Coupling Visual and Proprioceptive Signals." *Scientific Reports* 8, no. 1 (December 2018): 15819. https://doi.org/10/gfmpxb.

Nour, Matthew M., and Robin L. Carhart-Harris. "Psychedelics and the Science of Self-Experience." *British Journal of Psychiatry* 210, no. 3 (March 2017): 177–179. https://doi.org/10/gfv263.

Nozick, Robert. *Philosophical Explanations.* Cambridge, MA: Harvard University Press, 1983.

Overgaard, Morten S., and Asger Kirkeby-Hinrup. "Finding the NCCs Will Not Solve All Our Problems." *Philosophy and the Mind Sciences* 2 (19 July 2021). https://doi.org/10/gmg2bd.

Padmasambhava, and Karma Lingpa. *The Tibetan Book of the Dead: Awakening upon Dying.* Berkeley, CA: North Atlantic Books, 2013.

Papineau, David. "Naturalism." In *Stanford Encyclopedia of Philosophy*, edited by Edward N. Zalta, Summer 2021. Metaphysics Research Lab, Stanford University. https://plato.stanford.edu /archives/sum2021/entries/naturalism/.

Parr, Thomas, Andrew W. Corcoran, Karl J. Friston, and Jakob Hohwy. "Perceptual Awareness and Active Inference." *Neuroscience of Consciousness* 2019, no. 1 (January 1, 2019): niz012. https://doi.org/10.1093/nc/niz012.

Parr, Thomas, and Karl J. Friston. "Uncertainty, Epistemics and Active Inference." *Journal of the Royal Society Interface* 14, no. 136 (2017): 20170376. https://doi.org/10.1098/rsif.2017.0376.

Parr, Thomas, Giovanni Pezzulo, and K. J. Friston. *Active Inference: The Free Energy Principle in Mind, Brain, and Behavior*. Cambridge, MA: MIT Press, 2022.

Paul, L. A. *Transformative Experience*. Oxford: Oxford University Press, 2014.

Paul, L. A., and Sascha-Michael Benjamin Fink. *Was Können Wir Wissen, bevor Wir uns Entscheiden? Von Kinderwünschen und Vernunftgründen*. Translated by Jürgen Schröder. Reclams Universal-Bibliothek Was bedeutet das alles? Nr. 19654. Ditzingen, Germany: Reclam, 2020.

Persinger, Allan. "Foxfire: The Selected Poems of Yosa Buson: A Translation." 2013. https://dc.uwm.edu/etd/748.

Petitmengin, Claire. "On the Veiling and Unveiling of Experience: A Comparison Between the Micro-Phenomenological Method and the Practice of Meditation." *Journal of Phenomenological Psychology* 52, no. 1 (August 12, 2021): 36–77. https://doi.org/10.1163/15691624-12341383.

Petitmengin, Claire, Anne Remillieux, and Camila Valenzuela-Moguillansky. "Discovering the Structures of Lived Experience." *Phenomenology and the Cognitive Sciences* 18, no. 4 (2019): 691–730. https://doi.org/10.1007/s11097-018-9597-4.

Petitmengin, Claire, Martijn van Beek, Michel Bitbol, Jean-Michel Nissou, and Andreas Roepstorff. "Studying the Experience of Meditation through Micro-Phenomenology." *Current Opinion in Psychology* 28 (2019): 54–59. https://doi.org/10.1016/j.copsyc.2018.10.009.

Petitmengin, Claire, Martijn van Beek, Michel Bitbol, Jean-Michel Nissou, and Andreas Roepstorff. "What Is It Like to Meditate? Methods and Issues for a Microphenomenological Description of Meditative Experience." *Journal of Consciousness Studies* 24, no. 5–6 (2017): 170–198.

Pezzulo, Giovanni, Thomas Parr, and Karl Friston. "The Evolution of Brain Architectures for Predictive Coding and Active Inference." *Philosophical Transactions of the Royal Society B* 377, no. 1844 (2022): 20200531. https://doi.org/10/gnz5sb.

Pfeiffer, Franz, ed. *Deutsche Mystiker des vierzehnten Jahrhunderts. Bd. 1*. Neudr. d. Ausg. Leipzig 1845. Aalen, Germany: Scientia Verlag, 1962.

Picard, F., and A. D. Craig. "Ecstatic Epileptic Seizures: A Potential Window on the Neural Basis for Human Self-Awareness." *Epilepsy & Behavior* 16, no. 3 (November 2009): 539–546. https://doi.org/10/dttb44.

Picard, Fabienne, Didier Scavarda, and Fabrice Bartolomei. "Induction of a Sense of Bliss by Electrical Stimulation of the Anterior Insula." *Cortex; a Journal Devoted to the Study of the Nervous System and Behavior* 49, no. 10 (December 2013): 2935–2937. https://doi.org/10/ggxgfv.

Pigliucci, Massimo. "Prosochê as Stoic Mindfulness." In *Routledge Handbook on the Philosophy of Meditation*, edited by Rick Repetti, 371–382. London: Routledge, 2022. https://doi.org/10.4324/9781003127253-31.

Place, U. T. "Is Consciousness a Brain Process?" *British Journal of Psychology* 47 (1956): 44–50. https://doi.org/10/dvh5b5.

Pliushch, Iuliia, and Thomas Metzinger. "Self-Deception and the Dolphin Model of Cognition." In *Disturbed Consciousness*, edited by Rocco J. Gennaro, 167–208. Cambridge, MA: MIT Press, 2015. https://doi.org/10.7551/mitpress/9780262029346.003.0007.

Plotinus. *Plotinus: The Enneads*. Edited by Lloyd P. Gerson. New York: Cambridge University Press, 2018.

Plotinus. *Schriften in deutscher Übersetzung. Teilband 1: Schriften 1–38*. Translated by Richard Harder. Philosophische Bibliothek, Bd. 743a. Hamburg: Felix Meiner Verlag, 2020.

Plotinus. *Schriften in deutscher Übersetzung. Teilband 2: Schriften 39–54*. Translated by Richard Harder. Philosophische Bibliothek, Bd. 743b. Hamburg: Felix Meiner Verlag, 2020.

Popper, Karl R., and John C. Eccles. *The Self and Its Brain*. Berlin and Heidelberg: Springer Berlin Heidelberg, 1977.

Posner, Michael I. "Measuring Alertness." *Annals of the New York Academy of Sciences* 1129, no. 1 (May 2008): 193–199. https://doi.org/10.1196/annals.1417.011.

Putnam, Hilary. "Philosophy and Our Mental Life." In *Mind, Language, and Reality*, 291–303. Cambridge: Cambridge University Press, 1975.

Putnam, Hilary. "Psychological Predicates." In *Art, Mind, and Religion*, edited by W. H. Capitan and D. D. Merrill, 37–48. University of Pittsburgh Press, 1967.

Putnam, Hilary. "Why Functionalism Didn't Work." In *Inference, Explanation, and Other Frustrations*, 255–270. Berkeley, CA: University of California Press, 1992. https://doi.org/10.1525/9780520309876-012.

Pylyshyn, Zenon W. "Mental Imagery: In Search of a Theory." *Behavioral and Brain Sciences* 25, no. 2 (2002): 157–182. https://doi.org/10/dsk54w.

Pyszczynski, Tom, Sheldon Solomon, and Jeff Greenberg. "Thirty Years of Terror Management Theory: From Genesis to Revelation." *Advances in Experimental Social Psychology* 52 (2015): 1–70. https://doi.org/10/gmvjcd.

Rabjam, Longchen. "The Precious Treasury of Phenomenal Space." In *Great Perfection*, 31–164. Kathmandu: Vajra Books, 2014.

Ramm, Brentyn J. "Self-Experience." *Journal of Consciousness Studies* 24, no. 11–12 (2017): 142–166.

Ramm, Brentyn J. "First-Person Experiments: A Characterisation and Defence." *Review of Philosophy and Psychology* 9, no. 3 (September 1, 2018): 449–67. https://doi.org/10.1007/s13164-018-0388-1.

Ramm, Brentyn J. "Pure Awareness Experience." *Inquiry* 16, no. 1 (2019): 1–23. https://doi.org/10.1080/0020174X.2019.1592704.

Ramm, Brentyn J. "The Technology of Awakening: Experiments in Zen Phenomenology." *Religions* 12, no. 3 (March 13, 2021): 192. https://doi.org/10/gk5fdz.

Ramstead, Maxwell James, Anil Seth, Casper Hesp, et al. "From Generative Models to Generative Passages: A Computational Approach to (Neuro)phenomenology." Preprint. *PsyArXiv*, February 23, 2021. https://doi.org/10.31234/osf.io/k9pbn.

"Rang Rig Ye Shes." In *Rangjung Yeshe Wiki—Dharma Dictionary*, 2021. https://rywiki.tsadra.org/index.php?title=rang_rig_ye_shes&oldid=763086.

Ray, Reginald A. *Secret of the Vajra World: The Tantric Buddhism of Tibet*. Boston: Shambhala, 2001.

Renger, Almut-Barbara, ed. *Erleuchtung: Kultur- Und Religionsgeschichte Eines Begriffs*. Freiburg, Germany: Herder, 2016.

Repetti, Rick. "The Philosophy of Meditation." In *Routledge Handbook on the Philosophy of Meditation*, edited by Rick Repetti, 46–69. London: Routledge, 2022. https://doi.org/10.4324/9781003127253-4.

Repetti, Rick, ed. *Routledge Handbook on the Philosophy of Meditation*. Abingdon, UK, and New York: Routledge, 2022.

Revonsuo, Antti. *Inner Presence: Consciousness as a Biological Phenomenon*. Cambridge, MA, and London: MIT Press, 2009.

Rgyal-ba-g'yung-drung, Bru-sgom. *The Pith Instructions for the Stages of the Practice Sessions of the a Khrid System of Bon Rdzogs Chen (Great Completion) Meditation*. Translated by Geshe Sonam Gurung, and Daniel P. Brown. 2nd ed. Occidental, CA: Bright Alliance, 2017.

Rice, Collin, Yasha Rohwer, and André Ariew. "Explanatory Schema and the Process of Model Building." *Synthese* 196, no. 11 (2019): 4735–4757. https://doi.org/10/gn29wv.

Riedlinger, Thomas J. (Ed.). *The Sacred Mushroom Seeker: Essays for R. Gordon Wasson*. Ethnomycological Studies No. 11; Historical, Ethno-, and Economic Botany Series, Vol. 4, 1990. Portland, OR: Discorides Press.

Rilke, Rainer Maria. *Briefe aus Muzot 1921 bis 1926*. Edited by Ruth Sieber-Rilke and Carl Sieber. Leipzig: Insel-Verlag, 1935.

Rilke, Rainer Maria. *Letters of Rainer Maria Rilke 1910–1926*. Translated by Jane Bannard Greene and M. D. Herter Norton. New York: W. W. Norton, 1969.

Rilke, R. M. *Gedichte*. Edited by D. Bode. Reclam, 1997.

Rinpoche Thrangu, Khenchen. *Pointing out the Dharmakaya: Teachings on the Ninth Karmapa's Text*. Ithaca, NY: Snow Lion Publications, 2011.

Roberts, Thomas B. *Psychoactive Sacramentals: Essays on Entheogens and Religion*. San Francisco: Council on Spiritual Practices, 2001.

Roelfsema, Pieter R. "Solving the Binding Problem: Assemblies Form When Neurons Enhance Their Firing Rate—They Don't Need to Oscillate or Synchronize." *Neuron* 111, no. 7 (April 2023): 1003–1019. https://doi.org/10.1016/j.neuron.2023.03.016.

Rorot, Wiktor. "Bayesian Theories of Consciousness: A Review in Search for a Minimal Unifying Model." *Neuroscience of Consciousness* 2021, no. 2 (October 1, 2021). https://doi.org/10/gm7463.

Rosenthal, Norman E. *Super Mind: How to Boost Performance and Live a Richer and Happier Life through Transcendental Meditation*. New York: TarcherPerigee, 2016.

Routledge, Clay, and Matthew Vess. *Handbook of Terror Management Theory*. London, San Diego, CA, Cambridge, MA, and Oxford: Academic Press, 2018.

Sadaghiani, Sepideh, and Mark D'Esposito. "Functional Characterization of the Cingulo-Opercular Network in the Maintenance of Tonic Alertness." *Cerebral Cortex* 25, no. 9 (2015): 2763–2773. https://doi.org/10.1093/cercor/bhu072.

Sanders, Robert D., Giulio Tononi, Steven Laureys, and Jamie W. Sleigh. "Unresponsiveness ≠ Unconsciousness." *Anesthesiology: The Journal of the American Society of Anesthesiologists* 116, no. 4 (2012): 946–959. https://doi.org/10.1097%2FALN.0b013e318249d0a7.

Sandved-Smith, Lars, Casper Hesp, Jérémie Mattout, Karl Friston, Antoine Lutz, and Maxwell J. D. Ramstead. "Towards a Computational Phenomenology of Mental Action: Modelling Meta-Awareness and Attentional Control with Deep Parametric Active Inference." *Neuroscience of Consciousness* 2021, no. 2 (27 August 2021): niab018. https://doi.org/10/gmndz4.

Sangs-rgyas-rdo-rje, Klong-chen-pa Dri-med-'od-zer, and 'Jigs-med-gling-pa Rang-byung-rdo-rje. *Great Perfection: The Essence of Pure Spirituality*. Translated by Shyalpa Tenzin Rinpoche, Kathmandu: Vajra Books, 2014.

Schnell, Tatjana. "Existential Indifference: Another Quality of Meaning in Life." *Journal of Humanistic Psychology* 50, no. 3 (July 2010): 351–373. https://doi.org/10/cd5dh3.

Schnell, Tatjana. *The Psychology of Meaning in Life*. London and New York: Routledge, 2021.

Schooler, Jonathan W., Jonathan Smallwood, Kalina Christoff, Todd C. Handy, Erik D. Reichle, and Michael A. Sayette. "Meta-Awareness, Perceptual Decoupling and the Wandering Mind." *Trends in Cognitive Sciences* (June 2011): S1364661311000878. https://doi.org/10.1016/j.tics.2011.05.006.

Schultes, Richard Evans, and Albert Hofmann. "Plants of the Gods: Origins of Hallucinogenic Use." New York: Hutchinson, 1980.

Schuon, Frithjof. *The Transcendent Unity of Religions*. New York: Harper & Row, 1975.

Schütz-Bosbach, Simone, Jason Jiri Musil, and Patrick Haggard. "Touchant-Touché: The Role of Self-Touch in the Representation of Body Structure." *Consciousness and Cognition* 18, no. 1 (March 2009): 2–11. https://doi.org/10/dj4mvz.

Seager, William, ed. *The Routledge Handbook of Panpsychism*. Routledge Handbooks in Philosophy. New York and London: Routledge, Taylor & Francis Group, 2020.

"Seelengrund." In *Wikipedia*, February 16, 2021. https://de.wikipedia.org/w/index.php?title=Seelengrund&oldid=208869223.

Seli, P., M. J. Kane, T. Metzinger, et al. "The Family-Resemblances Framework for Mind-Wandering Remains Well Clad." *Trends in Cognitive Sciences* 22, no. 11 (2018), 959–961. https://doi.org/10.1016/j.tics.2018.07.007.

Sellars, Wilfrid. "Empiricism and the Philosophy of Mind." *Minnesota Studies in the Philosophy of Science* 1, no. 19 (1956): 253–329.

Serino, Andrea, Elisa Canzoneri, Marilena Marzolla, Giuseppe di Pellegrino, and Elisa Magosso. "Extending Peripersonal Space Representation without Tool-Use: Evidence from a Combined Behavioral-Computational Approach." *Frontiers in Behavioral Neuroscience* (2015). https://doi.org/10/ggqdjs.

Serino, Andrea, Jean-Paul Noel, Giulia Galli, et al. "Body Part-Centered and Full Body-Centered Peripersonal Space Representations." *Scientific Reports* 5, no. 1 (December 2015): 18603. https://doi.org/10/ggqdm3.

Seth, A. K. *Being You*. New York: Dutton, 2021.

Seth, A. K., and Tim Bayne. "Theories of Consciousness." *Nature Reviews Neuroscience*, 2022, 1–14.

Seth, A. K, and Jakob Hohwy. "Predictive Processing as an Empirical Theory for Consciousness Science." *Cognitive Neuroscience* 12, no. 2 (6 December 2020): 89–90. https://doi.org/10/gjvfb2.

Seuse, Heinrich. *Deutsche Schriften/Heinrich Seuse. Im Auftr. d. Württ. Komm. f. Landesgeschichte Hrsg. von Karl Bihlmeyer*. Edited by Karl Bihlmeyer. Frankfurt am Main, Germany: Minerva-Verlag, 1961.

Shear, Jonathan. "Eastern Methods for Investigating Mind and Consciousness." In *Blackwell Companion to Consciousness*, edited by Susan Schneider and Max Velmans, 697–710. Malden, MA: John Wiley & Sons, 2007.

Shear, Jonathan. "The Experience of Pure Consciousness: A New Perspective for Theories of Self." *Metaphilosophy* 14, January (1983): 53–62. https://doi.org/10/cjk2cx.

Sheehy, Michael R., and Klaus-Dieter Mathes, eds. *The Other Emptiness: Rethinking the Zhentong Buddhist Discourse in Tibet*. Albany: SUNY Press, 2020.

Sheng Yen. *Faith in Mind: A Commentary on Seng Ts'an's Classic*. Boston: Shambhala, 2006.

Sheng Yen. *The Method of No-Method: The Chan Practice of Silent Illumination*. Boston: Shambhala, 2008.

Siderits, Mark. *Empty Persons: Personal Identity and Buddhist Philosophy*. London: Routledge, 2003.

Siderits, Mark. *How Things Are: An Introduction to Buddhist Metaphysics*. Buddhist Philosophy for Philosophers. New York: Oxford University Press, 2021.

Siderits, Mark, Evan Thompson, and Dan Zahavi, eds. *Self, No Self? Perspectives from Analytical, Phenomenological, and Indian Traditions*. Oxford: Oxford University Press, 2011.

Signorelli, Camilo Miguel, Joanna Szczotka, and Robert Prentner. "Explanatory Profiles of Models of Consciousness—towards a Systematic Classification." *Neuroscience of Consciousness* 2021, no. 2 (October 1, 2021). https://doi.org/10/gm7465.

Slater, Mel. "Place Illusion and Plausibility Can Lead to Realistic Behaviour in Immersive Virtual Environments." *Philosophical Transactions of the Royal Society B: Biological Sciences* 364, no. 1535 (December 12, 2009): 3549–3557. https://doi.org/10/df44xc.

Smallwood, Jonathan, and Jonathan W. Schooler. "The Science of Mind Wandering: Empirically Navigating the Stream of Consciousness." *Annual Review of Psychology* 66 (2015): 487–518. https://doi.org/10/gd3vsv.

Solms, Mark. *The Hidden Spring: A Journey to the Source of Consciousness*. London: Profile Books, 2021.

Song, Xiaolan, and Xiao Wang. '"Mind Wandering in Chinese Daily Lives—an Experience Sampling Study." *PLoS One* 7, no. 9 (September 5, 2012): e44423. https://doi.org/10/f36wfv.

Spackman, John. "Meditation, Nonconceptuality, and the Reflexive Structure of Consciousness." In *Routledge Handbook on the Philosophy of Meditation*, edited by Rick Repetti, 137–149. London: Routledge, 2022. https://doi.org/10.4324/9781003127253-11.

Strauch, Philipp, Heinrich von Nördlingen, and Margaretha Ebner. *Margaretha Ebner und Heinrich von Nördlingen, ein Beitrag zur Geschichte der deutschen Mystik, von Philipp Strauch*. Freiburg im Bresgau and Tübingen, Germany: J. C. B. Mohr, 1882.

Strawson, Galen. "'Self-Intimation'." *Phenomenology and the Cognitive Sciences* 14, no. 1 (March 2015): 1–31. https://doi.org/10/gg3vk3.

Streib, Heinz, and Ralph W. Hood, eds. *Semantics and Psychology of Spirituality: A Cross-Cultural Analysis*. Cham, Switzerland: Springer International Publishing, 2016. https://doi.org/10.1007/978-3-319-21245-6.

Streib, Heinz, and Ralph Hood. "'Spirituality" as Privatized Experience-Oriented Religion: Empirical and Conceptual Perspectives." *Implicit Religion* 14, no. 4 (December 28, 2011): 433–453. https://doi.org/10/gf3qtx.

Struhl, Karsten J. "The Self." In *Routledge Handbook on the Philosophy of Meditation*, edited by Rick Repetti, 256–268. London: Routledge, 2022. https://doi.org/10.4324/9781003127253-20.

Suzuki, Daisetz Teitaro. *Zen Doctrine of No Mind*, 1949. http://archive.org/details/in.ernet.dli.2015.57108.

Tang, Yi-Yuan, Rongxiang Tang, Michael I. Posner, and James J. Gross. "Effortless Training of Attention and Self-Control: Mechanisms and Applications." *Trends in Cognitive Sciences*, 2022. https://doi.org/10.1016/j.tics.2022.04.006.

Tauler, Johannes, and Ferdinand Vetter. *Die Predigten Taulers*. Dublin: Weidmann, 1968.

Taves, Ann, and Egil Asprem. "Experience as Event: Event Cognition and the Study of (Religious) Experiences." *Religion, Brain & Behavior* 7, no. 1 (January 2, 2017): 43–62. https://doi.org/10/gmfft3.

Terry, Joseph. "The Philosophical Presuppositions of Christian Meditation." In *Routledge Handbook on the Philosophy of Meditation*, edited by Rick Repetti, 383–391. London: Routledge, 2022. https://doi.org/10.4324/9781003127253-32.

Thomas, Nigel J. T. "Mental Imagery." In *Stanford Encyclopedia of Philosophy*, edited by Edward N. Zalta, Fall 2021. Metaphysics Research Lab, Stanford University. https://plato.stanford.edu/archives/fall2021/entries/mental-imagery/.

Thompson, Evan. *Mind in Life*. Cambridge, MA: Belknap Press of Harvard University Press, 2010.

Thompson, Evan. "Self-no-self? Memory and Reflexive Awareness." In *Self, no self? Perspectives from Analytical, Phenomenological, and Indian Traditions*. Oxford, UK: Oxford University Press, 2011.

Thompson, Evan. *Waking, Dreaming, Being: Self and Consciousness in Neuroscience, Meditation, and Philosophy*. New York: Columbia University Press, 2015.

Thompson, Evan. *Why I Am Not a Buddhist*. New Haven, CT: Yale University Press, 2020.

Timalsina, Sthaneshwar. "The Phenomenology of Meditation." In *Routledge Handbook on the Philosophy of Meditation*, edited by Rick Repetti, 313–326. London: Routledge, 2022. https://doi.org/10.4324/9781003127253-26.

Tononi, Giulio, Melanie Boly, Marcello Massimini, and Christof Koch. "Integrated Information Theory: From Consciousness to Its Physical Substrate." *Nature Reviews Neuroscience* 17, no. 7 (2016): 450–461. https://doi.org/10/f8rbxc.

Trivers, Robert. "The Elements of a Scientific Theory of Self-deception." *Annals of the New York Academy of Sciences* 907, no. 1 (2000): 114–131. https://doi.org/10/fnn42d.

Trivers, Robert. *The Folly of Fools: The Logic of Deceit and Self-Deception in Human Life*. New York: Basic Books, 2011.

Tulver, Kadi, Karl Kristjan Kaup, Ruben Laukkonen, and Jaan Aru. "Restructuring Insight: An Integrative Review of Insight in Problem-Solving, Meditation, Psychotherapy, Delusions and Psychedelics," n.d. https://doi.org/10/gnwtsh.

Ueda, Shizuteru, and Ernst Benz. *Die Gottesgeburt in der Seele und der Durchbruch zur Gottheit: Die mystische Anthropologie Meister Eckharts und Ihre Konfrontation mit der Mystik des Zen-Buddhismus*. Edited by Wolf Burbat. Aktualisierte Neuausgabe. Freiburg im Bresgau and Munich, Germany: Verlag Karl Alber, 2018.

United Nations Office on Drugs and Crime. *World Drug Report 2022*. Vienna, 2022. https://www.unodc.org/unodc/en/data-and-analysis/world-drug-report-2022.html.

Urgyen, Tulku. *As It Is. Volume 1*. Hong Kong: Rangjung Yeshe Publications, 1999. https://www.overdrive.com/search?q=1183FDCF-36BD-4694-B679-8FDD0851B14D.

Urgyen, Tulku. *As It Is. Volume 2.* Hong Kong: Rangjung Yeshe Publications, 2000. https://www.overdrive.com/search?q=54BBFCEC-49D3-4D98-A577-FFB3C83E6BBA.

Urgyen, Tulku. *Vajra Speech: Pith Instructions for the Dzogchen Yogi.* Edited by Marcia Binder Schmidt and Michael Tweed. Translated by Erik Pema Kunsang. Hong Kong: Rangjung Yeshe Publications, 2001.

Venkataramiah, Sri Munagala. *Talks with Sri Ramana Maharshi.* Tiruvannamalai, India: Sri Ramanasramam, 2006.

Vignemont, Frédérique de, Andrea Serino, Hong Yu Wong, and Alessandro Farnè. "Peripersonal Space: A Special Way of Representing Space." In *The World at Our Fingertips*, edited by Frédérique de Vignemont, Andrea Serino, Hong Yu Wong, and Alessandro Farnè, 3–16. Oxford University Press, 2021. https://doi.org/10.1093/oso/9780198851738.003.0001.

Vignemont, Frédérique de, Andrea Serino, Hong Yu Wong, and Alessandro Farnè, eds. *The World at Our Fingertips: A Multidisciplinary Exploration of Peripersonal Space*. Oxford: Oxford University Press, 2021. https://doi.org/10.1093/oso/9780198851738.001.0001.

Vilas, Martina G., Ryszard Auksztulewicz, and Lucia Melloni. "Active Inference as a Computational Framework for Consciousness." *Review of Philosophy and Psychology*, August 10, 2021. https://doi.org/10/gmg2rn.

Vinding, Magnus. *Suffering-Focused Ethics: Defense and Implications*, Copenhagen: Ratio Ethica, 2020.

Voeneky, Silja, Philipp Kellmeyer, Oliver Mueller, and Wolfram Burgard, eds. *Cambridge Handbook of Responsible Artificial Intelligence: Interdisciplinary Perspectives*. Cambridge Law Handbooks. Cambridge: Cambridge University Press, 2022. https://www.cambridge.org/core/books/cambridge-handbook-of-responsible-artificial-intelligence/EF02D78934D18B9A22A57A46FF8FFAFC.

Vogt, Katja. "Ancient Skepticism." In *Stanford Encyclopedia of Philosophy*, edited by Edward N. Zalta, Summer 2021. Metaphysics Research Lab, Stanford University. https://plato.stanford.edu/archives/sum2021/entries/skepticism-ancient/.

Vöneky, S., P. Kellmeyer, O. Mueller, and W. Burgard (Eds.). *Cambridge Handbook of Responsible Artificial Intelligence: Interdisciplinary Perspectives*. Cambridge: Cambridge University Press, 2022. https://doi.org/10.1017/9781009207898.

Von Hippel, William, and Robert Trivers. "The Evolution and Psychology of Self-Deception." *Behavioral and Brain Sciences* 34, no. 1 (2011): 1.

Voss, Ursula, and Allan Hobson. "What Is the State-of-the-Art on Lucid Dreaming? Recent Advances and Questions for Future Research." In *Open MIND*, edited by Thomas K. Metzinger and Jennifer M. Windt. Frankfurt am Main, Germany: MIND Group, 2015. https://doi.org/10.15502/9783958570306.

Voss, Ursula, Romain Holzmann, Allan Hobson, Walter Paulus, Judith Koppehele-Gossel, Ansgar Klimke, and Michael A. Nitsche. "Induction of Self Awareness in Dreams through Frontal Low Current Stimulation of Gamma Activity." *Nature Neuroscience* 17, no. 6 (June 2014): 810–812. https://doi.org/10/sqt.

Voss, Ursula, Karin Schermelleh-Engel, Jennifer Windt, Clemens Frenzel, and Allan Hobson. "Measuring Consciousness in Dreams: The Lucidity and Consciousness in Dreams Scale." *Consciousness and Cognition* 22, no. 1 (2013): 8–21. https://doi.org/10/f4qm9r.

Waldschütz, Erwin. *Denken und Erfahren des Grundes: Zur Philosophischen Deutung Meister Eckharts*. Vienna, Freiburg im Bresgau, Germany, and Basel: Herder, 1989.

Wallace, B. Alan, and Brian Hodel. *Dreaming Yourself Awake: Lucid Dreaming and Tibetan Dream Yoga for Insight and Transformation*. Boston: Shambala, 2012.

Wangyal, Tenzin, and Mark Dahlby. *The Tibetan Yogas of Dream and Sleep*. Ithaca, NY: Snow Lion Publications, 1998.

Ward, Adrian F., and Daniel M. Wegner. "Mind-blanking: When the Mind Goes Away." *Frontiers in Psychology* 4 (2013). https://doi.org/10.3389/fpsyg.2013.00650.

Ward, Dave, Tom Roberts, and Andy Clark. "Knowing What We Can Do: Actions, Intentions, and the Construction of Phenomenal Experience." *Synthese* 181, no. 3 (2011): 375–394. https://doi.org/10.1007/s11229-010-9714-6.

Wasson, R. Gordon, Albert Hofmann, and Carl A. P. Ruck. *The Road to Eleusis: Unveiling the Secret of the Mysteries*. Berkeley, CA: North Atlantic Books, 2008.

Wasson, R. Gordon, Stella Kramrisch, and Carl Ruck. *Persephone's Quest*. New Haven, CT: Yale University Press, 2008.

Weber, Max, C. Wright Mills, and H. H Gerth. *From Max Weber: Essays in Sociology. Translated, Edited and with an Introduction by H.H. Gerth and C. Wright Mills*. London: Kegan Paul, 1947.

Wegner, Daniel M. "Ironic Processes of Mental Control." *Psychological Review* 101, no. 1 (1994): 34–52. https://doi.org/10.1037/0033-295X.101.1.34.

Wegner, Daniel M. *The Illusion of Conscious Will*. Cambridge, MA: MIT Press, 2018.

Wegner, Daniel M., David J. Schneider, Samuel R. Carter, and Teri L. White. "Paradoxical Effects of Thought Suppression." *Journal of Personality and Social Psychology* 53, no. 1 (1987): 5–13. https://doi.org/10.1037/0022-3514.53.1.5.

Weil, Simone. *Gravity and Grace*. London: Routledge, 1997.

Weil, S. *The Notebooks of Simone Weil. Vol. 2*. Translated by A. Wills. New York: Putnam's Sons, 1956.

Wiese, Wanja. *Experienced Wholeness: Integrating Insights from Gestalt Theory, Cognitive Neuroscience, and Predictive Processing*. Cambridge, MA: MIT Press, 2018.

Wiese, Wanja. "The Science of Consciousness Does Not Need Another Theory, It Needs a Minimal Unifying Model." *Neuroscience of Consciousness* 2020, no. 1 (January 1, 2020): niaa013. https://doi.org/10/gm95xr.

Wiese, Wanja. "Minimal Models of Consciousness: Understanding Consciousness in Human and Non-human Systems," preprint, 2023. https://philpapers.org/rec/WIEMMO-2.

Wiese, Wanja, and Karl J. Friston. "Examining the Continuity between Life and Mind: Is There a Continuity between Autopoietic Intentionality and Representationality?" *Philosophies* 6, no. 1 (February 21, 2021): 18. https://doi.org/10/gh4qkb.

Wiese, Wanja, and Karl J. Friston. "The Neural Correlates of Consciousness under the Free Energy Principle: From Computational Correlates to Computational Explanation." *Philosophy and the Mind Sciences* 2 (2021). https://doi.org/10/gmg2bh.

Wiese, Wanja, and Thomas K. Metzinger. "Vanilla PP for Philosophers: A Primer on Predictive Processing." In *Philosophy and Predictive Processing*, edited by Thomas K. Metzinger and Wanja Wiese. Frankfurt am Main, Germany: MIND Group, 2017. https://doi.org/10.15502/9783958573024.

Wikipedia contributors. "The Denial of Death." In *Wikipedia, the Free Encyclopedia*. Wikimedia Foundation, July 10, 2021. https://en.wikipedia.org/w/index.php?title=The_Denial_of_Death&oldid=1032213207.

Williams, Paul. *The Reflexive Nature of Awareness: A Tibetan Madhyamaka Defence*. Bd. 1. Psychology Press, 1998.

Williamson, Lola. *Transcendent in America: Hindu-Inspired Meditation Movements as New Religion*. New York: New York University Press, 2010.

Williford, Kenneth. "Self-Acquaintance and Three Regress Arguments." *ProtoSociology* 36 (2019): 368–412. https://doi.org/10/ggqrdk.

Windt, Jennifer Michelle. "Consciousness in Sleep: How Findings from Sleep and Dream Research Challenge Our Understanding of Sleep, Waking, and Consciousness." *Philosophy Compass* 15, no. 4 (2020): e12661. https://doi.org/10/gmdmzt.

Windt, Jennifer Michelle. "How Deep Is the Rift between Conscious States in Sleep and Wakefulness? Spontaneous Experience over the Sleep–Wake Cycle." *Philosophical Transactions of the Royal Society B: Biological Sciences* 376, no. 1817 (February 2021): 20190696. https://doi.org/10/gmh669.

Windt, Jennifer Michelle. "The Immersive Spatiotemporal Hallucination Model of Dreaming." *Phenomenology and the Cognitive Sciences* 9, no. 2 (June 1, 2010): 295–316. https://doi.org/10/bhg8q7.

Windt, Jennifer Michelle, and T. Metzinger. "Die phänomenale Signatur des Wissens: Experimentelle Philosophie des Geistes *mit* oder *ohne* Intuitionen?" In *Die Experimentelle Philosophie in der Diskussion*, edited by T. Grundmann, J. Horvath, and J. Kipper, 279–321. Berlin: Suhrkamp 2014.

Windt, Jennifer Michelle, and Thomas Metzinger. "The Philosophy of Dreaming and Self-Consciousness: What Happens to the Experiential Subject during the Dream State?" In *The New Science of Dreaming, Vol. 3: Cultural and Theoretical Perspectives*, edited by Deirdre Barrett and Patrick McNamara, 193–247. Westport, CT: Praeger Publishers/Greenwood Publishing Group, 2007.

Windt, J. M., T. Nielsen, and E. Thompson. "Does Consciousness Disappear in Dreamless Sleep? *Trends in Cognitive Sciences* 20, no. 12 (2016), 871–882. https://doi.org/10.1016/j.tics.2016.09.006

Windt, Jennifer Michelle, and Ursula Voss. "Spontaneous Thought, Insight, and Control in Lucid Dreams." In *Oxford Handbook of Spontaneous Thought: Mind-Wandering, Creativity, and Dreaming*, 385–410. New York: Oxford University Press, 2018.

Winter, Ulf, Pierre LeVan, Tilmann L. Borghardt, et al. "Content-Free Awareness: EEG-FcMRI Correlates of Consciousness as Such in an Expert Meditator." *Frontiers in Psychology* 10 (February 18, 2020): 3064. https://doi.org/10/gg9mjm.

Wittgenstein, Ludwig. *Tractatus Logico-Philosophicus: Logisch-Philosophische Abhandlung / Ludwig Wittgenstein*. Frankfurt am Main, Germany: Suhrkamp, 1966.

Wittmann, Marc. *Altered States of Consciousness: Experiences out of Time and Self*. Translated by Philippa Hurd. Cambridge, MA, and London: MIT Press, 2018.

Wittmann, Marc. "The Inner Sense of Time: How the Brain Creates a Representation of Duration." *Nature Reviews Neuroscience* 14, no. 3 (March 2013): 217–223. https://doi.org/10/ggtbvf.

Wolfe, Kathy, and Frank M. Ralls. "Rapid Eye Movement Sleep and Neuronal Development." *Current Opinion in Pulmonary Medicine* 25, no. 6 (November 2019): 555–560. https://doi.org/10.1097/MCP.0000000000000622.

Woods, Toby J., Jennifer M. Windt, and Olivia Carter. "Evidence Synthesis Indicates Contentless Experiences in Meditation Are Neither Truly Contentless nor Identical." *Phenomenology and the Cognitive Sciences*, 2022a. https://doi.org/10.1007/s11097-022-09811-z.

Woods, Toby J., Jennifer M. Windt, and Olivia Carter. "The Path to Contentless Experience in Meditation: An Evidence Synthesis Based on Expert Texts." *Phenomenology and the Cognitive Sciences*, 2022b. https://doi.org/10.1007/s11097-022-09812-y.

Woods, Toby J., Jennifer M. Windt, and Olivia Carter. "Silence in Shamatha, Transcendental, and Stillness Meditation: An Evidence Synthesis Based on Expert Texts." *Frontiers in Psychology* 11 (July 8, 2020): 1259. https://doi.org/10.3389/fpsyg.2020.01259.

Woods, Toby J., Jennifer M. Windt, Lydia Brown, Olivia Carter, and Nicholas T. Van Dam. "Subjective Experiences of Committed Meditators Across Practices Aiming for Contentless States." *Mindfulness*, June 22, 2023. https://doi.org/10.1007/s12671-023-02145-0.

Zedelius, Claire M., John Protzko, and Jonathan W. Schooler. "Lay Theories of the Wandering Mind: Control-Related Beliefs Predict Mind Wandering Rates in- and Outside the Lab." *Personality and Social Psychology Bulletin* 47, no. 6 (June 2021): 921–938. https://doi.org/10/gmxhk3.

Index

Note: Photos, figures, and tables are indicated by italicized page numbers.

Have you had an experience of pure awareness too?
Would you like to support our future research?

The revised version of our survey is now also
open to nonmeditators.
You can find it at mpe-project.info.